13 Questions

COUNTERPOINTS

Studies in Criticality

Shirley R. Steinberg
General Editor

Vol. 442

The Counterpoints series is part of the Peter Lang Education list.
Every volume is peer reviewed and meets
the highest quality standards for content and production.

PETER LANG
New York • Bern • Frankfurt • Berlin
Brussels • Vienna • Oxford • Warsaw

13 Questions

Reframing Education's Conversation: Science

Edited by
Lynn A. Bryan and Kenneth Tobin

PETER LANG
New York • Bern • Frankfurt • Berlin
Brussels • Vienna • Oxford • Warsaw

Library of Congress Cataloging-in-Publication Data

Names: Bryan, Lynn A. (Lynn Ann), editor. | Tobin, Kenneth George, editor.
Title: 13 questions: reframing education's conversation: science /
edited by Lynn A. Bryan, Kenneth Tobin.
Other titles: Thirteen questions
Description: New York: Peter Lang, 2018.
Series: Counterpoints. Studies in criticality; vol. 442 | ISSN 1058-1634
Includes bibliographical references and index.
Identifiers: LCCN 2016045990 | ISBN 978-1-4331-2780-9 (hardcover: alk. paper)
ISBN 978-1-4331-2779-3 (paperback: alk. paper) | ISBN 978-1-4331-4574-2 (ebook pdf)
ISBN 978-1-4331-4575-9 (epub) | ISBN 978-1-4331-4576-6 (mobi)
Subjects: LCSH: Science—Study and teaching.
Classification: LCC Q181 .T46 | DDC 507.1—dc23
LC record available at https://lccn.loc.gov/2016045990
DOI 10.3726/b11305

Bibliographic information published by **Die Deutsche Nationalbibliothek.**
Die Deutsche Nationalbibliothek lists this publication in the "Deutsche
Nationalbibliografie"; detailed bibliographic data are available
on the Internet at http://dnb.d-nb.de/.

The paper in this book meets the guidelines for permanence and durability
of the Committee on Production Guidelines for Book Longevity
of the Council of Library Resources.

29 Broadway, 18th floor, New York, NY 10006
www.peterlang.com

Printed in the United States of America

Table of Contents

Figures

Foreword

We Must Kill Our Darlings

SHIRLEY R. STEINBERG

In 1989, Joe Kincheloe conceptualized editing a book of questions to be answered by colleagues which focused on creating a new dialogue about education and schooling. Settling on 13 questions, he invited two educator/researcher/scholars to answer each question. Sending his proposal for *13 Questions: Reframing Education's Conversation* to Michael Flamini at Peter Lang Publishing, Joe wanted this book to create a focus on what educators did and didn't discuss with one another. After a short review, Michael responded with a contract for Joe's first edited book. Shortly after he received that contract, I met Joe in Dayton, Ohio on October 19, 1989, and he invited me to write a chapter answering one of the questions in his book. He also married me.

Working with all 26 authors on the chapters proved to be a larger task than Joe reckoned, and he asked me to co-edit the book. 1990 was the gestation period for those 13 questions and became the grounding for my own theoretical framework, my lifeworld of critical pedagogy, my first published chapter and book. I learned how to produce a book and was in constant contact with our gathering of radical and committed thinkers … the palaver created by our contributors was distinct, edgy, rigorous, and critical. The buzz was exhilarating, the book was adopted by many of our own authors, and within a couple of years, had been acknowledged as an innovative and topical book for teacher education. And, in the best of critical pedagogical traditions, *13 Questions* created a community of educators, who, for the most part, are still writing together, collaborating and continuing to change the course of the field.

Change being the operative word. Indeed, pedagogy cannot exist without change. Re-reading the original edition of *13 Questions,* I find that what was risky and subversive then, after three editions, is still in need of revision and new constructs. As Joe and I suggested in 1993, we must take a tentative position in our pedagogy. By doing this, we are able to demand that reform be addressed even as the new is proposed. Pedagogical arrogance becomes the norm when educators believe they have struck the *motherlode.* There is no motherlode in education. Education's conversation must be filled experimentation, demonstration, reflection, revision, change, 180° turns in direction, and, above all, humility. It is our obligation as educators to recognize when we must revise, renew, and, as William Faulkner and Allen Ginsberg advocated, even *kill our darlings* … we must destroy them at times, in order for critical pedagogy to survive. The only canon needed in education is *to not have a canon.* And that goes for radical pedagogies as well, nothing is sacred, nothing is permanent.

To say I am elated to be writing the foreword for the next incarnation of the original book would be beyond understatement. Lynn and Ken have created a gathering of fine authors and pedagoges in this compelling and most-needed volume. The coda? Five years after Flamini signed the original *13 Questions,* he left Lang to join St. Martin's Press. Somewhere in mid-Manhattan, Joe, Michael, and I dined to celebrate his new position. Michael handed us an envelope, "Joe & Shirley" written on the front. It contained a letter from Mr. Peter Lang, telling Michael he must not sign this book, *13 Questions,* it would never sell *and* there was no compelling argument to grant this contract. Michael read the letter and put it in his bottom desk drawer. The book became the top selling book on Peter Lang Publishing's list for over two decades.

Shirley R. Steinberg is the Research Professor of Critical Youth Studies at the University of Calgary, Werklund School of Education.

CHAPTER ONE

Of Eggs, Chickens, and Deep-Seated Ideologies

LYNN A. BRYAN AND KENNETH TOBIN

Joe Kincheloe and Shirley Steinberg described themselves as critical hermeneuts—not just when they wrote, but also (and especially) as they lived their lives. Many might consider Joe and Shirley to be troublemakers because they were determined to expose a core underbelly of everyday life—to all participants. For example, one of the things Shirley did in the community in which she, Joe and Ken Tobin lived, was to place a small notice on her mailbox: "free-range eggs—apply within." Within hours, the neighbors railed on social media about the impact of free-range chickens on property values, health, and the general welfare of the community. Of course, there were no eggs and no chickens. The presence of the sign was sufficient to unsettle deep-seated ideologies that rarely surfaced—precipitating anger, hate, and threats of violence. No doubt the intent of the sign extended well beyond causing trouble. Shirley and Joe sought to disrupt equilibria within the community as neighbors squabbled over axiological commitments, mainly related to neoliberalism and capitalism, especially manifest in property values. Similarly, many of the books edited and co-authored by Joe and Shirley were intended to make trouble—to create waves and thereby to reveal otherwise hidden values and commitments. Examples of such books include Joe's critique of McDonald's, *The Sign of the Burger: McDonald's and the Culture of Power* (Kincheloe, 2002a) and *Christotainment: Selling Jesus Through Popular Culture*, Shirley and Joe's co-edited volume (Steinberg & Kincheloe, 2009).

A primary goal of most of Shirley and Joe's authored and edited books was to expand what might be considered difficult topics, identify thorny issues, and reveal inequities. Discomfort was intended and convergence toward oversimplified solutions was never entertained. In many ways, the books mirrored the sign about free-range eggs and myriad conversations involving Shirley, Joe and colleague faculty and students. The status quo and its supporting ideologies and schema were not to be taken for granted and there were few, if any, taboos that were free of critical inquiry.

Of course, there is much more to Shirley and Joe than them unsettling taken-for-granted patterns of culture; they also were interested to learn from difference and practiced radical listening as they sought and listened attentively to written and spoken texts, while exploring the potential of others' ideas, and searched for structural underpinnings and sources of inequity. To some degree the two aspects we have addressed so far, calling "normal" patterns of questioning and listening to the perspectives of others to learn of their potential, create a context for the chapters in this volume.

THIRTEEN QUESTIONS

The origin of *Thirteen Questions: Reframing Education's Conversation: Science* traces back to one of Joe and Shirley's most influential books and a presidential plenary session at the 2014 annual meeting of NARST: A Worldwide Organization for Improving Science Teaching and Learning through Research. As president of the organization at the time, Lynn A. Bryan was responsible for organizing plenary sessions addressing the theme, "Awaking dialogues: Science education research, practices, and policies." The theme of the meeting was intended to instigate, provoke and otherwise rouse conversations about the "elephants in the room"—particularly historical and contemporary manifestations of social inequities in education—and challenge taken-for-granted beliefs, values, assumptions, and ideologies in science education. At the suggestion of a long-time friend and colleague, Deborah Tippins, Lynn read the second edition of *Thirteen Questions: Reframing Education's Conversation* (Kincheloe & Steinberg, 1996) as a possible source of inspiration. Barely beyond the first pages of the book, it became clear that this book would serve as a quintessential springboard for a plenary session. Consider Joe and Shirley's reflections in the preface about the first edition of their book:

> If a purpose of education is to provoke response and elicit action, we feel like we "musta done something right." Our editor, Michael Flamini, phoned us several months ago and said he had sent some returned volumes of *Thirteen Questions* to us for examination. Evidently one of the batches of books had some pages missing and, after the semester, a professor returned the books to Lang for credit. The situation as it stood was obviously

> forgettable; however, what Michael felt worth recording was the margin notes that students had written in a dozen or so of the volumes.
>
> When we received the books, we spent some time leafing through the familiar chapters marked with yellow, green, and pink highlighter with emotional expletives and inflamed statements ... Our discovery led us to believe that the students who responded with ire a la the margins were furious with our challenges to their own grand narratives of white, male curriculums and with our audacity in naming and demanding revelation of power structures within our educational society. Wow. *Thirteen Questions* (at least at the university that returned the books) was rocking the boat. (Kincheloe & Steinberg, 1996, p. xi)

Undoubtedly, *Thirteen Questions* (Kincheloe & Steinberg, 1996) was intended to disturb the status quo and catalyze changes in research, policy, curriculum, teacher education, and the roles of teachers and students. It expanded conversations about what is happening, why it is happening, and what supports the status quo. By selecting authors who embraced diverse theoretical perspectives, Joe and Shirley engineered a text that highlighted difference and contradictions while eschewing the idea that writers/scholars had a final word. The goal was not to obtain finality for 13 questions about education, but to disrupt comfort zones and unsettle taken-for-granted wisdom (i.e., to create turbulence by casting stones into a placid/serene pond).

Since 2014 marked just over two decades since the first edition of *Thirteen Questions*, it seemed fitting to revisit the original 13 questions with a diverse panel of scholars/educators/thinkers and from both retrospective and contemporary perspectives. Accordingly, a plenary panel was invited and cosponsored by the NARST Latina/o Research Interest Group, the Continental and Diasporic Africa in Science Education Research Interest Group, and the Equity and Ethics Committee. Four panel members spoke to a select number of the original 13 questions: Richard Milner, the Helen Faison Endowed Chair of Urban Education, Professor of Education, Professor of Social Work (by courtesy), Professor of Africana Studies (by courtesy), and Director of the Center for Urban Education at the University of Pittsburgh; Ana M. Becerra, Program Manager for Schools and Families at *Just Communities* Central Coast in Santa Barbara, California; Timothy Knight, founder and director of the Indianapolis O.K. Program for youth and former detective for the Indianapolis Metropolitan Police Department; and Shirley Steinberg, co-editor of *Thirteen Questions* and Research Professor of Youth Studies, formerly the Werklund Foundation Chair and the Director of the Werklund Foundation Centre for Youth Leadership Studies at the University of Calgary. The panel not only responded to the original 13 questions in the context of today's educational climate, but also reflected on the questions in terms of educational issues today vis-à-vis the dialogue from 20 years ago. Ken, a close colleague and friend of Joe and Shirley, served as the panel discussant.

At the end of the plenary session, Shirley invited us (Lynn and Ken) to edit a new version of the book, to be called *Thirteen Questions: Reframing Education's Conversation: Science.* When we opted to accept Shirley's invitation to edit a book concerning critical questions in science education, we too sought to disrupt and unsettle, with the purpose of expanding the conversations in science education, moving beyond the idea that research was producing an inviolate system of knowledge that could resolve stoic (i.e., static and resolute) problems in science education. Instead, we invited deep thinkers who embraced an array of sociocultural theories to identify and explicate what they considered to be the questions of the day—questions that scholars and practitioners might consider as they enacted science education in a dynamic and evolving universe threatened by issues of sustainability and disharmony. As editors, we embrace the goal of providing spaces for the authors to publish standpoints to provoke, expand, and enlighten readers about possibilities for being, acting, transforming, and enhancing the social and physical worlds we inhabit and for which we are stewards.

The questions of this volume share many of the questions from Joe and Shirley's original *Thirteen Questions.* However, in the development of this volume, we asked colleagues to examine the list of original questions and propose any questions that were not of the original 13 questions. From the responses of many dozens of colleagues, we generated the list of questions for this volume, replacing a few of the original questions, combining a few of the original questions, and keeping many of the original questions. Thus, this volume is organized into 13 sections, each guided by one of the 13 questions.

This volume does not provide final words on science education; instead it serves as a step forward in an ongoing journey of ever-changing façades. We anticipate that the chapters of this book will disrupt and stimulate, providing a turbulent soup in which fresh ideas are born, nurtured, grow, and flourish.

REFERENCES

Kincheloe, J. L. (2002a). *The sign of the burger: McDonald's and the culture of power.* Philadelphia, PA: Temple University Press.

Kincheloe, J. L. (Ed.). (2002b). *Teachers as researchers: Qualitative paths to empowerment* (2nd ed.). New York, NY: RoutledgeFalmer.

Kincheloe, J. L., & Steinberg, S. R. (Eds.). (1996). *Thirteen questions: Reframing education's conversation* (2nd ed.). New York, NY: Peter Lang.

Steinberg, S. R., & Kincheloe, J. L. (Eds.). (2009). *Christotainment: Selling Jesus through popular culture.* Boulder, CO: Westview Press.

SECTION ONE

The Science Curriculum

What Are the Basics and Are We Teaching Them?

CHAPTER TWO

The Science Curriculum at the Elementary Level

What Are the Basics and Are We Teaching Them?

CHRISTINA SIRY

When I was asked to contribute to this book by addressing the question of what the basics are for teaching science at the elementary school level, and whether or not these are being taught, I began by first taking a step back to consider why science as a discipline is (or should be) taught at the elementary level. I write this from the standpoint of a former elementary school "science specialist", grounded in critical sociocultural theoretical frameworks. I begin with this question of "why" because it is one that I am often confronted with in my own research, and it is a question that I argue all science education researchers should engage with reflectively. In particular, I contend that as a field we ought to consider: what are the needs and purposes of teaching science at the elementary level? Is it to create scientists? Is it to support an informed citizenry? Or perhaps it is to facilitate scientific literacy for either of these viewpoints? The answers to these questions are of course complex, especially as they have been asked by science educators for generations, and also given that the answers have changed as what is valued by society has shifted over time. With that in mind, in the sections that follow, I reflect on the purpose and potential foci for elementary school science teaching, guided by international research findings, critical theoretical perspectives, and my own experiences working in the field of elementary science education for the past 20+ years.

WHY TEACH SCIENCE IN ELEMENTARY SCHOOL?

> There are many who do not believe it amounts to much one way or the other what children do in science in the elementary school. I do not agree, for upon the whole, I believe the attitude toward the study of science is, and should be, fixed during the early years of life. (Dewey, 1910, p. 123)

There has been much written in the field of science education about teaching children positive attitudes toward science at an early age, and young children have been situated in the literature as being "natural scientists" (e.g., Head Start, 2010, p. 18), "scientific learners" (e.g., Gopnik, 2012, p. 1625), and "emergent inquirers" (e.g., Hedges, 2014, p. 39). Undoubtedly, interacting with young children often reveals curiosity and fascination with new discoveries, especially when engaging with science questions. Science itself is a human endeavor, one that is particular to inquiring into phenomena of the natural world. In curricular debates regarding science at the elementary levels, there tends to be a split between those who emphasize primarily science content and those who emphasize primarily the processes of science, but in both of these perspectives, positive attitudes towards science are often held as a central underpinning, as also emphasized by John Dewey in the above quote.

Science as a specific focus for a school discipline typically lacks the clarification of what, specifically, is meant by the term "science" in thinking of school science—many have preconceived notions of textbooks, magnifying lenses, and lab experiments—all of which come from clearly confined cultural contexts and expectations. Recently, the Science Council in the UK (2015) took a step back and defined the word *science*, which they contend is the first time this has been done by scientists for clarification of the term. Their perspective is that "science is the pursuit and application of knowledge and understanding of the natural and social world following a systematic methodology based on evidence" (http://www.sciencecouncil.org/definition). Related to this is the Next Generation Science Standards ([NGSS]; NGSS Lead States, 2013) in the US which have clarified that "in the K-12 context, 'science' is generally taken to mean the traditional natural sciences: physics, chemistry, biology, and (more recently) earth, space, and environmental sciences" (p. 103), and further that science is "both a body of knowledge and an evidence-based, model and theory building enterprise that continually extends, refines, and revises knowledge" (Next Generation Science Standards, 2013).

Taking these perspectives together, science can be perceived as an approach to investigating the known, and the unknown, and to an understanding of a body of knowledge that evolves over time. Further, science can be situated as being central to engaging in processes of knowledge construction—of the known and the unknown—which certainly extends far beyond the realm of science in and of itself, in order to support learning, reflecting, and understanding of the social and natural worlds in which we participate. Yet we find in many schools that the primary focus in elementary grades is on literacy and numeracy (Appleton, 2007), and that science is not consistently taught at the elementary level. When science is taught, it is common to be in a fragmented, reduced manner, focusing almost exclusively on developing vocabulary (e.g., Newton & Newton, 2000) and/or descriptive regurgitation of information. This is unfortunately not restricted to particular isolated regions of the world as there is a globalized focus in elementary school of accountability. The pressures of such accountability position teachers to teach only that which will be tested and relies on rote memorization of facts, which marginalizes a subject such as science (Berg & Mensah, 2014). As a former researcher and teacher in the United States, I was confronted with this situation time and time again. Currently I live in the European country of Luxembourg, and continue to be confronted with the same types of experiences regarding the de-valued role of science at the primary school level. As an example, my own daughter recently came home from her first day of sixth grade with a note to parents from her teacher, which stated that: "Due to the standardized tests later this year [in French, German and math] the following subjects will not be taught during the first two trimesters: art, music, history, geography, science, and Luxembourgish." Imagine for a moment the experiences of children who are in classes all day long, only being instructed in three subjects, every day, all day … one mustn't wonder then why my daughter no longer wants to go to school. While this is perhaps merely one example of one teacher's reaction to standardized testing pressures, those of us working in elementary science education internationally are likely more than aware of this situation in numerous classrooms. An education system that only values the things that are tested will not educate and support children to be able to construct complex understandings of knowledges—instead, what is valued is reduced to what is tested: isolated facts removed from any context, presented as known information to be memorized, or pre-packaged, scripted curricula, created by corporations and designed for teaching to tests (Au, 2011), which is a far cry from the perspective of science as "both a body of knowledge and an evidence-based, model and theory building enterprise that continually extends, refines, and revises knowledge" (http://www.nextgenscience.org/three-dimensions) and similar explanations of the discipline of science as represented in policy documents.

In considering why science is taught at the elementary level, it is prudent on the one hand to acknowledge the damaging culture of accountability that continues to shape classroom practices and policies, marginalizing and reducing science to something that can be measured and is fact-based; which is far removed from the ideals of why science should even be taught in the first place. On the other hand, it is an opportunity to reflect on the immense possibilities that science as a field can provide at the elementary level for engaging children in enacted, embodied processes of knowledge construction. For example, the work that my research team has been doing examines the ways in which children's interactions around science can serve to facilitate dialogic encounters that support the emergence of complex understandings and questions. This research has revealed that supporting children with extended time for open-ended approaches to investigating can increase children's agency as learners (e.g., Siry, Wilmes, & Haus, 2016), and that building science investigations from children's own questions can lead to new conceptual understandings (e.g., Siry & Kremer, 2011). I contend that science education as a *practice* is an entwined mixture of content, process, and attitude, which is deeply contextual, and that none of these can be pulled apart from the other. With a grounding in dialectical theoretical perspectives, science to me is both the *process* as well as the *product*—the act of doing science and the development of conceptual understandings; each of which facilitates the other recursively and together is embedded in the attitudes that are shaped in this. As such, a question such as "why" teach science at the elementary level can be approached with an understanding of working towards the development of content, processes, and critical, positive attitudes towards the discipline. Yet this still leaves the big question of what, specifically, should be taught in elementary school, which I address in the next section.

WHAT SHOULD BE TAUGHT IN ELEMENTARY SCHOOL SCIENCE?

"What" in regards to curriculum is a problematic consideration, as it encompasses not only the intended goals and objectives, but also the specifics of the subject matter, the scope and sequence of these, and the instructional strategies, among other considerations (Gehrke, Knapp, & Sirotnik, 1992). Science as a school subject is impacted in practice by what is valued; valued by policy makers, valued within the curriculum itself, and valued by those who implement these policies—the teachers. Further, the actual implementation of what happens in science classrooms is a complex relationship between the student, the teacher, the curriculum, the context, and the policies (Rodriguez, 2015). Certainly there have been a multitude of policy reforms focusing on the teaching of science in the recent past (e.g., American Association for the Advancement of Science, 1993, 2001; NGSS Lead States, 2013; Rutherford & Ahlgren, 1991), but at the interface between these actual policies and their implementation lies the decision making of the teachers, and this is in essence the curriculum as enacted. In addressing the central question of "what" I attempt to bridge between the printed and the performed curriculum, but it is necessary to acknowledge the possible discrepancies between what exists in teachers' guides and what is lived and practiced in classrooms.

The debate over what, specifically, should be in elementary curricula to be taught in science has focused on content of science and the process of science for over 100 years (e.g., Dewey, 1910). Visions for science curriculum have included a variety of perspectives on what should be taught at the elementary school level, but quite consistent over the past several decades has been a discourse with foci on considering scientific content as thematic and focused on "big ideas" (e.g., Harlen, 2015), and scientific skills as process-based and focused on "habits of mind" (e.g., Gehrke et al., 1992). In short, we have learned from international research findings that there is a benefit for working towards emphasizing knowledge, skills, and values of science with children at the elementary level. What is often missing, however, is the focus on the contextualized ways in which children and teachers can learn together, in order to emerge with culturally and socially relevant understandings of the known, and the unknown.

What are the basics of a well-conceived science education is a controversial question in many ways—is it that the basics are an objective body of facts? Turning to the NGSS illustration mentioned above of science as the traditional natural sciences: physics, chemistry, biology, as well as the earth, space, and environmental sciences, once could perceive of this as a focus on knowing a specific canon of knowledge, even if the NGSS clarifies that the focus ought to be on a small number of core ideas in the sciences. In times of immense knowledge literally at our fingertips, the memorization of specific facts is in many ways superfluous, yet this is often the focus of elementary school science teaching practices, as students are to be prepared to memorize information that will be evaluated with tests. I contend that instead our focus should be on working to support students in creating knowledge and valuing the knowledges that they have and that evolve over time. Students ought to be supported in creating their own conceptual frameworks for understanding science phenomena and connections, and make meanings that connect with their own experiences and perspectives. If we consider such meanings as not things that are static, but rather as constructs that "come-to-be" in relations (Vagle, 2015, p. 9), then it is critical that we focus on these contextualized relations; between students, students and teachers, students and subject matter.

Above I asked if the "basics" are an objective body of facts. Perhaps one can also ask if the basics include a generalized approach for engaging in science as well. Turning again to the NGSS as an example of a standards-based conception of a discipline, once could conceive of science as a "theory building enterprise", which turns attention to the action of engaging in science and the "doing" of science (e.g., Siry, Ziegler, & Max, 2012). The doing of science in a classroom situation is a relational one (Cavicchi, 2014), as participants engage collectively in approaching science phenomena as something that is done. This relational aspect of science education turns the attention to the value of the process of science as a central basic grounding for science education. The argument for integrating content and process is certainly not new—it is one that has been in the literature on elementary science education for over 100 years (e.g., Dewey, 1910), and had been persistently argued for generations. David Hawkins (the director of the "elementary science study" in the early 1960s) emphasized the necessity of bridging what he coined "messing about" in science (1965) as he illustrated the value of balancing open-ended investigation time with more structured discussion opportunities. The connections between doing science, talking science, and learning science are deeply integrated and inseparable. Why then do we continually find that classroom practices return to knowledge acquisition alone? This persistent focus on knowledge acquisition in science has been suggested to be damaging to those concerned (Bryce, 2010), and it is far removed from any relational, contextualized meaning-making that students are more than capable of creating at the elementary level. This relational meaning-making was recognized in a review of the curriculum materials in science education from almost 50 years ago: "… the student is constantly making discoveries and organizing the processes and content which he is learning into some sort of fabric meaningful to him" (O'Hearn, 1966, p. 1). If we flip the focus from concrete certainty to be recast as a focus on uncertainty, there emerges a simple focus on valuing difference (Osborne, 1997), and an emphasis also on the focus on the unknown.

Today we are stuck in a pre-internet paradigm in elementary school science, where facts needed to be easily recalled to be useful. Simple facts are easy to find online, but the relationships between these are more difficult. The Internet gives students the ability to retrieve simple facts and thus the starting point to analyze these facts, their contexts, and their relationship to children's contexts. While those simpler facts may be easy to test, and thus measure, the purpose of the potential next step is missing–the development of scientific understandings in a relational manner. However, such contextualized understandings and foundations for an appreciation of uncertainty and the unknown is quite difficult to measure. If we work more on facilitating students' abilities to investigate the unknown, we are supporting them to create their own knowledges. In not teaching students to think and apply knowledge, we also limit the scope of who is reached, as there is a need to embed notions of engagement, equity, and diversity, to make science learning accessible to a wide range of students (Rodriguez, 2015). Further, we need to consider policy and curriculum as a lived text (Vagle, 2015)—to consider curriculum as it might be lived, not as it has been designed and planned, in order to emerge with a more contextual, relational view of science, and a school population supported in thinking, learning and being able to discover the unknown. In doing so, we can move towards the "re-imagining" of science education in order to move away from the "authoritarian knowledge structures" (Tytler, 2007, p. 67) to incorporate approaches to classroom investigations that lead to more diverse approaches to thinking about knowledge as well as about learning.

HOW SHOULD SCIENCE BE TAUGHT IN ELEMENTARY SCHOOL?

> Scientific knowledge is not a universal "metanarrative" from which one might eventually expect to be able to deduce a reliable answer to every meaningful question about the world. It is not objective but reflexive: The interaction between the knower and what is to be known is an essential element of the knowledge. And like any other human product, it is not value-free, but permeated with social interests. (Ziman, 2000, p. 327)

There has been significant research regarding the low confidence of elementary teachers to teach science (e.g., Appleton, 2007), suggested to emerge in part from teachers' own perceived lack of content knowledge (e.g., Keys & Bryan, 2000). Additionally, it has been elaborated that teachers' past experiences in science education have a strong impact on teachers' thinking about science education as a discipline (e.g., Bryan & Abell, 1999). The low confidence in teachers is not surprising if we think of teaching science as requiring an understanding of curricular objectives, pedagogical approaches for teaching in a "hands-on" manner, and the entire body of knowledge that is science. But, if we reframe what these basics are we can restructure teachers' roles as well as the purposes of the curriculum. In doing so, teachers can be supported to deconstruct their own notions of science with an emphasis on reconstructing to have an understanding of uncertainty, which is the heart of the scientific endeavor. If teachers begin to understand science as a search for meaning in essence, then this meaning making can be framed as locally bounded, and as individual and collective—content and process. Science is a set of cultural practices and discursive relationships focused on "constructing explanations, defending and challenging claims, interpreting evidence, using and developing models, transforming observations into findings, and arguing theories" (Kaartinen & Kumpulainen, 2002, p. 190). Reframing the focus of science as a school content area to move away from discreet, isolated facts, and instead to be on the discursive and cultural processes at play in the act of science, can ideally support a contextualized, relational teaching of science at the elementary level. As such, the interactive nature of learning—between the knower and the known as John Ziman (2000) elaborates—can be moved to the foreground of science teaching and learning.

As Madeleine Grumet noted in the original *Thirteen Questions* book over 2 decades ago, the notion of relation is one that is basic to education writ large

(Grumet, 1995), yet it is often what is left out of the conversation of what "the basics" actually are. Relationships between the students and the material, and also human relationships, can become central as "the relational processes that negotiate trust are inseparable from the intellectual processes that initiate experimenting" (Cavicchi, 2014, p. 188). Relational, contextual approaches can connect science to students' lifeworlds, and there is sufficient evidence from research of the value of reframing science education as participation in communities (e.g., Roth & Lee, 2004) and as a cultural way of knowing (e.g., Meyer & Crawford, 2011), so that cultural practices of science can be woven together with the cultural experiences of students. Returning to the opening quote from John Dewey from over 100 years ago, attitudes are a critical part of science education, and I suggest that positive attitudes for children as well as their teachers can emerge if science is framed as something that is close to participants' lives. In supporting teachers to find contextualized approaches to engage in science with their students can lead to a contextualized science based on diversity and difference, one that is not only focused on memorizing and regurgitating information, but one that is focused on teaching and learning science as a complex, contextualized experience.

> When our schools truly become laboratories of knowledge-making, not mills fitted out with information-hoppers, there will no longer be a need to discuss the place of science in education. (Dewey, 1910, p. 127)

REFERENCES

American Association for the Advancement of Science. (1993). *Benchmarks for science literacy*. New York, NY: Oxford University Press.

American Association for the Advancement of Science. (2001). *Atlas of science literacy: Project 2061*. Author.

Appleton, K. (2007). Elementary science teaching. In S. K. Abell & N. G. Lederman (Eds.), *Handbook of research on science education* (pp. 493–535). Mahwah, NJ: Lawrence Erlbaum Associates.

Au, W. (2011). Teaching under the new Taylorism: High-stakes testing and the standardization of the 21st century curriculum. *Journal of Curriculum Studies, 43*(1), 25–45. doi:10.1080/00220272.2010.521261.

Berg, A., & Mensah, F. M. (2014). De-marginalizing science in the elementary classroom by coaching teachers to address perceived dilemmas. *Education Policy Analysis Archives, 22*(57), 1–31. doi:10.14507/epaa.v22n57.2014.

Bryan, L. A., & Abell, S. K. (1999). Development of professional knowledge in learning to teach elementary science. *Journal of Research in Science Teaching, 36*, 121–139. doi:10.1002/(SICI)1098-2736.

Bryce, T. G. K. (2010). Sardonic science? The resistance to more humanistic forms of science education. *Cultural Studies of Science Education, 5*, 591–612. doi:10.1007/s11422-010-9266-6.

Cavicchi, E. (2014). Learning science as explorers: Historical resonances, inventive instruments, evolving community. *Interchange, 45*, 185–204. doi:10.1007/s10780-015-9235-9.

Dewey, J. (1910). Science as subject matter and as method. *Science, New Series, 31*(787), 121–127.

Gehrke, N. J., Knapp, M. S., & Sirotnik, K. A. (1992). In search of the school curriculum. *Review of Research in Education, 18*, 51–110. doi:10.3102/0091732X018001051.

Gopnik, A. (2012). Scientific thinking in young children: Theoretical advances, empirical research, and policy implications. *Science, 337*, 1623–1627. doi:10.1126/science.1223416.

Grumet, M. R. (1995). The curriculum: What are the basics and are we teaching them? In J. Kincheloe & S. Steinberg (Eds.), *Thirteen questions: Reframing education's conversation* (pp. 15–21). New York, NY: Peter Lang.

Harlen, W. (Ed.). (2015). *Working with big ideas of science education.* Trieste, Italy: The Science Education Programme (SEP) of IAP.

Hawkins, D. (1965). Messing about in science. *Science and Children, 2*, 5–9.

Head Start. (2010). *Child development and learning framework: Promoting positive outcomes in early childhood programs serving children 3–5 years old.* Washington, DC: Office of Head Start. Retrieved from http://eclkc.ohs.acf.hhs.gov/hslc/tta-system/teaching/eecd/Assessment/Child%20Outcomes/HS_Revised_Child_Outcomes_Framework(rev-Sept2011).pdf

Hedges, H. (2014). Young children's "working theories": Building and connecting understandings. *Journal of Early Childhood Research, 12*(1), 35–49. doi:10.1177/1476718X13515417.

Kaartinen, S., & Kumpulainen, K. (2002). Collaborative inquiry and the construction of explanations in the learning of science. *Learning and Instruction, 12*, 189–212. doi:10.1016/S0959-4752(01)00004-4.

Keys, C. W., & Bryan, L. A. (2000). Co-constructing inquiry-based science with teachers: Essential research for lasting reform. *Journal of Research in Science Teaching, 38*, 631–645. doi:10.1002/tea.1023.

Meyer, X., & Crawford, B. (2011). Teaching science as a cultural way of knowing: Merging authentic inquiry, nature of science, and multicultural strategies. *Cultural Studies of Science Education, 6*, 525–547. doi:10.1007/s11422-011-9318-6.

Next Generation Science Standards. (2013). *Three dimensions*. Retrieved from http://www.nextgenscience.org/three-dimensions

Newton, D., & Newton, L. (2000). Do teachers support causal understanding through their discourse when teaching primary science? *British Educational Research Journal, 26*, 599–613. doi:10.1080/713651580.

NGSS Lead States. (2013). *Next generation science standards: For states, by states*. Washington, DC: The National Academies Press.

O'Hearn, G. (1966). *A review of new science curriculum materials* (Bulletin 134). Madison, WI: WI Department of Public Instruction.

Osborne, M. D. (1997). Balancing individual and the group: A dilemma for the constructivist teacher. *Journal of Curriculum Studies, 29*, 183–196. doi:10.1080/002202797184125.

Rodriguez, A. J. (2015). What about a dimension of engagement, equity, and diversity practices? A critique of the Next Generation Science Standards. *Journal of Research in Science Teaching, 5*, 1031–1051. doi:10.1002/tea.21232.

Roth, W.-M., & Lee, S. (2004). Science education as/for participation in the community. *Science Education, 88*, 263–291. doi:10.1002/sce.10113.

Rutherford, F. J., & Ahlgren, A. (1991). *Science for all Americans.* Oxford, England: Oxford University Press.

Science Council. (2015). *What is science?* Retrieved from http://www.sciencecouncil.org/definition

Siry, C., & Kremer, I. (2011). Children explain the rainbow: using young children's ideas to guide science curricula. *Journal of Science Education and Technology, 20*, 643–655. doi:10.1007/s10956-011-9320-5.

Siry, C., Wilmes, S., & Haus, J. (2016). Examining children's agency within participatory structures in primary science investigations. *Learning, Culture, and Social Interaction.* doi:10.1016/j.lcsi.2016.01.001.

Siry, C., Ziegler, G., & Max, C. (2012). "Doing science" through discourse-in-interaction: Young children's science investigations at the early childhood level. *Science Education, 96*, 311–326. doi:10.1002/sce.20481.

Tytler, R. (2007). *Re-imagining science education: Engaging students in science for Australia's future.* Retrieved from http://research.acer.edu.au/aer/3

Vagle, M. D. (2015). Curriculum as post-intentional phenomenological text: Working along the edges and margins of phenomenology using post-structuralist ideas. *Journal of Curriculum Studies, 47*, 594–612. doi:10.1080/00220272.2015.1051118.

Ziman, J. (2000). *Real science: What it is and what it means.* Cambridge: Cambridge University Press.

CHAPTER THREE

The Science Curriculum

What Are the Basics and Are We Teaching Them?

FEMI S. OTULAJA AND MESHACH B. OGUNNIYI

The world is so much different today than it was when the book, "*Thirteen Questions: Reframing Education's Conversation*" was published in 1994 by Joe Kincheloe and Shirley Steinberg (1994). It may be a surprise to us to know how many people have died and/or been killed in wars in which knowledge or application of basic science has played a significant part, for example, in the design and production of the weapons of warfare, in the dissemination of the news related to the war, in the process of rescuing solders and the victims of such wars, and in trying to save their lives at the hospital or on the field. Of course, it would be foolhardy to argue that wars and other human tragedies, in which the products of science (including technology) have been used, are necessarily caused by this human enterprise. Like any currency, science is neither good nor bad. It depends on the way it has been used.

The question is what science knowledge is of most worth and should be taught to students. The same question has been discussed even more vociferously before. The argument regarding the launching of Sputnik by Russia in 1957 was that the Russian science (and technology) curriculum was perhaps superior to that of the U.S. This spurred frantic new science curricula reform in the U.S., the West, and subsequently the developing world. In Africa, the curriculum reforms were largely part of the response to the curriculum reforms happening in the West. In addition, it was in response to the challenges posed by hard-won political independence. As African countries gained political independence, they became aware of their inadequacies in many areas. There were great shortages with respect to human and material resources, qualified science teachers, science laboratories, textbooks, enrolment in science; etc. African countries soon realized that political independence without economic independence was a hollow achievement (Ogunniyi, 1986, 1995).

The major focus of the curricular innovations and related textbooks that resulted from this reform has been to study science not as a completive body of knowledge to be transmitted to students, but as an inquiry. This focus was followed by a concern for both teachers and students to understand the nature of science (NOS), in essence, to become scientifically literate. In terms of the products of science, a number of science curricula specified certain conceptual schemes and assumptions—e.g., assumptions about diversity, change, continuity, interactions, organization and limitations of science. Other curricula emphasized scientific processes such as observation, classification, measurement, prediction, problem identification, formulating of hypotheses, experimentation, gathering data, interpreting and analysing data, synthesizing data, drawing valid inferences, etc. Students were encouraged to behave as scientists in terms of making decisions, allowing for serendipity, undertaking scientific inquiries, expanding scientific knowledge, and communicating their findings (Ogunniyi, 1997).

But teaching science as an inquiry or learner-centred activity especially in an examination-driven education system has its own challenges. In most African countries, assessment maintains strangle-hold effects on the entire education system (Ogunniyi & Rollnick, 2015). Also, in an era where change is the only constant, we are confronted with the educational issue of preparing today's children for that change. For teachers, this means knowing and teaching more than what can traditionally be found in the regular science textbooks. For learners, it means knowing more than what they encounter in the science classroom. Amidst the avalanche of scientific information available on the internet, how does the teacher guide his/her students to decipher between what is or what is not relevant? The implications for curriculum development and instructional practice of adequately dealing with all of this information, though not the focus of this chapter, are certainly worthy of closer consideration.

The issue of making informed choices becomes even more complex in an era of information explosion. How does the teacher or learner know what to select? When the need to make informed choices is juxtaposed against the gap between school science and mass scientific illiteracy existing in the society outside the school walls, the task of what basic science to teach or learn becomes a Herculean endeavour. As a result, the average citizen of today, particularly in Africa, is relatively naïve about how to tackle the diverse environmental problems of our time—the pollution of our environment, the rapid erosion of the ozone layer resulting from the excessive use of pesticides, ignorance, disease and poverty of unparalleled proportion, drought, flooding, the swelling population of refugees, the antiscientific movements, resistance to antibiotics, overpopulation, to list but a few. In light of all such challenges, what role should the science curriculum play? What science basics should be taught?

There seems to be a credibility gap, far greater than science and science educators would like to admit, existing between the science curriculum we teach and the lived experiences of our students. The same is also true between the intellectual interests of the scientist and those of the science teacher or the society at large. The basic scientist is not primarily concerned about whether or not her/his findings would contribute to human social welfare. Rather s/he is driven by creativity and passion for discovery and the freedom to explore more and more towards an ultimate reality. Of course, we in science education (including the curriculum planners and policy makers) frequently are driven more by pragmatic considerations than by the inherent beauty and elegance of scientific pursuit for its own sake.

Often, most curricula are driven by political rationality than scientific and/or technical rationality. The caveat of course, is that the development of science curricula for pragmatic consideration alone is short-sighted in light of the fact that we are preparing students for a constantly changing world and a world of work. In our view, it is unjustified to engage students with scientific concepts which are in the process of becoming obsolete anyway. Indeed, the basics of school science are anything but current. Also, if we expect every curriculum to be immediately relevant for here and now, we do the students a great disservice. After all, what is considered irrelevant today may turn out to be relevant in the future. It is our view that apart from covering the essential scientific concepts and principles, every curriculum should have a futuristic or visionary element to it.

From the foregoing, it may be interesting to know how many children have been caught in the curricular melee and debate and who are going to be subjected to whatever dominant discourse is in vogue. Yet, such a dominant discourse might have precluded their voices and perspectives (including their ways of knowing, thinking, doing and being). As a result, the official curriculum, rather than stimulate the children's intellectual interests, has actually motivated them to resist the curriculum. In this regard, the important questions to ask, among others include: Whose science? Whose curriculum? Should what is basic to the science curriculum in the U.S. be equally basic in other countries or cultures? Was the "Science for All" really science for all peoples? Was the "Common Core" curriculum really about a common core for all, irrespective of the diversities of the nation-states, their goals and aspirations? Was the science curriculum of 1994 and the current one inclusive of the non-mainstream or second language students? Is the new science curriculum dubbed the "Next Generation Science Standards" (NGSS) (NGSS Lead

States, 2013) really for the next generation of students in all non-Western cultures worldwide? What and whom does NGSS preclude, distort and/or suppress? Whose voices, perspectives, and ways of knowing, doing, and thinking are being considered? Is NGSS not another effort to perpetuate the same old Eurocentric dominance of the past in a growing multicultural context in the US? We can go on and on, but it suffices to say that no curriculum (including the science curriculum) is culture-free. Regarding the core ideas of NGSS, why are they core ideas in the first place? Or in what sense are these core ideas—i.e., essential, fundamental of what? Are they core in an inclusivist or exclusivist sense? Are they core to the extent that no matter whatever change may be occurring in the society, they will still be appropriate? How, for instance, will students exposed to such core ideas be prepared enough for the changes taking place in their society? Will the core ideas help students become aware of the social aspects of science—e.g., perception of the cultural context in which science thrives, recognition of the role that science could play in the society or the broad perspectives of culture, society and its history? Would they become more aware of the strengths and limitations of science in tackling diverse human problems? The same questions arise when we think of the cross-cutting concepts. What are these concepts and why are they so designated? Are they consistent with the nature of science? Are they cognizant of indigenous sciences and the cultural environment and worldviews of the students? Would they contribute to students' scientific literacy? Looking at the science and engineering practices espoused in/by NGSS, what are they and how are they articulated in the curriculum? What is the distinction between the practices in science and in engineering, and how are they reflected in the NGSS? Where is the nature of science, which for long has provided the platform and the driving force for scientific inquiry and the way science is taught? Where is the aspect dealing with scientific literacy—a major goal of science education? Where is the aspect that recognizes the educational and cultural values of the indigenous wisdom which has for centuries shaped the way of life of indigenous people? What sort of pedagogies could be mobilized to integrate indigenous science and engineering with school science? In short, will the students be more scientifically literate than has been the case? We can go on and on.

As Alan Bishop (1990) contends about the myths of a culture-free mathematics curriculum, that school mathematics is one of the colonialists' weapons of oppression and subjugation of indigenous knowledges, so do we contend that the basics of the school science curriculum has also been used as weapons of oppression and subjugation against indigenous peoples, even decades after most African countries had gained independence. The question is: how will the African countries own the science that is taught in their schools? Can teachers who have been groomed in Western science teach science that is culturally sensitive and relevant to non-Western students?

THE HEGEMONY OF THE BASICS VERSUS THE GOAL OF SCIENTIFIC LITERACY

The fact that for many decades now, after political independence, and most likely, for the foreseeable future, African nations still continue to follow the Eurocentric science curriculum is an indication of the apparent relational disconnection with the needed indigenization of the curriculum. The consequence of this has been the perpetuation of the dominant voice (discourse) and the disenfranchising and alienation of local students or learners from school science early in their educational pursuit. The basics of the science curriculum in African schools and the ways the science is being taught is totally disconnected from the context in which science is taught.

One often wonders, in what ways have the current or previous science curricula been implemented in African schools? Are they really reflective of the wills and aspirations of the indigenous peoples of Africa? How suitable or sensitive is science pedagogy for the African classroom environment? Africans used to "do" science before the advent of Eurocentric science. Africans taught and learned science in various ways (e.g., through drama, music, apprenticeship, storytelling, symbolic representations, songs, idioms) before Eurocentric science. African science had been borrowed or literally stolen, repackaged and sold back at huge costs to African lives. The questions are: Were the pyramids built using Eurocentric science? Were mummification of the Pharaohs and nobles of Egypt based on modern scientific processes? To what extent has the knowledge of the ancient University of Timbuktu been reflected in the African science curricula? The questions that we have raised here and similar ones that will arise in the readers' minds are meant for all to ponder. Some have easy answers while others are complex and difficult to answer—all certainly warrant scholarly consideration.

THE DISCONNECTEDNESS

As stated earlier, the basics of the science curriculum in Africa are still the same as the ones introduced by the colonial masters half a century ago. African peoples' interconnectedness to their spirituality and the elements of nature that sustain them are missing in their curricula. The contents and concepts of science curriculum are still the same even though many changes have taken place. The science curriculum, which promised to bring indigenous people out of the so-called dark ages, of what was thought to be false beliefs and superstition, turned out to separate indigenous people from their roles as custodians of the elements of nature. The Eurocentric science curriculum and the ways the basics of science have been taught brought in its wake diverse environmental problems associated with scientific and industrial revolution, notably, global warming, environmental pollution, greenhouse effects, flooding, drought, desertification, etc.

The connections Africans have to their own knowledge base and relationships with other cohabiting earthly and planetary "beings" that had sustained them for centuries before the advent of colonizers have since been devalued and made insignificant in the Westernized science classroom. Many of the African's "superstitions" and taboos turn out to be safeguarding measures to protect and manage the resources within this one universe we share with other non-human beings. Science is only recently realizing that these resources are finite. The basics of science curriculum, as we know it, are yet to fulfil its grandiose claims of solving human problems. For every problem it seems to have solved, it has created more—for example, the problem of urbanization with its concrete jungles. The fact that Western science has failed to solve diverse social problems such as hunger, poverty and resurgence of diseases that were supposedly "conquered," are some vivid examples. Science's claim to humanity without spirituality has also contributed to the destruction of indigenous people's self-image. In other words, the average non-Western indigene has suffered irreparable damages, such as, loss of self-identity, the essence and components of his/her being, and s/he has remain marginalized, even today.

This loss of self-image is much more prominent in the teachers of Westernized science who often enforce Eurocentric science as the only worldview and knowledge. Most of these teachers of science, having been schooled only in Westernized science, do not see the science in indigenous knowledge and the indigenous knowledge that is existential in science. They look down on their own indigenous knowledge, their local wisdom, which they often refuse to learn and yet use in the backrooms. They remain numb to the realization and consciousness that science, as we know it, is multicultural, an amalgamation of other peoples' sciences repackaged. When a person is subjugated and oppressed, his/her way of knowing and being are suppressed following which his/her relational principles are distressed and end up being repressed. The basics of any curriculum and its pedagogy should be to connect and/or reconnect the essence and components of self and being in holistic ways to all the elements of nature that sustain the self—the food we eat and where it comes from, the air we breathe, the water we drink and their interconnectedness to the hereafter, the recycling of all being.

THE IMPOSED CURRICULUM

When the basics of the science curriculum are prescribed or enforced overtly and covertly, they serve the self-interests of some people over others. To accomplish the self-interests of Western culture, the essence of schooling is not meant to emancipate other non-Western people and engage their worldviews in the spirit of knowledge sharing; it was to inculcate them through the written and hidden curriculum to the bidding of the dominant discourse's only view of the world. By imposing the basics of the science curriculum on the will and being of the people of indigenous communities, the dislike of Westernized science was imminent. Hence, it is no wonder why indigenous students are failing Westernized science.

In the first place, the fact remains that the basics of the science curriculum, the policies and pedagogies for enacting them, have always been written in the language and discourse of the colonial masters. In most parts of sub-Saharan Africa, English language is the dominant and official language of discourse in the former Anglo-Saxon colonies' (Nigeria, Ghana, Gambia, Senegal, Sierra Leone, Liberia, Tanzania, Zimbabwe, Botswana, Swaziland, etc.). It is French in the former territories of France (Togo, Benin, Central African Republic, Cote d'Ivoire, Chad, Madagascar, etc.). It is still Portuguese that is valued in the former colonies of Portugal (Angola, Mozambique, Cape Verde, etc.), and it is still Spanish in Equatorial-Guinea, the former colony of Spain. And in South Africa, despite the 1976 uprising against the imposition of Afrikaans as the language of learning and teaching in Black schools, the invented language of the apartheid segregators is still the second dominating language of academic discourse next to English language. This is despite the fact that these languages and cultural groups are the minorities, in population, among the larger African population with nine additional official languages.

In the second place, the conceptualization of what should constitute the basics of the science curriculum has never involved the people upon whom the curriculum is imposed, neither was their knowledge base engaged in this process. The science curriculum has always been wilfully conceptualized without the participation and contributions of indigenous people and is enforced as the dominant discourse without regards to the peoples' needs. Neither can we remember a gathering where even the Ministry of Education officials called for a local referendum on the curriculum to be enacted in our schools. In many cases, teachers are handed the curriculum with make-shift training on how to use it and are told to go and teach. We are yet to see any nation in Africa where any local, indigenous, language is made the official language of academic, scientific and/or commerce discourses.

It is apposite to say that during our days as pupils, concerted efforts were made to disconnect us from our local language and ways of knowing and being. Social, cultural and symbolic violence were meted out by our teachers, trained up as European apologists (colonialists). Corporal punishments, in one form or another, were unleashed on us (which is still being done) to dissociate us from using our home languages in the school—and even in the name of being "civilized" at various homes. There were cultural and social incentives for making caricatures of those who struggle to master the colonial master's language; hence, many dropped out of school as a result of not being able to cope with the language demands of the basics of curricular activities. The results were seething angers and resistances to having our beings discounted as humans; for to be able to speak the colonial master's language eloquently grants the indigene some social, cultural and symbolic statuses as "cultured," and recognized as the "Black-White person" and the cognitive symbolism of being "smart."

Deliberative curriculum theory holds that the stakeholders who are invested in the development of a particular curriculum should engage in the development process through a process of mutual deliberation (Curry, 1992). The science curricula in many African countries during the independence era of the 1960s and 1970s and even afterwards have been controversial for many reasons. Some of the controversies include: the poor preparation of teachers who were expected to implement the new curricula that required a radically different instructional approach from the usual; the imposition of curricula in which teachers have not been actively involved; the demand of certain curricula to include the indigenous knowledge in the science curriculum (e.g., South Africa, Lesotho, Swaziland and, to some degree, several other countries) without adequate preparation or textbooks to teach such curricula; etc.

Even the new science curricula in Africa, for various reasons, are not problem-free. The new South African science curriculum exemplifies this. Indeed, the public outcry against the new curriculum is not unrelated to the top-down approach and the haphazard ways in which the curriculum was developed and implemented. This led Jansen (1997) to state unequivocally ten reasons why the new South African science curriculum would fail.

Besides the reasons given by Jansen (1997), the new curriculum was introduced at a time when the education system was still reeling from a plethora of reforms as described. In addition, individual teachers, many of whom were products of the old apartheid education system, all of a sudden, found themselves having to implement a curriculum that required of them a completely new way of thinking and relating to learners. To make matters worse, there was very little support in terms of resources; and even in the tertiary education institutions, few understood the demands of the new curriculum. Critics of the new curriculum argue that the national curriculum was introduced too early and without adequate preparation (Baxen, 2001). As will be seen later, the introduction of the new curriculum has had far reaching effects on teacher education.

The attempt to indigenize the new science curriculum in South Africa has been criticized for the same reason of poor consultation with other stakeholders, particularly the teachers. Whatever might be the justification for including indigenous knowledge (IK) in the science curriculum, it should have been realized that most teachers have been schooled largely in Western science and hence were unfamiliar with the IK worldview. The new curriculum demands new instructional approaches and goals in terms of contextualization and indigenization. However, these contradict the countless assessment protocols associated with the new curricula. The lack of consultation by the curriculum planners seems to underrate teachers' roles in curriculum planning and implementation. In addition, there was lack of clarity on how a science-IK curriculum could be implemented. Science curricula which emphasizes alternative goals for classroom pedagogy in a context where conceptual goals predominate, is notoriously difficult to implement unless a well-planned and supportive teacher training programme is in place.

Despite the mammoth problems surrounding science curricula in Africa, some giant leaps have been made to reform the old curricula, post-1994, in South Africa. But moving beyond the rhetoric of curriculum reforms the issue of preparing a cadre of teachers who are able to fulfil the aspirations and mandates of the new curricula in Africa is not negotiable. It certainly warrants a closer consideration by teacher training institutions and other stakeholders. The conundrum: Good teachers prepare good learners while bad teachers drive learners out is equally applicable to the curriculum as well. It is a well-known fact that the quality of teachers capable of implementing the intentions of any curriculum is determined by the quality of training they received at higher education institutions.

Any new science curricula in most African countries invariably require new instructional strategies that go beyond the chalk-and-talk approach within which most teachers have been groomed. Furthermore, most of the curricula envision teachers who are not only knowledgeable in their respective disciplines but who are also able to function effectively in a holistic, learner-centered, well-integrated, activity-based, and multicultural classroom. Also, the new curricula aim at developing the full potential of each learner to become a scientifically literate productive member of a democratic society. The challenge of achieving this aim of course rests heavily on the teachers. As knowledge brokers, of scientific knowledge and social values, teachers have the unique role of preparing students for a scientifically and technologically-driven economy. Furthermore, these science curricula envisage teachers who are capable of mediating learning experiences gained in the classroom with experiences that learners encounter in the world outside the school walls (Department of Basic Education, 2002). These expectations might seem unachievable, but in reality they depict the challenges faced by a science teacher teaching in the current multicultural African classrooms. In other words, teaching goes far beyond the transmission of a litany of scientific terms; it is a transaction between people seeking to share mutual experiences and values critically important in a modern society.

The kind of teachers envisaged by the new curricula in African countries are certainly different from what most teachers entering into, or are already in the teaching profession, are familiar with. Nevertheless, most of these curricular statements construe teaching as a complex and multifaceted activity. Whatever training that teachers might have, it is the school contexts, the teachers' understanding of such contexts, and the beliefs they hold about teaching and learning and education as a whole that would ultimately shape their instructional practices. Unless teachers have functional understandings of what teaching and learning entail and the socio-cultural context in which such endeavours are embedded, they might fail to move beyond the performative act of knowledge dissemination (Ogunniyi & Mushayikwa, 2015).

We can still remember in our students' days how we studied science (Physics, Chemistry, Life Sciences, Natural and Earth Science, and Agricultural Science) that was quite removed from our daily experiences, even though many of us were from the farming communities. We knew so much about the flora and fauna and their dietary and medicinal uses. We knew about the seasons, days and nights, the phases of the moon, and the stars in the sky, yet our stocks of knowledge were rolled-up and bounded as voodoo or pseudo-sciences rather than "real" or authentic indigenous sciences. As much as only the approved basics of the science curriculum were enacted in the classroom so would the minds of learners be considered "tabular rasa," with no prior knowledge that was significant to the science classroom discourses. Then and even now, most learners in African classrooms are still regarded as empty vessels to be filled with accepted standard accounts of Western science.

The above scenario still depicts the way that science is taught in many African science classrooms. Though the contents and emphases of the curricula have changed, the modes of instruction have remained virtually the same as it was at the independence era, even in South Africa two decades after the demise of the apartheid system of education. It seems to us that a credibility gap exists between what the new science curricula envisage and what is enacted in the science classroom. As students our needs were not considered part of the equation, so to speak. The indigenous knowledge we brought into the classroom was rarely seen as a source of learning and a basis for the exchange of ideas in a fair and equitable way. We were made, and learners are still being made, to accept that the indigenous taxonomy of flora and fauna based on careful observation, which in turn depends on the sense of sight, smell, hearing, touch and taste we knew before we started schooling were considered not valid. From the foregoing, we can see that, as in the normal biological practice, criteria used for description and classification are based largely on observation of animals and plants. Of course, this practice is not limited only to the indigenous communities; common names used to designate plants and animals by Westernized science are also based on observation in terms of color, taste or other characteristics e.g., cuckoo, the migratory bird named from its call, cuckoo-pint, the wild arum, popularly known as wake-robin or lords-and-ladies, cuckoo-flower or lady's smock, and so on. The same power of observation is used for innumerable chemical substances in terms of their appearance, color, smell, taste, touch, and sound. Other criteria used for classification include their ecological characteristics as were practiced in indigenous communities before the advent of the Westernized science.

These same principles used by our forefathers for describing flora and fauna in terms of their color, taste, ecological niches, or in associating with the names of the discoverer is not only a scientific practice but a commonly used indigenous practice as well. Indeed, the use of Greek and Latin words in modern science (including astrophysics), to describe diverse phenomena are vivid exemplars of this common practice. However, due to the hegemony of Western science that has occupied our thinking patterns for many decades, we now need to relearn the names of many flora and fauna we once knew from childhood. As in science, the indigenous peoples use broad categories which in turn are then sub-divided into smaller more specific functionally-related categories, i.e., something reminiscent of the Swedish scientist, Carolus (Carl) Linnaeus (1707–1778), though quite distinct from it. Hermeneutical analysis of the nomenclatures revealed that indigenous peoples have been using observed behaviours and characteristics of the flora and fauna in giving them their local names. This method, which is acclaimed as a Western scientific practice, was actually borrowed from indigenous methodologies (Hountondji, 1997). Worse still, indigenous non-Western learners have for decades been forced to learn the new scientific nomenclatures and taxonomies in languages that were totally foreign to us.

The faulty perception that the so-called "savages" are innocent of rational thought has for long been debunked in serious scholarship. Even up till the early 20th century, it was thought that indigenous students could neither grasp complex scientific terms nor categorize such complex terms in the description of plants and animals, their environments, the ecosystems and the biomes in which they existed. However, given the right atmosphere, learners from indigenous communities have performed as well as their Western counterparts. Such students have not only been able to master the names and taxonomies of diverse phenomena but have also been able to determine their physical and chemical properties. However, our own indigenous knowledge was devalued outright from the onset of our schooling and was tossed out of the window, so to speak, in conformity with the conundrum of "throwing out the baby with the bath water."

To compound the challenges of indigenous students, they are still not even allowed to code switch in the science classroom. To do so was considered an anathema to their cognitive development. In fact, indigenous students were punished for deploying that practice inside and outside the classroom. So, most of them ended up learning by rote without being able to link what they were learning to their life world experiences. Likewise, their ways of thinking and learning were taken from them as their time was preoccupied with learning the so-called "modern" way by learning Western science. They were deprived of their common practice of learning from nature. Instead, they learned vicariously and in contrived learning environments when in actual fact they were surrounded by nature as their natural laboratory. Even today, it is not uncommon to see a biology teacher abstractly drawing the roots of plants, different types of vegetation and various ecological niches on the chalkboard when there is an abundance just outside the classroom window where students could go to make learning real to them.

ARE WE TEACHING THE SCIENCE CURRICULUM BASICS?

There is no doubt that to a large extent the basics of the science curriculum are still being taught, in the same old ways as they were several decades ago when we were students. To a large extent science is still being taught with renewed vigour under the guise that Eurocentric and Westernized science is the only worldview to human understanding of nature and knowledge. As stated earlier, most science curricular have been the direct copies or slightly modified copies of science curricula from Western countries. However, these curricula have not been redesigned to reflect the language, ways of knowing, thinking and doing of indigenous peoples. In South Africa for instance, since independence in 1994, the apartheid curriculum has been revised four times with the goal of enacting outcomes-based curriculum that the West has long abandoned. While the basics of the science curriculum have been revised to include the incorporations of indigenous knowledge, the teaching is yet to include and access learners' indigenous knowledge or cultural wisdom.

The revised South African National Curriculum and Assessment Policy Statement (http://www.education.gov.za/Curriculum/NationalCurriculumStatementsGradesR-12.aspx) is expected to produce the next generation of engineers, scientists, technologists and medical professionals with a Westernized worldview. These experts are expected to usher African nations into global prominence through the enactment of the basics of Westernized science curriculum. In addition, these neo-colonial elites are expected to make it possible for African countries to become globally competitive and to become powerbrokers in their axes i.e., the North-South, South-South or North-North axes. However, it seems to us that policy makers and education stakeholders in Africa are spellbound to forget that global competitiveness is determined and defined by the same powers that colonized their forebears and therefore, are not likely to relinquish their stranglehold—a set goal to keep Africa as the source for their raw materials and receptacles for manufactured goods. Hence, the second scramble for Africa has since been underway and it is not for liberation but cognitive and economic enslavement; not for reconnecting Africans with their knowledge base.

The basics of the Eurocentric science curriculum continue to be taught, even right to the letter of the textbooks. But as stated earlier, all these will lead to the production of individuals with a colonized mentality i.e., elites with well-packed minds rather than elites with well-formed minds. Such elites tend to lose their sense of social identity in that they have lost their interconnectedness to their own language and cultural wisdom. They often have lost their indigenous sciences and their African ways of knowing, thinking, doing, and being. At best, as science teachers, they merely reproduce what they have been taught or what have been written in the textbooks. So, their teaching of the basics is purely didactic with little or no innovation. These teachers often fail to access the background knowledge of their African learners of science and as such disenfranchise them. Any little wonder then why education stakeholders wonder why African students are not performing, let alone, excelling in science. The questions remain: Whose science and for whom? The continued decline of student enrolment in science is not unconnected to the same issue of what science is being taught or learned in African science classrooms. It has been shown that these learners of science are more likely to engage more in the science they are learning if it resonates with their cultural identity (Ogunniyi, 1996, 1997, 2004, 2007a & b).

RENEGOTIATING THE BASICS OF SCIENCE

With the current resurgence of multicultural classrooms becoming the norm in previously monocultural societies worldwide and, most likely, lasting into the foreseeable future, even Western science teachers are facing challenges in the classroom. The science classroom is becoming much more ethnically, culturally, religiously and linguistically diversified. So, to hold on to the notions that the basics of the science curriculum will or should continue to serve the self-interests of the Western worldview will be foolhardy. There is therefore the need to renegotiate the basics of the science curriculum and the teaching of a Westernized science curriculum to include the ways of knowing of other cultures. Without this authentic renegotiation and genuine transformation, multicultural science classrooms will continue to post challenges for science teachers. In addition, non-mainstream, English as second language students will continue to suffer unnecessarily. They will perform woefully and be disadvantaged from active participation in science discourses in their classrooms. Among other consequences is also the fact that non-Western students will remain in perpetual conflicts between their home-based funds of knowledge (indigenous worldview) and school science (Eurocentric worldview). Mitigating the home-based knowledge with school science calls for integration of the basics of science (science curriculum) with other worldviews. Making science more culturally relevant, in contents and pedagogy, is beneficial to all students.

The educational policies of many indigenous nations have embraced the inclusion of indigenous knowledge, rather than the integration of indigenous knowledge with science. Here in South Africa, the National Curriculum Statement (NCS) document, implemented in 2003, and the recent Curriculum and Assessment Policy Statements (CAPS) document, implemented in 2011, both provide for the need to value indigenous knowledge in teaching and learning by, saying,

> … up until then [1960] the Western world had only valued logical, mathematical and specific linguistic abilities, and rated people as "intelligent" only if they were adept in these ways. Now people recognise the wide diversity of knowledge systems through which people make sense of and attach meaning to the world in which they live. Indigenous knowledge systems in the South African context refer to a body of knowledge embedded in African philosophical thinking and social practices that have evolved over thousands of years. The National Curriculum Statement … has infused indigenous knowledge systems into the Subject Statements. It acknowledges the rich history and heritage of this country as important contributors to nurturing the values contained in the Constitution. As many different perspectives as possible have been included to assist problem solving in all fields. (NCS, 2003, p. 14)

As we look at what is and what might be in the years ahead, it behooves us to look within Africa for potential solutions to the vexing problems of educating South Africans in and about science. Within this endeavor the final section of our chapter examines the potential of Ubuntu-driven science.

UBUNTU-DRIVEN SCIENCE CURRICULUM?

Ubuntu is a worldview theory that is central to African indigenous people. *Ubuntu* literally means, "I am because of you"; "I am what I am because of who we all are." It is the essence of being human. It is the belief in the universality of the bond of sharing the space and the resources that connects all humanity with all non-humanity in existence in the universe. Humans will not be humans without the non-humans elements in existence that sustains us. *Ubuntu* construes people's knowledge, perceptions and behavioural patterns in accordance with indigenous societal norms. It is a way of thinking and doing that constructs humans as being endowed with such virtues as kindness, respect, sympathy, empathy, cooperative spirit, concern for each other's welfare, and having a sense of moral responsibility. It also emphasizes human interrelatedness, interdependence, reciprocity, synchronicity and the inseparability of a person or his/her ideas from that of the community in which s/he is reared or lives in. It is more about "us" not "me." Your joy or pain is mine as mine is yours. In such a setting, any personal claim to a particular idea is therefore unwarranted. It is all about our shared values, mutual responsibilities and respect to each other and the elements of nature that sustains us.

Though an Nguni word (a Niger-Congo language group), *ubuntu* is a philosophy and an ideology that is universal to African indigenous communities throughout the continent. It is called different names by different indigenous communities. For example, it is called *botho* in the Sotho language, *umuntu* in Chewa/Nyanya language, *unhu* in Shona language in Zimbabwe, *hbret* or *meth be bar* in Tigrigna. It is called *omoluabi* (a personification of *agbajo-owo* (collectiveness) undergirding humanness, virtues, etc.) in Yoruba language. It is about shared values, mutual responsibility and mutual respect. In sum, *ubuntu* is the underlying philosophy that shapes and drives the African intellectual interests and his/her way of life as a whole. It forms the basis for resolving conflicts and for undertaking daily activities. It has universal application. Although *Ubuntu* emphasizes communality rather than individuality that is prevalent in Western ideologies; it equally recognizes the importance of individuals in the community. The African term *umuntu* or man/woman, in a generic sense, or *ntu*, a person or human being, construes an African as a tripartite being, with body, soul and spirit. To remove the spirit aspect, as Western science has done, is to leave man dispirited and disconnected from him/herself and all else.

Ubuntu-driven science basics and curriculum will recognize other worldviews and incorporate them in the teaching and learning of the sciences. It will accord recognition to other elements of nature, their uses and more importantly, their conservation. It will remove the disconnection that non-mainstream indigenous students find between the ways they do things at home, e.g., in terms of cooperation, seeking the welfare of others, putting the interests of others before their own self-interests that are totally deemphasized in their science classrooms. It will restore their sense of fulfilment in their relationships with others that is rooted in interdependence, interrelatedness and collectivism, which Western science has replaced with competitiveness and go-it-alone individual attainment. The instructional strategies that will help in making science more relevant to second language indigenous students will include indigenous methods of solving practical problems in the science classroom and as it is done in the community. Indigenous human archives will be invited to share age-old indigenous ways of doing (cultural wisdom) in the science classroom. In addition, students will engage in the use of cultural symbols, myths, metaphors, poems, proverbs, storytelling, arts, drama, songs and music and indigenous technologies in everyday life and games in the science classroom. The interconnectedness that the curriculum (science and/or other subjects) was supposed to engender will no longer be a pipe dream.

REFERENCES

Baxen, J. (2001). Teachers' perspectives on outcomes-based education in South Africa. *International Journal of Education Reform, 10*(1), 2–13.

Bishop, A. J. (1990). Western mathematics: The secret weapon of cultural imperialism. *Race and Class, 32*(2), 51–63. doi:10.1177/030639689003200204.

CAPS. (2011). *Curriculum and assessment policy statements*. Pretoria: Department of Education.

Curry, L. (1992). Deliberative curriculum inquiry and its application in the education of health service administrators. *Journal of Health Administration Education, 510*, 519–526.

Department of Basic Education. (2002). *Revised national curriculum statement*. Pretoria: Department of Education.

Hountondji, P. (1997). *Endogenous knowledge: Research trails*. CODESRIA Book Series.

Jansen, J. D. (1997). Essential alterations? A critical analysis of the states syllabus revision process. *Perspectives in Education, 17*, 1–11.

Kincheloe, J. L., & Steinberg, S. R. (Eds.). (1994). *Thirteen questions: Reframing education's conversation*. New York, NY: Peter Lang.

NCS. (2003). *National curriculum statement. Grades 10–12. Life Sciences*. Pretoria: Department of Education.

NGSS Lead States. (2013). *Next generation science standards: For states, by states*. Washington, DC: The National Academies Press.

Ogunniyi, M. B. (1986). Two decades of science education in Africa. *Science Education, 70*, 111–122. doi:10.1002/sce.3730700205.

Ogunniyi, M. B. (1995). The development of science education in Botswana. *Science Education, 79*, 95–109. doi:10.1002/sce.3730790107.

Ogunniyi, M. B. (1996). Science, technology and mathematics: The problem of developing critical human power in Africa. *International Journal of Science Education, 18*, 267–284. doi:10.1080/0950069960180301.

Ogunniyi, M. B. (1997). Science education in a multi-cultural South Africa. In M. Ogawa (Ed.), *Report of an international research program on the effects of traditional culture on science education* (pp. 84–95). Mito: Ibaraki University Press.

Ogunniyi, M. B. (2004). The challenge of preparing and equipping science teachers in higher education with knowledge and skills to integrate science and indigenous knowledge systems for learners. *South African Journal of Higher Education, 18*, 289–304. doi:10.4314/sajhe.v18i3.25498.

Ogunniyi, M. B. (2007a). Teachers' stances and practical arguments regarding a science-indigenous knowledge curriculum, Part 1. *International Journal of Science Education, 29*, 963–985. doi:10.1080/09500690600931020.

Ogunniyi, M. B. (2007b). Teachers' stances and practical arguments regarding a science –indigenous knowledge curriculum, part 2. *International Journal of Science Education, 29*, 1189–1207. doi:10.1080/09500690600931038.

Ogunniyi, M. B., & Mushayikwa, E. (2015). Teacher education in South Africa: Issues and challenges. In B. Adegoke & A. Oni (Eds.), *Teacher education systems in Africa in the digital era* (pp. 71–90). Dakar: *CODESRIA*.

Ogunniyi, M. B., & Rollnick, R. (2015). Preservice science teacher education in Africa: Prospects and challenges. *Journal of Science Teacher Education, 26*, 65–79. doi:10.1007/s10972-014-9415-y.

CHAPTER FOUR

Science Education and Transformative Activist Stance

Activism as a Quest for Becoming via Authentic-Authorial Contribution to Communal Practices

ANNA STETSENKO

There is a surge of novel ideas and approaches relating activism to science and science education. Researchers even talk about "activism turn" in academia (e.g., Wittel, 2012), echoing earlier descriptions of "cultural turn," "discursive turn" and other radical changes in perspectives and approaches that took place in recent decades. There is as yet no sufficient clarity in explaining the reasons for the rapid rise of scholarly work connected to academic activism. As Andreas Wittel notes, two developments stand out as possible explanations—the crisis of capitalism and the crisis of universities and higher education. Even as we still do not fully grasp the scope of the activism turn including its underlying reasons and possible ramifications, it is important to explore various meanings of activism in academia and, for the purposes of the current chapter, its meanings in and implications for science education. As pertains specifically to science education, the relationship of activism to social inquiry and political action and motivation has been recognized (e.g. Bencze & Alsop, 2014), even though for various reasons it is still on the periphery of mainstream approaches (Carter, Castano Rodriguez, & Jones, 2017). Summarizing many important trends in this field, Jesse Bazzul (2015) writes that "if science education is going to play an integral part in solving 21st-century problems such as climate change and social inequality [...], it must become *a site of political contestation*" (p. 1). Important dimensions of such political contestation include interrogating and rejecting accepted rules that govern how participants are positioned vis-à-vis educational projects and how they are excluded along the lines of social hierarchy and power. Further, there are suggestions to re-position activism as a possible aim of science education, for example, in "broadening of the aims of science education to include social justice and sociopolitical action" (Reiss, 2007, p. 21).

While sharing these positions, what I want to explore in this brief chapter is how science education is *always already* a site of political contestation even when it *appears* to be confined within the classroom walls (which in fact it never is) such as when students learn, for example, about evolution or laws of gravity, or even just learn about and practice multiplication table. My argument is that even when science education appears to be "merely" about teaching and learning some established "facts of science," it actually never is only about that but also, and quite centrally, about activism in a broad and quite fundamental sense. That is, science education is also, at once, about activism understood as the right and ability to know for oneself within what is one's own—unique, authentic and authorial, as well as ever-emerging and shifting—quest to make a difference in the world. Activism thus understood is about agency and ability to take a stand on what is going on in one's community practices and the wider world, to contest existing status quo, and to stake a claim to how things could be otherwise including what should and can be done to achieve a different, and a better world.

In this broad sense, activism is about agentive and activist strivings and struggles of becoming, that is, about human worth and significance and, most critically, the right for each and every one of us to be agentive *actors* and de facto activists of our own lives and of society at large, who contribute to community practices and bring them into realization, thus themselves coming into being, while rebelling against the status quo, contesting its rules, and righting its injustices and entrenched biases. In the context of science education, activism in this broad sense is about each learner *authoring* knowledge and science, claiming it, making it one's own, all while becoming an agentive actor within the practices of "doing" science and what these practices actually stand for—namely, the ongoing and uninterrupted human pursuits within the broader world in the full richness and complexity of its cultural-historical, sociopolitical, and ethical-valuation dimensions. Importantly, the emphasis in this conceptualization is on the processes and practices at *the intersection* of individual quests of becoming with the wider social practices and processes embodied in what we traditionally—and quite narrowly—think of as "science" and "science education".

According to this position, teaching-learning and science education are not about transmission of knowledge and facts (a point well understood in many recently developed frameworks and articulated already by Dewey, e.g. 1916/1966), and not even about participation in community practices (as in the recently influential sociocultural framework, e.g. Lave & Wenger, 1991; for a detailed critique, see Stetsenko, 2008) but more radically, about agentive, authorial, authentic and activist *contribution* by each learner, and each teacher, to what are the workings of science. Importantly, this approach hinges on the concept of what science is all about that is different from extant models. In particular, science in this approach is understood, firstly, as an ongoing and open-ended human endeavor fully immersed in worldly contexts and everyday realities of lives, continuously unfolding through generations, historically specific, and filled with human significance and relevance within situated quests of humanity. Secondly, the endeavor of science is highlighted to be reliant for its realization on sociocultural practices, interactivities and their tools yet also reliant, and no less critically, on each and every learner taking up and realizing these endeavors in her or his own quests of becoming that are uniquely positioned, authentically conceived, and authorially carried out. Thirdly, science is understood to be open-ended, indeterminate, forever un-finalized and contested, and therefore resisting any final definition or canonization, just as the world and life itself, wherein science is immersed, are.

At the level of practical implications for classroom practices and science curriculum, this approach suggests that science education needs to be about arranging for the processes in which students—and teachers too!—can agentively connect and contribute to investigative projects of science as a human endeavor, while relying on the tools of activism and knowledge in coordinated pursuits of self- and world-realization as the pathways to transcending the status quo in sciences and beyond. This broad conceptualization of activism presupposes a conception of human development and subjectivity—that is, the processes of being-knowing-doing—that is radically different from the traditional views still dominant in mainstream approaches within the range of disciplines from psychology, philosophy, and sociology to education and pedagogy. It also presupposes, concomitantly, a conception of what is science education and science itself that is radically different from the traditional models aligned with canons of objectivism and positivism and even from the more recent alternatives such as postmodernist, critical, and sociocultural approaches. These more recent accounts still carry with them some of the remnants of "the old legendry" about science—the legendry that needs to be more resolutely contested than has been achieved so far. Contesting traditional assumptions simultaneously across all the spectrum of positions pertinent to knowledge, teaching-learning, subjectivity, agency, science, and science education is not an easy task. Indeed, working out a challenge to these assumptions requires much conceptual and theoretical work at the level of the deeply-seated, foundational assumptions of a philosophical caliber including about ontology, epistemology, and ethics. At stake is no less than a challenge to traditional worldviews and models that through many centuries have shaped accounts of these matters. Yet this is what is required to allow for establishing and realizing an ontologically and epistemologically justified and conceptually warranted connection between sciences and activism.

Importantly, I believe that all educators and teachers are faced, and indeed grapple, with the questions of this magnitude, whether or not they directly acknowledge and reflect on this. Each teacher entering science education classroom takes on a "philosopher hat" and has to address fundamental questions about human being-knowing-doing. This is because teachers and educators are always searching for the best ways to engage students, understand their concerns, connect to their struggles, spur their interests, rally their motivation, arrange for social interactions, explain complex conceptual matters, resolve conflicts, and so on. These are not mundane issues. Instead, these are monumental questions of cosmic proportions and the answers to them rely on the deep, fundamental understandings of human beings, their strivings and quests, their ways of being-knowing-doing—in all the gamut of the most intricate and complex aspects pertaining to the human condition, society, nature, knowledge, and development.

Such broad questions of ontology and epistemology are often supposed to be the province of a "big philosophy" to be addressed by a select coterie of elite specialists who work in the footsteps and emulate giants no less than those like Kant, Hegel or Marx. Most teachers and science educators do not count themselves among such elites and stay away from formulating their own answers to these cosmic philosophical questions and their own worldview-level conceptions because they consider themselves to be engaged in putatively less significant, allegedly "minor" topics related to education and classroom management. Yet these topics are no less deep and complex than any of those addressed in the "big philosophy." A push for a more democratic science education has to include the challenge to an established division of labor where we have elitist approaches to philosophy on one hand, and rank-and-file theorists and practitioners of classroom practices and curriculum development, on the other. To think otherwise would mean to subscribe to old-fashioned approaches and their status quo that overlook the profound significance of teaching-learning and of collaborative endeavors and communal practices including what only appears to be day-to-day, mundane classroom processes and interactions. To emphasize again, teachers and science educators are in the business of no less than what, at its core, is a deep philosophy and a worldview-level theorizing. These endeavors of teaching-learning may be called, in the words of Isabelle Stengers (2006) who conveyed Alfred North Whitehead's insight, "a cosmic adventure, because the manner [the learners] will experience new possibilities, feelings and ideas, or stubbornly keep to their abstractions, to their judgment about what matters and what does not, is indeed a cosmic stake" (p. 15).

TRANSFORMATIVE ACTIVIST STANCE

Novel approaches to integrating activism into science education can be fruitfully advanced on the grounds of what I have termed a transformative activist stance (TAS; see e.g., Stetsenko, 2015, 2016). This perspective has been developed as a direct challenge to, firstly, the mechanical worldview that portrays human beings to be passively adapting to the world and—as concerns teaching-learning and knowledge—to be merely processing information received from outside, dispassionately and neutrally, much in similarity with how computers operate. Secondly, TAS aspires to resolve some of the impasses in recent theories that capitalize on the role of culture, mediation, and social interaction in development and education. While building upon and integrating many important insights stemming from these theories, TAS suggests steps to move beyond their underpinnings in the relational worldview and its notions about mind, knowledge, and learning as situated, relational, contextualized, embodied, enacted, and dynamic. The shift beyond these notions is achieved by introducing transformative worldview premised on onto-epistemology that centrally integrates human agency and activism in highlighting human capacity to transform and transcend the status quo.

The TAS builds and expands upon a number of theoretical positions such as Vygotsky's cultural-historical theory, Bakhtin's dialogical theory, Freire' critical pedagogy, as these are further integrated with insights from recent works in feminist, ecological and critical approaches such as critical race theory. Central to transformative ontology and epistemology is the notion of the dual and ceaseless dynamics at the *shifting nexus* of people collaboratively changing their world while simultaneously bringing it—and themselves—into existence through the very process of agentively contributing to collaborative and communal transformations of existing realties. Although the notion of transformation is very familiar to scholars in the critical tradition and especially honored in Marxism, its radical implications at all levels of theorizing, including about the process of knowledge production and learning, have not been fully addressed (for details, see Stetsenko, 2014, 2016) and TAS aspires to do just that. The central notion in TAS is that people contribute to and thus change the world and its community practices—while the world is reciprocally changing people through cycles of recursive interactions—in individuals and communities struggling for a sought-after future that they themselves envision, imagine, and commit to. Posited in opposition to passive adaptation, while also supplanting a more active yet not fully agentive process of participation in community practices, the activist contributions to ongoing transformations of the world gain the status of ontologically and epistemologically central processes.

The transformative ontology and epistemology are both premised on the notion that agency embodied in activist stance is inherent in human development in its interrelated dimensions of being-knowing-doing. Agency is posited to be associated with, and emblematic of, people collaboratively moving beyond the status quo (i.e., the presently "given" reality), via individual agentive and activist contributions to this process, while relying on interactivities and cultural tools of creating social change predicated on a sought-after future. In this dialectically recursive and dynamically co-constitutive approach, people can be said to realize their development *in the enactment of changes that bring the world, and simultaneously their own lives, including their selves and minds, into reality.* These transformative processes are situated in shared contexts of communal history enacted by collective practices while relying on their resources, tools, spaces, and collaborative interactivities. Yet the active and indeed activist role of people themselves in realizing these processes while acting on commitments to a sought-after future, to what they themselves deem important and worth struggling for, is placed at the center (see Vianna & Stetsenko, 2011).

The focus on agency and activism expands and moves beyond many established approaches including the tenets of Vygotsky's theory. These tenets were centrally focused on the present communal practices and, perhaps especially, their past history. A critical expansion offered by TAS concerns the relevance of a *forward-looking* activist positioning vis-à-vis the future and of a commitment to social change in order to bring this future into reality. The critical constituent of human development, mind, and teaching-learning is posited to consist in taking stands and staking claims on ongoing events, conflicts and contradictions in view of the goals, commitments, and aspirations for the future—the process of *making up one's mind* as literally a process through which human subjectivity, and processes of teaching-learning, come about and which they are made up of.

An additional and no less critical specification is that this capacity for agency, along with other expressions of human subjectivity, are understood to be fully social, that is, developed and realized in acting collaboratively and cooperatively with the cultural tools, and within the communal spaces, of the world in its ongoing historicity and evolving quests shared with others. The processes such as mind, knowledge, identity, and teaching-learning including in science education, are understood to be simultaneously social and individual contributions to the future-oriented dynamics of transformative shared practices of communal life in their activist, that is, world-changing and history-making status.

The concept of activism in the broad sense of authoring community practices, including practices of doing science and producing knowledge, finds a central place and plays a ubiquitous role within this perspective. *Activism* conveys the sense that all individuals and communities are immersed within, and themselves are always contributing to, not just the neutral contexts or environments that somehow peacefully "surround" them. Instead, human acts of being-knowing-doing are parts and parcels of the unfolding drama and struggle that is the world itself infused with conflicts and contradictions, dilemmas and challenges—that even in their daily expressions and everyday contexts are always about the struggle for transformation of the world (cf. Freire, 1970/2005).

In this view, in continuation of many critical and sociocultural approaches, yet with a number of revisions made to their canonical postulates, the processes of being-knowing-doing do not just somehow happen to people. Instead, they are creatively, agentively, and collaboratively carried out, organized, and performed, that is, *worked out*—as an ongoing effort and a continuous, open-ended striving at becoming by each learner to be an actor of community practices. In this approach, activism premises human development understood as a continuous and uninterrupted *striving for contribution to communal forms of life* that is simultaneously the condition for self-realization and realization of fellow human beings (see Stetsenko, 2013). What is highlighted by TAS is that in their acts of being-knowing-doing, people can and in fact *always do challenge the taken for granted realities and rules*—while transcending them in collectively moving forward and jointly *co-authoring* these practices and, simultaneously, themselves. Underlying these notions is a shift away from the sociobiological ethos of adaptation, and its attendant values of competition and survival, to the ethos of interactivity, collaboration and solidarity. Further, what is acknowledged and strongly emphasized in this conception, most critically, is that solidaristic communities are only possible if activism of *each and every individual*—as an ability to form one's unique stance, position, and voice that make contributing to communal practices possible—is socially and culturally co-constituted, nurtured, supported, and sustained via collective practices and tools especially in education. This can be achieved by a communally organized provision of sociocultural structures, mediators, shared spaces, and cultural tools that individuals can agentively and creatively take up and carry out, in novel and authentic ways, from their unique positions and authorial stances that are themselves formed in this very process of "taking up" the world and its cultural practices and tools. The transformative onto-epistemology displaces naturalist explanations that exclude human dimensions to instead open ways to theorize and account for emancipatory agency and activism as central to all processes of being-knowing-doing including as these are enacted in science education.

IMPLICATIONS FOR ACTIVISM IN SCIENCE EDUCATION

To reiterate, the suggestion in this chapter is that activism in science education has to do with the right of each learner to know for oneself within an authentic and activist quest for agency embodied in the process of co-authoring community practices and knowledge—a quest that is achievable through making a difference via unique and irreplaceable contributions to communal practices including science and its investigative practices and projects. Accordingly, the core implication from the TAS for the practices of science education is that there is no knowledge and no teaching-learning that exist prior to, and can be separated from, each individual learner's transformative, activist engagement with the communal world.

The premise that education is always about what lies behind teaching and learning of seemingly neutral "facts" has been addressed, for example, in the vast research on the "hidden curriculum" (for a recent overview, see Kentli, 2009). This research pertains to the unarticulated dimensions that often fall outside the formal curriculum including the transmission of norms, values, forms of control, privileges and power hierarchies through the classroom practices. The "hidden curriculum" research helps to uncover how school practices, typically differing by social class and ethnicity, reproduce social systems and structures by tacitly preparing students for particular roles and relationships in society. Yet less addressed, perhaps one could say hidden, within the "hidden curriculum" research is a dimension of activism in the broad connotation of this term as outlined in the introduction. This dimension of activism pertains to personal struggles of becoming through co-authoring the world in contesting its status quo by each learner who agentively participates in and contributes to the ongoing dynamics of classroom practices and those wider social dynamics that lay hidden behind these practices.

On the other side of the spectrum of research on science education, there has been much progress that highlights how knowledge and teaching-learning are deeply personal pursuits of meaning and identity development, rather than purely cognitive processes of mental computation akin to computerized calculations (for details, see Vianna & Stetsenko, 2011). Yet this research rarely engages with the topics of how science works and how knowledge is also a producer and product of social relations at the structural levels. The TAS opens ways to simultaneously address the processes at the individual and collective levels of social practices through the concept of contribution that bridges the divide between them. In this emphasis, the focus on identity development and personal becoming is understood to be entailed by and even "build into" the seemingly neutral classroom practices such as teaching and learning of science—*along and together with* the collective power dynamics pertaining to reproduction of social hierarchies also enacted and built into these very practices. These two dimensions of classroom practices, including in science education, are understood to be neither complementary nor additive. Instead, they are seen as indivisibly merged and mutually co-constitutive because individual and collective layers of social dynamics act as inalienable forces and facets in the same process of these dynamics realizations, as highlighted by the notion of individual contribution to collective practices.

Any lesson about, for example, evolution or the forces of gravity is also a lesson about science and knowledge production—about how knowledge is made, how it came about and how it has been and can be enacted in the present to play a role in the world and its community practices in view of the present challenges, conflicts, and conundrums. This process is also about who each learner is and could be vis-à-vis these processes as they stretch from the past into the present and already project into the future (see Stetsenko, 2010a). These lessons are about *science and knowledge as human endeavors, strivings and pursuits* that are embodied and implicated in each and every fact of science, each and every bit of knowledge, and also, quite critically, in each and every act of knowing and understanding by learners. Importantly, in this approach, knowledge and science are not about elitist endeavors conducted in the ivory tower by exceptional individuals standing on pedestals at some putatively unreachable Olympian heights (as science is still typically portrayed in the "great men" tradition). Instead, knowledge and science are but parts and parcels of collaborative projects carried out in the real world of activist struggles and strivings for answers—and often new questions—emerging within the ever-unfolding dynamics of history and society in which each and every learner makes a difference and matters.

That is, science education is not separate from the learners' life projects—themselves ineluctably social—that embody strivings and struggles for how learners want to be and what they want their world to be, as interrelated facets of teaching-learning. In this approach, any understanding and knowledge building—any act of knowing including in the context of science education—is but part and parcel of the larger projects of our becoming, inclusive of identity development of ourselves as actors in the social world of communal practices shared with others. Each act of knowing, in other words, is a complex, contested, and non-neutral endeavor loaded with human significance, including identity dimensions such as emotions and passions. In this endeavor of knowing, much is at stake and it itself is at stake in many ways and in many aspects of life well beyond the immediate context of learning and the classroom walls. For example, the learners' goals and strivings towards the future, how they imagine the horizon of possibilities and what they struggle for, in fact color and shape their understandings and all of the learning dynamics in the present (for details, see Stetsenko, 2014). Importantly, while being deeply personal, these processes are not about personalized quests somehow confined to isolated individuals and defined in terms of solo dynamics separated from the larger world. Rather, teaching-learning and knowing are inextricably parts and parcels in the dynamic processes of an active and indeed activist striving for one's authenticity, authorship and place in the *shared world of communal practices.*

This is about activism as an individual quest that is not extracted from the typically structurally-based, collective solutions of major social and environmental problems including social injustices. Instead, their intimate relationship and intricate blending as facets of the same process is brought to the fore. This is important in order to avoid connotations of social reforms and activist transformations as top-down impositions of "truths" and directions on passive individuals who have to comply and fit in with these impositions. For broad activism and social reforms to work, each individual and each community have to always be in the process of exploration, search, contestation, and dissensus with the status quo and even these reforms themselves.

ROUSING ACTIVIST AGENDAS TO LIFE: CONTESTING TRADITIONAL MODELS OF SCIENCE

I fully endorse the goal of educators as activists and advocates for science and science education to play a role in the solution of major social and environmental problems including social injustices. My suggestion is to expand this position to include an emphasis on activism in the broader sense—that is, on activism as an ability to take a stand on, and stake a claim to, how things are and how they can be changed, as a broad and foundational characteristic of the processes of being-knowing-doing in their most basic ontological and epistemological dimensions. From this it follows that science education has to be about connecting to, engaging with, and even more critically, spurring these activist projects of becoming—in their interrelated dimensions of being-knowing-doing—in thus rousing not only minds to life (to use an excellent expression by Tharp & Gallimore, 1988) but also these activist projects and pursuits themselves. This is not about indoctrination or imposition of alien agendas on unsuspecting students. Instead, science education needs and can be, to use Tom Barone's (2006) expression, about *artfully* coaxing learners "into collaborative interrogation of stale, tired, taken-for-granted facts" (p. 227), while learners carry out the work *to assert themselves in the actualization of their potential.*

This position can be sustained only on the condition that the established canons of objectivity are radically challenged. As Stephen Jay Gould (1996) wrote decades ago, "scientists must give up the twin myths and canons of objectivity and of knowledge as inexorable march toward truth that can be established once and for all" (p. 55). These canons are the pillars that support exclusionary practices and stay in the way of implementing an activist positioning in science education by all participants, that is, teachers and learners alike. Yet this has to be done without falling into the extremes of uncommitted relativism and denial of science and knowledge as such, so that no knowledge claims are closer to reality than any of the others. That is, on one hand, it is important to reject quaint epistemologies according to which the purpose of science is to describe, reveal and explain the world "as it is" and, accordingly, the purpose of science education is to deliver (or transmit) knowledge in a canonical form of established "facts" or "truths" in a top-down fashion, unchanged and unchallenged, from science via teachers to students. On the other hand, challenges to traditional models of science have to avoid dismissing the value of knowledge and science. Giving credit to science and acknowledging its success does not need to go hand in hand with accepting "facts of science" as some sort of "higher" truths produced by elite authorities that can be neither questioned nor contested. It is of utmost importance to advance critiques of science and its methods alternative both to groveling before their supposed superiority on one hand, and to dismissals of science that throw out the baby of knowledge with the bathwater of objectivism, on the other.

Much has been written about these canons in sociocultural and critical scholarship especially in arguing for the right to diverse perspectives and dispersed points of view, for the value of knowledge especially from marginalized positions and for the cultural, social and political embedding of science and knowledge. Strong critiques of objectivist, value-neutral models have been developed, especially consistently in feminist scholarship (e.g., Harding, 2004) and science studies (Latour, 1987). These critiques have helped to expose contradictions, mythologies and dead ends of traditional models, showing how the supposedly neutral conduct of science is in fact rooted in history and entangled with cultural and political dimensions, including the reciprocal relationships, co-constitutions and bindings among evidence, methodology, normative assumptions, political interests, instrumentalities, variables, and models of reality (cf. Burman, 1994).

Yet many critical post- and anti-positivists in the social sciences are still often deferential to the natural sciences models and often tacitly accept their superiority, even while developing alternative approaches (as in crypto-positivism, see Kincheloe & Tobin, 2009). In this deference, there is sometimes a residual subordination of qualitative, sociocultural, and critical approaches to those of natural sciences in the acceptance of ideals of certainty and reliability of hard facts and evidence. This is exemplified, among other things, in a belief in incontrovertible authority of mathematics that proclaims that numbers are the ultimate test of objectivity, combined with a naïve under-appreciation of how contingent, open-ended, malleable and uncertain *the very endeavor of science*, any science without exception, is. A reverential attitude towards science, especially physics and more recently neurosciences and genetics, is partly understandable given how much has been achieved in applications of scientific discoveries and advances. Indeed, after only four hundred years of modern science, its record is remarkable and scientific knowledge, as Marcelo Gleiser (2010) puts it, is now stretching "all the way from the inner confines of atomic nuclei to galaxies billions of light-years away" (p. xvi).

In an alternative definition of science and science education in line with the premises of TAS, there can be no unquestioned assumptions, unchallenged axioms, fixed "evidence," hard "facts," rigidly accepted rules and unshakeable arguments. Instead, all science and all knowledge production, though all of its steps, is always *inherently* indeterminate, open-ended, and therefore, contestable—"an uncertain process of exploration, deconstruction, refashioning and interrogation" (Burke & Jackson, 2007, p. 196). This position amounts to accepting that knowledge is *normally* defined through conflict (cf. Burke & Jackson, 2007) and, I would emphasize in addition, that *contestations, contradictions and dissensus are the very practices that science and knowledge are made of.* This is a strong epistemic stance according to which contestability is a form of a new quasi-rule and quasi-universality (cf. Barnett, 2005) where a process of questioning, interrogating and moving beyond the status quo *is the very fabric of knowing*. In science thus understood, no one has the answers to the questions that arise. As Marcel Gauchet (2002) puts it, "this is the ultimate reason why learning is so difficult. In searching for knowledge, we have to confront a situation which tells us that we can never know" (2010). Yet the search continues and can provide legitimate answers under the condition that we search and find these answers for ourselves. As I would re-formulate Gauchet's position, in searching for knowledge, we have to confront a situation which tells us that we have to know for ourselves, have to find our own answers—certainly not from scratch, not in discarding previous discoveries, and not without the tools that science and society have to offer. Yet even in "standing on the shoulders of others" (to paraphrase an old saying while avoiding a reference to "giants," for explication see Stetsenko, 2010b), much authentic and authorial work is needed—the work of finding the balance, seeing farther, and ultimately figuring everything on our own, even in standing on such shoulders yet while critically taking up, contesting, expanding, and transforming what our predecessors have achieved.

MOVING FORWARD: ACTIVISM AS AUTHORING

The ultimate lesson from the foregoing discussion is that big ideas and answers to the challenges we together and each of us are facing today—and not as lonely individuals but as society members and agents-activists of social change—are not and cannot be handed down from on high. *The answers to these challenges have to be figured out by each individual learner qua actor of social communal practices and agent of the world- and history-in-the-making*. Embedded in this notion is also that we need to constantly challenge ourselves in our quests for being-knowing-doing while taking a stand vis-à-vis ongoing struggles and conflicts in society, in figuring how to contribute to resolving them, including through the tools of science. I believe it is important that science education takes up the challenge of engaging in teaching-learning of science as a collaborative endeavor and a collective, open-ended quest for becoming—through activism—by students and teaching together exploring how the world is through the prism of how it came to be and how it could be, while simultaneously figuring out how they themselves can and already do matter in these processes by making activist contributions (however modest and imperfect) to them.

In the spirit of self-reflection and to illustrate this position, let me comment on employing my approach of TAS in interpreting the topic of activism in science education. Advancing and naming TAS as an original approach might be seen as immodest, too ambitious or less preferable than a humble following in the footsteps of those who are typically described as "giants" such as in the already mentioned "great men" tradition (for a critique of this tradition, see Stetsenko & Arievitch, 2004). Also, the act of naming an approach always carries the risk of essentializing and setting ideas in place rather than leaving "*no-place* where everything is possible" (see Sandoval, 2000, p. 141). Given the transformative gist paramount in the TAS, however, such connotations can be avoided on both counts. With the emphasis on change and transformation, and on activism and contribution, this approach itself is open-ended and subjected to constant amendments, revisions, and transformations—since it stands for a kind of thinking that never finds itself at the end even though it posits an endpoint of where it strives to arrive and commits to its realization. Finally and most critically, introducing this approach as an original position on matters at the worldview-level is a conscious and deliberate act that echoes the central premise of the transformative activist stance itself—that we all, each and every one of us, matter and have the right to co-authoring the world in all of its aspects and dimensions, through our agentive, authentic, and authorial contributions.

REFERENCES

Barnett, R. (2005). Recapturing the universal in the university. *Educational Philosophy and Theory, 37*, 785–797. doi:10.1111/j.1469-5812.2005.00158.x.

Barone, T. (2006). Making educational history: Qualitative inquiry, artistry, and the public interest. In G. Ladson-Billings & W. F. Tate (Eds.), *Education research in the public interest: Social justice, action, and policy* (pp. 213–230). New York, NY: Teachers College.

Bazzul, J. (2015). Towards a politicized notion of citizenship for science education: Engaging the social through dissensus. *Canadian Journal of Science, Mathematics and Technology Education.* doi:10.1080/14926156.2015.1051670.

Bencze, J. L., & Alsop, S. (2014). *Socio-political activism and science & technology education.* Dordrecht: Springer.

Burke, P., & Jackson, S. (2007). *Reconceptualising lifelong learning: feminist interventions.* London: Routledge.

Burman, E. (1994). *Deconstructing developmental psychology.* New York, NY: Routledge.

Carter, L., Castano Rodriguez, C., & Mellita Jones, M. (2017). *Sociopolitical activism and transformative learning: Expanding the discourse about what counts in science education.* In L. A. Bryan

& K. Tobin (Eds.), *13 Questions: Reframing Education's Conversation: Science* (pp. xx-xx). New York: Peter Lang

Dewey, J. (1916/1966). *Democracy and education. An introduction to the philosophy of education.* New York, NY: Free Press.

Freire, P. (1970/2005). *Pedagogy of the oppressed* (Rev. 30th anniversary ed.). New York, NY: Continuum.

Gauchet, M. (2002). *La démocratie contre elle-même.* Paris: Gallimard.

Gleiser, M. (2010). *A tear at the edge of creation: A radical new vision for life in an imperfect universe.* New York, NY: Free Press.

Gould, S. J. (1996). *The mismeasure of man* (2nd ed.). New York, NY: W.W. Norton. (Originally published in 1981).

Harding, S. (2004). A socially relevant philosophy of science? Resources from standpoint theory's controversiality. *Hypatia, 19*(1), 25–47. doi:10.1111/j.1527-2001.2004.tb01267.x.

Kentli, F. D. (2009). Comparison of hidden curriculum theories. *European Journal of Educational Studies, 1*(2). Retrieved from http://www.ozelacademy.com/EJES_v1n2_Kentli.pdf

Kincheloe, L. J., & Tobin, K. (2009). The much exaggerated death of positivism. *Cultural Study of Science Education, 4*, 513–528. doi:10.1007/s11422-009-9178-5.

Latour, B. (1987). *Science in action: How to follow scientists and engineers through society.* Cambridge, MA: Harvard University Press.

Lave, J., & Wenger, E. (1991). *Situated learning: Legitimate peripheral participation.* New York, NY: Cambridge University Press.

Reiss, M. (2007). What should be the aim(s) of school science education? In D. Corrigan, J. Dillon, & R. Gunstone (Eds.), *The re-emergence of values in science education* (pp. 13–28). Rotterdam: Sense Publishers.

Sandoval, C. (2000). *Methodology of the oppressed.* Minneapolis, MN: University of Minnesota Press.

Stengers, I. (2006). *Whitehead and science: From philosophy of nature to speculative cosmology.* Retrieved from http://www.mcgill.ca/fi les/hpsc/Whitmontreal.pdf

Stetsenko, A. (2008). From relational ontology to transformative activist stance: Expanding Vygotsky's (CHAT) project. *Cultural Studies of Science Education, 3*, 465–485. doi:10.1007/s11422-008-9111-3.

Stetsenko, A. (2010a). Teaching-learning and development as activist projects of historical becoming: Expanding Vygotsky's approach to pedagogy. *Pedagogies: An International Journal, 5*, 6–16. doi:10.1080/15544800903406266.

Stetsenko, A. (2010b). Standing on the shoulders of giants: A balancing act of dialectically theorizing conceptual understanding on the grounds of Vygotsky's project. In W.-M. Roth (Ed.), *Re/structuring science education: Reuniting psychological and sociological perspectives* (pp. 53–72). New York, NY: Springer.

Stetsenko, A. (2013). Theorizing personhood for the world in transition and change: Reflections from a transformative activist stance. In J. Martin & M. H. Bickhard (Eds.), *The Psychology of personhood: Philosophical, historical, social-developmental, and narrative perspectives* (pp. 181–203). Cambridge: Cambridge University Press.

Stetsenko, A. (2014). Transformative activist stance for education: Inventing the future in moving beyond the status quo. In T. Corcoran (Ed.), *Psychology in education: Critical theory~practice* (pp. 181 198). Rotterdam: Sense Publishers.

Stetsenko, A. (2015). Theory for and as social practice of realizing the future: Implications from a transformative activist stance. In J. Martin, J. Sugarman, & K. Slaney (Eds.), *The Wiley handbook of theoretical and philosophical psychology: Methods, approaches, and new directions for social sciences* (pp. 102–116). New York, NY: Wiley.

Stetsenko, A. (2016). *The transformative mind: Expanding Vygotsky's approach to development and education.* New York, NY: Cambridge University Press.

Stetsenko, A., & Arievitch, I. M. (2004). Vygotskian collaborative project of social transformation: History, politics, and practice in knowledge construction. *The International Journal of Critical Psychology, 12*(4), 58–80.

Tharp, R. G., & Gallimore, R. (1988). *Rousing minds to life.* Cambridge: Cambridge University Press.

Vianna, E., & Stetsenko, A. (2011). Connecting learning and identity development through a transformative activist stance: Application in adolescent development in a child welfare program. *Human Development, 54,* 313–338. doi:10.1159/000331484.

Wittel, A. (2012). Digital Marx: Toward a political economy of distributed media. *TripleC (Cognition, Communication, Co-operation): Open Access Journal for a Global Sustainable Information Society, 10*(2). doi:10.1163/9789004291393_004.

SECTION TWO

Power and Science Education

Who Decides the Forms Science Education Has Taken and Who Should Decide?

CHAPTER FIVE

What Can Teachers Do to Restructure Power Dynamics in Science Classrooms?

Exploring the Personal and Social Transformative Power of Science Learning through a Funds of Knowledge Approach

SARA TOLBERT, ANGELA CALABRESE BARTON, AND LUIS C. MOLL

> This is what our scooter started off as, with plain old Razor scooter. Kids can actually make a change. Look where it is now, look at all the work. We were making an electric scooter. So we made a solar-powered scooter. These, we put, this is the lights. This is the tape. MSU! This is the solar power that we made it run on.
>
> I feel like super cool. People will love it. They'll say, "Who made this?" It was me. Then they'll ask me like, "The tiny person always in the background did this?" I'll say, "Yeah, I did that." They're like, "Oh my gosh. Look, everybody. This girl did this giant scooter. We all love it. It's awesome and cool." … Little kids can do ginormous work!
>
> —JENNIFER, 11 YEARS OLD, INTERVIEW

The opening transcript comes from an interview we had with Jennifer (all names are pseudonyms), an African American female in the 6th grade, when asked about what others would think of the solar powered scooter she prototyped. Her comments emphasize a vision of doing science as dynamic and an avenue to take action in, for, and beyond her community. Learning science holds personal and social transformative power for Jennifer by extending beyond knowledge acquisition into action taking in order to "make a change." Her quote reveals an empowered sense of self in science—a contrast to stereotypical views of young minoritized females in science.

Jennifer's story is not an isolated case, despite the narratives told by mainstream media. Many youth have powerful stories about when, where, how, and why science matters to them. Many of their stories position doing science as part of their lives, in ways that are congruent with their lived histories. Nevertheless, despite multiple national reform efforts, opportunity and interest gaps continue to persist in school science, disproportionately impacting minoritized youth. Yet, a growing body of research indicates that many minoritized youth, like Jennifer, who are not performing well in traditional school settings, are excelling in science in other spaces (Deschenes & Malone, 2010). This poses the question of how might young people's lives matter in school science? How might we break down the traditional power dynamics in science classrooms to position youth with power, agency, and authority? How might we create opportunities for students to engage in science in ways that are meaningful to both the discipline and to their lives? *What can teachers do to restructure power dynamics in science classrooms and why does it matter*?

THE POWER OF LABELING

Fundamental to sustaining or challenging the status quo in education is how teachers (and others in the hierarchy) come to represent students. The power to label students, which defines their possibilities for learning, and frames their social histories and family circumstances as limiting or empowering, helps legitimize the nature of the practices of schooling. Deficit thinking occurs when teachers attribute the educational challenges and barriers faced among marginalized youth and their families to perceived deficiencies in students' intellectual abilities, motivation, culture, language, behavior, etc., rather than to the structural and systemic oppression that limits their educational opportunities. We recognize that systemic oppression (e.g., largely racialized, gendered, and classed) manifests as inequitable economic and education opportunities for minoritized communities. This is a local and global lived reality that teachers, communities, academics, and youth must work to disrupt. We also recognize, however, that the dominant (hegemonic) view among teachers and administrators of racially and economically minoritized students is one that mischaracterizes youth as deficient and lacking persistence, and positions their families as unconcerned with their children's schooling and education (Gorski, 2013).

WHAT ARE FUNDS OF KNOWLEDGE?

The central idea of funds of knowledge is that all families, although our work concentrates on those in the working class, can be characterized by the social practices they have developed and knowledge they have produced and acquired in the living of their lives. The social history of families, and their productive or labor activities in both the primary and secondary sector of the economy, are particularly salient because they reveal experiences (e.g., in farming, construction, auto repair, household maintenance, entrepreneurship) that generate much of the knowledge household members may possess, display, elaborate, or distribute to others through social relations or networks. A funds-of-knowledge approach defines as the leading activity (e.g., for disrupting inequitable classroom power dynamics) the analysis of households as dynamic, social historical entities (González, Moll, & Amanti, 2005). Foregrounded in this work is the importance of teachers coming to know, through ethnographic-like observations and interviews, not only the children but their parents and other household members by documenting first-hand the social practices and knowledge base that constitute household life.

The goal has been to help teachers become resourceful practitioners when addressing issues of diversity in their classroom and in combating damaging stereotypes of low-income children and families. Furthermore, we have also emphasized the importance of teachers collaborating in the analysis of the pedagogical potential of funds of knowledge by generating case studies of their own classrooms. This is, to be sure, a laborious process. But as Linda Darling-Hammond (2007) proposed, teachers must not only avoid trivializing the connection between school-based knowledge and everyday knowledge, but also overcome any deficit assumptions about the students or their practices. As she points out, the capacity to understand the other and make connections to classroom learning are central challenges in contemporary schooling, especially given the changing demographics of schools. It does not come naturally, but must be "developed through study, reflection, guided experience and inquiry" (p. xxi). This process captures the intent of the funds of knowledge approach.

Two caveats are important here. Our primary claim is that teachers documenting or knowing about this family-related knowledge can become an important resource for teaching and learning in classrooms. It represents, one could say, an opportunity for teachers, as essential to their pedagogy, to establish a resource-based representation of families and students, especially those who are often assumed to be lacking any such assets. It is a way, then, of getting to know students and their familial context, beyond the narrow confines of classrooms. However, as we emphasize herein, children may also produce knowledge (e.g., about computers) that is independent of the adults in their lives, their social practices, or their social relations but related to the children's peer networks or primary interests or activities. Teachers may certainly have more access to the latter funds of knowledge than to the former (e.g., Moll, Soto-Santiago, & Schwartz, 2013).

Other researchers who have built on a funds-of-knowledge approach have pointed out that such detailed household analysis is not always feasible, that one must also develop alternative methods of study that similarly challenge deficit perspectives. Lew Zipin and colleagues, for example, have made important alterations to the original funds-of-knowledge model by replacing the household observations, which they deemed as untenable in their particular circumstances of study, with pedagogical attempts that build on the lived experiences and knowledge resources of the students themselves in their own lifeworlds beyond school, with the goal of making their "community curricular" (Zipin, 2013). In this case, central to the students' work is the development of clay animations that capture detailed stories about community life, both the good and the bad, which are then presented at a school-wide exhibit of their work. There is also a strong effort here to establish relations of trust, especially among the students and between the students and the teacher, in providing their public depiction and analysis of their lifeworld experiences.

In this chapter, we explore, through two narratives, how teachers and students draw from personal, community, and familial funds of knowledge to restructure inequitable power dynamics from within school and non-school science learning spaces. Through the lenses of these two narratives, we can see how science learning experiences can be/are collectively negotiated and how youth and community funds of knowledge become mediating tools in this process. Our first example illustrates how the teacher (Andy) taking seriously the simple dictum of "get to know your students," can alter the nature of the relationships between teacher and students and how they mutually come to re-define their engagement in academic work. The second example depicts two friends (Emily and Jennifer), both remarkably resourceful learners, it turns out, and their perseverance in gaining the knowledge necessary to alter an electric scooter, especially for nighttime use.

ANDY'S LEARNING TO TEACH

Andy, at the time of the study from which his story was documented, was a 23-year-old White male preservice teacher enrolled in a 1-year Masters/credentialing program who had recently completed his 14-week student teaching requirement. During his student teaching experience, he was hired by the district to teach a credit recovery course for high school students who had not passed their biology class during the academic year. The course was held at La Esperanza High School, which was located in the same district where he had been student teaching, but in a more economically oppressed migrant farming community. Students at La Esperenza are predominantly Latin@ (approximately 96%), 90% of students are on free/reduced lunch, and over 30% of the students are identified as English learners. Andy admitted he had had little experience working with predominantly Latin@ students and struggled as a credit recovery teacher at La Esperanza, becoming resentful toward his students for what he perceived as their "lack of motivation." However, an intervention on the part of his teacher supervisor, who encouraged Andy to get to know his students, helped Andy change course midway through the semester. Andy's story illustrates how connecting with students' personal and community funds of knowledge can have transformational effects on both teachers' and students' views of science and science education, and of each other.

Toward the beginning of his teacher education program, Andy, a White middle class male from a suburb in southern California, positioned "inner-city" students as not capable of seeing or understanding the "big picture" because they had not been exposed to the green spaces that he and students like him (White, middle class, from suburban homes) had experienced. Andy also felt strongly that students were responsible for their own success in secondary science. As a science teacher, Andy felt it was his role to provide standards-based learning opportunities for students, but, ultimately, it was up to the students to engage or not. During the spring semester of his program, Andy took a position as a credit recovery teacher in a high school located in a migrant farming community. Andy described it as "a high school unlike mine … ninety plus percent Hispanic, you know."

He recalled how he initially went into teaching the credit recovery biology course with the idea that all students would be intrinsically motivated to do well in the course: "I went into it expecting like, okay, these kids flunked. Now they're [going to be] really motivated to recover these credits. They're going to be on the ball." A couple of weeks into the course, he became increasingly frustrated with what he perceived as a lack of motivation on the part of his students and decided that he would establish a more disciplined classroom environment through individual behavior contracts: "If they [students] break the contracts then I don't want them in the class anymore because it's going rough for me." Andy decided to share his plan with his university supervisor.

> … I said, "what do you think?" And she [my university supervisor] said, "They don't work. They'll sign the contract and their parents will sign the contract and then they'll break the contract or you're going to fall through on what you say you're going to do with the contract." I said "yeah, that sounds about right." I would either kick them out of the class because I know they're going to break the contract or I'd be lenient and what's the point in doing it if I'm going to be lenient about it. She really tried to encourage me to take it another way … [She said] get to know these kids and find out where they're from … what's important to them right now.

Andy took his supervisor's advice and restructured his approach to teaching and learning in the credit recovery course. In his view, her advice "altered [his] whole year." Following the mantra of "get to know these kids," Andy asked students to share with him their perspectives on what issues were most important to them in the world. Andy recalled that all of his students responded that family was the most important thing to them. In response to this, Andy began to explore designing units that blurred boundaries between home and school contexts. For example, during a unit on genetics, students interviewed family members about their genetic traits. He remarked how students' attitudes toward the class changed significantly through these more responsive experiences, "seeing their eyes get a little wider and how they stopped talking to their neighbors and stopped texting on their cellphones ... It took the time and effort of me asking them, 'What's important to you?'"

For the final project which was required to be 30% of students' grades, Andy continued to build on his funds of knowledge approach to teaching and learning, creating an open-ended inquiry project in which students would investigate a science question of their choice, formulate a hypothesis, write up their findings, and present to the class. To scaffold this experience, Andy began taking the students on nature walks outside in the wetlands during class, encouraging them to make observations and ask questions about what they saw.

> A lot of the questions that La Esperanza High kids came up with to investigate wasn't stuff that they noticed on our walk but things that they noticed about their community. Walking to school, they had seen eagles flying around the wetlands and walking to school they saw squirrels running around—and they weren't any strangers to ticks ... One girl [Gabriella] thought that ticks were there just to make your day horrible. Just to bite you. They were nasty and ugly just to bother you. They sucked your blood for that reason—to just to get on your bad side. *That was her hypothesis*: Ticks bit you because they're mean. She did her research and found out that ticks suck your blood because they need to nourish their eggs and lay eggs, and without a blood male, they can't do that. As a mother, she totally tied into that and did a great presentation at the end of the year about why ticks suck your blood, and no it's not to bother you, they need it for their babies, and that they even secrete something that makes it painless when they're sucking your blood and then they fall off of you after they've sucked your blood. She kind of went from thinking that they're disgusting to no they're not so bad, they're trying to get by and have their babies, which is pretty cool.

In this example, Andy facilitates an opportunity in which students are experts in science, capable and knowledgeable investigators of the natural world. Through the resulting blurred boundaries between home, community, and school science learning contexts, an opportunity emerges for Gabriella, as a young woman and mother, to engage a marginalized discourse in science: "a feeling for the organism" (Calabrese Barton & Osborne, 2001; Fox Keller, 1983). Andy recalled other significant changes in how students oriented themselves to the science class, once it became apparent that Andy was authentically interested in engaging their interests and experiences. He talked about how students who were unwilling to speak in front of the group became willing to present their work: "Even the girl who only spoke Spanish and refused to speak any English—and hardly spoke Spanish out loud—gave a presentation and I thought that was great at the end of the year."

Andy also shared how "knowing students" transformed his approach to "discipline" from one of authoritarianism toward an ethic of mutual respect and understanding:

> I had to break up a fight one day, and I asked one [of the students] to sit back down in his seat and one to go outside, and I pulled out a chair and told the [rest of the] class "you guys need to be working on this, and I'm going to go outside" and we had a talk. I sat down next to him. I said, I want to talk to you and he shakes his head no and so I sat there for five minutes. He finally started talking to me, and I think it was because of the exchanges that we had [before in class] about where he came from, who he was, and what was important to him. So that's gone a long way … you need to know the students' names. If possible you need to have some sort of communication with their home. Your student needs to know that you're support for them. You're not some guy up on a pillar … At the end of the year it did end up working, and I asked students to do some narrative evaluations of me. I asked them a few questions, and one of the boys that was writing the contract for said "at the beginning of the year, I hated you." That was his first sentence. Then he goes, "I've really appreciated the chances that you've given me because I've been a screw-up in this class and I wasn't right to you and I wasn't right in this class and you could have taken me out if you wanted to, but you let me stay and I appreciate that."

By the end of the program when Andy had completed teaching the credit recovery course, it was clear that he had learned a lot about the power of getting to know students:

> I guess that every student given the right sort of teaching can be equally successful in the classroom … I found that the most important strategy discovered this year was just getting to know students on a personal level. Building a relationship with them because at that age, it's so important to feel like that they can interact with adults in some sort of equitable way. That it's not a completely top down operation … It altered my whole year this year. It totally took me from a slippery slope of negativity and trying to fight the small wars with students and to be a little more copacetic with students, understanding that they're coming from a different place than I am … We don't all come from the same background … Just getting to know them and feeling comfortable and engaging with them.

This symbolic power of labeling was patently clear in Andy's example; a change in his perceptions, motivated no doubt by the personal success he had in altering the classroom participation in ways that engaged the students, provoked a helpful, productive, and perhaps sustainable change in pedagogy. When Andy positioned his students as "unmotivated," he essentially put additional constraints (beyond those imposed by structural inequalities) on his own agency to teach science. As

Russell Bishop (2011) has found in his work with secondary teachers, a discursive repositioning (or re/authoring) of minoritized students can occur if/when teachers are presented with "an opportunity to draw explanations and subsequent practices from alternative discourses that offer them solutions instead of reinforcing problems and barriers ... [thereby] discursively reposition[ing] themselves from discourses that limit their agency to those where they can be agentic" (pp. 48–49). Establishing relations of trust, and especially affirming that the student experiences would be respected, were also critically important in Andy's re-constitution of his classroom dynamics. Through being offered an opportunity to "get to know" the lifeworlds of his Latin@ students, Andy was able to take on a more agentic positioning to science instruction, with his students then being able to (re)orient themselves as "insiders" to science, curious and concerned investigators of the natural world.

JENNIFER AND EMILY'S SOLAR POWERED ELECTRIC SCOOTER

Jennifer and Emily joined Get City (GC) in autumn of 6th grade in 2012. Jennifer joined because she wanted to use the computers that she knew GC had. She thought that "science is boring," and often complained that she was not allowed to build things or use technology in school science. However, she felt that she was good at using the computer and the Internet, and GC would give her a chance to learn new skills. She was also very proud of her technology skills, as she explained her ability to jailbreak her mother's iPhone, "I am good at jailbreaking stuff. I am actually good at jailbreaking. Like going on an iPad or iPod or whatever, it's like an unlock. You can jailbreak it and get inside of it." More often than not, outside of school, Jennifer felt that no one recognized her interest or ability with technology. For example, when teachers asked questions about computer skills and the Internet during class, Jennifer was the first one to raise her hand to answer the questions, even though she was rarely called on to help.

Emily, on the other hand, joined GC because Jennifer joined. The two were best friends and had been from an early age. They both came to the local community club most days after school because both had mothers who worked in the after school hours, and their schools did not provide after school programs. Emily felt that GC gave her something to do with her friend that was interesting.

In the vignette we share below, we describe how Jennifer and Emily designed a scooter in which attached LED lights were powered by a solar panel and rechargeable batteries. They designed the scooter because their peers were unable to play with their scooters or travel to each other's homes after dark due to the limited number of streetlights in their neighborhood. After they learned how to use solar energy technology and a three-dimensional modeling computer program, they tried to design a three-dimensional model of her scooter and then build a prototype. They shared their prototype with their friends at school and community.

Why a Solar Powered Electric Scooter? Jennifer and Emily initially decided on building an electric scooter because they wanted a way to travel to each other's homes after dark. They each owned a razor scooter, in fact Emily stated that she had four of them, and scooters were cool—everyone has one:

Researcher: Why did you try to make something new using the scooters?
Emily: Scooters are cool. Scooters are just cool; they're cool.
Researcher: How did you know that?
Jennifer: A lot of kids own scooters.
Emily: Yeah, like every member. I used to play with scooters. I have four of them. At first, I didn't have any idea about we would add besides lights when we were going to do a scooter. We all just thought of it.

The girls did not settle on the idea of a scooter right away. At first they wanted to make an "iTech phone" which was a phone that had a portable charger. However, they decided that would be too expensive, and that even though a lot of kids had phones, they did not. The girls started brainstorming different ideas. One idea that they talked about was how they needed better ways to get to each other's homes after dark. Jennifer's family did not have a car, and even if they did, her mother worked most evenings. Emily also felt that they lived close enough to each other that they should be able to go visit on their own. Their idea was simple. First, Emily explained:

> Yeah, it's a scooter with … It's kind of like the same type of scooter, but it kind of goes a little bit faster. The brakes are more automatic. You can just push a button, and it stops right there where you want to stop. The bars have a little bit more cushion to the regular one. Usually when you're at the bike and you get tired, you're able to just put it back on the bars, and it feels way better.

And then Jennifer elaborated:

> You roll on your scooter. You're just doing this all day long. Your arms get tired. Around the block you're like, "How am I supposed to get home?" You have to ride it again. Stop for a moment. Then you're like, "I need a break from my bike that I can just sit down, put away my legs." We can make that scooter better, more popular better.

As can be seen in these descriptions, the girls were interested in engineering a scooter that would get them home quickly, easily, and safely.

The Initial Design. The two girls went out and collected observational data about who uses scooters and when. They even interviewed a few of their peers about their initial ideas. They noticed that many different ages of people used scooters, and that most people used the razor scooter.

They initially sketched out their idea by hand (see Figure 5.1). Their sketch began with the razor scooter. And, it included many different concerns they felt that their peers would have: It would be electric so that they could move quickly, especially when they needed to get somewhere fast. As Emily said about the importance of speed: "speed is important to me … the razor is one of the ones where you have to, like, push yourself. They are not like the motorized. So as far as speed and everything like that, we were looking more just for operations for getting things moving."

The scooter would also have a phone dock (for the "ten-, twelve-year olds who probably do have phones") as well as "soft handle bars" to make the ride comfortable. The scooter would be adjustable in size to meet a wide range of ages,

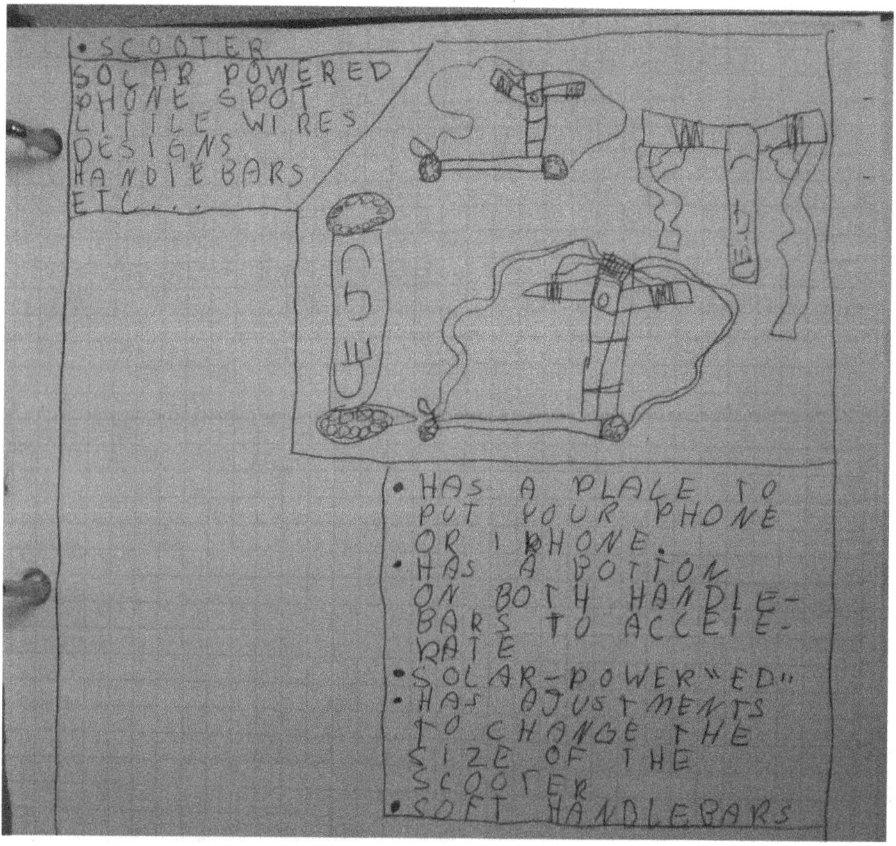

Figure 5.1: Sketch of Electric Scooter.
Source: Author

and would have a button on the handle bars to allow the user to "accelerate" in case you need to "get away" from something. As they further described in their interview:

> Well, it kind of entertains kids, and adults can ride it. For little kids, there's streamer things on it, so you could just attach the streamers or take them off if you don't want them on or not. You can get them in different colors. It kind of helps parents like when they're going somewhere, and they leave their kid with someone at home, and they're like a troubled kid. You want to buy something for them so it entertains them so the people who watch them don't have as much trouble as having to be in a house all day hearing them scream, yell, and break things. It keeps them entertained, and it also gives them exercise because your leg is moving, and it's pushing you. You're pushing yourself.

Not only were the girls interested in the technical features, the quote above also shows that the girls cared about how the scooter looked, and in keeping kids entertained. These concerns all factored into their designs.

The two girls shared their sketch up with their peers outside of their after-school club in order to get feedback on their design, and to see if others would find it as useful as they thought it was. Later when they surveyed their community members about the solar powered scooter they noted that most people said that they would use the scooter for "7–9 months of the year, but I don't really think anybody would choose a lot of 10 - 12 because that's getting to the winter and it'd be a lot of snow, no sidewalk" (Jennifer, feedback cycle). The girls also noted that everyone but one person said that they would buy a solar powered light up scooter: "Only one person chose 'no,' which was my grandpa."

They also thought about the role scooters played in transportation in their community. They noted in their survey that most people said that people would like the scooter because it would be "green" and would "save on gas" but they noted that they did not "really think people want to save on gas, 'cause they don't have a car.'"

The survey data presented a conundrum for the girls in terms of the issue of winter transportation, especially as they thought about the younger children who might use their scooter. As they wrote in their notebook: They needed "special wheels so the scooter has a chance against the snow or mud" and it would "come with padding so if you fall and you fall on the scooter the fall isn't so hard." They also decided at this point to add "lights" because their peers noted that people needed to be able to "see you."

Moving from a Sketch to a Prototype. The girls used the feedback they received from their peers and decided to make a sketch up of it in a 3-D modeling program (Google Sketch Up). Neither girl had used this program but both girls were excited to try it out. Emily "likes everything about computers!" and "it's the thing that is one of the best things, it's one of my hobbies and I'm really good at it."

Jennifer also felt that computers were her "best subject" and that "I love computers and I want to spend time on them." In fact, Jennifer noted that "one thing in my family that my grandma, she works for her own business [with computers]. She fixes computers and programs for people." And that she actually started using a computer when she was "four years old" when her older sister put her "in front of the computer. She taught me a lot about computers and stuff."

The two girls were the first in their afterschool club to complete a three-dimensional sketch up (see Figure 5.2). Asked why they wanted to spend more time revising their three-dimensional computer model, Jennifer stated she wanted to "show the experts her intelligence and creativity." Jennifer and Emily, in fact, saw the possibility of using her three-dimensional computer model to change how people viewed them.

However, moving from the 3-D sketch up to a physical prototype proved to be challenging. The first thing they did was to bring in Christmas lights from home to wrap around their scooter. But, these lights were "too long" and "too expensive." They decided to solve this problem by "slapping some stickers on it" for decoration and visibility. But, the girls did not have any stickers that they liked. When they started digging around the supplies that were available in their afterschool club they found some university themed duct tape and a small string of LED lights that

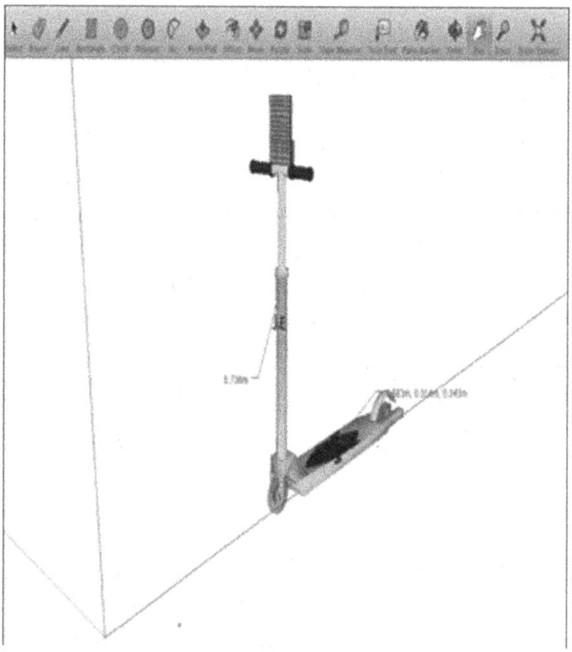

Figure 5.2: Three-Dimensional Model of the Scooter.
Source: Author

one of their afterschool mentors had affixed to an old set of shoes. This gave them the idea of both decorating their scooter in a way that people in their community might really like (since they lived near the university), and the lights would be efficient. However, they still felt challenged with how to attach the lights because the string of lights was shorter than they wanted and they did not know how to attach the lights to the solar panel.

In fact, Jennifer described the "wiring was the most important part because you have to wonder about where the wires go, how they connect and what we're supposed to do with them." They had to figure out how to "connect the wires and bolts, because you have to take them apart and put back together and stuff. And, find the right tools."

The girls drew from so many different sources of ideas in their designs. For example, the girls went onto YouTube to find solutions for how to solve their problem. Again, as Jennifer explained: "There's a lot of stuff I got on YouTube … that we got on YouTube for hooking up the things and copper ball and everything we got off YouTube. I'm actually really good with YouTube actually." But the girls soon ran into a new problem. The scooter did not work the way they wanted. The solar panel did not keep the scooter charged the way they wanted or needed. They had already learned in school that the sun's angle changes and that there was less daylight in the winter. Jennifer explained:

> Yeah, but in our science class we're learning about how the sunrise and the sunset and the sun angle and stuff can be very different. In the winter, the daytime, the daylight is way shorter because in the winter it's really dark. Probably like at 5 or 4 it gets super dark. It starts getting really dark, and the sun starts to fade away more and more. Even in the morning it's dark.

Merging their understandings developed at school of the different angles of the sun across the seasons, therefore, also played an important role in their designs.

The two girls shared their design with a local science education expert and an engineer at "feedback cycle day" in their club. They described this day as "really important" because, as Emily stated, "Well, we just got to, we were able to learn more stuff about our scooter. Actually, before we met the experts, they were actually, I just looked down here and they were the ones who told us to have a solar panel that faces up and down and we needed a bigger one … and that it needed to be 45 degrees, yeah, because before we had it smaller." But, even after solving this problem with the help of some outside experts, they ran into further problems because the solar panel did not work in the dark:

> Emily: Okay, we put the batteries because … [overlapping]
>
> Jennifer: … We put the recharger batteries so it would light up this light because what happens if the solar panel runs out.

When the girls finally got their design to work they were ecstatic. As Jennifer explained, "I'm pretty proud that the scooter is really cool. I would totally buy it. I would probably be on my knees to my mom. 'Please buy it. Please. It's Christmas. Please.'" Jennifer even said that when her sister saw her scooter she said, "Oh my God Jennifer! You did that? Awesome!"

The girls also brought the completed project to school to share with their friends. They shared their scooter at recess: "Recess, yeah, 'cause how to have fun with this, helped us think of this 'cause we're kids and recess" and also over Kik (a social media platform). Jennifer summed up their feelings when she said that what the scooter would say about them is:

> It will probably say about me she knows how to have fun. She can get down and smart when she really needs to. This girl can be fun. She could build things. She could make the world a different place and help everybody else learn how to have the type of fun she has and stuff.

The community-based work of Emily and Jennifer conceptualized and conducted a project that adapted a scooter by attaching LED lights powered by a solar panel and rechargeable batteries. It is an example of how the students' extracurricular

Figure 5.3: Jennifer's Scientific Movie.
Source: Author

interests in the topic, their perceived need for transportation after dark, and the funds of knowledge found in their immediate and digital communities, combined in providing information to assist them with their project. This work also represents an example of making the community curricular, in the sense of their strategy of seeking knowledge from different sources that would help them conduct and effectively complete their project. The girls' interest in computers, in particular, motivated their search for solutions and they cleverly used all resources available to them, ranging from expert advice provided in the after-school program to the use of YouTube displays to gather information. Thus, the knowledge needed to complete their project was distributed in their immediate and digital community, and the girls displayed the wherewithal to locate it, thus reaffirming for themselves, their peers and adults, their smarts, agency and competence. Science learning became a vehicle for personal and social transformative power as the girls used their knowledge to design a solution to a problem *they had identified.*

As community insiders, Jennifer and Emily, in on-going ways, positioned themselves as a part of, or inside, their community, rather than outside of it. In schools, teachers and students are rarely asked to identify with community. The very notion that one can position oneself as inside or outside of community stands in stark contrast to the driving norm in science education, where the focus is on generalized, "across the board" scientific knowledge and practice. However, in taking on local problems, in community spaces, the girls used their insider knowledge and status to frame the problem space, and to work towards community sustaining solutions. Through integrating their funds with scientific and technical data, the girls also came to view themselves as experts.

REFLECTION

Every educator (and every educational researcher) addresses, in one way or another, deliberately or not, relations of power. After all, these relations are part and parcel of, and help constitute, the social and material structures of schooling. Earlier in this chapter, we discussed how the power to label students is the power to reproduce larger societal inequities. As Pierre Bourdieu (1989) has suggested, the power to make groups, especially in education, is political power *par excellence.* Richard Valencia (2010) reminds us that this practice of labeling, or deficit thinking, "amounts to a form of oppression—that is the cruel and unjust use of authority and power to keep a group of people in their place" (p. 9). When educators are supported with sustained opportunities to critically reflect on the power of labeling, how they have come to label students, and how the ways that they label students contribute to a reproduction of the larger economic and political inequities in society, they can begin to discursively reposition themselves as having agency over their practice (vs. blaming students and families) while they discursively reposition students as engaged knowledge producers and learners (Bishop, 2011; Tolbert, 2015).

In this chapter, we set out to explore the question, *What can teachers do to restructure power dynamics in science classrooms (and why does it matter)*? Interrogating power dynamics as a part of science learning over space and time is an important equity-minded project. Young people's experiences in science often take place in what Patricia Collins (2000) calls a "matrix of oppression," the structure that operates with race, class, gender and other forms of oppression. They are often positioned as "outsiders" to science based on where they live, what they look like, or a perceived lack of capabilities (or interest) in making contributions to science investigations. Normative discourses and practices (of those in power—in science/western society—mainly white, male and middle class) position minoritized youth in real and symbolic hierarchies, leaving actors to confront contrary narratives in their experiences. This often plays out in the struggle between the ways in which school positions and values certain ways of knowing and doing science and the cultural meaning-making practices youth from non-dominant communities bring with them from personal and sociohistorical experiences. These intersections impact conceptions of what it means to be scientific, who can participate in science and how or if science might become more accessible and meaningful to young people.

REFERENCES

Bishop, R. (2011). *Freeing ourselves.* Rotterdam: Sense Publishers.

Bourdieu, P. (1989). Social space and symbolic power. *Sociological Theory,* 7(1), 14–25. doi:10.2307/202060.

Calabrese Barton, A. & Osborne, M. (Eds.). (2001). *Teaching science in diverse settings: Marginalized discourses and classroom practice.* New York, NY: Peter Lang Publishing.

Collins, P. H. (2000). What's going on? Black feminist though and the politics of postmodernism. In E. Pierre & W. Pillow (Eds.), *Working the ruins: Feminist poststructural theory and methods in education* (pp. 41–73). New York, NY: Routledge.

Darling-Hammond, L. (2007). Foreword. In C. D. Lee (Ed.), *Culture, literacy, and learning: Taking bloom in the midst of the whirlwind* (pp. xvii–xxi). New York, NY: Teachers College Press.

Deschenes, S., & Malone, H. J. (2010). *Year-round learning: Linking school, afterschool and summer learning to support student success.* Retrieved from Harvard University, Harvard Family Research Project: www.hfrp.org

Fox Keller, E. (1983). *A feeling for the organism: The life and work of Barbara McClintock.* New York, NY: W.H. Freeman & Company.

González, N., Moll, L. C., & Amanti, C. (Eds.). (2005). *Funds of knowledge: Theorizing practices in households, communities, and classrooms.* Mahwah, NJ: Erlbaum.

Gorski, P. (2013). *Reaching and teaching students in poverty: Strategies for erasing the opportunity gap.* New York, NY & London: Teachers College Press.

Moll, L. C., Soto-Santiago, S. L., & Schwartz, L. (2013). Funds of knowledge in changing communities. In K. Hall, T. Cremin, B. Comber, & L. C. Moll (Eds.), *International handbook of research on children's literacy, learning, and culture.* Oxford: John Wiley & Sons, Ltd.

Tolbert, S. (2015). "Because they want to teach you about their culture …": Analyzing effective mentoring conversations between culturally responsible mentors and secondary science teachers of indigenous students. *Journal of Research in Science Teaching, 52,* 1325–1361. doi:10.1002/tea.21240.

Valencia, R. (2010). *Dismantling contemporary deficit thinking: Educational thought and practice.* New York, NY: Routledge.

Zipin, L. (2013). Engaging middle years learners by making their communities curricular: A funds of knowledge approach. *Curriculum Perspectives, 33*(3), 1–12.

CHAPTER SIX

STEM Education as "Trojan Horse"

Deconstructed and Reinvented for All

LARRY BENCZE, MICHAEL J. REISS, AJAY SHARMA, AND MATTHEW WEINSTEIN

A powerful and widespread movement in education focuses on encouraging integration and/or interrelationships among fields of science, technology, engineering and mathematics (STEM). While the movement is not universally accepted, many stakeholders, including school administrators, business leaders, government education officials, teacher educators, teachers, students, and others, avidly support various forms of STEM education. Indeed, it seems increasingly to be the accepted norm in many jurisdictions. Nevertheless, it is our contention that, like the proverbial Trojan horse, STEM education initiatives sometimes (or often) project idealized images of prosperity, wellbeing and status that may result from their implementation that, however, hide serious STEM-related dangers for the wellbeing of individuals, societies and environments such as dramatic dislocation and disruption from severe climate change. Although constructors and proponents of the STEM Trojan horse are many and varied, we also believe that much of this subterfuge appears attributable to vast global networks of individuals, organizations, technologies, infrastructure, etc. supporting neoliberal capitalism. This is a somewhat nebulous ideology, but it often appears to involve interventions by governments and other entities in economic markets and in societies more generally to promote private sector gains. While we acknowledge that power is distributed across such global networks, it also appears to us that much of the impetus for these networks can be traced to relatively few wealthy individuals and groups. In this chapter, we provide elaboration and justification for these claims and suggestions for some alternatives that we hope may help realize improvements in wellbeing for all.

STEM EDUCATION AS TROJAN HORSE

Although STEM education appears to have achieved the status of generalized "common sense" in many educational contexts around the world, there seems to be much about this movement (acknowledging its diversity) that can be called into question. On the one hand, there are aspects of proposed STEM integration and/or interrelationships that seem reasonable. Many proponents suggest, for instance, that there often are significant connections among professional scientists, technologists, engineers and mathematicians (Rennie, Venville, & Wallace, 2012). Of course, many of the instruments, tools, etc. used in fields of science have been developed by technologists and engineers. Similarly, many approaches to data-collection, analyses and representation have been derived, to a great extent, from fields of mathematics. There also is the argument that involvement of engineers in the work of scientists can help expedite translation of abstract knowledge from the sciences (e.g., about the structure of DNA) into useful products and services (e.g., genetically-modified organisms for more efficient agriculture) (Etzkowitz & Leydesdorff, 2000). Moreover, there are arguments for epistemological and ontological relationships between science and engineering, in the sense that conduct of science and engineering often co-influence each other and, therefore, may have characteristics of the other in them (Roth, 2001). For example, translation of phenomena (e.g., trees) into representations of them (e.g., drawings of trees), which is said to be a major process of science, can affect translation of revised representations (e.g., molecular pathway drawing) into revised phenomena (e.g., engineered trees), which is said to be a process of engineering. On the other hand, despite some apparent similarities and relationships in professional practice, there does appear to be resistance to integration of and/or interrelations among STEM fields in education. Some argue, for example, that there is some "muddling" in STEM education projects of epistemological and ontological differences among the four STEM fields. In many cases, it appears that teachers of science, teachers of technology, teachers of engineering, and teachers of mathematics are now considered, simply, "STEM teachers." There is research suggesting, however, that different fields of endeavour in the sciences—such as chemistry, as compared to physics (Thalos, 2013)—are based on different knowledge generation principles and practices. Similarly, think of how we arrive at truth in mathematics. We establish that the number of primes is infinite, that the square root of 2 is irrational and that the sum of the internal angles of a flat triangle add up to 180 degrees in ways that have little in common with ways in which knowledge is constructed in fields of science, technology or engineering. A scientist might head off into the field, gather samples of triangles, bring them back into the laboratory, cultivate them in Petri dishes, measure their three internal angles, sum them and then compute means and some measure of variableness between triangles. Perhaps partly because of such differences, school systems have frequently separated instruction in the four STEM fields. Moreover, in light of such philosophical and methodological differences, there appears to be have been some *antagonism* between science and technology education, with science frequently enjoying higher status while technology has been stigmatized as only appropriate for "less able, concrete thinkers" (Fensham & Gardner, 1994, p. 168). It has long been suggested that school science has been given high status in schools because, to a great extent, scientists have taken advantage of the traditional claim that abstract knowledge they generate is more important than technological knowledge, the former considered more generalizable and, therefore, more powerful than the latter (Fensham, 1993).

Given apparent long and deep traditions of separation of teaching of science, technology, engineering and mathematics in schools and, moreover, apparent special status for science as compared to technology and engineering, it seems clear that STEM education promoters need to work to "sell" linkages among these historically isolated fields of education. Related to this, because of traditions of isolation from one another, teachers of different STEM fields may have to learn to "play nice" with each other and, in a sense, be on one discursive team. Breaking traditions of and structures promoting their isolation and bringing teachers of the four STEM fields together seems to require significant resources and resourcefulness. This, in turn, raises questions like, "Who or what is behind STEM education?" "What have they to gain from such associations?" and "How do they appear to be attempting to achieve their ends?" We think clues to answering such questions lie in analyzing which fields of endeavor are *excluded* from many STEM education initiatives. It seems to us that articulation (of STEM fields) occurs simultaneously with disarticulation. By focusing on who is not included, we can start to make visible the ideological work of this stew. Missing in action, it seems, are fields of humanities, sociology, history, politics, anthropology and possibly many others (Zeidler, 2016). Perhaps we can gain clues about sources of power behind STEM initiatives by thinking about what omitted fields like these would reveal about STEM fields.

In examining websites, curriculum policy and instructional materials, STEM education brochures, publications about relevant research, etc. through insights from the humanities and social sciences, it is apparent to us that a—perhaps *the*—driving force determining much of STEM education is support for *neoliberal capitalism*. *Neo*(new)-liberalism marks a return, or reference to, economic liberalism of the 18th century, in which priority was given to private sector freedoms from, especially, government interference in economic affairs. Neoliberals are not merely trying to restore a previous order, however. Their version of free markets is not about a *laissez faire* state; but, rather, involves government intervention in markets, particularly in terms of establishing laws that enable financiers and companies to freely pursue profit-generation activities. Part of the worldview of neoliberals has been an emphasis on perpetual entrepreneurialism (lifelong learning), in which everyone is supposed to see themselves as constantly trying to increase their personal human capital in order to survive in a world in which the market, rather than the state, provides for social needs, and in which there is little security (McMurtry, 2013).

Although its origins are complex and somewhat controversial, many suggest neoliberalism gained traction after the second world war with formation of the Mont Pèlerin Society (MPS), a think tank consisting of economists and philosophers including Friedrich Hayek, Karl Popper, George Stigler and Milton Friedman, which sought to resurrect and renew ideologies supporting personal and market freedoms—suggested in light of harms linked to fascist and communist regimes in Europe (Harvey, 2005). Its efforts have, to say the least, been extremely successful. Now affiliated with the MPS, apparently, is a vast network of neoliberal think tanks (e.g., Atlas Network, Heritage Foundation), banks (e.g., World Bank), financiers (e.g., Filiz Sahenk, Wee Cho Yaw), corporate owners (e.g., Koch Brothers, Bill Gates), media barons (e.g., Rupert Murdoch), university departments (e.g., Department of Economics, University of Chicago), transnational corporations (e.g., Disney), transnational trade organizations (e.g., World Trade Organization), conventions (e.g., World Economic Forum, Davos) and many others (Ball, 2012). Indeed, neoliberalism's capture of supporters worldwide appears so extensive that its network seems like The Borg of Star Trek™ fame, threatening to assimilate everything in its path, stating "resistance is futile" (https://en.wikipedia.org/wiki/Borg_(Star_Trek)). About neoliberalism, two of its most famous supporters, Margaret Thatcher of Great Britain and Ronald Reagan of the USA, famously said, respectively, that: "There is no alternative" and it is "The New World Order."

To gain support for their agendas, neoliberals have sometimes resorted to military force—as famously enacted in the Chilean dictatorship of Augusto Pinochet (in office from 1974 to 1990). However, it seems they often have preferred battles won by heavily financed *persuasion*. Accordingly, a key element in neoliberal tactics is education of populations in ways congruent with their purposes. Stephen Ball (2012), for instance, in his book *Global Education Inc.*, identifies a vast network of individuals and groups that, funded by numerous millionaires and billionaires, like Eli Broad, the Walton Family, and Bill Gates, have been placing enormous pressure on education systems to enact perspectives and practices of the so-called

Global Education Reform Movement (GERM), including: *curriculum standardization, international competitiveness, testing and reporting* and *emphases on "core" literacies* (e.g., language(s), mathematics, science and information technology) (Sellar & Lingard, 2013). A major aspect of this in recent years has been promotion of STEM education. In what has become known as the "neo-Sputnik" race, governments, companies, financiers, universities, transnational trade organizations and others in the USA, the EU and other "Western" nations have become increasingly concerned about competition over the last decade with countries like China and India and, accordingly, have been urging education systems to focus on identification and education of prospective STEM workers to assist in such economic struggles (Pierce, 2013).

Although vast and complex networks of pro-capitalist entities like those noted above likely use—not, necessarily, consciously—an array of strategies for achieving their ends regarding STEM education, we suggest that a significant tack is *subterfuge*. Like a magician, if people can be convinced to focus their attention on certain, perhaps common, phenomena, they will not notice what the magician is attempting to conceal. Like the story of the proverbial Trojan horse (see Figure 6.1 and https://en.wikipedia.org/wiki/Trojan_Horse), it seems that promoters of STEM education often encourage acceptance by portraying it as a wonderful solution to

Figure 6.1: Trojan Horse.
Source: https://commons.wikimedia.org/wiki/File:Giovanni_Domenico_Tipeolo,_Procession_of_the_Trojan_Horse_in_Troy._1773..jpg

a range of problems in ways that may distract recipients from its hidden dangers. Apart from evidence from analyses of STEM education promotional materials, etc., as described below, it should first be mentioned that this tack appears congruent with capitalists' sales techniques for many or most other commodities such as cars, cosmetics and financial futures (Bencze & Carter, 2015). Consumers are routinely distracted from noticing or addressing personal, social and environmental problems (e.g., harms from climate change) linked to for-profit products and services (e.g., cars) by idealized abstractions such as "sexy," "classy" and "powerful" associated with them.

As with marketing of many commodities, STEM education seems to be sold largely on the basis of a set of images to which consumers might identify. Particularly important to STEM education "sales" appears to be its image as a solution to many economic problems. Apparently aligned with the nature of global economic competitiveness, it seems that STEM sales campaigns rely on a common neoliberal tack; namely, disaster capitalism (Klein, 2007). In other words, in order to infuse neoliberal priorities, capitalists may capitalize on natural or manufactured disasters. STEM education promoters often, for instance, appear to adopt salvationary rhetoric—claiming that STEM education should allow individuals and jurisdictions (e.g., states/provinces, countries) to be saved from economic disaster; and, indeed, to prosper, especially in terms of jobs and associated products and services, in the face of increased international economic competitiveness from other countries, like India and China (Pierce, 2013). By dazzling the public and even policy makers with the idea that STEM education is the route to economic security and the solution/avoidance of said disasters, these publics, policy makers etc., are led away from neoliberalism's and even STEM education's contradictions and problems. Warnings of dire consequences associated with purported needs for more STEM workers, for instance, appear exaggerated. With increasing technological advances, it may well be that fewer STEM professionals will be needed than claimed (Frey & Osborne, 2013); and, moreover, there is the suggestion that calls for more STEM workers is intended to glut the market and, consequently, minimize costs for wages and benefits (Pierce, 2013). Indeed, there have been overt statements indicating that relatively few STEM professionals are needed to help direct technological innovation and production; and, moreover, many STEM jobs will be increasingly "precarious" (part-time and impermanent). The U. S. National Research Council (NRC, 2011), for example, which helped facilitate development of that country's *Next Generation Science Standards* (NGSS) (NGSS Lead States, 2013), stated in its document, *Successful STEM Education*, that relatively few STEM professionals are needed to direct many jobs:

> [t]he primary driver of the future economy and concomitant creation of jobs will be innovation, largely derived from advances in science and engineering … 4 percent of the nation's workforce is composed of scientists and engineers; this group disproportionately creates jobs for the other 96 percent. (p. 2)

Meanwhile, claims that jobs are tied to particular nations also seem illusionary. Competition, often managed by transnational trade agreements (e.g., the proposed Transatlantic Trade and Investment Partnership), is increasingly between financiers and corporations with minimal ties to any one nation (McMurtry, 2013). Jobs in such environments often are moved to nations with the best legal arrangements for profit generation, such as in terms of minimal wages, worker benefits and environmental protections. At the same time, because of companies' legal rights to minimize costs associated with their commodities (e.g., for labour, materials, transportation, marketing and reparation for negative side-effects of commodities), many of them are linked to a range of personal, social and/or environmental harms (Mirowski, 2011), including from: cigarettes (Barnes, Hammond, & Glantz, 2006), agro-toxins (Hileman, 1998), fast foods and other manufactured foods (e.g., Weber, 2009), highly toxic household cleansers (e.g., Leonard, 2010), environmentally disruptive biotechnologies (Kleinman, 2003) and, perhaps especially significantly, burning of petroleum products that accelerates global climate change (Klein, 2014).

In light of such problems, perhaps technological innovation and engineering that is so prominent in STEM education initiatives is more about encouraging repeating cycles of production, consumption, and disposal of commodities that are not, necessarily, meant to benefit living and nonliving environments. Many such STEM-associated problems are not strongly addressed in STEM education projects (Gough, 2014). Where attention is given to such problems, it seems solutions often involve "techno-fixes"; that is, applying STEM techniques with little attention, as noted above, to perspectives and practices from the humanities and social sciences. For example, in the U.S.A.'s recently published NGSS (NGSS Lead States, 2013), a secondary school standard for *Earth and Space Science* suggests: "Examples for limiting future impacts could range from local efforts (such as reducing, reusing, and recycling resources) to large-scale geoengineering design solutions (such as altering global temperatures by making large changes to the atmosphere or ocean)" (www.nextgenscience.org/pe/hs-ess3-4-earth-and-human-activity). Here, the NGSS suggest that widespread chemical alteration of the oceans and atmosphere is to be treated as a perfectly acceptable solution to human impacts (there is a weird contradiction, of course, in that the goal here is to reduce the human footprint, but one permitted answer is to wildly expand them). Moreover, perhaps mirroring companies' tendencies to employ reputable scientists to dispute science findings that would discredit many of their potentially-harmful commodities (Oreskes & Conway, 2010), school students routinely are invited to view potential problems linked to STEM fields as controversial (Levinson, 2013).

TOWARDS EDUCATION FOR SOCIAL AND ENVIRONMENTAL JUSTICE

Although STEM education has a vast and deep set of supporters and although there appear to be some epistemological, ontological, and practical merits in integration and/or interrelationships in education among the four STEM fields, we suggest that educators, policy makers and others be wary in accepting a virtual STEM Trojan horse, an apparently convenient construction intended to advance neoliberal capitalist agendas, often at the expense of the wellbeing of many individuals, societies and environments. As university-based science education scholars feeling immersed in a deep and extensive milieu dominated by neoliberal capitalist perspectives and practices that we find problematic, we work to help educate citizens about dangers of aspects of STEM education and, moreover, provide alternatives that we hope may lead to more democratic control of education and, perhaps, increase worldwide social justice and environmental sustainability.

Socio-Political Dissent and Conflict. For at least the last 40 years, scholars and others have been encouraging science educators to teach science in terms of its relationships with fields of technology, societies and environments (STSE) (Pedretti & Nazir, 2011). Such approaches recognize that science does not occur in isolation of other professional fields (e.g., technology and engineering), other members of societies and their living and non-living environments. In our analyses of STEM education initiatives, however, much of this recognition seems to be dissolving. Ironically, in integrating STEM topics, a new isolationism seems to have resulted. In other words, just as STEM brings together (articulates) science with math, engineering and technology, it disconnects (disarticulates) these same fields from political, social, and humanistic inquiries. We strongly believe, however, in reversing this "erasure" and re-establishing "STSE" education—or, perhaps, with addition of engineering and mathematics, "STEMSE" education. Problems faced by humanity appear to need citizenry prepared to attempt to bring about a better world—in part, through more *critical* and *holistic* uses of STEM knowledge, skills, dispositions, etc.

As briefly noted above, it is apparent to us that many attempts over the last several decades to integrate science (and/or technology) with humanities and social sciences have urged students to treat potential personal, social, and environmental problems as *controversial.* Perhaps the most popular of these movements has been education regarding *socio-scientific issues* (SSIs), in which students often are invited to consider evidence and arguments that they may use for making reasoned personal choices about controversies (Levinson, 2013). Dana Zeidler, Sadler, Applebaum, and Callahan (2009), for example, who have had significant influences on the nature of SSIs education, summarize the approach this way:

> Central to this approach is the concerted effort to provide opportunities for students to reflect on issues in order to evaluate claims, analyze evidence, and assess multiple viewpoints regarding ethical issues on scientific topics through social interaction and discourse. (p. 75)

Although such personal-choice approaches have been popular and appear to lead to numerous educational benefits for students, many scholars suggest that SSIs (or "STEMSE") education needs to promote models of citizenship that extend beyond personal choice, giving much more priority to actively addressing, in more collectivist ways, the many problems faced by humanity (Bencze & Alsop, 2014). Many such problems and corresponding actions appear to us to be congruent with ecojustice education, which prioritizes enlightening students about various problems of social justice (e.g., androcentrism, individualism) and environmental sustainability (e.g., anthropocentrism, commodification) and, as argued above, preparing them to engage in socio-political actions to address problems they identify (Martusewicz, Edmundson, & Lupinacci, 2014).

There are, undoubtedly, many possible approaches to "STEMSE" education promoting actions to address ecojustice concerns. From our previous research and practice, we provide some examples that may serve as models for further development of such approaches. One set of these, *A Habermasian Contestation Approach to EcoJust STEM-SE Action Education* (exemplified by nuclear physics education; adapted from Reiss, 2003), draws on Habermas's (1983) perspectives on communicative action:

- **Research the roles played by such scientists** as Henri Becquerel, Ernest Rutherford, Marie Curie and Lise Meitner. [Lise Meitner played a crucial role in the discovery of nuclear fission but was not awarded the Nobel Prize with Otto Hahn in 1944 for this research.]
- **Plot a map of the distribution of nuclear power stations** around the globe and suggest reasons for the results observed. [Nuclear power stations are expensive to build and require considerable engineering expertise. In some countries public support for new nuclear power stations is lacking; in others, e.g. Germany and Japan, it has changed dramatically in recent years.]
- **Plot graphs of decreases in radioactivity in vegetation** in Cumbria, UK in the years after Chernobyl and compare the results with government predictions. [It has taken orders of magnitudes longer for the radioactive levels to return to safe levels than had been predicted. Science is not always a certain subject.]

- **Explain how carbon dioxide emissions from electricity-producing stations in France fell by two-thirds during the 1980s.** [Expansion of the French nuclear power industry. Over two-thirds of French electricity is generated in this way, a higher percentage than in any other country.]
- **Write to both pro- and anti-nuclear power organisations asking them the same specific questions**, e.g. "How safe is nuclear power?" and "How important is nuclear power for electricity generation?". [Helps students to consider the significance of sources of scientific knowledge and enables them to consider the extent to which such knowledge is value-free.]
- **Examine the medical evidence for and against an increase in the incidence of leukaemia** around certain nuclear power stations. [Controversial. Can help students to appreciate how difficult it may be to see if technologies are safe or not. In addition, to what should the safety of nuclear power stations be compared?]
- **Design and use a questionnaire to investigate fellow students' knowledge of and attitudes towards nuclear power.** [A good learning experience, developing and reinforcing knowledge about nuclear power. The work on attitudes can introduce students in science lessons to both quantitative and qualitative approaches to the gathering and interpretation of data.]
- **Role play a government meeting trying to decide whether to extend a country's nuclear power programme or to scrap it.** [Role plays don't appeal to all teachers and students and can polarise arguments. Alternatives include discussion in small groups. Done well, though, role plays can enhance empathy and understanding of the position of others, especially if one role plays a view different from one's own. Should be followed by de-briefing.]
- **Write an imaginary letter** from one of the service persons or indigenous people on test islands like Bikini Atoll. [Too little writing in science is in such a genre. Also helps pupils realise that issues of sickness and death resulting from nuclear explosions aren't restricted to Japan in the Second World War. NB controversy over use of depleted uranium shells in the Gulf War.]

Habermas advanced an emancipatory agenda (its moral dimension) while attending carefully to power in acts of communication. In school science, its adoption would mean students listening carefully and respectfully to views of others (their teachers, textbook authors, their peers), but then challenging them through rigorous enquiry. In a classroom setting, a teacher would want to search for possible consensus, without forcing agreement on students (which is often not possible for contentious issues). Such a consensus should be based largely on reason and genuine debate and take into account established practices of ethical reasoning. At the same time, it should be open to criticism, refutation and possibilities for change.

Furthermore, consensus should not be equated with majority voting. It needs to be borne in mind that while a consensus may eventually emerge, there is a time when what is more important is simply to engage in valid debate in which the participants respect one another and seek for truth through dialogue. As noted above in *A Habermasian Contestation Approach to EcoJust STEM-SE Action Education*, however, after students negotiate personal positions regarding merits of nuclear power, this approach then encourages them to develop and implement plans of action to communicate their perspectives to those who may affect change, thus benefiting all concerned (Reiss, 2003).

The approach in *A Habermasian Contestation Approach to EcoJust STEM-SE Action Education* provides considerable guidance, although also with significant choice, for students in their explorations and actions to address perceived ecojustice problems—in this instance, connected to the subject of nuclear physics. Guidance can be helpful for certain students, particularly those not familiar/comfortable with freedom of choice in schooling. It is apparent, however, that students also given significant control over decisions about representing phenomena of the world (e.g., graphs of car uses) and uses of such representations to imagine new kinds of phenomena that they then try to realize (e.g., new car uses) can become very committed to educational matters, including ecojustice concerns and actions (Wenger, 1998). However, since students seldom have such choices in schooling, the STEPWISE (Science & Technology Education Promoting Wellbeing for Individuals, Societies & Environments [www.stepwiser.ca]) approach illustrated in Figure 6.2 was developed. It is designed to provide students with apprenticeship lessons and activities intended to encourage and enable them to eventually self-direct research-informed and negotiated action (RiNA) projects to address power-related problems in STSE relationships (Bencze, 2017). Students may, for instance, be concerned about private sector advertising for fast foods that their secondary research suggests have ingredients linked to medical problems and their primary research suggests are commonly used by youth at particular times of day. Drawing on their findings, they may then develop advertising campaigns to enlighten fellow students about possible harms and healthy alternatives.

The general approach shown in Figure 6.2 appears to have been successful in helping several students to self-direct RiNA projects to address STEMSE problems of their concern. Finding that many fellow students were discarding plastic water bottles in garbage bins, a group of tenth-grade students developed a water bottle label that encourages users to recycle the empty bottles (Krstovic, 2014). Although most students seemed highly motivated to engage in various aspects of such RiNA projects, it also was felt that their actions were relatively narrow (e.g., limited to aspects of local contexts, such as encouraging the school principal to provide more recycling bins). To help broaden their research and actions, therefore, teachers turned to actor network theory (Latour, 2005). Pierce (2013),

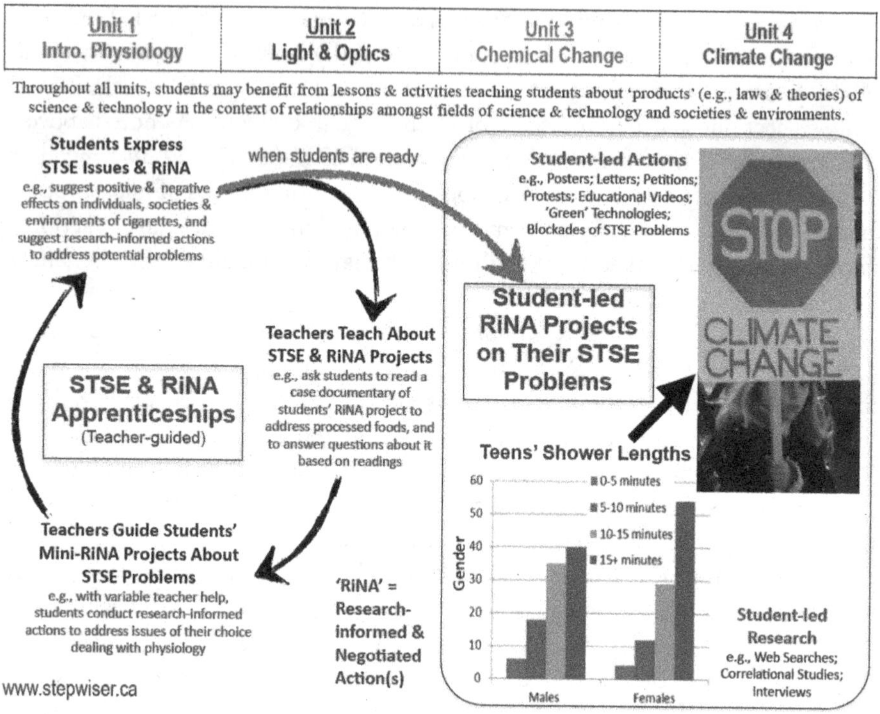

Figure 6.2: The STEPWISE Approach.
Source: J. L. Bencze, Author of stepwise.ca

for instance, suggested that helping students to develop actor network maps to depict STSE relationships can help expose them to living, nonliving, and symbolic entities (actants) to which they might not otherwise be exposed. In considering medications, for instance, students may not be aware of roles of governments in developing regulations that allow companies to test drugs in limited situations that, consequently, may lead to adverse side effects in some people (Angell, 2004). Accordingly, in work based on the framework in Figure 6.2, a teacher encouraged and enabled students to construct actor network maps to represent their analyses of STEMSE relationships regarding commodities they routinely purchase. A focus on such maps seemed, indeed, to enlighten youth about often-unseen actants and, as indicated below by eleventh-grade students' choices for a re-design (innovation) of a men's cologne, also appeared to motivate them to incorporate into their technology designs the wellbeing of a broader spectrum of living things and institutions:

> The cologne itself is made from primarily pure and natural ingredients these include: Bergamot essential oil, cedar wood essential oil, lemon, cinnamon sticks, green tea leaves, and absolute vodka. Many of our ingredients, if not grown in Canada do not contain harsh

> chemicals so the people working in other countries to harvest cinnamon for example, would not be exposed to dangerous fumes. … [T]he production process of cork is less harmful to the environment then making a glass or plastic lid, the cork is stripped off the trunk of the tree every 9 or 10 years, this does not kill the tree. … Our product would appear more attractive to a consumer because they are able to read and identify all seven ingredients and maybe even have majority of them in their household. (March 21, 2014) (Bencze & Krstovic, 2017, p. 211)

Aligned with ecojustice education foci, these students used STEM knowledge and skills and incorporated matters of social justice, such as in terms of clarity of consumer choices via labelling, and for matters of environmental sustainability, such as in terms of use of tree bark without killing the trees.

Towards Networked Ecojustice Education. Although we have been encouraged by students' uses of attitudes, skills and knowledge associated with STEM fields (e.g., research, negotiation and innovation) to bring about a better world, their actions often are, overall, relatively narrow; e.g., frequently limited to educating friends, families and others. Moreover, it is apparent that such more altruistic uses of STEM expertise, etc. has been limited to a few teachers' cases, particularly where a range of contextual factors appear to align (e.g., a supportive principal and teaching colleagues). This suggests at least two possibilities for improving our work: i) students could benefit from apprenticeships emphasizing *networking* of their research and actions; and, ii) supporters of critical STEMSE education and RiNA projects could benefit from efforts to further network (distribute) relevant pedagogical perspectives and practices. In other words, like neoliberal capitalists, as described above, those promoting ecojustice goals need to work to encourage a wide range of people, institutions and structures to align in collective support of social justice and environmental sustainability. Critique and disarticulation of hegemonic practices associated with STEM education, albeit necessary, may not be sufficient on its own. This is because liberatory politics needs to involve both deconstruction and reconstruction (Freire, 1997). That is, educators will need "a strategy whose objective is, through a set of counter-hegemonic interventions, to disarticulate the existing hegemony and to establish a more progressive one thanks to a process of re-articulation [merging] of new and old elements into different configuration of power" (Mouffe, 2008, para. 18).

As science education scholars working in university-based teacher education contexts, we can contribute to the formation of an alternative network and infrastructure that support activist ecojustice goals for STEMSE education. One such approach, called STEAMD (science, technology, engineering, arts, mathematics and democracy) (Weinstein, unpublished work), starts with creation of critical consciousness surrounding our capitalist milieu. Neoliberalism is placed front and center at the level of global, national, local and discipline-specific effects. Students are given an introduction to the history and impacts of neoliberalism (and patriarchy, capitalism and colonialism) in their foundation coursework. This is then reinforced in discipline-specific ways through alternative framings of science, engineering and knowledge more generally in their coursework (methods classes and more general classes on multiculturalism, in which links between science/STEM and colonialism are taught as well as some foundations of sociology and philosophy of knowledge). This approach involves repeated uses of alternative lenses for understanding science and engineering. For engineering, the instructor draws heavily on the work of Richard Sclove (1995) on democracy and technology and Ivan Illich's (1973) *Tools for Conviviality*, both of which emphasize local, easy-to-maintain, and low-impact technological interventions. For science, in addition to the work of Philip Mirowski (2011) and Melinda Cooper (2008) on neoliberalism and science, students benefit from reading Rebecca Skloot's (2011) account of the development of "HeLa" cells—the first cell line to be successfully produced and commercialized, involuntarily extracted from a dying African American cancer patient, Henrietta Lacks—to help them make curricular the exploration of science as an intersection of nature, economics, colonial/racial social orders, and power.

Running parallel to this social justice focus is an exploration of how teachers are themselves resisting neoliberal rationalities. In the local context, this foregrounds the work of the Bad-assed Teachers Association (BATS [www.badassteacher.org]), a network made possible by social media such as Twitter™ and Facebook™, and which works both on questions of social justice and corporate takeover of teaching. A few students have conducted practice teaching in BAT classrooms and report to their colleagues regularly. It also is possible to have students dialogue with BAT personnel, such as one of the BAT leadership who is also a regional science teacher (and is Chair of a secret science education BAT group on Facebook™). A related approach may be to add a blogger network, such as one that includes young upstarts like EduShyster (http://edushyster.com) and old guard activists like Mike Rose, all of whom are active in creating a counter articulation of education. On a larger scale, there also is the Network of Teacher Activist Groups (TAG [www.teacheractivistgroups.org/]), a U.S. national coalition of grassroots teacher organizing groups—including: Association of Raza Educators (San Diego/Oakland), Education for Liberation Network, Educators' Network for Social Justice (ENSJ) (Milwaukee), Metro Atlantans for Public Schools (MAPS), and New York Collective of Radical Educators (NYCoRE). TAG aims to disarticulate hegemonic practices associated with neoliberal standardization of curricula, high stakes testing and teacher accountability.

Such articulations of preservice teachers with groups like the BATS and TAG, in the USA and elsewhere, may be crucial to networking ecojustice approaches in science/STEM education, as they point student-teachers to where they need to go and how they need to act after finishing their teacher education programs. As

Ernesto Laclau (2001) asserted, "no overall historical transformation is possible unless the particularism of the struggles is superseded and a wider 'collective will' is constituted" (p. 8). According to Ernesto Laclau and Chantal Mouffe (2000), this would require acts of political articulation in which different local struggles strive to engage in common platforms connected by relations of equivalence. That is, educators and educator groups working to disarticulate STEM and re-articulate an alternative progressive vision of school science should benefit from struggles to form global movements against STEM—coordinated mobilization that has been labeled by scholars as grassroots globalization or "globalization from below" (Appadurai, 2000).

Overall, we critical science education scholars feel we are merely preparing student-teachers for a life of struggle over the very terms of the job of (science) teaching. From our point of view, the critical seed to plant is that this struggle has to be about more than preserving the professionalism of teachers. It must be mindful of social justice and environmental sustainability and critical of projects that may create worlds of massive deprivation and ecological disaster; sadly, often framed as social justice, but actually arguing for problematic aggressive individualism.

IMAGINING CRITICAL AND ACTIVIST STEM EDUCATION

Given the claim that all knowledge—and, presumably, curriculum innovations—serves someone's (or a group's) interests (Habermas, 1972), it seems clear that the relatively sudden and comprehensive acceptance of various STEM education initiatives represents a mechanism for achieving particular, pre-determined, ends. In our analyses, and that of others, there is considerable evidence to suggest that it is very difficult to tease out from among the many living, nonliving, and symbolic entities and name anyone responsible for STEM education. Nevertheless, it is apparent that, in answer to the fundamental question addressed by this chapter ("Who decides the forms science education has taken, …"), driving much of STEM education in many places in the world is a relatively small group of people and groups with significant financial power and interests. More particularly, calls for integration and/or interrelationships among STEM fields, which represent significant challenges to traditional perspectives and practices, often seem to appeal to citizens' apparently-conditioned priorities for material and symbolic comforts (e.g., as for-profit goods and services) associated with neoliberal capitalists' pursuit of wealth, often regardless of potential adverse effects on the wellbeing of individuals, societies, and environments.

In this chapter, we have suggested that those who may consume (implement) STEM education initiatives consider the extent to which such projects may represent a *Trojan horse*; that is, a kind of commodity that is attractive on the outside, but is, perhaps, hiding dangers within. On the surface, STEM initiatives may promise jobs, including for more women, minorities and the poor (Eisenhart et al., 2015), and economic prosperity in the face of international economic competitions. Indeed, there has been sudden increase in relative proportion of wealth of companies officially linked to Asian countries as compared to those linked to the USA and EU. At the same time, however, it is apparent that many financiers and corporations (which often control STEM fields) have limited connections to individual nation states; and, consequently, have made transnational agreements that prioritize movement of jobs and production etc. to places in the world with the "best" conditions for profit-generation—including in terms of minimal salaries, benefits, job-security, materials' costs and liability for personal, social and/or environmental harms often linked to their goods and services.

To address many of the personal, social and environmental harms that appear to be linked to neoliberal capitalists' influences on and uses of STEM fields, we suggest that science education needs to educate students about such relationships and, as well, help them to develop expertise, confidence and motivation for eventually self-directing research into such associations and, where they perceive problems, develop and implement socio-political actions to try to bring about a better world. In this chapter, we have, indeed, reported some successes in this regard. At the same time, however, such critical and activist science education seems to be restricted to relatively rare contexts in which several mutually-supportive conditions exist. Moreover, it seems that STEM education initiatives seem particularly inhibitory for such forms of science education. We recommend, accordingly, that educators and others interested in promoting critical and activist science education "borrow a page from capitalists' playbook"; in the sense of working to rally large, comprehensive, networks of living, nonliving, and symbolic entities in ways prioritizing social justice and environmental sustainability. Rather than an elite few deciding the forms science education takes, perhaps a much broader representation of the public will be influential. In this chapter, we offer cases in science teacher education that seem to offer hope for greater ecojustice in science education through establishment of citizen networks. Nevertheless, we are conscious of needs for continuous work in this regard.

ACKNOWLEDGEMENTS

This chapter came together, in part, through efforts of the journal, *Cultural Studies of Science Education*, which co-sponsored a research brainstorming workshop (June, 2014) in association with the University of Luxembourg and Springer Publishers. Our collaboration emerged from discussions at that workshop. We also would like to acknowledge and thank the editors of this volume, Lynn A. Bryan and Kenneth Tobin for the invitation to participate in this important collection following in the footsteps of pioneering authors and editors, Joe Kincheloe and Shirley Steinberg.

REFERENCES

Angell, M. (2004). *The truth about the drug companies: How they deceive us and what to do about it.* New York, NY: Random House.

Appadurai, A. (2000). Grassroots globalization and the research imagination. *Public Culture, 12*(1), 1–19. doi:10.1215/08992363-12-1-1.

Ball, S. J. (2012). *Global education, inc.: New policy networks and the neo-liberal imaginary.* Abingdon: Routledge. doi:10.4324/9780203803301.

Barnes, R. L., Hammond, S. K., & Glantz, S. A. (2006). The tobacco industry's role in the 16 Cities Study of Secondhand Tobacco Smoke: Do the data support the stated conclusions? *Environmental Health Perspectives, 114,* 1890–1897. doi:10.1289/ehp.9385.

Bencze, J. L. (Ed.). (2017). *Science & technology education promoting wellbeing for individuals, societies & environments.* Dordrecht: Springer.

Bencze, J. L., & Alsop, S. (Eds.). (2014). *Activist science & technology education.* Dordrecht: Springer. doi:10.1007/978-94-007-4360-1.

Bencze, J. L., & Carter, L. (2015). Capitalists' profitable virtual worlds: Roles for science & technology education. In P. P. Trifonas (Ed.), *International handbook of semiotics* (Vols. 1 & 2, pp. 1197–1212). Dordrecht: Springer. doi:10.1007/978-94-017-9404-6_57.

Bencze, L., & Krstovic, M. (2017). Science students' ethical technology designs as solutions to socio-scientific problems. In J.L. Bencze (Ed.), *Science & technology education promoting wellbeing for individuals, societies & environments* (pp. 201-226). Dordrecht: Springer.

Cooper, M. (2008). *Life as surplus: Biotechnology and capitalism in the neoliberal era.* Seattle, WA: University of Washington Press.

Eisenhart, M., Weis, L., Allen, C. D., Cipollone, K., Stich, A., & Dominguez, R. (2015). High school opportunities for STEM: Comparing inclusive STEM-focused and comprehensive high schools in two US cities. *Journal of Research in Science Teaching, 52,* 763–789. doi:10.1002/tea.21213.

Etzkowitz, H., & Leydesdorff, L. (2000). The dynamics of innovation: From national systems and "Mode 2" to a triple helix of university–industry–government relations. *Research Policy, 29*(2), 109–123. doi:10.1016/s0048-7333(99)00055-4.

Fensham, P. J. (1993). Academic influence on school science curricula. *Journal of Curriculum Studies, 25*(1), 53–64. doi:10.1080/0022027930250103.

Fensham, P. J., & Gardner, P. L. (1994). Technology education and science education: A new relationship? In D. Layton (Ed.), *Innovations in science and technology education* (Vol. V, pp. 159–170). Paris: UNESCO.

Freire, P. (1997). *Pedagogy of the oppressed* (New rev. 20th-anniversay ed.). New York, NY: Continuum.

Frey, C. B., & Osborne, M. A. (2013, September 17). *The future of employment: How susceptible are jobs to computerisation?* "Machines and Employment" Workshop, University of Oxford. Retrieved April 25, 2014 from http://www.oxfordmartin.ox.ac.uk/downloads/academic/The_Future_of_Employment.pdf

Gough, A. (2014). STEM policy and science education: Scientistic curriculum and sociopolitical silences. *Cultural Studies of Science Education, 10*, 445–458. doi:10.1007/s11422-014-9590-3.

Habermas, J. (1972). *Knowledge and human interest* (J. J. Shapiro, Trans.). London: Heinemann.

Habermas, J. (1983). *Moralbewusstsein und kommunikatives handeln*. Frankfurt am Main: Suhrkamp Verlag.

Harvey, D. (2005). *A brief history of neoliberalism*. Oxford: Oxford University Press.

Hileman, B. (1998). Industry's privacy rights: Is science shortchanged? *Chemical & Engineering News, 76*(17 August), 36. doi:10.1021/cen-v076n033.p036.

Illich, I. (1973). *Tools for conviviality* (1st ed.). New York, NY: Harper & Row.

Klein, N. (2007). *The shock doctrine: The rise of disaster capitalism*. New York, NY: Henry Holt.

Klein, N. (2014). *This changes everything: Capitalism vs. the climate*. Toronto, ON: Simon & Schuster.

Kleinman, D. L. (2003). *Impure cultures: University biology and the world of commerce*. Madison, WI: University of Wisconsin Press.

Krstovic, M. (2014). Preparing students for self-directed research-informed actions on socioscientific issues. In L. Bencze & S. Alsop (Eds.), *Activist science and technology education* (pp. 399–417): Dordrecht: Springer. doi:10.1007/978-94-007-4360-1_22.

Laclau, E. (2001). Review: Can immanence explain social struggles? *Diacritics, 31*(4), 3–10. doi:10.1353/dia.2004.0008.

Laclau, E., & Mouffe, C. (2000). *Hegemony and socialist strategy: Towards a radical democratic politics*. London: Verso.

Latour, B. (2005). *Reassembling the social: An introduction to actor-network-theory*. Oxford: Oxford University Press.

Leonard, A. (2010). *The Story of stuff: How our obsession with stuff is trashing the planet, our communities, and our health – and a vision for change*. New York, NY: Free Press.

Levinson, R. (2013). Practice and theory of socio-scientific issues: An authentic model? *Studies in Science Education, 49*(10), 99–116. doi:10.1080/03057267.2012.746819.

Martusewicz, R., Edmundson, J., & Lupinacci, J. (2014). *Ecojustice education: Toward diverse, democratic, and sustainable communities* (2nd ed.). New York, NY: Routledge. doi:10.4324/9781315779492.

McMurtry, J. (2013). *The cancer stage of capitalism: From crisis to cure*. London: Pluto.

Mirowski, P. (2011). *Science-mart: Privatizing American science*. Cambridge, MA: Harvard University Press. doi:10.4159/harvard.9780674061132.

Mouffe, C. (2008). Critique as counter-hegemonic intervention. *Transversal*. Retrieved from http://eipcp.net/transversal/0808/mouffe/en

National Research Council [NRC] (2011). *Successful K-12 STEM education: Identifying effective approaches in science, technology, engineering, and mathematics*. Washington, DC: National Academies Press. doi:10.17226/13158.

NGSS Lead States. (2013). *Next generation science standards: For states, by states*. Washington, DC: The National Academies Press. doi:10.17226/18290.

Oreskes, N., & Conway, E. (2010). *Merchants of doubt*. London: Bloomsbury Press.

Pedretti, E., & Nazir, J. (2011). Currents in STSE education: Mapping a complex field, 40 years on. *Science Education, 95*, 601–626. doi:10.1002/sce.20435.

Pierce, C. (2013). *Education in the age of biocapitalism: Optimizing educational life for a flat world*. New York, NY: Palgrave Macmillan. doi:10.1057/9781137027832.

Reiss, M. J. (2003). Science education for social justice. In C. Vincent (Ed.), *Social justice, education and identity* (pp. 153–165). London: RoutledgeFalmer.

Rennie, L., Venville, G., & Wallace, J. (Eds.). (2012). *Integrating science, technology, engineering and mathematics: Issues, reflections and ways forward.* New York, NY: Routledge. doi:10.4324/9780203803899.

Roth, W.-M. (2001). Learning science through technological design. *Journal of Research in Science Teaching, 38*, 768–790. doi:10.1002/tea.1031.

Sclove, R. (1995). *Democracy and technology.* New York, NY: Guilford Press.

Sellar, S., & Lingard, B. (2013). The OECD and global governance in education. *Journal of Education Policy, 28*, 710–725. doi:10.1080/02680939.2013.779791.

Skloot, R. (2011). *The immortal life of Henrietta Lacks.* New York, NY: Broadway Paperbacks.

Thalos, M. (2013). The lens of chemistry. *Science & Education, 22*(7), 1707–1721. doi:10.1007/s11191-012-9443-y.

Weber, K. (Ed.). (2009). *Food, Inc.: How industrial food is making us sicker, fatter, and poorer-and what you can do about it: A participant media guide.* New York, NY: Public Affairs.

Wenger, E. (1998). *Communities of practice.* Cambridge: Cambridge University Press.

Zeidler, D. L. (2016). STEM education: A deficit framework for the twenty first century? A sociocultural socioscientific response. *Cultural Studies of Science Education, 11*, 11–26. doi:10.1007/s11422-014-9578-z.

Zeidler, D. L., Sadler, T. D., Applebaum, S., & Callahan, B. E. (2009). Advancing reflective judgement through socioscientific issues. *Journal of Research in Science Teaching, 46*, 74–101. doi:10.1002/tea.20281.

SECTION THREE

Science Teachers Under Suspicion

Is It True That Science Teachers Aren't as Good as They Used to Be?

CHAPTER SEVEN

Obsessed with Accountability?

Science Teachers Under the Microscope

SHAKHNOZA KAYUMOVA AND DEBORAH J. TIPPINS

In this essay, our aim is to unpack and problematize some of the ways in which science teachers are made visible in the public STEM discourse and positioned in the literature as self-entrepreneurs needing continuous self-improvement. In the original book *Thirteen Questions: Reframing Education's Conversation* (Kincheloe & Steinberg, 1996), Deborah Britzman (1996) argues that the continuous quest for improving and scrutinizing teachers comes from romanticized and nostalgic historical notions of what it means to be a teacher. What happens when the past meets the present? Fast forward to the present, it is fall 2017. In our teacher preparation program, we have just finished introducing the new teacher accountability system, edTPA, to a group of prospective science teachers. The edTPA is a performance-based, subject-specific, evaluation system formerly known as TPA (Teacher Performance Assessment), administered in teacher preparation programs for teacher certification, nationwide. The look on students' faces, like that of deer caught under the headlights, and our own uneasiness with increasing unsubstantiated scrutiny over science teachers' practices, and teacher education programs, provokes us to revisit Britzman's description of teachers under suspicion in light of contemporary overt public targeting of educators under the mantra of achievement discourses and accountability policies. Curiously, we sense—although not with absolute certainty—that the narratives that depict romanticized roles of science teachers as "potential saviors" has given way to a story of de-professionalization of teaching in ways we could not have anticipated in the modern era. Shaped by changing contexts and understandings about what it means to be a STEM teacher and to do STEM, in this essay we use vignettes from practicing and pre-service teachers to explore the ways in which accountability discourses are structuring instructional practices of teachers, particularly in relation to science education, at the dawning of the year 2016. While doing so, we reflect on some of the situated tensions in science teacher education, both historically and in the present age, that mediate "fix it" approaches within science teacher preparation. In order not to be trapped in this dominant discourse of teacher accountability, we rethink change in science education in terms of the conditions of possibilities for teachers to be positioned as empowered intellectuals.

TEACHERS AND THE CONTEXT OF STEM

One day, in a rural school district located in what is known as the Blackbelt Region of Georgia, students in Jasmin Wideen's 10th grade Biology class entered the room excitedly discussing a nearby forest fire, the result of a forced burn. Recognizing the students' interest in the fire, Jasmine quickly made the decision to put aside her lesson on the structure, function and biochemistry of cells to build on the teachable moment with a lesson on forest succession. Because Jasmin was a teacher at a public school that had not achieved Adequate Yearly Progress (AYP), a supervisor from the state was seated at the back of the room to observe the lesson. Adequate yearly progress (AYP) is an accountability measure defined by the U.S. No Child Left Behind Act to determine academic performance of schools, school districts, and states, based on all students' academic performance levels in the state standardized test results (US Department of Education, 2009). As soon as it became obvious that Jasmin had "deviated" from her planned lesson, the supervisor moved to the front of the classroom and directed Jasmin to stop the forest succession lesson immediately, exclaiming "this topic is not a 10th grade Biology standard—you need to stick to the Standards if you want your students to do well on the end of the course test." Later, sharing this incident in a frank conversation with other teachers in her professional learning community, Jasmin lamented:

> When I became a teacher three years ago, I was full of energy and optimism, you might even say I was an idealist. I wanted to inspire students in creative ways, not only to value science but to love learning. But I feel like the obsessive drive toward school accountability has stripped me of my creativity and inspiration. I am asked to teach to the tests to ensure that our school ranks favorably when compared with other schools, districts and even teachers. In my attempts to make science relevant to the lives of my students, I somehow feel that I am viewed as a failure as a teacher.

Jasmin's story is not unlike those of many prospective and practicing teachers we work with who feel the pressures of a narrowing curriculum and pedagogy in light of a neoliberal education reform agenda. In 1993, when Shirley Steinberg and Deborah P. Britzman posed the question "Teachers under suspicion: Is it true that teachers aren't as good as they used to be?" the world's population was at 5.6 billion compared to today's estimated 7.3 billion people. Nelson Mandela and F. W. de Klerk of South Africa had received the Nobel Peace Prize; a fire resulting from the raid of the Branch Davidion religious compound in Texas resulted in the deaths of 72 men, women and children; and police officers were convicted on civil rights charges in the Los Angeles beating of Rodney King. *Schindler's List*, *The Piano* and *Philadelphia* topped the list of movies that year, reflecting the public's interest in issues of equity and social justice. In education, this was a time of increasing concern about financing of schools and colleges at all levels, particularly in light of the fact that college tuition and costs were exceeded by only those of housing. Some states began to rely on lottery proceeds to support public education, and president Bill Clinton proposed linking community service to funding for college tuition. Issues such as school violence, the effects of war on children, religious, ethnic and racial questions and changing demographics were at the forefront of educational discourse. Minority enrollment accounted for 31% of the public school population, and calls for Spanish-language courses and programs for second language learners were on the rise. Fast forward to the year 2016 when Steinberg and Britzman's original assertion of teachers under suspicion is just as relevant today as it was 20 years ago, perhaps even more so for science teachers who are "under the microscope," held accountable for goals of STEM reforms and workforce preparation. We live in a world where the Human Genome Project, which seemed an impossibility in 1996, has created the possibilities of life extension and expansion. Today's youth, through virtual reality, can experience simulated complex worlds and robot raccoons. Matt Damon, in his role as a Martian, makes real the possibilities of examining the surface of Mars in our lifetimes, and nanotechnology has opened the door to new ways of treating disease or engineering solar panels, along with ethical concerns for regulating rapid increases in biotechnology (drones are just one example). In education, we are witnessing a new era of accountability, value added measurements of teacher effectiveness, scripted curricular demands, and large scale testing, the beginnings of which can be traced back to 1993 reform movements such as the Texas Assessment of Basic Skills (TABS) which provided more than 20 million dollars in bonuses to schools and teachers who showed improvements in high test scores. At this time schools could lose their accreditation and be placed on a watch list for failure to show significant improvement in test sores (Sacks, 1999). Additionally, the minority enrollment in public schools is projected to be more than 50%, reflecting a shift to a non-White majority student population. Ever increasing diversity in public education and disparities in test scores among students brings us to another familiar discourse of our times: the achievement gap.

TEACHERS AND THE ACHIEVEMENT GAP

"One of the most common phrases in today's education literature is the achievement gap. The term produces more than 11 million citations on Google. 'Achievement gap,' much like certain popular culture music stars, has become a crossover hit," said Gloria Ladson-Billings (2006, p. 3) in her presidential address at the 2006 annual meeting of the American Education Research Association. In the United States context, the achievement gap signifies disproportional differences in test scores among children of color compared to their White counterparts. Ladson-Billings reconceptualized the gap as an educational debt, and argued that the gap "manifests itself in the distrust and suspicion about what schools [and teachers] can and will do in communities serving the poor and children of color" (p. 9). As scholars and science teacher educators committed to equitable and socially just science education, we have grappled with similar challenges of how to prepare, support, and empower teachers to enact science teaching that is equitable and rigorous in educational settings and conditions which are inherently inequitable (Kayumova & Tippins, 2017). Together with our teacher candidates and practicing teachers we have been asking questions such as: *How do we structure formal and informal science practices in ways in which attention is given to situated and lived experiences of children, and subsequently be aware of the ways in which power and positioning shape curricular decisions?* Yet, we have noticed that, both in research and public discourse, teachers are constantly positioned as ones who lack enough content and pedagogical knowledge and skills, and whose practices and beliefs need to be fixed congruent with the new STEM initiatives and reforms. This is especially the case for teachers practicing science and mathematics at the early childhood, elementary, and middle grades levels who are constantly disempowered by deficit discourses. When deficit discourses are coupled with the top-down STEM reforms, we are observing resistance and uneasiness across the board. In an illustrative example, Isabel described her struggle:

> One week before the start of the new school year we were called into a faculty meeting during which our principal announced that our school would seek to become a STEM-certified school this year. He went on to explain that this would require 100% participation of all teachers who would be expected to document how they were making STEM a central part of their curriculum. As I sat there listening to the plans for becoming a STEM-certified school I couldn't help but wonder, "What does this mean for me as a kindergarten teacher?" It may sound simplistic, but as an early childhood teacher, I have always believed in educating the "whole child" by uncovering the unique talents and interests each of my students carry with them to school. It is essential for my kindergarten students to explore their world in ways which transcend what I see as a narrow focus on STEM related knowledge and careers. Who knows—the next musician, artist or journalist might be sitting in my class right now. I am feeling somewhat unsettled about this new STEM reform, maybe because I had no part in the decision-making process and am struggling to see how it fits within my philosophy of early childhood education.

As Isabel's comments suggest, in many cases teachers are faced with a top-down approach and have not been given the professional respect they deserve. The above quote highlights the tensions that Isabel is experiencing in light of her school's quest for STEM certification, and more broadly highlights the discourses that are at play structuring public education in relation to national economic goals and reforms. Isabel's pedagogical understandings of what counts as quality teaching and learning, and the justifications she provides for particular ways of teaching and goals of education, seems to be overlooked in favor of larger reforms. For example, public and private schools in Georgia can apply for state STEM certification, an official recognition bestowed on institutions who meet a specific set of criteria by documenting how they are providing students with 21st century skills needed for careers in STEM industries and empowering them as technological innovators and problem solvers. A school seeking STEM certification must demonstrate that 100% of students and teachers are involved in school-wide STEM initiatives. Isabel's comment reflects the stance of teachers who are perplexed at the visions of change and who seek to understand and challenge policies that constrain and narrow their practice. Her comments remind us that Britzman (2012) described teaching as:

> a contest between what the technological society demands of its teachers and the personal visions of those teachers, their often repressed notions of what is desirable and what ought to be ... It is a sense of being watched, and viewing the self through the eyes of others. (p. ix)

We have been working with science teachers in multiple states and in different contexts and based on our observations we can tell that teachers feeling immense pressure and confusion are not patterns of sporadic and intermittent occurrences limited to one school, district, or the state. Unfortunately, Jasmine's and Isabel's voices are not only repressed, but they also echo continuous tension prelavent in current education reforms. On the one hand teachers are forced to follow the standards. On the other hand, they are witnessing that teaching to standards has not resulted in "achievement for all." As Gloria Ladson-Billings (2006) in her presidential address argues, an "all-out focus on the 'Achievement Gap' moves teachers toward short-term solutions that are unlikely to address the long-term underlying problem" (p. 4). The discourses of achievement are setting a new norm in the everyday lives of teachers in which learning is conceived as the means and ends of technical and codified skills. Subsequently, teachers are held accountable for perceived lack of knowledge and skills, underachievement of diverse learners, school closures, and often accused of not being able to adopt and adapt to a changing STEM world. And yet simultenousely teachers are expected to follow prescribed scripts, without any flexibility and allowance of time to address the needs of diverse learners. Every time there is a professional development or local science teacher conference, we hear more and more science teachers expressing disdain that it is very difficult to enact desired science teaching under all the pressures of accountability, reduced instructional time, and constantly changing standards and expectations. Ken Tobin (2011) argued that most of the contradictory, fixated, and unachievable expectations, with and about teachers, heavily rest on the instrumentalist and neoliberal notions of autonomous and rational individuals. As other contributors to this book will describe, neoliberalism is realigning science education to fit within a STEM discourse of national global competitiveness. In the context of current geopolitical, national, and international discourse of globalization and knowledge economy, teachers and schools are expected not only to teach, but also prepare skilled workers for future jobs. With new value added measures of evaluation introduced into the system of education, each teacher is expected to be an effective, self-governing, and self-entrepreneur professional who meets the goals and objectives of education reforms under any given condition (Ball & Olmedo, 2013). Moreover, the narratives about holding teachers responsible for educational outcomes is further concealing the unequal and inequitable social, economic, political, and curricular structures which are at the root of a problem.

TEACHERS AND ACCOUNTABILITY DISCOURSES

Stephen Ball (1990) argues that the field of education is not only subject to social, political, and economic discourses, but also part of "the selective dissemination" and "social appropriation of discourse" (p. 3). Ball gives an example of how in 1970s, the sociology of education "informed and reinforced the 'problem of working-class underachievement'" (p. 4). Ball states that the education studies and research findings of the time constructed working-class families as underachieving and deprived, furthering deficit perspectives about families and children based on their social class. Today, we are troubled to see a similar trend: the field of science education, similar to other fields, is also structured by social, political, and economic discourses of present age. One such structuring has to do with instrumentalization of knowledge economy and the need for skilled STEM workers. Tobin (2010) explains that with the dissemination of accountability discourses related to the instrumental nature of STEM education for a future work force, "policies and associated rules enforced in a hierarchy in which institutional power [is] greatest at the state and school district levels and students and their teachers [have] little power to make curricular choices" (p. 6).

Take for instance, edTPA is one of the regulatory-practices lodged in discourses of accountability impacting teacher education programs in recent years. edTPA is perhaps even more significant to science education than previous accountability systems given that it is subject specific and required for teacher candidates to receive certification in states which choose to participate. edTPA encompasses many different aspects and values of what is perceived to be quality teaching, with 27 different versions in fields covering many subjects, including science, in grades K-12. As part of the edTPA process, prospective teachers submit documents as evidence of their competence in three areas: (a) planning for instruction and assessment, (b) instructing and engaging students in learning, and (c) assessing student learning. Prospective teachers submit instructional plans, teaching materials, student work samples, and unedited videotape segments as part of the documentation process. edTPA in many ways is an extension of the value-added evaluation systems (VAMS), the Teacher Advancement Program (TAP) run by the National Institute for Excellence in Teaching (2012), and the Classroom Assessment Scoring System (Class) (Pianta, Karen, Paro, & Hamre, 2008). The VAMS are designed to gauge the academic influence of a particular teacher upon students while accounting and correcting for other elements that might affect student achievement (Goldhaber, Walch, & Gabele, 2014). VAMS have been used in several states to inform administrative decisions regarding dismissal of low performing teachers and awarding of achievement bonuses (Myers, 2014). Similarly, the TAP program was developed to evaluate, develop and reward accomplished teaching (Toch, 2008). The use of these teacher accountability and evaluation systems has been criticized because of the wide variation in teacher test scores every year and the difficulty in accounting for all of the factors which might impact student achievement (Darling-Hammond, Amrein-Beardsley, Haertel, & Rothsetin, 2012). Moreover, when accountability and evaluation policies are based on standardized test scores and rely on certain prescribed and narrow skill sets, teachers' instructional decisions and practices are reduced to the discrete and fixed performances, further moving away from creativity and critical thinking.

To illustrate for instance how edTPA evaluation system is re-structuring pre-service teachers' instructional decisions, we share one recent observation. It was Friday morning and Jake and Melanie, together with other secondary science student teachers in their cohort, were meeting at the university to work on lesson planning as part of the "planning for instruction and assessment tasks" required for edTPA. Typically the group of student teachers would be in their classrooms during this time, but the vast amount of time needed to complete the documentation in relation to each of the three edTPA tasks forced the science education faculty to set aside edTPA "work days" in which the student teachers leave their classroom field placements and meet with each other and their course instructors at the university. Jake exhibited a strong sense of doubt and concern as he shared his lesson plan ideas with Melanie:

> My teacher, Ms. Arnold, doesn't base her middle school science curriculum on standards. She explained to me how she felt it was more important to identify the interests that capture the imagination of her students, and build the curriculum around them. She also thinks that it is important for middle schoolers to have some ownership in what they will learn and how they will be held accountable for it. Right now students in the class are

> learning about pollinators and are engaged in citizen science projects. I haven't been able to find any standards that fit, and I know our edTPA lessons must be based on standards. When I discuss this with Ms. Arnold it is clear that she believes standards should not drive the curriculum for middle school science.

In this excerpt, Jake is expressing his dilemma associated with edTPA's requirement for lessons to be standards-based and a child-centered approach advocated by the mentor teacher. He is placed in a position of uncertainty when working with a teacher such as Ms. Arnold who passionately embraces a philosophy of teaching that is more than standards and is driven by student motivation and interests. Jake notes that in the literature, he rarely comes across a study lauding teachers like Ms. Arnold. In science education research, teachers (i.e., their beliefs, practices, and knowledge) are often constructed as needing to be changed, fixed, or altered. Teachers are positioned as unable, scared, lacking knowledge, skills, agency, and bottom-line ineffective. When such a positioning is contrasted with the accountability and achievement discourses in which a teaching job is conceived today, we regretfully have to say that good teaching is confined to narrowly defined practices such as meeting standards, exceeding benchmarks, and managing overcrowded classrooms with diverse children. David and Julie Gorlewski (2015), in their critique of edTPA, highlight several other concerns in relation to this evaluation system. They point out that the detailed rubrics that are integral to the process may reflect subtle forms of bias, even though they create "a perception of objectivity" (p. 21). Learning to teach is a messy and complex process that cannot be captured well by a set of stringent rubrics. The use of rubrics in the edTPA system also begs the question of what counts as a knowledge when measuring teacher quality. Gorlewski and Gorlewski argue that when the complexity of teaching is reduced to a set of discrete subcategories there is a technical orientation to learning to teach, one in which prospective teachers merely check off tasks they have completed. If we want rigorous and rich science practices to be a part of everyday learning, as science educators we must consider the ways in which we can best support classroom teachers such as Jasmine, Isabel, Jake, and Melanie to weave different practices together, whether it be science, math, art, and/or poetry. We are left wondering whether these future teachers, under the guise of edTPA, will be *empowered* to take on challenge of supporting diverse children's learning and identities in their future roles as science teachers. How can we ask teachers to empower their students by drawing on students' every day lives, experiences, and their cultural and linguistic backgrounds if we do not enact similar pedagogy in colleges of education? How can we ask teachers to position diverse learners as capable if we often position teachers as lacking knowledge and skills? How can we ask teachers to view science practices as dynamic entities in the making if all what we hold them accountable for is a static knowledge? How can we transform education if we are positioning energetic, optimistic, and idealist teachers like Jasmine and Isabel as failures in the system of accountability? We are also left wondering about a language of possibility and empowerment that speaks to teachers' and students' diverse and complex experiences, both equally and equitably. In the last section of our essay, we share our imaginations and visions of hope about the future of science teacher education.

IMAGINATIONS AND VISIONS OF HOPE

The motivation for this chapter stemmed in part from our own roles as science teacher educators who are committed to socio-spatial and ecological justice, and who value authentic learning, lived experiences, and diversity. We asked, through this essay, how we might resist various aspects of the current neoliberal accountability discourses and move forward with creative imaginations that are at the heart of any innovation in science education for future generations. As John Dewey noted in a volume written with James A. McLellan (1895), there is a need for teachers to "understand, and even be able to criticize, the general principles upon which the whole education system is formed and administered" (p. 14). He further argues that a teacher "is not like a private soldier in an army, expected merely to obey, or like a cog in the wheel, expected merely to respond to and transmit external energy" (p. 15). Instead, Dewey positioned a teacher as an "intelligent medium of action" (p. 15). Yet, as science teacher educators we are equally perplexed about how to enable current and future science teachers as empowered intellectuals. How do we empower teachers in the context of accountability? We are not suggesting that there should not be any accountability measures in schools. Instead, we are problematizing the ways in which accountability discourses are used to identify, reprimand and/or position teachers in the spectrum between the margins of public "saviors" and "failures." We are problematizing the point that the accountability discourses are structuring "a system of governance" instead of "(mutual) responsibility" (Biesta, 2004, p. 235). Maxine Greene (1997), in her essay *Teaching as possibility in dark and constraining times*, suggests that teaching in difficult times "is a matter of awakening and empowering … to name, reflect, to imagine, and to act … matter of enabling … to remain in touch with dread and desire." Greene argues that it is an imagination that allows teachers to "think of things as if they could be otherwise; it is the capacity that allows a looking through the windows of the actual towards alternative realities" (p. 2). As science teacher educators, we are concluding our chapter by imaging those alternatives:

- We imagine that one-day teacher accountability will be conceptualized as a mutual "responsibility of teachers to themselves as professionals, to their colleagues and professional associates, and to pupils, parents, and … [to] the democratization of education" (Biesta, 2004, pp. 235–236).

- We imagine that one day evaluation systems will take into account the different contexts in which science learners, both teachers and students, produce knowledge.
- We imagine that one day, teachers will be given a professional respect, the respect that is given to doctors, engineers, and scientists.
- We imagine science teachers and students creating partnerships to join together in designing assessments that are part of the learning process.
- We imagine that one day, teachers will be positioned as socio-politically active intellectuals.
- We imagine that one day, we will come together as a community of educators, teachers, teachers educators, scientists, engineers, artists, musicians, parents, and students to reclaim education as spaces of collective transformation.
- We imagine a science education for people of all ages rooted in communities, planetary health and democratic practice.
- We imagine science education programs which call on prospective and practicing teachers to interrogate what constitutes traditional knowledge in light of new contexts and questions.
- We imagine one day there will be neither an achievement gap nor an education debt, and STEM fields will become equal and equitable spaces filled with people from diverse backgrounds.

We imagine …

REFERENCES

Ball, S. J. (Eds.). (1990). *Foucault and education: Disciplines and knowledge*. New York, NY: Routledge.

Ball, S. J., & Olmedo, A. (2013). Care of the self, resistance and subjectivity under neoliberal governmentalities. *Critical Studies in Education, 54*(1), 85–96. doi:10.1080/17508487.2013.740678.

Biesta, G. J. (2004). Education, accountability, and the ethical demand: Can the democratic potential of accountability be regained? *Educational Theory, 54*, 233–250. doi:10.1111/j.0013-2004.2004.00017.x.

Britzman, D. P. (1996). Teachers under suspicion: Is it true that teachers are not as good as they used to be? In J. L. Kincheloe & S. R. Steinberg (Eds.), *Thirteen questions: Reframing education's conversation* (2nd ed., pp. 69–76). New York, NY: Peter Lang.

Britzman, D. P. (2012). *Practice makes practice: A critical study of learning to teach*. Albany, NY: Suny Press.

Darling-Hammond, L., Amrein-Beardsley, A., Haertel, E., & Rothsetin, J. (2012). Evaluating teacher evaluation. *Phi Delta Kappan, 93*(6), 8–15. doi:10.1177/003172171209300603.

Goldhaber, D., Walch, J., & Gabele, B. (2014). Does the model matter? Exploring the relationship between different student achievement-based teacher assessments. *Statistics and Public Policy, 1*(1), 28–39. doi:10.1080/2330443x.2013.856169.

Gorlewski, D. A., & Gorlewski, J. A. (2015). Producing professionals: Analyzing what counts for edTPA. In K. E. O'Hara (Ed.), *Teacher evaluation: The charge and the challenges.* New York, NY: Peter Lang.

Greene, M. (1997). Teaching as possibility: A light in dark times. *Journal of Pedagogy, Pluralism and Practice, 1*(1), 1–10. doi:10.1057/9780230100893_8.

Kayumova, S., & Tippins, D. (2016). Toward re-thinking science education in terms of affective practices: reflections from the field. *Cultural Studies of Science Education, 11*, 567–575.

Kincheloe, J. L., & Steinberg, S. R. (Eds.). (1996). *Thirteen questions: Reframing education's conversation* (2nd ed.). New York, NY: Peter Lang

Ladson-Billings, G. (2006). From the achievement gap to the education debt: Understanding achievement in US schools. *Educational Researcher, 35*(7), 3–12. doi:10.3102/0013189x035007003.

McLellan, J. A., & Dewey, J. (1895). *The psychology of number and its applications to methods of teaching arithmetic.* New York, NY: D. Appleton.

Myers, J. (2014). ASA issues statement with recommendations for using VAMs for educational assessment. *American Statistical Association News.* Retrieved from http://www.amstat.org/policy/boardstatements.cfm

National Institute for Excellence in Teaching. (2012). *The effectiveness of TAP: Research summary 2012.* Santa Monica, CA: Author. Retrieved from http://www.tapsystem.org

Pianta, R. C., Karen, M., Paro, L., & Hamre, B. K. (2008). *Classroom assessment scoring system (CLASS) manual, pre-K.* Baltimore, MD: Paul H. Brookes Publishing Company.

Sacks, P. (1999). *Standardized minds: The high price of America's testing culture and what we can do about it.* New York, NY: Harper Collins.

Tobin, K. (2011). Global reproduction and transformation of science education. *Cultural Studies of Science Education, 6*(1), 127-142.

Toch, T. (2008). Fixing teacher evaluation: Evaluations pay large dividends when they improve teaching practices. *Educational Leadership, 66*(2), 32–37.

CHAPTER EIGHT

En Route to Becoming a "Good" Teacher

An Attempted Hijack of Pedagogical Autonomy?

GIULIANO REIS

WHO ARE YOU CALLING OLD?

As I write this chapter, the book *Thirteen Questions: Reframing Education's Conversation* (Kincheloe & Steinberg, 1996), has been around for about two decades. After having come across the book—notably by recommendation of Lynn Bryan and Ken Tobin—I decided to adopt it for a graduate course I teach on curriculum studies. As part of the course assignments, students had to write weekly responses to the ideas presented in each chapter. Surprisingly, a few of my students revealed a certain frustration with the fact that the book was outdated and unrelated to the Canadian education context. In other words, and although my students really engaged in the discussions that we had in class, some of them wished that the book had been published more recently (maybe within the last five or ten years?) and that it dealt more specifically with the predicaments of the Canadian school curricula. Even though I thought their point was valid, I too felt a bit frustrated.

Indeed, the book offers a glimpse of the past of education in the U.S. and the enrichment of my students' perceptions of today's education matters in Canada was absolutely desired. To that end, the course syllabus made provision for each chapter to be complemented with an extra reading of their choice—Canadian-oriented if they chose. Likewise, the suggested bibliography listed a fair number of works by Canadian scholars that students could use to write their final paper. What bothered me the most was that those students had likely lost perspective of the opportunity that the book presented them to appreciate the changes—if any—that had taken place outside the Canadian educational scenario over the course of 20 years. More importantly, how else could the students see the ways in which Canada differs—or not—from their neighbours south of the border? If the ideas in the book were obsolete and out of sync with the reality of those in my class, how come they seemed to speak with propriety about the topics covered? (Add to that the fact that my class was composed of graduate students from a variety of nationalities: from Asia and Africa to Middle East and South America.) Are curriculum-related complications of race, democracy, politics, power, socioeconomic class, media, etc., so time-sensitive and geographically-bound to certain periods of the U.S. history that they are now deemed irrelevant to the critical analysis of contemporary education elsewhere? Hasn't Canada (or any other country for that matter) seen its own share of these same problems?

This is not to say that the educational circumstances of all countries are the identical or—worst—that a one-size-fits-all solution might exist to eradicate today's issues around instruction, curriculum and education policy. Though a number of difficulties (and successes) in the field have survived the test of time, a gold standard in education is likely not to exist (see also Grant, 2015). Ultimately, the same "13 questions" that Joe Kincheloe and Shirley Steinberg once asked years ago remain as current as ever.

This chapter is about one of those questions that appear to linger in the collective imaginary of teachers, parents, and even students: Is it true that science teachers aren't as good as they used to be? (Translation: Do people trust that science teachers know what they are doing?). I adopt a personal angle to reflect on the significance of a positive sense of "pedagogical autonomy" for the construction of one's (my own in this case) identity as a "good" science teacher. Thus, I discuss two specific moments in my teaching career: i) a parent-teacher interview gone wrong and ii) a textbook selection gone wild. Fortunately, I believe that those events have prompted me now (as a university professor) to apply myself in developing the confidence of my students (i.e., pre-service science teachers) in their own "pedagogical instincts." In addition, this chapter invites the readers to (re)consider the inevitability of a value-laden orientation to science curricula and the undesirability of any attempts to sanitize (i.e., neutralize) school science teaching and learning.

SHORT CONTEXTUALIZATION OF MY NARRATIVE

In order for the reader to better understand the impact that the experiences reported here have had on my teaching career, I believe that a short contextualization of my narrative is required. During part of my last years as a biology student in Brazil, I held a small undergraduate research assistantship position in the cellular biology laboratory of my department. Although I felt it was a valuable learning experience—a publication even came out of my research (Name, Reis, & Báo, 2007)—I had a hard time adjusting to the bench work type of environment. Besides, the location of the lab (basement of the building) associated with the lack of windows helped to create a lingering sense of imprisonment and isolation from the outside world. I wanted (needed?) a higher pay and believed that there was more to life than spending hours in a dark room examining dissected rice weevils' testes under an electron microscope. Whatever the real motivation was to quit the lab, the fact that I had seen my mother teach high school math for decades got me intrigued into considering switching jobs—only to the dismay of my parents.

I began teaching science at a traditional local middle-class private Catholic elementary school before I had obtained my B.Ed. degree, which was possible—not to say common—in the country back then. Overall, the low national average salary of teachers combined with long hours of work both at school and home made teaching fairly unattractive those days—an unfortunate reality that is still palpable to many (e.g., Alves & Pinto, 2011). Nevertheless, from my perspective, things never looked better: I was making more money, and I had more access to sunlight and human interactions than when I was working in the university lab. Soon after, I moved to teaching to an upper class private Christian school at the same time that I secured a tenured position with the local public school board. I was working day and night.

A few years later, I would decide to get a Master's degree in science education—which crushed any hope that my parents might have had for their son to become something more than a prospective underpaid and overworked (science) teacher. Ever persistent, after completing my first graduate degree, I made the decision, with the incentive and support from my wife, of course, to pursue a Ph.D. in North America. I believed that it would not only open doors to better working conditions, but also justify all the years invested in studying English. During my search, I came across my future supervisor, who welcomed my family and me to the province of British Columbia a few months into the new year (see Lee et al., 2010). Canada has been our home ever since.

In sum, my teaching career is not the result of chance. Regardless of the well-known struggles inherent to the profession, I made a conscious decision to become a teacher; and I always aspired to be a "good" one, able to be a positive influence in the life of my students.

LEARNING THE HARD WAY

It was my first year teaching a "real" class—and I wanted my Grade 5 students to take science seriously by making them study hard for tests. That would prove that I was a good teacher. I was hoping to make an impression.

By the time that the first parent-teacher interviews took place, I had already had to explain my teaching methods to a few concerned parents. Apparently, my students were performing more poorly than they expected. During that evening, it came the turn of the father of one of my lower-grade students to sit down with me. He started the conversation by noting that he had a doctorate in biology and was in the habit of consistently teaching science to his son at home. To conclude, he indicated that he believed my knowledge of the natural sciences was lacking. According to him, his credentials as a science expert and family tutor were enough to explain his son's below average performance: my teaching simply confused his son. It was as if his son was learning two conflicting fifth grade "sciences" at the same time: one at home and one in school (in my class, to be exact). To support his argument, the father compared "his" definition of ecological community against "mine," also found in the textbook adopted for my classes. I replied by saying that it was not the case that his definition was incorrect, but I felt that it was too advanced for an 11 year-old to understand. (Luckily for me, I had taken a course on ecology at the university just a few months before.) I tried to reason with him, but he seemed inflexible. I was getting tired and started losing my patience. I considered it presumptuous on his part to rub his academic qualifications in my face and to impose his assessment of my teaching. The conversation got heated. The next thing I knew, we were both standing when a school counsellor intervened to break up the argument (it was very close to getting physical). I looked around and saw the other parents in the room staring at us with their jaws dropped. I could not help but think about how I would explain to my soon-to-be wife that I had been fired from my job over a discussion with a parent. I have never engaged in an argument like this ever again—PhD or not. I learned the hard way.

Looking back at what happened that night, I now believe that what upset me the most was the fact that the father called my pedagogical autonomy into question. In other words, he doubted my ability to decide what and how to teach my students, his son included. How could I have been that far off my original plans of leaving a positive impression? Was I really as bad a teacher as that parent seemed to believe?

As it turned out, I did not get fired. The father—and later I found out that the mother too—were notorious in the school community for often being rude and confrontational to teachers. Nevertheless, it troubled me that someone outside of the immediate universe of my classes would presume to know better than me (a.k.a. the science teacher) about how to teach my own classes. (Funnily enough, I felt like the main character of the movie "Back to the Future," who never took it lightly when called "chicken.")

PICKING THE "RIGHT" CHOICE

A few years after the parent-teacher conference incident, I found myself teaching science in Grade 8 at a different private institution. There, teachers were given a chance to select a new textbook for their classes every two or three years. In between that time, we had to stick to whatever book we had chosen in order to help the students' families save money on school supplies. That way, books could be passed down to younger siblings or immediate relatives who attended the school.

The time to choose a new book in science had arrived. I remember being in one of the school's offices going through a number of examination copies that had been sent by the major publishers, when one of the counsellors approached me. She wanted to tell me that one specific book in the pile was not to be chosen. I do not remember if I had seen the book before, but I was certainly curious to hear the reasons why she thought the book was out of the race for adoption. She opened it to a page with a cartoonish reproduction of an advertisement for a lingerie company (a representation of the ad is shown in Figure 8.1) and explained that the

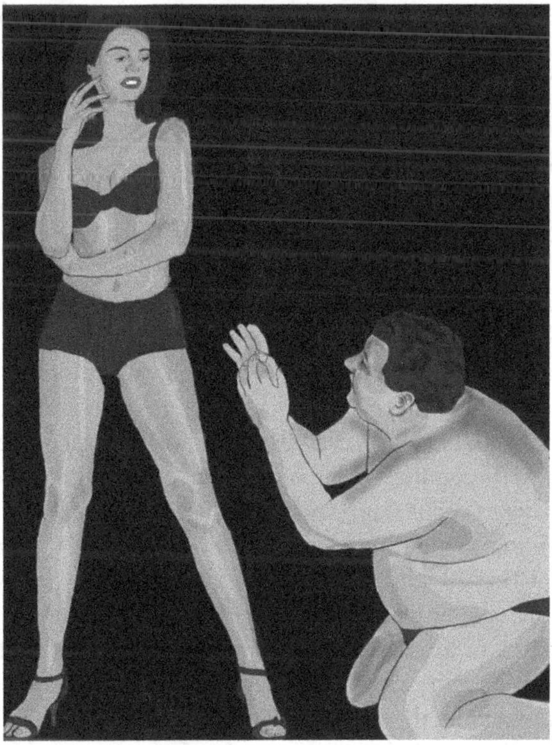

Figure 8.1: Representation of a Portion of a Lingerie Advertisement Published in Brazilian Magazines in the 1990s.
Source: Maria-Luiza Reis (2017)

book was "inappropriate." I was baffled. For whom was the book inappropriate? Why? All because of one picture? She also mentioned the school's Christian orientation, but that did not help to ease my perplexity. I did not ask any questions, even though I really wanted to. (I had learned my lesson and could not afford to lose my job over a textbook selection). Inside myself, I believed that the potential sociocultural approach of the book to matters of science and society was exactly what "my" students needed. It was a good example of how science could be used as a catalyst for important discussions on how to deconstruct the hidden subjective values of media outlets, the health risks linked to increasing obesity rates amongst the general population, the negative social consequences (like suicide and excessive plastic surgery) of mass promotion of certain stereotypes of beauty, the importance of preventing eating disorders, the economic impact on the city caused by malnutrition, etc. Likewise, it would have been great to learn what my students made of the picture. Would they have thought that it was offensive? How would they relate it to their own realities? Regardless of my disagreement with the counsellor's resolution, I did what I had to do (to keep my job) and selected a different science textbook for the next year. Once again, it troubled me that someone foreign to my teaching philosophy would presume to know better than me what was best for my own classes. (Metaphorically, my inner Marty McFly was being called "coward!" one more time.)

What Does Pedagogical Autonomy Have to Do with It?

Certainly, one could take me as someone who dislikes having my teaching called into question. What if I was indeed teaching science incorrectly to my students? What if I was actually better off not showing my "innocent" teenage students a picture of a half-naked model looking down on her supplicant overweight partner? Could it be that both the parent and the counsellor were right? After all, I had been a teacher for a couple of years and had much to learn.

On the other hand, I knew I was forever to be an "unfinished" science teacher—i.e., in the process of "becoming" (Freire, 1970/2005, p. 84). I had made many mistakes and had no false hopes about stopping for as long as I taught. What about the long hours I had put into planning my lessons? Didn't (or shouldn't) I have my students' best interest in my heart? Didn't (or shouldn't) I know better what I was doing?

Maybe I don't take criticism well. Or maybe I don't like being told what to do without a chance of a rebuttal. That parent did not want to know why I was teaching his child a certain way. He went to the meeting that evening to tell me what I was doing incorrectly. The counsellor did not ask for my opinion on the science textbook. She simply informed me that I was not to choose it. In a way, both of them appeared to have mistrusted my pedagogical autonomy. They put my professional ability to decide how to conduct my classes under suspicion. Hence, their actions have affected (even if only for a brief moment) my identity, my self-perception as a "good" teacher, one who truly cares for the education of his students.

RE-CONSIDERING THE QUESTION

When I expressed my willingness to contribute to this book, it was mainly because the editors wanted essays that were "reflective and very personal" (L. Bryan and K. Tobin, personal communication, March 17, 2015). They believe (and I agree) that the personal experiences that shaped each author's response to the proposed questions are powerful in that they offer concrete possible ways of thinking about contemporary issues in science education. In line with their expectations, I have dared to try to write my response from a personal angle. As such, I feel now obligated to adjust the original question in order to make it more "mine": Is it true that I am not as good a science teacher as those before me? What if I am? What if I am not? Does it matter?

I don't pretend to have a strong historical perspective of science teaching over any given period of time. Even after nearly 20 years since I started teaching science, my career is sufficiently short for me not to draw any reliable comparisons between myself and older generations of fellow science teachers. Besides, as someone who adheres to a sociocultural perspective on education, I believe that science teaching is contextually situated. As a result, I must modify the original question once more: Am I as good a science teacher as I can be? What if I am? What if I am not? Does it matter? Unlike the other questions, this one demands that I consider what I can—or cannot—realize within the boudaries of my immediate professional and personal context. For instance, what risks did I take when I let that parent's remarks get to me? Why didn't I insist in adopting that "inappropriate" textbook?

I do believe that I have done everything within my reach to be a "good" teacher to my students at the time of the two episodes introduced so far. That does not mean that I succeeded in actually being that type of teacher—but I certainly tried. Even when I failed, I knew I had done so. I rexamined what went wrong and I sought new ways to improve. I searched for new resources—books, activities and ideas from more experienced colleagues—to teach topics considered abstract (like the concept of ecological communities) and found ways to bring in important discussions into my classroom without using censored materials. I didn't give up on striving to be as good a science teacher as I could be. What else could I have done? Quit?

BRIDGING MY EXPERIENCE

In all the years that I have been working in the teacher education program of my university, I learned that there are essentially two types of pre-service science teachers: those who lack the confidence to teach science and those who are over-confident about their skills to teach it. In the first group, there are the students who typically do not have a strong science background—i.e., they have not studied much science after graduating from high school. The members of this group need to build their self-reliance as science teachers. They must appreciate their teaching potentialities in order to learn the science content that they yet have not mastered. My expectation is that they will flourish once they start believing that they can teach science.

As for the second group, they are usually the students who have at least the minimum number of university science credits to be considered admissible to our program. They know science, but do not know how to teach it beyond the traditional lecture format. In this case, my approach has always been to refine their understanding of the complexities of teaching by exposing them to different teaching strategies. Eventually, they can customize the ones they like to fit their needs.

One common message I try to get across to both groups of students is that they must trust their pedagogical autonomy—their teacher instinct—especially when nothing else seems to work. They have to be able to rely on their authority as professionally trained teachers to get the job done. This is a way of preparing them to deal with their own share of suspicious "expert" parents and co-workers. If they don't have conviction on their "teaching gut," who will?

Unbeknownst to many student teachers at the beginning of their professional journey is the reality that they will be constantly required to make important decisions on their own in order to cover the curriculum—it is simply impossible to teach everything. Their pedagogical autonomy will be often put to the test. Otherwise, who else is to decide how much emphasis to put on certain parts of the science curriculum? For example, should the fact that childhood obesity is only mentioned once in Ontario's high school science curriculum document (Ontario Ministry of Education, 2008) be considered an indication that this subject is irrelevant? Should science teachers shy away from discussing it in the classroom? Of course not! Especially considering that it is a growing epidemic problem amongst school-aged children and young adults in developing countries (e.g., Poobalan & Aucott, 2016).

Another example of the intricacies of being as good of a science teacher as one can be came about in the Canadian national print media in 2012. On October 3rd of that year, the newspaper *National Post* published an article entitled, "B.C. Teachers' Union Defends Anti-pipeline Teaching Materials." Briefly, the proposed Northern Gateway pipeline project was a $7.9 billion dollar undertaking that would see the building of two pipelines stretching 1177 kilometres from Bruderheim (Alberta) to Kitimat (British Columbia). These pipelines—the majority of which would be buried underground—would allow Canadian oil producers to reach the emerging markets of Asia as well as facilitate the transport of toxic liquid hydrocarbons used as thinning agents for heavy oils, including bitumen.

The British Columbia Teachers' Federation (BCTF) reacted by offering various teaching materials on the controversial megaproject as part of its social justice program. The resources were developed by a group of BCTF members and included lesson plans, a fact sheet and a poster that stressed the risk of a major oil spill (Figure 8.2). The Natural Resources Minister and part of the public opinion counter-reacted by accusing BCTF of promoting "unbalanced" teaching materials, to the point where they were considered an attempt to brainwash Canadian students (Reynolds, 2012).

If it is true that one aspect of science education is to help students participate in a fuller and more informed manner in public debates about socio-scientific issues (SSIs) (e.g., Roth & Désautels, 2002), then science teachers must exercise (pedagogical) autonomy when selecting materials to engage their students in practices that deepen their understanding of those same SSIs. The poor reception of such decisions by certain segments of the public leave teachers feeling disempowered and off track to achieving teaching excellence (see also Apple, 2013). Worse, they might start thinking that making decisions on their own—and which they believe speak to the best interest of their students' development as Canadian citizens—is beyond their immediate responsibility of being "impartial" science teachers. It appears that these professionals believe that their teaching should be as unbiased as possible, much like science itself.

In a recent study (Reis, Tippett, Scott, & Oliveira-Jayme, 2014), I investigated the perceptions of pedagogical autonomy held by two groups of elementary pre-service teachers at my institution. For data collection purposes, participants were asked to comment on the original news article about the BCTF-public opinion dispute. The preliminary results suggest that teachers tend to acknowledge the importance of a public debate about the pipeline project while also recognizing the need to be impartial when teaching about this particular issue. Put differently, participants seem to believe that the exercise of their pedagogical autonomy when deciding what socio-scientific matters to bring into their classrooms needs to be counter-balanced by a more "detached" presentation to students. In sum, those soon-to-be teachers admit to expecting to have limited pedagogical flexibility to decide what and how to teach the curriculum.

While a neutral approach to science teaching might sound appealing, it is likely unrealistic and naïve (e.g., Alsop & Fawcett, 2010). The decision to teach dispassionately (i.e., without taking any sides) is as partial, biased, and sided as any other. In striving for an illusory non-politicized science teaching, we run

Figure 8.2: British Columbia Teachers' Federation Poster.
Source: British Columbia Teachers' Federation (2016); https://www.bctf.ca/SocialJustice.aspx?id=27811&libID=27801

the risk of questioning teachers' pedagogical autonomy to decide how they can improve the quality of the education that they provide. Likewise, this would be to deny the existence of political aspects of science teaching that permeate public discussions in/about school science. After all, the responsibilities of educators to their students extend far beyond the delivery of the content (Gardner, Jones, & Ferzli, 2009).

SUMMING UP

I started my reflection by inviting readers to consider that the questions raised in this book are in no way exclusive to the maladies and triumphs of education in North America. Likewise, I argued that a positive sense of pedagogical autonomy has been fundamental for the construction of my identity as a "good" science teacher. In support of my arguments, I described two incidents where my identity as a "good" science teacher was put under suspicion. Next, I considered some of the lessons I took away from those events before proposing that the original question be adjusted to echo the personal vein of my narrative. Then, I explained how those early events in my teaching career had impacted my work as a university professor involved with the professional development of new (and potentially "good") science teachers. Finally, I resorted to an issue reported in the Canadian media to demonstrate how the theme of pedagogical autonomy affects science teachers more broadly and to exemplify the inevitability of a value-laden orientation to science teaching and learning. This last episode also suggests that pre-service science teachers have deeply ingrained convictions about the neutrality (fragility?) of science teaching, which are likely in part to be caused by years of (dangerous) indoctrination in schools by many experienced teachers. Such a sanitized approach to science education can weaken teachers' self-reliance on their pedagogical autonomy and compromise the quality of their teaching over time.

"Good" science teachers understand the social, political, cultural and physical environments in which their students are asked to learn (Cochran, DeRuiter, & King, 1993). However, certain expectations from parents and sectors of the public opinion for the "right execution" of the science curriculum not only contribute to reduce teachers' perceived pedagogical autonomy, but also create the impression that science teaching can be impartial. I am not advocating here that teachers should decide whose viewpoints to omit or include in classroom discussions about important SSIs. Quite the contrary, I believe they should present all perspectives as opposed to not debating them altogether. Nevertheless, I can only wish that my position on the topic, even though personal, can stimulate readers to consider the perils of letting/pushing teachers to lose conviction in their own ways of being as "good" a teacher as they can be.

REFERENCES

Alsop, S., & Fawcett, L. (2010). After this nothing happened. *Cultural Studies of Science Education, 5*, 1027–1045. doi:10.1007/s11422-010-9298-y.

Alves, T., & Pinto, J. (2011). Payment and characteristics of the teaching work in Brazil: A contribution of data from school census and PNAD. *Cadernos de Pesquisa, 41*, 606–639. doi:10.1590/S0100-15742011000200014.

Apple, M. (2013). Controlling the work of teachers. In D. Flinders & S. Thornton (Eds.), *The curriculum studies reader* (pp. 167–181). New York, NY: Routledge.

British Columbia Teachers' Federation. (2016). *Oil pipeline poster.* Retrieved May 2016 from https://www.bctf.ca/uploadedFiles/Public/SocialJustice/Issues/EnvJustice/EnbridgePoster.pdf

Cochran, K., DeRuiter, J., & King, R. (1993). Pedagogical content knowing: An integrative model for teacher preparation. *Journal of Teacher Education, 44*, 263–272. doi:10.1177/0022487193044004004.

Freire, P. (1970/2005). *Pedagogy of the oppressed.* New York, NY: Continuum.

Gardner, G., Jones, M., & Ferzli, M. (2009). Popular media in the biology classroom: Viewing popular science skeptically. *American Biology Teacher, 71*, 332–335. doi:10.2307/20565328.

Grant, K. (2015). From lists to images: Exploring the concept of the good teacher in teacher education. *Journal of the Canadian Association for Curriculum Studies, 13*, 36–59.

Kincheloe, J., & Steinberg, S. (Eds.). (1996). *Thirteen questions: Reframing education's conversation.* New York, NY: Peter Lang.

Lee, Y.-J., Ardenghi, D., Gholam-Reza, E., Hsu, P.-L., Oliveira-Jayme, B., Mijung, K., … Reis, G. (2010). Celebrating science education from people for people: A tribute to Wolff-Michael Roth. *Cultural Studies of Science Education, 5*, 787–805. doi:10.1007/s11422-010-9287-1.

Name, K., Reis, G., & Báo, S. (2007). An ultrastructural study of spermiogenesis in two species of Sitophilus (Coleoptera: Curculionidae). *Biocell, 31*, 229–236.

Ontario Ministry of Education. (2008). *The Ontario curriculum Grades 11 and 12: Science.* Toronto, ON: Queen's Printer for Ontario.

Poobalan, A., & Aucott, L. (2016). Obesity amongst young adults in developing countries: A systemic overview. *Current Obesity Reports, 5*, 2–13. doi:10.1007/s13679-016-0187-x.

Reis, G., Tippett, C., Scott, J., & Oliveira-Jayme, B. (2014, September). Pre-service teachers' perceptions of pedagogical autonomy through the examination of a public socio-scientific debate in the media. In *Proceedings from the European Science Education Research Association Biannual Conference* (ESERA), Nicosia (Cyprus). Retrieved from http://www.esera.org/media/eBook_2013/Strand 13/Giuliano_Reis_20Feb2014.pdf

Reis, M.-L. (2017). Representation of a portion of a lingerie advertisement published in Brazilian Magazines in the 1990s. Ottawa: Private Collection.

Reynolds, C. (2012, November 5). Why are schools brainwashing our children? Protesting oil pipelines, celebrating polygamy: Is the new "social justice" agenda in class pushing politics at the expense of learning? *Maclean's.* Retrieved from http://www2.macleans.ca/2012/10/31/why-are-schools-brainwashing-our-children/

Roth, W.-M., & Désautels, J. (Eds.). (2002). *Science education for/as socio-political action.* New York, NY: Peter Lang.

SECTION FOUR

Science Students Under Suspicion

Is It True That Students Are Less Interested in Science Than They Used to Be and Perform More Poorly in Science Than They Used to?

CHAPTER NINE

Problematizing "Problematic" Students in Lower Track Classes

TANG WEE TEO

I begin my chapter with a narrative describing a science lesson that I had observed in a Grade 8 lower track classroom in a typical public middle school located in Singapore, so as to situate my discussion within the social context in which the "problems" were identified. The lesson was on the topic on electricity and the teacher, Jason, was about to begin the lesson.

> "Please cooperate. Everybody should be in their own places. No walking around." Jason was pleading with his students to be quiet and to settle down for the past three minutes. In his bid to get attention, he clapped his hand and asked, "Are you ready?" He greeted the students, but only a few of them at the front of the class reciprocated. He repeated, "Okay, class … everybody settle down. Shh, keep quiet first!" Jason abruptly begun the lesson with a comment and question, "In the last lesson, we learnt about the hazards of electricity. I'm sure you have heard of this saying: 'Electricity is a good servant but a bad …?'" Without waiting for a response, he repeated, "Listen again, 'Electricity is a good servant but a bad what?'" A boy blurted out, "Bad ending!" followed by another boy who shouted out, "Bad enemy!" Jason acknowledged, "Bad enemy. Good try." Jason did not give up and cued the students, "Okay, I give you a hint. It starts with the letter 'm'." A student, Ming Teck, who was seated at the back of the classroom shouted out, "Teacher, speak louder!" Jason defended himself in saying, "So you need to … please be quiet first. Then you can listen. Don't shout at the same time. Put up your hand if you want to answer. Electricity is a good servant but a bad …? Bad what?" Without putting up his hand, Ming Teck demanded again, "Teacher, speak louder." In frustration, Jason said, "Okay, I speak very loud already. Here I'm trying to talk as loudly as I can, so please cooperate. Put up your hand [Jason

> puts up his right hand] if you want to answer. Like fire, electricity is a good servant but a bad …? Starts with the letter 'm'." A boy blurted out, "Macau!" The other students near him laughed. For the next 30 seconds, the students suggested many possible answers and laughed over their peers' responses, much to the dismay of Jason who eventually blew his top, "Okay, class. Listen. I want you to be serious and focus here. As much as I would like you to have an enjoyable lesson in class, I also want you to cooperate with me, so that I don't have to shout so loudly. Already, I am at the top of my voice already. Okay? If you are like that, this quiet, then I don't have to shout. Okay?!"

In Singapore, Grade 7 students are streamed into the Express, Normal Academic (NA), or Normal Technical (NT) tracks based on their academic performance at the Grade 6 national examinations (Ministry of Education, 2015). The NA and NT streams, collectively referred to as the "lower tracks" in this paper, typically make up 40% of the student cohort each year. While the lesson in the above narrative was not representative of over 150 lessons I have observed in seven other Grade 7 and 8 lower track classrooms in two middle schools over two academic years, my experience in Jason's class certainly resonated with my own teaching experience as a preservice teacher 14 years ago. Constant interruptions from students who blurted out irrelevant or random remarks, background noises from screeching chairs and students' small group talks, and quiet resistance from those who slept in class were a common sight. Urban science education scholars would most probably agree that they had seen similar classrooms in other parts of the world. I could draw parallels to the students and classrooms described in Kenneth Tobin, Rowhea Elmesky and Gale Seiler (2005) City High School science classroom and also Jeanne Oak's (1985) and Reba Page's (1991) lower track classrooms. My research consultant, Sonya Martin, who had the experience teaching in urban science classrooms in New York, USA, was observing Jason's class with me and she too, was surprised at the familiarity of the classrooms situated in different parts of the world. While one may be interested to know: What makes the problems that I see in these classrooms so persistent and prevalent? I would like to take a step back and ask another question: What makes the issues that I see in lower track classrooms "problems" to begin with?

"PROBLEMATIC"?

This paper seeks to problematize the labeling of science learners in the lower academic tracks as "problematic" with the aim of opening up dialogue about the genesis of this stigmatic and sweeping term. According to Ismail Masturah and Aik Ling Tan (2006), Singapore students in lower track classrooms often have been described as having short attention span, low self-esteem, lower motivation, poorer study habits, poorer self-concept, and unable to unpack academic expectations and requirements. I had the experience of talking with friends and colleagues who had limited knowledge about students in lower track classes and referred to them as "problematic." Much to my dismay and surprise, a colleague who is also a parent of a child in secondary school, referred to students in the lower track classes as "those students with [a] learning disability." I believe that he was probably not using the term "learning disability" in the same way as the World Health Organization (2011) to refer to an impairment in information processing that results in difficulties in listening, talking, writing, spelling, and so on. Rather, he was most probably referring to "bad behaviors" interpreted as a form of "learning disability." I believe that he is not alone in his thinking about students in lower track classes. His views could be in part due to the (unqualified) direct correlation individuals make between academic achievement with behaviors, attitudes, motivation, interests, and so on. I argue that the view of lower academic track students as "problematic" should not be unquestionably accepted as the "truth". Rather, this mindset requires a critical reexamination of the term "problematic" by interrogating what constitutes a "problem," what turns an issue into a "problem," and for whom are the lower track students a "problem."

IT'S ALWAYS ABOUT OTHERS

In my two-year, large-scale study examining science learners in Singapore lower track classrooms, I interacted with many students, teachers, and school administrators. I also observed many of these classrooms to understand how science teaching and learning took place. In analyzing data belonging to classrooms that were described by teachers as more "problematic" than others, it became apparent to me over time that teachers and students saw others, and not themselves, as responsible for the lessons that did not proceed as planned.

According to Jason, he found that the main problem with the lower track students was their lack of motivation to learn and low self-expectations. He said, "They are not academically inclined so they don't like to set high expectations of themselves." This mindset had in some way, shaped how he responded to students in the classroom. He said that if students approached him with questions during lessons, "I can help them, but I try not to because otherwise it will disrupt the flow of the lesson." According to him, "[T]he students don't ask many questions actually. They are not that inquisitive." His views were, however, contradicted by his students. One girl in his class shared, "When I call him [Jason] to come [over], he [Jason] said, 'Why should I come?'" She also claimed that Jason did not give students enough time to copy the notes on the PowerPoint slides. When students asked for more time, Jason would refuse to comply; Jason's lessons also were regarded by students as boring.

Whose "fault" is it? In this case, the "problem" lies in how the individuals thought s/he was being treated. Jason viewed the student as a "problematic" student who was disinterested to learn. Based on his preconceived perception about the students in that class, he did not think that students would be inquisitive and interested to learn. Hence, he probably did not think that the student would ask anything sensible that justified his effort to snake around the other tables to reach her. As such, he turned down the student's request to walk to her table and attend to her queries. He even confronted the student and questioned why he, as a teacher, should accede to her request to walk over to her. On the other hand, the girl reasoned that since she was not allowed to walk around in class, the teacher should go to her. By turning down her request and not allowing her to walk to him, she felt that the teacher was being unreasonable and not fulfilling his basic duty as a teacher.

Jason's power struggle with students occurred on more than one occasion. Returning to the opening narrative, Ming Teck, who was sitting at the back of the classroom, had asked Jason to speak louder as he had difficulty listening to Jason. Instead of raising his volume, Jason defended himself in saying, "You need to … please be quiet first. Then you can listen." The teacher-student contestations led to constant struggles between Jason and his students. The narrative about Jason and his two classes showed that students in lower track classrooms became "problematic" in the minds of individuals who held different views about who should take responsibility for learning. Clearly, there was no intention on the part of Jason or his students to relent and slowly move towards balancing the power relationship that could allow for open negotiations and mediation of learning. Typically, individuals saw that the other persons should take responsibility for ensuring that learning takes place and did not view learning as a partnership between the pedagogue and learner. In this case, what is "problematic" is the *incommensurability of perspectives of stakeholders* who should equally be responsible for the outcomes of education. It is always easier to put the blame on others because people are tangible targets that appear accessible in front of oneself.

Communication Breakdown. Jason's students had implicitly expressed to Jason that they could not and did not want to participate in the classroom discourse. Unfortunately, Jason did not pick up the cue and continued to carry out what he set out to do. This case alluded to a problem, which a teacher in another school had identified—the lack of understanding between teachers and students.

According to this teacher, she had difficulty understanding why students in a lower track class, in which she was a substitute teacher, did not follow her instructions. She felt that she could not understand what students wanted to do and kept instructing them to follow her way. In the end, the students were frustrated with her. Similarly, Jason did not understand that his students did not have the language proficiency to give him the answer and since there was no such "English common saying" in the first place, the students would not have heard it and hence, were unable to give him the expected answer.

DECENTERING THE "PROBLEM"

Thus far, I have discussed how individuals' stances may be reasons why lower track students are viewed as "problematic." There are many other "problems" about lower track students that teachers have shared with me. This includes having poor English and mathematics literacy skills, lack assessment literacy skills, poor memory, short attention span, anger management issues, learning disabilities (e.g., dyslexia), low self-esteem, poor attitude towards learning, behavioral disorder, family issues, inability to recognize the relevance of what is taught, poor foundations, and poor study habits. In this section, I propose an alternative view that shifts the focus from students to something that is more abstract, yet systemic enough to result in the pervasiveness of "problems." Through decentering the "problem" of individuals to the more complex and intertwined structural and systemic sociocultural factors and forces, I raise attention to the hegemonic structures, including practices and processes that exist in classrooms to entrap lower science achievers. In doing so, I argue that my observations of the so-called "problems" of lower track students in Singapore are in effect, manifestations of broader sociopolitical and sociocultural factors that they may have little control over.

Classroom Constraints. According to the Ministry of Education Digest 2015, the average class size for Grade 7 and 8 classes were 35.5 and 36.5, respectively. In comparing the eight case study classes I had observed, five classes had above average class size. In fact, three of the NT classes had 40 students. It did not help that each classroom was approximately 32 feet in length by 24 feet in width. This means that each student had about 19 square feet (or 4.3 by 4.3 feet) of space, most of which were taken up by the furniture, leaving little space for teachers and students to move around and physically engage in some group learning activities. The classroom conditions were not ideal for learning, especially in the Singapore climate, which according to the National Environmental Agency (2015), has an average humidity of 84% high temperatures ranging 31–34 degree Celsius throughout the year even if there were many ceiling and wall fans installed. As such, the sightings of students falling asleep, resting their heads on the table, or dozing off were common in all the classrooms I observed.

Imposing Social Norms. Surely, the stereotypical views that many individuals shared about lower track students could be responsible for the pervasiveness of the "problem." When a view is imprinted in the minds of a sizable group of individuals, the view becomes pervasive. Socially and culturally, a good learner is perceived to be someone who conforms to normative views and expectations in terms of behaviors, attitudes towards learning, and academic achievements. As such, when students do not listen attentively in science lessons, show little interest in the science subject, and perform poorly in science tests and examinations, they are collectively regarded as a "problematic" group that needs more teacher control.

In the opening narrative, Jason had wanted students to be quieter so that he could be heard at the back of the class. He had imposed his views about proper classroom decorum—that is, to keep quiet and listen when others are talking. However, the students did not abide as they could not engage with the lesson and hence, chose to participate in their own small group conversations. The students, in their open resistance against Jason, decided that they would "act out" and "out of line" (Teo & Osborne, 2012, p. 545). To Jason, students in that class were "problematic" because they did not conform to school culture norms that demanded students to give respect to teachers as authoritative figures.

Family Problems. According to one teacher in my research study, students could be acting out because of their frustration and emotional problems due to family baggage, so they would vent their frustrations in school. These personal problems were apparently more frequent in the lower track classes in both of my case study schools. One teacher research participant shared her knowledge about the personal baggage of her lower track students:

> [S]ometimes when we start to scold or discipline them for example when they are late, they will start talking about not staying with parents, or they work until very late. When they woke up to come to school the parents are still asleep and things like that. (…) [B]ecause a lot of NA [students], like my class, [there are] a lot of single parent families, and a lot of children are not staying with any of the parents, or one of the parent pass away; got one of the parent in prison.

Another teacher research participant commented that more students in NT classes came from lower socioeconomic status backgrounds. The lack of family support for some students was one cause of their "out of line" behaviors. One teacher anecdotally shared her conversation with a parent:

> Teacher: Your child is smoking.
> Parent: I know, I know.
> Teacher: Your child scolded me.
> Parent: I know, I know.
> Teacher: Your child is late [for school].
> Parent: I know, I know.
> Teacher: Your child is not in school [but] playing somewhere.
> Parent: I know, I know.

According to this teacher, the parent was not concerned about the education of her child and hence, as a teacher, there was little that she could do to help the student as she saw that there were boundaries that she should not cross as a teacher in meddling with her students' family affairs. While this may be her stance as a teacher, the personal problems that the lower track students bring into the classroom were evidently a source of the "problems" manifested in the form of resistance and defiance against authority and learning.

HARNESSING ALTERNATIVE VIEWS

Through decentering the "problem" from the individual, I have foregrounded structural and sociocultural factors as possible reasons for the prevalence and persistence of the "problems" observed in lower track classes. I hope that through presenting this alternative view, I have convinced the readers that seldom do problems originate from individuals—a view which Jason held. This possibly resulted in his failure to see beyond students' actions to understand the reasons why they acted they way they did, and to realize that problems of science teaching and learning are manifestations of broader factors such as imbalanced power relationships, the lack of conducive environment for active learning, the imposition of social norms, and lack of control over one's family issues.

REFERENCES

Masturah, I., & Tan, A. L. (2006). Snapshots from the normal (technical) world: A case study of low-track students in Singapore. In W. D. Bokhorst-Heng, M. Osborne, & K. Lee (Eds.), *Redesigning pedagogy: Reflections of theory and praxis* (pp. 73–88). Rotterdam: Sense.

Ministry of Education, Singapore. (2015). *Education statistics digest 2015*. Retrieved February 15, 2016 from https://www.moe.gov.sg/docs/default-source/document/publications/education-statistics-digest/esd-2015.pdf

National Environmental Agency. (2015). *Local climatology*. Retrieved September 4, 2015 from http://www.nea.gov.sg/weather-climate/climate-information/local-climatology

Oaks, J. (1985). *Keeping track: How schools structure inequality*. New Haven, CT: Yale University.

Page, R. N. (1991). *Lower track: A curricular and cultural perspective*. New York, NY: Teachers College Press.

Teo, T. W., & Osborne, M. (2012). Using symbolic interactionism to analyze a specialized STEM high school teachers' experience in curriculum reform. *Cultural Studies of Science Education, 7*, 541–567. doi:10.1007/s11422-011-9364-0.

Tobin, K., Elmesky, R., & Seiler, G. (2005). *Improving urban science education: New roles for teachers, students, and researchers*. Lanham, MD: Rowman & Littlefield.

World Health Organization. (2011). *World report on disability*. Retrieved September 4, 2015 from http://www.who.int/disabilities/world_report/2011/report.pdf

CHAPTER TEN

Confronting Prevailing Narratives of Student Engagement and Participation in Science Classrooms

SONYA N. MARTIN

When asked to contribute to this book by exploring how interest affects student performance in science, I found myself reflecting on how my views about the value of student interest in connection with student achievement have changed over the last 20 years. Serving as a science teacher in the Philadelphia public school system in the northeastern United States in the late 1990s, I was convinced that if I could capture the interest of my students through dynamic and well-designed science lessons, then my students' active participation and engagement in science learning could be reflected in their academic achievement. These notions were reinforced by participation in teacher education programs and professional development courses that suggested a strong link between student interest in science and positive attitudes towards learning, increased motivations to study science, and increased self-efficacy for performing well on science assessments. In addition, my school district touted hands-on inquiry activities and the creation of student-centered learning environments as essential features for enhancing student interest in science—which would lead to sustained motivation for learning science and achieving at high levels.

As a science teacher, I engaged middle and high school students in action-research projects designed to develop participatory structures that sought to privilege their voice and interests in an effort to expand science learning opportunities and increase student achievement. Later, as a science teacher educator and researcher, I continued to focus on understanding how to develop effective science learning environments by inviting teachers and students to participate in dialogues with a goal of increasing student participation and engagement in school science. A central aim of these dialogues was to promote positive attitudes towards learning science and to increase student interest and motivation for pursuing science as a career—all while increasing academic achievement. My work was particularly focused on improving interest and achievement in science for students who were typically marginalized in school science, including students who were racial, cultural, and linguistic minorities and who were often economically disadvantaged.

During the first 15 years of my career as a science teacher and science teacher educator, I never questioned the connection between student interest in science and student academic success. Certainly I encountered plenty of students who failed to transform their interests in science and positive attitudes about learning science into academic success. However, many students did show increased interest, and their grades reflected improved understanding. For those that continued to fail, I could generally point to myriad other factors impacting these students' achievement—including systemic failure of the school system to adequately educate students to be able to read and write on grade level. Alternatively, I could point to the cultural misalignments that existed between the mostly White, middle class teachers and their students that resulted in combative and unproductive learning environments. Finally, I could also identify ways that social and institutional racism and various other forms of discrimination prevented many students from making positive connections to school or science, all of which could explain why some of my students failed to excel in assessments on school science even when they eagerly participated in learning activities and demonstrated great interest and motivation for learning about science. However, my experiences during the last six years while living and teaching in the Republic of Korea (hereafter referred to as Korea) and conducting research in science classrooms in Korea, Taiwan, and Singapore, have forced me to reconsider the supposed connections that exist between student interest, classroom participation, and achievement in science.

In the sections that follow, I first provide some context for my critique of the current discourse in science education that suggests student interest in science and school positively impacts on student participation and engagement and on students' achievement in science. I offer examples from research in Korea and other Asian countries to raise some questions about hegemonic assumptions that currently exist regarding student interest, engagement, and science achievement. I use these examples to reflect on how the prevailing discourse on engagement has narrowed the scope of what research is being conducted around this topic, and in doing so, has obscured opportunities for learning across difference, and I raise questions about the ways in which the hegemony of the generalizability of research limits researchers from questioning whether what is learned in one context is actually applicable to another. I conclude this chapter by introducing some questions to reframe how researchers examine the issues of student engagement, participation, and interest, and I offer implications for how an expanded view of these issues can advance science education research.

AN OVERVIEW OF ENGAGEMENT AND PARTICIPATION IN SCHOOL AND SCIENCE

Internationally, student engagement has been widely discussed in various educational contexts for more than three decades (Harris, 2011). Engagement is a complex construct that researchers struggle to understand, but, historically, most studies have conceived of engagement as including several different components, including cognitive, psychological, and behavioral factors (Fredericks, Blumenfeld, & Paris, 2004). Participatory practices used to describe behavioral engagement include following class rules, asking and answering questions, and actively engaging with both the teacher and peers in class activities. Alternatively, research on cognitive and psychological engagement has tended to explore less observable factors (Appleton, Christenson, Kim, & Reschly, 2006) and instead has attempted to identify internal indicators that impact on student engagement (Fredericks et al., 2004). Some researchers have further divided the construct of cognitive engagement into cognitive and academic engagement. Students who are academically engaged are more likely to maintain time on tasks, accumulate credit hours needed to graduate from school, and complete in-class assignments and homework. This kind of engagement is generally differentiated from research on cognitive engagement, which focuses attention on students' motivation to learn, amount of time invested in learning, and students' perceptions about the value of school and value of education for meeting future goals. A significant number of cognitive engagement studies have focused on identifying students' self-regulation strategies to help meet learning goals (Harris, 2011). More recently, researchers have also begun to consider psychological or affective aspects of engagement by attempting to measure students' feelings, attitudes, and perceptions towards their educational environment, relationships with school and teachers/peers, and attitudes and perceptions towards specific content (Mitchell & Carbone, 2011).

Studies examining student engagement in educational research have been extensive, as evidenced by a quick search in Google Scholar for the term "student engagement," which resulted in nearly 2.4 million hits. One reason engagement has been so targeted as a topic for investigation is that it is considered to be both malleable (Mitchell & Carbone, 2011) and positively correlated with improvements in student academic achievement, school attendance, and participation in class (Fredericks et al., 2004). These factors have contributed to engagement becoming a key component of educational reforms, including in science education, where much research has aimed at improving student science achievement by focusing on finding ways to engage students as science learners by increasing their interest in science. Such studies may reflect the notion that positive emotions could serve as a mediator between cognitive engagement and academic achievement by enhancing students' ability to self-regulate learning that requires them to engage in effortful tasks over a period of time.

In recent years, more studies have emerged exploring the effect of student and teacher emotions on the teaching and learning of science (see Milne & Otieno, 2007; Olitsky, 2007), and some of these studies have also explored physiological measures of emotion (see Ritchie et al., 2015). The inclusive nature of these studies may offer researchers a more expansive view of the issue of student engagement by providing more opportunities for students to offer their own interpretations of how interest or enjoyment may impact on their motivations to learn or pursue science. Such studies may also enhance the reliance of student self-report data because the data can be triangulated with other measures, such as physiological measures, interviews, and classroom observations. While this area of research on engagement is promising and has seen some growth in recent years, a recent review of the literature found that between 2001 and 2011, less than 10% of articles published in the three top-ranked science education journals (*Journal of Research in Science Teaching, Science Education,* and *International Journal of Science Education*) dealt with emotion or affect in science teaching and learning (see Fortus, 2014).

Currently, the majority of the studies examining student engagement fail to consider the impact of sociocultural factors, such as gender, social class, race/ethnicity, or ability on students' behavioral, academic/cognitive, or affective engagement in school. That the "vast majority of theoretical frameworks and research on the topic of engagement are couched within the Western world" (Maslak, Kim, & McLoughlin, 2010, p. 256) also means that a narrow perspective about engagement, informed by Western and Anglophone perspectives, currently dominate the discourses on the ways in which engagement impacts participation, interest, and achievement (Harris, 2011). My search for publications about studies examining the connections between students' interest, engagement, and participation in science in Asian countries (in both English and native language publications) have confirmed not only that there is a need for this research, but also that what can be learned from exploring these issues in non-Western contexts has the potential to inform and expand what is currently known in the field.

In the sections that follow, I offer examples from research in Korea and other Asian countries that differ from the prevailing narrative that students' enjoyment and interest in science are key factors for promoting cognitive engagement and high achievement in science. Specifically, I introduce what is known as the East Asian *Achievement Paradox* or *Achievement Disparity*, which refers to the phenomenon seen in many East Asian countries where researchers have identified a disparity between students' high achievement in math and science and students' reported lack of enjoyment and interest for learning math and science (Song, 2013). Building from my discussion of this phenomenon, I argue the need for science education researchers to critically question some of the hegemonic assumptions that underpin research on student engagement and participation in science. I conclude the chapter by raising some questions for the field regarding the necessity of reconsidering the Western assumption that there are concrete, predetermined ways in which student "engagement" is demonstrated in classrooms. In particular, I suggest that there needs to be a stronger emphasis on the contextualized ways in which participation can be demonstrated, and I offer the issue of student silence as but one example of how researchers can begin to expand our perspective about what it means to participate and be engaged in science learning in different contexts.

EXAMINING STUDENT ACHIEVEMENT IN MATH AND SCIENCE IN EAST ASIAN COUNTRIES

The high performance of students from countries in East Asia on international literacy, mathematics, and science assessments is well established in the literature. Comparative analysis of the Trends in International Mathematics and Science Study (TIMSS) assessments from 1995 to 2007 (Gonzales et al., 2008) indicate that students from Taiwan, Korea, Japan, Hong Kong, and Singapore routinely rank in the top five of all participating countries. In addition, student performance on the Programme for International Student Assessment (PISA) from 2003 to 2012 shows similar results, with students from East Asian countries scoring in the top five positions in all content areas. Results from the TIMSS and PISA exams indicate that within the Organization for Economic Co-operation and Development (OECD) countries, students from East Asian countries (i.e., those from the territories and regions of China, Hong Kong, Japan, Singapore, South Korea, Macau, and Taiwan) consistently score well above the OECD average for all content areas and there are no measurable gender differences in math or science scores for students in Korea, Japan, Taiwan, or China (OECD, 2010, 2013e). In addition, while a comparison of PISA scores of students in East Asian countries from 2003 to 2012 show a steady increase in the achievement gap between students from wealthy and socioeconomically disadvantaged families, the gap has remained much smaller in these countries compared to students in Western countries. Results from the 2012 PISA mathematics exam found that students in China, Hong Kong, Japan, Korea, and Macau had an above average mathematics performance compared to the OECD average and that students from the lower socioeconomic classes were as likely to have high performance on the exam as peers in higher socioeconomic classes (OECD, 2013a, 2013e).

If we examine Korea's 2007 PISA results independently, we see that Korea has one of the narrowest distributions of ability regardless of gender or social class (OECD, 2010). This means that in Korea, large percentages of students from poor families routinely score in the high and middle range on both PISA and TIMSS (OECD, 2010). When we consider the results of the 2007 TIMSS math assessment, we see that the worst performing eighth grade students (those in lowest 10th percentile) in Korea, Taiwan, Singapore, and Japan significantly outperformed all other low achievers in the OECD countries and even outperformed the highest achievers (those in the top 90th percentile) in six countries (Gonzales et al., 2008). In science, Korea's lowest-performing students outperformed all other low performers in the OECD, with the exception of Japan (Gonzales et al., 2008). A more recent comparison of mathematics scores on the 2012 PISA exam found that 13% students from Korea, Hong Kong, Macau, Shanghai, Singapore, and Vietnam, whose families were in the bottom quarter of each country's socioeconomic class, performed among the top 25% of students across all participating countries and economies (OECD, 2013d). By comparison, analysis of the 2009 PISA data revealed that only about 7% of economically disadvantaged students in the USA excel in school despite their background (OECD, 2011).

Usually students who are female or who are living in conditions of poverty show lower achievement on math and science assessments than male students or students in higher socioeconomic classes. These data suggest that not only do students in East Asian countries tend to have high achievement in math and science, but also that females and students living in poverty are not predicted to perform poorly as a result of differences in gender or social class status. Even with closer examination of this phenomenon, I have found little discussion about these achievement differences in the science education literature. The few references I found in the PISA and TIMSS reports merely mentioned these differences, but did not interrogate the implications of these findings more generally. These data are interesting to me as an urban science education researcher because I am more familiar with studies asserting that gender and socioeconomic class are significant predictors of how successful students are likely to be on math and science assessments. In the next section, I begin to explore some of the data measuring East Asian students' cognitive and affective engagement.

A DISPARITY BETWEEN SCIENCE ACHIEVEMENT AND STUDENT ENGAGEMENT IN EAST ASIA

The vast body of research linking student interest and enjoyment in math and science to student performance on assessments would suggest that, given their high performance, students in Korea and in East Asian countries should express high levels of interest in the subjects of math and science. Instead, PISA and TIMSS assessments show that students in Korea, Taiwan, China, Singapore, and Japan report scores that are lower than the OECD average for interest and enjoyment in studying science (Martin, Martin, & Foy, 2008). In fact, among all countries participating in the 2007 TIMSS exam, students in Korea, Taiwan, and Japan reported the greatest disinterest (first through third, respectively) for learning science or enjoying science. The same exam found that students in Hong Kong, Korea, and Taiwan reported the lowest levels of self-confidence (first through third, respectively) with regard to their ability to learn and do well in science (Gonzales et al., 2008). Referred to as the *East - Asian disparity* (Song, 2013), this achievement without enjoyment paradox is disconcerting for educators and government officials who are concerned that students' poor perceptions of math and science could have negative consequences for STEM (science, technology, engineering, and mathematics) innovation and development in the future.

When we look at more general measures of affective engagement related to school, we see similar trends in the data. For example, a recent study measuring student health and school engagement found Korean students ranked last out of all 65 countries in the study when asked if they were "happy in school." Korean students' positive response rate of 60% is considerably lower than the OECD average of 80% (OECD, 2015a). In addition, for more than six years, Korean students have ranked at the bottom of the OECD's Better Life Index (Jeong, 2016). This low happiness indicator may be evident in Korea's suicide rate for young people, which is the highest among all OECD nations (OECD, 2013c). In fact, suicide has been the leading cause of death in Korea among people aged 10–19 for several years and is the fourth leading cause of death in Korea for all age groups (Korean National Statistics Office [KOSTAT], 2015). This trend runs counter to that of most other countries in the OECD, where the majority of countries have seen an overall decrease in the number of suicides for young people over the last twenty years (OECD, 2013d). On the contrary, Korea's suicide rates have risen sharply since 1990 (OECD, 2015b) and the suicide/suicide attempt rates among young women are higher in Korea than they are in all other OECD nations (Noah, Acciai, & Firebaugh, 2016). A recent study conducted in Korean K-12 schools (as reported in Jeong, 2016) found that 26% of high school, 20% of middle school, and 17% of elementary school students reported having considered suicide in the last year.

These numbers are bleak and suggest that many students who may be high achievers on math and science assessments may also be poorly engaged at an affective level. Interestingly, Korea was the only country in the Asian region that reported below average feelings of happiness at school. In contrast, several countries in Asia reported very positive feelings, including Indonesia, Thailand, Malaysia, and Singapore (1st, 4th, 6th, and 12th, respectively). Of these countries, only Singapore's students have consistently high levels of achievement on the PISA and TIMSS, while Indonesian students score near the bottom for all content areas in both exams (Gonzales et al., 2008; OECD, 2014). The data presented thus far have been fairly stable over the last 15 years, but there have been no comparative studies exploring how these differences in students' perceptions of school or success may influence students' achievement in science. In the section that follows, I examine more closely some issues related to academic and cognitive engagement for students in Korea and in East Asian countries to draw attention to some areas where these students tend to demonstrate different patterns of engagement than is commonly reported in the international literature.

ACADEMIC AND COGNITIVE ENGAGEMENT IN EAST ASIAN COUNTRIES

Introduced earlier in the chapter, academic engagement is a common construct used to measure students' engagement with school and schooling. An example of academic engagement includes factors like school attendance and students' perceptions of belonging to the school or community. Cognitive engagement generally refers to students' motivation to learn. The amount of time students invest in learning has been often used to measure a students' cognitive engagement. In many studies, researchers have positively correlated high measures of student affect (interest, attitudes, and enjoyment) with high academic and cognitive engagement. Thus we may expect that students who report feeling unhappy at school would also have lower than average school attendance rates and that they may also spend less time engaged in *doing school.* However again, we find this is not the case in Korea and Japan and in some other East Asian countries.

Studies examining school attendance rates as a measure of engagement have shown that when comparing the average number of days and hours that students spend in school varies greatly from country to country (OECD, 2013b) and that spending increased amounts of time in school does not necessarily result in higher achievement on international assessments. For example, Finnish students have routinely scored in the top five on the math, science, and literacy PISA exams since 2003, but they receive fewer hours of instruction and attend fewer days of school than students in any of the East Asian countries (OECD, 2012). Both the TIMSS and PISA learning environment data show that some students in China, Taiwan, Japan, and Korea endure the longest school days in the world (an average of 10–12 hours per day for 5–6 days per week). Students in these countries generally attend public school from about 8am until 5pm with a lunch or study break of 1–2 hours. However, after the end of the normal school day, students are compelled to attend compulsory after-school classes offered at the public school as an extended evening program subsidized by the government or, for those who can afford it, to attend private education institutes.

These institutes are commonly referred to as "cram schools," and they serve the purpose of preparing students for the college admission examination. According to the 2015 Survey of Private Education Expenditure conducted by KOSTAT, 81% of elementary school students, 69% of middle school students, and 50% of general high school students attended some form of private education after the completion of the regular school day. For nearly 10 years, the Korean government has been attempting to regulate the operating hours of cram schools by imposing a 10 p.m. curfew for the schools. Even while government officials have raided some cram schools in an attempt to minimize intense student study, reports show these efforts have not been very successful in curtailing after-school exam preparation programs (Choi & Choi, 2016). One response to these regulations by families has been to simply have students receive one-on-one tutoring in their homes after the cram schools close. This has been especially true for students from higher socioeconomic levels. As a result, some critics argue these policies have increased inequities in education among Korean students (Choi & Cho, 2015). Researchers in other East Asian countries report similar practices with regard to hours spent in study, the relative high cost of supplementary private education for students, and the ineffectiveness of government policies to regulate or minimize the number of hours students are expected to spend studying beyond the end of the normal school day. While the percentages above suggest participation in cram school decreases from elementary to high school, other studies indicate that many high school students prefer to use one-on-one study guide programs, e-learning programs, and tutors, all of which allow for more flexible scheduling. However, many researchers lament that we actually know very little about the real impact of "shadow education" or private education on student learning in Asian countries.

While the reasons for the unprecedented growth of private education in Asian countries since the 1990s is somewhat varied, Percy Kwok (2004) identified several factors contributing to the persistence of private education in Hong Kong that are shared across the region, including a tendency towards smaller family size that allows for increased educational expenditures, large daytime school class sizes, diminished ability of public school teachers to teach effectively due to heavy administrative responsibilities that impede them from developing innovative lessons like those offered in pay-for-access cram schools, increased access to and advances in information and communication technologies allowing for greater connectivity, and intense competition among a large number of students for a limited number of employment positions in a post-graduate market. And while many reports about cram schools are negative, a recent study in Taiwan (Liu, 2012) found that not only did cram schools help to improve academic achievement, but they could also expand equitable access to improved educational outcomes for students who were from lower socioeconomic backgrounds and/or who were female. Cram schools provided the students in this study with better learning opportunities than were available by attending only public school. Mark Bray (2010) argues that researchers need to better understand how cram schools are influencing education in general. However, he indicates that currently researchers have limited access to data and they need more appropriate measures for collecting data that can accurately describe the types of private education experiences students have in different countries. Bray also suggests that for researchers to be able to effectively interpret the findings from cross-country comparative studies, they need to develop cross-cultural teams who can help to contextualize what is learned from the research. Regardless, it is clear that the examination-oriented pedagogies that are prevalent in Asian countries are not likely to disappear any time soon.

And when viewed from the perspective of academic engagement, it is clear that students in East Asian countries appear to be highly engaged as evidenced by the large amounts of time invested in participation in educational activities. At the same time, many of these students report a limited interest and enjoyment in learning and high levels of affective disengagement from school. Some studies have shown that getting good grades can be an important motivator for helping students to maintain interest and to exert energy to meet goals over a period of time. However, responses to the 2012 PISA survey on students' attitudes towards school found that only 58% of Japanese students said they enjoy getting good grades, followed by students from Korea (79%) and Vietnam (86%) who represented the next two lowest responses in the study (OECD, 2014). Those unfamiliar with the education systems of East Asian countries may wonder, for what are these students studying? In Korea and other countries in East Asia, students focus on preparing for a single exam offered once a year that is taken in the final year of high school. The score from this one exam is used to determine to which college a student can apply. In Korea, where 90% of youth (OECD, 2012) attend tertiary education, gaining admittance to college is not as important as attending a specific college. Admittance to the top universities is incredibly competitive, as only those students who score in the top 1% are eligible to apply for a spot in one of the top universities. In 2014, this meant that less than 2% of more than 650,000 students who took the exam qualified for first-tier admission universities. However, these 2% then needed to pass several more highly competitive university and department admissions examinations before they could successfully enroll in specific programs of study, such as teacher education or medical school.

For students in Korea and in other East Asian countries, the pathway to success is very narrow and there is little room for making mistakes on an exam that is offered only once a year. For those who do poorly, they may spend several years studying for and re-sitting the once-a-year exam in an attempt to improve their chances of being admitted to a better quality university. In countries where college graduates face unemployment rates of 40% for full-time jobs (Ministry of Employment and Labor, 2017), scoring high on math and science exams can ensure access to high status universities where post-graduation name recognition and social networks can position students more favorably in the job market. Thus, for students in these countries, the single most important measure of their success is their performance on a single exam. In Korea, the exam is taken so seriously that during the 30-minute English listening portion of the test, the government enforces a no-flight period for all airplanes, both commercial and military. With so much pressure facing these students, it is easier to appreciate that students may express dissatisfaction with school, be disinterested in receiving high grades, and experience the kinds of depression or anxiety that could contribute to high suicide levels. In the next section, I explore the issue of behavioral engagement, especially in the context of science classrooms.

BEHAVIORAL ENGAGEMENT AND PARTICIPATION IN SCIENCE IN EAST ASIAN COUNTRIES

Brenda McMahon and David Zyngier (2009) argue that because so many studies characterize engagement as something that is done to students, it has resulted in the alteration of school policies, curriculum, and teaching pedagogies that make engagement and participation central for all students. Additionally, many studies have examined affective engagement by considering how students' perceptions and attitudes towards science impact students' motivations to learn science, to participate in science classes, and to decide to pursue science study in the future (for reviews of research on attitudes towards science, see Osborne, Simon, & Collins, 2003; Tytler & Osborne, 2012). A common assertion from such studies is that enjoying science, having positive personal views about science learning, and having an interest in science can be strong predictors of students' active participation in science learning activities, which may lead to increased motivations to pursue science learning opportunities over time (Ainley & Ainley, 2011). Student-centered pedagogies are widely considered to be instrumental in developing students' interest and enjoyment in science and promoting conceptual understanding and high achievement in science. However, when comparing student science achievement among countries with differing patterns of classroom participation, researchers observed that science achievement for Korean and Japanese students, relative to their participation in interactive classroom activities, ran counter to the general narrative asserting that active student participation is necessary for high achievement (Preznel, Seidel, & Kobarg, 2012).

Using the 2006 PISA results, researchers (Seidel et al., 2007, as reported in Preznel et al., 2012) found that in comparison to 30 other OECD nations, students in science classrooms in Japan and Korea reported the lowest frequency of interactive teaching patterns, including explaining ideas in class, stating their opinions, engaging in class debate, or discussing topics. However, students in both Japan and Korea scored highest among all countries with regard to scientific competency. The same report found that students in countries with higher levels of interactive participation (such as Australia, Italy, Mexico, Portugal, Turkey, the US, and New Zealand) scored lower than students in classrooms characterized as having medium to low interaction patterns. So although Japan and Korea report the most teacher-centered classroom learning environments, students in these countries scored higher on the 2006 PISA science exam than students in countries who experienced medium or high levels of student-centered teaching (Preznel et al., 2012).

Research has shown that active student participation can foster students' abilities to engage in scientific inquiry (National Research Council, 1996); to develop skills, attitudes, and values associated with scientific thinking (Duschl, Schweingruber, & Shouse, 2007); and to construct scientific understanding through social interaction and dialogue (Lemke, 1990). Other researchers have found that active student participation and engagement in science class are critical in supporting students to acquire specific language needed to communicate about science concepts (Bransford & Donovan, 2005) and that interactive teaching can foster student learning (Hofstein & Lunetta, 2004), especially with regard to students' cognitive development and motivational learning outcomes (Lazarowitz, Baird, & Bowlden, 1996), and even increased student achievement (Martin, 2006). Various instructional strategies, such as collaborative learning, inquiry, modeling, and argumentation are all predicated on the notion that learning occurs when students are actively involved in a process of meaning and knowledge construction through interaction and dialogue with the teacher, peers, and material resources (Duit & Treagust, 1998). If this is the case, then what is happening in science classrooms that tend to be fairly teacher-centered and students are reported to rarely have opportunities to engage in small group interactions or to participate in hands-on activities?

In the last two decades especially, there has been a growing number of studies focused on the promotion of pedagogical strategies based on constructivist notions that promote student-centered classrooms as a way to engage students in science. These strategies focus attention on increasing students' behavioral enactment of practices such as "talking science" and "doing science" (Olitsky & Milne, 2012). Indeed, science educators who are critical of the TIMSS and PISA assessment data have asserted that students in Asian countries who perform well on these assessments are limited in their ability to be creative and they lack critical problem-solving skills. Critics have questioned the value of the TIMSS and PISA assessments by insinuating that the "kind of science knowledge" measured by these tools is not as valuable as knowledge gained while engaging in student-centered inquiry-based science classrooms. However, the new PISA (OECD, 2014) assessment was designed to measure students' creative problem-solving abilities by having students apply knowledge to solve complex and novel tasks. Many researchers found it surprising that students in Singapore, Korea, and Japan scored in the top 10% of all OECD nations.

More than 20% of the students in these three countries achieved the highest problem-solving ability (Levels 5–6), meaning they could "systematically explore a complex problem scenario, devise multi-step solutions that take into account all constraints" (OECD, 2014, p. 30). The new assessment also found that students in Hong Kong, Japan, Korea, Macao, Shanghai, and Taiwan performed significantly better on problems that require understanding, formulating, or representing new knowledge than students in other OECD nations. In addition, these countries had the smallest percentage of students measured to have the lowest problem solving abilities (Levels 1–2), ranging from 12% of students in Taiwan and only 7% in Korea. When viewed from this perspective, the PISA and TIMSS data from East Asian countries would suggest that classroom participation might not play as large a role in shaping students' science learning as many mainstream researchers have previously thought. Unfortunately, the dominant discourse in science education about student engagement, interest, participation, and achievement cannot effectively explain the varying disparities occurring between these factors in many East Asian countries.

REFRAMING PARTICIPATION IN SCIENCE CLASSROOMS FROM A NON-PARTICIPATORY PERSPECTIVE

My interest in understanding what drives a vast majority of students from East Asian countries to be able to master certain forms of science knowledge, regardless of gender or social class or whether they are taught using teacher-centered strategies, has led me to conduct several studies in Korea focused on exploring science teaching and learning in various schools and classroom settings. Over the last four years, my colleagues and I have implemented surveys designed to characterize student preferences for enacting different participatory practices in science classrooms, and we have conducted video ethnographies to examine teacher and student interactions in traditional teacher-centered classrooms and in classrooms in innovative schools designed to promote student-centered learning (see Park, Martin, & Chu, 2015). We have engaged teachers and students in interviews using video from their own classrooms in order to ask them to describe and reflect upon their interactions in an effort to better understand how these participants would characterize their own behavioral practices in the science classroom.

In our studies, we set the video recordings to capture several minutes before and after the start and end of class. These videos reveal animated, loud, and boisterous students who can be seen running, shouting, hitting, and jumping around the room—but as soon as the bell rings, the class falls silent and students can be seen sitting with very limited movement for nearly 45 minutes. In a typical class, we generally see fewer than 10 students out of 35–40 who vocalize answers to teachers' questions, and the majority of these responses are simple affirmations of "yes" or "no." However, when the same students in this class respond to survey items we designed to measure a broad possibility of participation practices in science classrooms, more than half of all students rate themselves as being highly participatory. Through surveys, interviews, and video analysis, students and teachers indicated a variety of active participation strategies being enacted in the science classroom that they felt were more valuable than talk. Examples included active listening, taking notes, maintaining eye contact with the teacher during the lecture, reading silently along in the book while the teacher reads aloud, computing answers to problems, remaining awake or alert, and refraining from speaking during class.

Our studies have shown that both teachers and students hold different perceptions of what it means to participate or engage actively in learning science. For example, many students who characterize themselves as being "silent in class," also identified themselves as being "highly participatory." When watching videos of science lessons with students from silent classrooms, students pointed out numerous examples of their "participatory practices," including moments when they were "listening hard," maintaining eye contact with the teacher, taking notes, and remaining still in their seats. These were all important examples of these students' "active" participation and engagement in the science classroom (Kim, 2015). These experiences have forced me to reconsider what it means for students to *be engaged* in school or science or even what it means for students to *participate* in learning science. Specifically, my experiences in Korean classrooms have led me to question and critically reflect on my own cultural biases that privilege talk and gestures/movement as evidence of engagement, participation, and learning in science classrooms. My work has forced me to question the primacy of talk as evidence for student participation and engagement, and instead I have begun to reframe my research to consider student silence and stillness as important evidence of student participation and engagement.

Currently, my colleagues and I (Chien, et al., 2016) are expanding this study to include students in Taiwan where we are working to identify and categorize different types of student silence, and we are trying to develop new methodologies for measuring participatory practices that are not always evident through observation. Our video analysis of different science classrooms has revealed various kinds of "silent but participatory students." By re-defining active participation in ways that encompass a wider notion of activity (including thinking, listening, observing, reflecting, theorizing, or interpreting without speech), we hope to learn more about how students are processing information internally to form their understanding of science concepts without engaging in overt talk and teacher-student or student-student interaction. However, this is challenging because a review of the literature shows that engagement has historically been measured using self-report survey instruments or observational tools that focus on measuring overt, observable practices that can be construed as "participatory actions." This has led researchers to focus on the identification of outward manifestations of student practices as they engage in academic or social activities in school (Zyngier, 2008). In our research, we are considering what methodologies are needed to support data collection and interpretation that can expand our understanding of what is happening in these classrooms and that can allow us access to the private thinking of "silent" learners.

CAN STUDENTS BE ENGAGED, INTERESTED, PARTICIPATORY AND HAVE HIGH ACHIEVEMENT IN SCIENCE?

It would seem that researchers could learn a lot about what factors account for student learning by exploring how students manage to achieve at such high levels even while reporting limited affective and behavioral engagement in science learning. Instead, I have found that my colleagues from Western educational contexts tend to discount East Asian students' high achievement on these exams as being suspect with regards to what "kind of learning" it measures. At the same time, these colleagues are engaged in research designed to increase the achievement of students in their countries on these and similar assessments. This has led me to question what it really means for students to *know or do science?* What forms of science knowledge are important? Why are some forms of knowledge or ways of knowing and representing science knowledge more valuable than others? Why are students' high achievements in Asian countries discounted while at the same time, educators in Western countries are seeking ways to increase student achievement on similar assessments?

Alternatively, I find there is tendency by science education researchers in Korea and other East Asian countries to be focused on finding ways to enhance student interest and enjoyment and to increase participation practices in science classrooms, but they expect students to do so *without sacrificing high scores* on international and national assessments. There is no doubt that the achievement without enjoyment paradox is disconcerting for parents and educators who are concerned about the health and well-being of an entire generation of students. In addition, government officials in these countries are worried about the potential impact on their country's economy, as many believe that students' limited classroom participation and low interest and enjoyment in math and science may serve to diminish the number of students who will choose to pursue STEM related careers in the future. Publications in domestic and international journals by Korean researchers over the last fifteen years reflect many of the governments' policy changes and funding initiatives designed to expand the use of inquiry teaching, to engage students in collaborative group work, and to participate in hands-on experiments. More recently, there have been publications describing the use of flipped classrooms, and the implementation of modeling and argumentation strategies to increase student enjoyment and interest in learning science. These pedagogical strategies all position student talk and activity as central for learning—which does not seem compatible with teachers' and students' expectations for good science teaching and learning in these contexts. This concern is generally reflected in the conclusions and implications sections of these papers, which tend to remark that while the strategies used in the research did improve students' behavioral participation in class and their enjoyment and interest for learning science, the pedagogies implemented were not sustainable due to the pressure felt by students and teachers who need to cover large amounts of material in preparation for the college entrance exam.

Given that the education systems of Korea and other East Asian countries have produced high levels of student achievement with relative equity among learners (especially compared to many Western educational contexts), might we be able to learn something positive from the teaching and learning occurring in these classrooms that could be applied in other contexts? For example, what can we learn about student perseverance and resilience? What can account for some students' ability to overcome their negative perceptions of school and learning in order to achieve at such high levels? What motivates these students to persist in their educational studies under such challenging conditions? With regard to affect, what sort of factors may account for differences in students' affective engagement with school or science? How does affective engagement impact learning for students in different educational contexts? What can we learn about how interest and enjoyment shape student learning? What factors maximize student learning in different contexts? How can understanding high achievement by students in East

Asian countries inform researchers' understandings about achievement of students of Asian descent educated in Western contexts? What contributes to the patterns of high achievement and relatively limited affective engagement with school and science for students in East Asian countries, but not in the wider Asian region? How does the way these students experience science learning inform how they view science as a discipline?

MOVING FORWARD TO REFRAME STUDENT ENGAGEMENT AND PARTICIPATION

In this chapter, I have drawn heavily from reports describing student performance on the TIMSS and PISA exams. Unfortunately, comparative assessments based on quantitative survey data fail to offer perspectives about the wider educational contexts in which these students toil. Research that considers the socio-historical development of different education systems and the political, economic, and cultural factors that have shaped how teaching and learning are practiced in different countries would provide us more opportunities to consider the complex connections between engagement in school and learning and participation in school science (Martin et al., 2014). However, my review of the literature produced few studies addressing these kinds of concerns and almost none that were exploring the math and science achievement disparity experienced by students in East Asian countries. In fact, a cursory review of the literature focusing on the learning of Asian students provides a large number of papers that position these learners as being exemplars of the "model minority": students who are successful in school and who demonstrate high achievement in all categories, especially math and science (Ng, Lee, & Yoon, 2007). Alternatively, there is a substantial body of work that devalues these students' achievements by qualifying Asian students' learning by using broad-based stereotypes to describe students as being passive, dependent rote memorizers who are prone to plagiarism and who are lacking in creativity or critical thinking ability (Saravanamuthu, 2008).

In either case, it seems these studies present an over-generalization of Asian students (both in Western and East-Asian educational contexts) that position more than a billion people as being homogenous and studiously ignores the dynamic and changing nature of cultures that are continuously influenced by changes in politics, society, culture, and economy (Ryan, 2010). Discourses about "Western" and "Eastern" educational philosophies are generally described in binary terms, such as "student centered" and "teacher centered" or "individualistic" and "communal" (Ryan & Louie, 2007). This practice creates dichotomies that ignore the complexities and diversity of evolving political, social, cultural, and economic factors that are continuously shaping how science and education unfold in each context. Unfortunately, there is little room in mainstream academia for questioning or critical reflection on the part of researchers in different contexts about what science teaching and learning practices are "normal" or expected or even what kinds of practices should be cultivated or accepted. In addition, there has been limited space for researchers from Asian countries to contribute to the ongoing discourse about science education. A recent analysis (Martin & Chu, 2015) of all papers published in the four top-ranked journals in the field of science education between 2010 and 2014 found that only 127 of 1,309 (9.7%) papers included authors who represented an institution in an Asian country. Of those 127 papers, only 12 (0.92%) were single-authored by a researcher from an institution in Asia, and close analysis of the 127 papers published revealed that about half of the papers published were unrelated to science education in Asia at all. The few that offered descriptive analysis of the local science education context were generally being reported as an international cross-comparative study where the focus was on a shared issue rather than a particular learning context.

That there is limited representation of Asian learning contexts in the science education literature means that researchers may be limited in their opportunities to learn about what is happening outside of the research published by mainstream contributors. The same could be true of science education research in African and Central and South American countries as well. This lack of representation means we are missing a chance to learn from one another—which may prevent us from asking new questions or considering alternative perspectives. I know from my own experience that I could not have begun to question the primacy of verbalized speech as a requirement for helping students to learn science until I had spent considerable time teaching and learning in contexts where student speaking is not the norm. Using Korea as a lens to make arguments for the deeply contextualized ways in which students engage in and participate in science has taught me that studies grounded on a firm (Western) notion of what it means to "participate" do not allow for more and different ways of thinking about student engagement in science. My experience has taught me that engagement is personally- as well as collectively-understood and demonstrated or performed. And it was only by embedding myself in a different context that I was able to recognize that my (Western) assumptions of participation were not universally shared. While this statement may appear to some researchers to be self-evident, the literature on student engagement and participation would suggest that there are universal participation practices that are not only easily observed, but also easily interpreted.

However, my collaborations with my peers and my reflections on my experiences in classrooms in Korea and Taiwan have helped me to recognize the need to contextualize the "evidence" that can be seen and to try to suspend expectations of what it means to be an active student in science in order to better learn from how students see themselves as learners. Unfortunately, the majority of engagement research has tended to omit students' perspectives about what it means to be engaged as learners. This has resulted in an objectification of students' experiences that serves to position students as being either "engaged and successful" or "unengaged and deficient" (McMahon & Zyngier, 2009). As such, relatively few quantitative studies exploring different constructs of students' engagement in science have contributed to a fairly robust assertion that student enjoyment, interests, and attitudes about science are significant factors for determining students' achievement in science. Most of these studies neglected to include students in identifying the factors most salient for engaging them as learners. Such research limits our understandings of engagement to a binary, either-or perspective. When researchers fail to consider broader views about what it means to be engaged or to participate in school or science, they limit the possibilities for considering relationships between engagement and positive social and academic outcomes that are more complex than simple cause and effect notions of engagement and learning (Harris, 2011).

Sharing our different perspectives and lived experiences can lead us to different conceptions about teaching and learning that can be valuable for helping us question and problematize norms in ways that can expand our thinking and our learning. For me, it has required a reconsideration of the role of silence as an active choice rather than a passive action. When we reframe silence as a conscious activity, we can begin to think about the ways in which silence can be a kind of performative act that students may employ as learners in the classroom. To help expand my thinking about the role of silence as a performative act that may mediate science learning, I have begun to consider more fully Lev S. Vygotsky's (1986) work about the role of "private speech" for supporting "inner" vocalization and internalization of concepts rather than verbalization. As I revisit Vygotsky's work, I am beginning to reframe how I think about *talking science*, and instead, I have begun to consider what are the implications for students' science learning when they are *thinking science*.

My examination of these issues has been enhanced and enriched by my ability to work closely with colleagues, teachers, and students with emic perspectives about science education in Korea and Taiwan. It is in our discussions across our differences that we have gained some new insights and have been able to reconsider what questions we need to ask in order to probe more deeply into the achievement paradox. Because we seek to learn about how students' internalized vocalizations about science inform their conceptual understanding, we are also working to develop new ways to collect and interpret data that involve students as co-researchers. By providing students with opportunities to express alternative notions about participatory practices, we seek to understand how these practices may influence their science learning experiences in ways that are respectful, inclusive, and locally contextualized. Hopefully, contributions from our collective research will enable us to begin to confront some of the prevailing narratives that exist and instead allow us to reframe our individual and collective views about what it means to for students to be engaged or to participate in science. In doing so, I hope my colleagues and I can better understand how students' participatory practices, which vary in different contexts, may contribute to students' science learning in ways that go beyond what is currently understood.

ACKNOWLEDGEMENT

This work was supported by the Ministry of Education of the Republic of Korea and the National Research Foundation of Korea (NRF-2016S1A3A2925401).

REFERENCES

Ainley, M., & Ainley, J. (2011). Student engagement with science in early adolescence: The contribution of enjoyment to students' continuing interest in learning about science. *Contemporary Educational Psychology, 36*, 4–12. doi:10.1016/j.cedpsych.2010.08.001.

Appleton, J. J., Christenson, S. L., Kim, D., & Reschly, A. L. (2006). Measuring cognitive and psychological engagement: Validation of the student engagement instrument. *Journal of School Psychology, 44*, 427–445. doi:10.1016/j.jsp.2006.04.002.

Bransford, J. D., & Donovan, M. S. (2005). Scientific inquiry and how people learn. In M. S. Donovan & J. D. Bransford (Eds.), *How students learn: History, mathematics, and science in the classroom* (pp. 397–420). Washington, DC: National Academies Press. doi:10.17226/10126.

Bray, M. (2010). Research shadow education: Methodological challenges and directions. *Asia Pacific Education Review, 11*, 3–13. doi:10.1007/s12564-009-9056-6.

Chien, Y.T., Chun, H.J., Martin, S., Chu, H.E., Ahn, W., & Chang, C.Y. (August, 2016). *Toward an understanding of students' verbal and non-verbal participatory practices in the science classroom.* Oral presentation for the 5th bi-annual meeting of the East-Asian Association for Science Education (EASE), Tokyo, Japan, August 26-28, 2016.

Choi, J., & Cho, R. M. (2015). Evaluating the effects of governmental regulations on South Korean private cram schools. *Asia Pacific Journal of Education.* doi:10.1080/02188791.2015.1064356.

Choi, H., & Choi, A. (2016). Regulation private tutoring consumption in Korea: Lessons from another failure. *International Journal of Educational Development, 49*, 144–156. doi:10.1016/j.ijedudev.2016.03.002.

Duit, R., & Treagust, D. (1998). Learning in science – From behaviorism towards social constructivism and beyond. In B. J. Fraser & K. G. Tobin (Eds.), *International handbook of science education* (pp. 3–25). Dordrecht, The Netherlands. Kluwer Academic Publishers. doi:10.1007/978-94-011-4940-2_1.

Duschl, R. A., Schweingruber, H. A., & Shouse, A. W. (Eds.). (2007). *Taking science to school: Learning and teaching science in grades K-8.* Washington, DC: National Academies Press. doi:10.17226/11625.

Fortus, D. (2014). Attending to affect. *Journal of Research in Science Teaching,* 51(7), 821–835. doi:10.1002/tea.21155

Fredericks, J. A., Blumenfeld, P. C., & Paris, A. C. (2004). School engagement: Potential of the concept, state of the evidence. *Review of Educational Research, 74*, 59–109. doi:10.3102/00346543074001059.

Gonzales, P., Williams, T., Jocelyn, L., Roey, S., Kastberg, D., & Brenwald, S. (2008). *Highlights from TIMSS 2007: Mathematics and science achievement of U.S. fourth- and eighth-grade students in an international context* (NCES 2009–001 Revised). Washington, DC: National Center for Education Statistics, Institute of Education Sciences, U.S. Department of Education.

Harris, L. (2011). Secondary teachers' conceptions of student engagement: Engagement in learning or schooling? *Teaching and Teacher Education, 27*, 376–386. doi:10.1016/j.tate.2010.09.006.

Hofstein, A., & Lunetta, V. N. (2004). The laboratory in science education: Foundations for the twenty-first century. *Science Education, 88*, 28–54. doi:10.1002/sce.10106.

Jeong, E. J. (2016, May 3). South Korean youngsters back at the bottom of OECD happiness rank. *The Hankyoreh*. Retrieved from http://english.hani.co.kr

Kim, E. J. (2015). *Understanding student and teacher interactions in Korean high school science classrooms from a structure/agency dialectic perspective: Implications for pedagogy and research.* Masters of Physics Education thesis, Seoul National University, Seoul, Republic of Korea.

Korean National Statistics Office [KOSTAT]. (2015). *2015 Statistics on the youth*. Korean Statistics. Retrieved May 1, 2016 from http://kostat.go.kr/portal/korea/index.action

Kwok, P. (2004). Examination-oriented knowledge and value transformation in East Asian cram schools. *Asia Pacific Education Review, 5*(1), 64–75. doi:10.1007/bf03026280.

Lazarowitz, R., Baird, J. H., & Bowlden, V. (1996). Teaching biology in a group mastery learning mode: High school students' academic achievement and affective outcomes. *International Journal of Science Education, 18*, 447–462. doi:10.1080/0950069960180404.

Lemke, J. (1990). *Talking science*. New York, NY: Ablex.

Liu, J. (2012). Does cram school mattering? How does to cram schools? Evidence from Taiwan. *International Journal of Educational Development, 32*, 46–52. doi:10.1016/j.ijedudev.2011.01.014.

Martin, I. V. S., Martin, M. O., & Foy, R. (2008). *TIMSS 2007 international science report: Findings from IEA's Trends in International Mathematics and Science Study at the fourth and eighth grades.* Chestnut Hill, MA: Boston College.

Martin, S. (2006). Where practice and theory intersect in the chemistry classroom: Using cogenerative dialogue to identify the critical point in science education. *Cultural Studies of Science Education, 1*(4), 693–720. doi:10.1007/s11422-006-9031-z.

Martin, S., Choe, S.-U., Kim, C.-J., & Kwak, Y. (January, 2014). Employing a socio-historical perspective for understanding the impact of ideology and policy on educational achievement in the Republic of Korea. In J.V. Clark (Ed), *Closing the achievement gap from an international perspective: Transforming STEM for effective education* (p. 229–250). The Netherlands: Springer.

Martin, S., & Chu, H. E. (2015). Asia-Pacific Science Education (APSE): Expanding opportunities for publishing science education research. *Asia-Pacific Science Education, 1*(3), 1–18. doi:10.1186/s41029-015-0006-9.

Maslak, M. A., Kim, J., & McLoughlin, A. S. (2010). Educational engagement in China: A case from the Northwest. *International Journal of Educational Development, 30*, 254–262. doi:10.1016/j.ijedudev.2009.09.003.

McMahon, B. J., & Zyngier, D. (2009). Student engagement: Contested concepts in two continents. *Research in Comparative and International Education, 4*, 164–181. doi:10.2304/rcie.2009.4.2.164.

Milne, O., & Otieno, T. (2007). Understanding engagement: Science demonstrations and emotional energy. *Science Education, 91*, 523–553. doi:10.1002/sce.20203.

Ministry of Employment and Labor. (2017). *Youth Employment Statistics.* Retrieved from http://laborstat.molab.go.kr/

Mitchell, I., & Carbone, A. (2011). A typology of task characteristics and their effects on student engagement. *International Journal of Educational Research, 50*, 257–270. doi:10.1016/j.ijer.2011.05.001.

National Research Council [NRC]. (1996). *National science education standards.* Washington, DC: National Academy Press. doi:10.17226/4962.

Ng, J. C., Lee, S. S., & Yoon, K. (2007). Contesting the model minority and perpetual foreigner stereotypes: A critical review of literature on Asian Americans in education. *Review of Research in Education, 31*, 95–130. doi:10.3102/0091732x07300046095.

Noah, A., Acciai, F., & Firebuagh, G. (2016). Understanding the contribution of suicide to life expectancy in South Korea. *Demographic Research, 35(22), 617–644.* doi: 10.4054/DemRes.2016.35.22.

OECD. (2010). *PISA-2009 results: What students know and can do: Student performance in reading, mathematics, and science* (Vol. I). Paris: OECD Publishing.

OECD. (2011). *Lessons from PISA for the United States: Strong performers and successful reformers in education.* Paris: OECD Publishing. doi:10.1787/9789264096660-en.

OECD. (2012). *Education at a glance 2012: Highlights.* Paris: OECD Publishing.

OECD. (2013a). Defining and measuring equity in education. In *PISA 2012 Results: Excellence through equity (Volume II): Giving every student the chance to succeed.* Paris: OECD Publishing. doi:/10.1787/9789264201132-6-en.

OECD. (2013b). *Education at a glance 2013: OECD indicators.* Paris: OECD Publishing. Retrieved October 7, 2014 from www.oecd.org/edu/eag.htm

OECD. (2013c). *Suicides.* OECD Factbook 2013: Economic, environmental, and social statistics. Paris: OECD Publishing. Retrieved May 1, 2016 from http://www.oecd-ilibrary.org/sites/factbook-2013-en/

OECD. (2013d). *Making Mental Health Count: The social and economic costs of neglecting mental health care.* OECD Health Policy Studies. Paris: OECD Publishing. Retrieved May 1, 2016 from https://www.oecd.org/els/health-systems/MMHC-Country-Press-Note-Korea.pdf

OECD. (2013e). *Japan Country Note: Results from PISA 2012.* Paris: OECD Publishing. Retrieved May 1, 2016 from https://www.oecd.org/pisa/keyfindings/PISA-2012-results-japan.pdf

OECD. (2014). *PISA-2012 results in focus: What 15-year-olds know and what they can do with what they know.* Paris: OECD Publishing.

OECD. (2015a). *How's life? 2015 measuring well-being.* Paris: OECD Publishing. doi:10.1787/how_life-2015-en.

OECD. (2015b). *Health at a Glance 2015: OECD Indicators.* Paris: OECD Publishing. http://dx.doi.org/10.1787/health_glance-2015-en

Olitsky, S. (2007). Promoting student engagement in science: Interaction rituals and the pursuit of a community of practice. *Journal of Research in Science Teaching, 44*, 33–56. doi:10.1002/tea.20128.

Olitsky, S., & Milne, C. (2012). Understanding engagement in science education: The psychological and the social. In B. J. Fraser, K. Tobin, & C. J. McRobbie (Eds.), *Second international handbook of science education* (pp. 19–33). Dordrecht: Springer. doi:10.1007/978-1-4020-9041-7_44.

Osborne, J., Simon, S., & Collins, S. (2003). Attitudes towards science: A review of the literature and its implications. *International Journal of Science Education, 25(9), 1049–1079.* doi: 10.1080/0950069032000032199.

Park, J., Martin, S., & Chu, H. E. (2015). Examining how structures shape teacher and student agency in science classrooms in an innovative middle school: Implications for policy and practice. *Journal of the Korean Association for Science Education, 35*, 773–790. doi:10.14697/jkase.2015.35.4.0773.

Preznel, M., Seidel, T., & Kobarg, M. (2012). Science teaching and learning: An international comparative perspective. In B. J. Fraser, K. G Tobin, & R. Campbell (Eds.), *Second international handbook of science education* (pp. 667–678). Dordrecht: Springer. doi:10.1007/978-1-4020-9041-7_44.

Ritchie, S. M., Hudson, P., Bellocchi, A., Henderson, S., King, D., & Tobin, K. (2015). Evolution of self-reporting methods for identifying discrete emotions in science classrooms. *Cultural Studies of Science Education* [online first], 1–17. doi:10.1007/s11422-014-9607-y.

Ryan, J. (2010). The Chinese learner: Misconceptions and realities. In J. Ryan & G. Slethaug (Eds.), *International education and the Chinese learner* (pp. 37–26). Hong Kong: Hong Kong University Press. doi:10.5790/hongkong/9789888028450.003.0003.

Ryan, J., & Louie, K. (2007). False dichotomy? "Western" and "Confucian" concepts of scholarship and learning. *Educational Philosophy and Theory, 39*, 404–417. doi:10.1111/j.1469-5812.2007.00347.x.

Saravanamuthu, K. (2008). Reflecting on the Biggs-Watkins theory of the Chinese learner. *Critical Perspectives on Accounting, 19*(2), 138–180. doi:10.1016/j.cpa.2005.12.005.

Song, J. (2013). The disparity between achievement and engagement in students' science learning: A case of East-Asian regions. In D. Corrigan, R. Gunstone, & A. Jones (Eds.), *Valuing assessment in science education: Pedagogy, curriculum, and policy* (pp. 285–306). Dordrecht: Springer. doi:10.1007/978-94-007-6668-6.

Tytler, R., & Osborne, J. (2012). Student attitudes and aspirations towards science. In B. J. Fraser, K. Tobin, & C. J. McRobbie (Eds.), *Second international handbook of science education* (pp. 597–626). Dordrecht, The Netherlands: Springer. doi:10.1007/978-1-4020-9041-7_41.

Vygotsky, L. S. (1986). *Thought and language*. Cambridge, MA: The MIT Press.

Zyngier, D. (2008). (Re)conceptualizing student engagement: Doing education not doing time. *Teaching and Teacher Education, 24*, 1765–1776. doi:10.1016/j.tate.2007.09.004.

SECTION FIVE

Science Teacher Education

What Is Good Science Teaching and How Do We Teach People to Be Good Science Teachers?

CHAPTER ELEVEN

Professional Duties and Challenges of Novice Teachers

Level of Consciousness in Facing the Pedagogical Risk

MAURÍCIO PIETROCOLA

Teaching science is a complex, deeply contextual and situational activity, requiring teachers to draw upon a vast repertoire of practices and deep knowledge about science, students, and pedagogy. Rather than viewing science teaching as a technical activity to be enacted without regard to specific institutional, political, and social contexts, I frame science teaching as a persistently problematic activity in which teachers continually redesign, negotiate, and relearn throughout careers and within specific contexts. Nowadays, preparing science teachers for the task is an education project of great complexity and importance.

Over the last 50 years, the academic model for the preparation of science teachers has undergone a series of changes. At first, this model focused on the conceptual content of the sciences. The content-focused model was followed by models that supplemented content with educational knowledge and pedagogical practices. More recently, pedagogical models have included specific knowledge for science teaching. The notions of *pedagogical content knowledge* (PCK) by Lee Shulman (1987), *didactic engineering* by Guy Brousseau (1988) or even *didactic transposition* by Yves Chevallard (1988) suggest that teaching strategies should be adapted to specific disciplinary content such as natural sciences. These strategies would take into account epistemological, ontological and axiological aspects of the sciences and therefore differ from the teaching strategies in other fields of knowledge.

However, even with the advancement of pedagogical models in education courses, science teachers' practices still seem very distant from what is idealized as good quality science teaching. The manner in which young teachers enter the school environment and start their professional activity remains a challenge. How can we facilitate novice science teachers' professional integration of practices so that their teaching is consistent with the pedagogical models experienced in their teacher preparation? How do we enact teacher preparation in ways that afford teachers transforming the quality science teaching beyond the more traditional pedagogical models? These are contemporary questions with tough answers.

In the vast majority of countries, a teacher's activity is regulated by partial or full certification through college courses. Therefore, the quality of these courses is directly related to the quality of teachers' performance. Many surveys have studied the relationship between teacher education programs and the performance of teachers in schools. The difficult relationship between theory and practice has been a recurring theme in educational scholarship. To some scholars, the problems are addressed using a model of education that encompasses at least three steps (e.g., Wideen, Mayer-Smith, & Moon, 1998): (1) Future teachers initially take college courses where they learn theories and methods and develop teaching skills; (2) While, they teach students, future teachers become acquainted with the school environment, its demands, and its modus operandi; and (3) The now novice teachers must apply the theoretical knowledge obtained in their college courses in a professional capacity. Usually, the success in this final step comes solely from the teachers' efforts and responsibilities. The beginning of professional school activity requires novice teachers to overcome two types of discontinuities: one related to knowledge, as it implies transposing the knowledge and practices from an idealized context to a real life context (of the school), and another related to the changes of social contexts with different rules and commitments, as the novice teachers transition from being college students to becoming professional educators in a school environment.

Overcoming these two discontinuities is a determining factor in the success of the teachers as well as the pedagogical models they incorporate throughout their professional careers. Studying the history of teacher education seems to be an important aspect when attempting to understand the best methods of facilitating their development of knowledge and skills for teaching. Over three decades ago, Kenneth Zeichner and Robert Tabachnick (1981) stressed that future teachers regress to more traditional models as they assume their professional duties. According to Zeichner and Tabachnick:

> It now has become commonly accepted within the teacher education community that students become increasingly more progressive or liberal in their attitudes towards education

> during their stay at the university and then shift to opposing and more traditional views as they move into student teaching and in-service experience. (p. 7)

Simon Veenman (1984) came to similar conclusions in researching beginning teachers. Difficulties and even barriers in the transition from teacher education to practice are well documented (Feiman-Nemser & Buchman, 1989). Many researchers use expressions such as "reality shock" or "practice shock" (e.g., Dann, Müller-Fohrbrodt, & Cloetta, 1981). Adra Cole and Gary Knowles (1993) interpreted this shock as a consequence of student teachers' idealized picture of teaching and learning being shattered when facing the reality of schools. The development of the student teachers' attitudes when observed over several years presents a "U-shaped curve": student teachers begin with very traditional behaviors, stemming from their previous experience in education courses. Then, certain innovative attitudes gain strength during the preservice courses that weaken after graduation and the beginning of their service as teachers. It seems that the social demands of the school end up creating a very stressful environment in which novice teachers struggle to maintain the enactment of more progressive and innovative pedagogical models. The tendency is for there to be a shift from being a moderately successful and generally conformist student to being a pedagogically conservative teacher (Lacey, 1995).

Novice teachers often experience a disconnection between the ideas they learn in teacher education programs and the traditional ways of teaching expected by other faculty members, parents, and the students themselves. Hanns-Dietrich Dann and colleagues (1981) call this "discrepancy experiences" during the socialization process of the teachers. One source of this problem lies in part on the teacher education programs, which are frequently shown to be fragmented, in the form of a separate collection of courses in which knowledge is presented in a theoretical fashion and without a clear connection to the actual practice of teaching. Such programs often underestimate the dominant role of actual practice in shaping the development of the teachers (Lortie, 1975). Without knowing for certain how to implement theory learned in college on the practical situations in school, novice teachers end up regressing to pedagogical models experienced as pupils. In this case, both old school teachers and teacher educators from pre-service courses may not display the best examples of good teaching.

SCHOOL SOCIALIZATION

One of the turning points in a teacher's professional life is the start of her/his teaching career. Schools are settings of great social complexity, where individual practices are mediated by demands of many kinds—rules and commitments of epistemological and psychological nature, human values, etc. Hence, it should come as no surprise that teachers' socialization is of critical importance to their professionalization. Socialization is defined here as:

> ... the process by which people selectively acquire the values and attitudes, the interests, skills and knowledge—in short the culture—current in groups to which they are, or seek to become a member. (Merton, Reader, & Kendall, 1957, p. 287)

The literature on teacher socialization emphasizes that the school is the dominant factor in shaping teaching practices, and that school cultures discourage more progressive and modern courses of action. We know very little about the degree to which specific arrangements and strategies in teacher education may counterbalance the tendency of teachers to socialize compared to well established practices in schools (Brouwer & Korthagen, 2005). Becoming a teacher is a process that involves learning to teach and involves continuous emotional production and mediation. Beginning teachers work hard to maintain control amidst feelings of frustration, anger, shame, and bewilderment (Wideen et al., 1998).

The challenge for science teacher educators is to understand the limits and possibilities of changes in teachers' practices and enact programs to overcome the previously described historical trends and associated problems. We assume that the success beginning teachers experience from teaching in an innovative way is in part due to their awareness of the risks and conflicts to assume in classroom practices. The problems I would like to analyze deal with the teachers' experiences from their careers as educators. It is important to discuss the commitments, duties, and constraints involved in these pedagogical practices as belonging to a historical cultural set of practices of teaching and learning. Using this sociocultural framework, the school is conceived as a cultural space made up of a series of moments of action of specific behaviors and resources fitting the culture of teaching and learning. It could be seen as a cultural system belonging to the educational world where symbols and meanings were shared by a community in a coherent way, as proposed by Geertz (1973). But the actors in the educational world enact culture by reproducing patterns in the system while also transforming it, because those actors (teachers and students) could exercise agency as they appropriate the schemas and resources in different ways; and there is always the possibility that schemas and resources from other cultures will mediate what is intended (Sewell, 2005).

The challenge for a teacher is to incorporate schemas and resources from an educational structure and address events that disrupt what is intended by teaching in ways that are cognizant of what is happening in the moment and being open to enacting changes in ways that are contingent and emergent while responding to the needs of learners. That is, teachers must adopt the ways of researchers so as to be anticipatory while teaching in ways that are timely and appropriate to the needs of students. It is interesting that, when asked, teachers can vividly recall events from the first day, month, and even the first year of their lives as teachers. Normally, these are moments in which disruption puts teachers in the position to use schemas and resources that they first learned when they themselves were students. These moments are memorable to them because the emergence of different types of pedagogical situations prompts decision making for which they are otherwise unprepared. Without conscious awareness, they enact what they learned from being in with others as students in their lives as K-12 students. This tacit knowledge affords them seeing and acting fluently to cope with the social complexity of a classroom.

As time passes, teachers adapt themselves to performing in new social contexts—the school—and become experts. Day after day, the practical activities in a school reach a level of routine, which emerges from a necessity to enact specific practices in a classroom without sufficient time to reflect on the complex interactions that unfold continuously. The necessity to act fluently is a condition for successful teaching and learning, requiring practical skills that can be enacted without conscious awareness in the moment. In many cases, the necessity to act fluently results in a return to old practices the teachers experienced when they themselves were students.

In trying to explain the roles of social actors and activity in the face of social life, Anthony Giddens (1984) discusses some aspects of consciousness present in any kind of social practice. He highlights the importance of understanding the human awareness of our own actions in social life. In his words, "actors have, as an inherent aspect of what they do, the capacity to understand what they do while they do it. The reflexive capacities of human actors are characteristically involved in a continuous manner with the flow of day to day conduct in the context of social activity" (Giddens, 1984, p. xxii/xxiii). Giddens called this "knowledgeability," i.e., everything actors understand (or believe they understand) about their actions and the actions of others in the context of action and the others involved in it. This knowledgeability allows for the production and reproduction of actions. Giddens presents three types of consciousness humans have about what they do: *discursive consciousness*, *practical consciousness*, and *unconsciousness*.

Using Stephen Toulmin's idea of "articulateness," Giddens defines the notion of *discursive consciousness* as a type of reflection about what we do that could be a topic of conversation (to talk about). In Giddens' words: "'Consciousness' … presumes being able to give a coherent account of one's activities and the reasons for them" (Giddens, 1984, p. 45). However, this reflexivity operates only partially on a discursive level. Much of what people know about what they do is because what they do is on the level of practical consciousness and consists primarily of tacit knowledge. Giddens calls the tacit process of reflexivity "practical consciousness." Recurrent situations or sets of recurrent situations are put in action without the need for deep reflexive thinking. In this sense, Giddens says that, "practical consciousness consists of all things which actors know tacitly about how to 'go on' in the context of social life without being able to give them direct discursive expression" (1984, p. xxiii).

Giddens concludes the idea of practical consciousness with the notion of routine, which sheds light on courses of action being taken automatically, without a level of discursive consciousness. The majority of day-to-day social activities are made during routines, because the nature of social life is very recursive. Normally all social life activities are immersed within these three levels of consciousness. Therefore, to Giddens, social life implies the transformation of repetitive activities in routines. He calls this process *routinization*: "The habitual, taken-for-granted character of the vast bulk of the activities of day-to-day social life; the prevalence of familiar styles and forms of conduct, both supporting and supported by a sense of ontological security" (Giddens, 1984, p. 376). Ontological security is a necessity of social life and should be understood as "a confidence or trust that the natural and social worlds are as they appear to be, including the basic existential parameters of self and social identity" (Giddens, 1984, p. 375). Routinization is a fundamental process to offer ontological security to social agents.

Anthony Elliot (2011) details the meanings of the concepts of routines and ontological security in Giddens' work:

> Giddens makes the concept of routinization central to the constitution and reproduction of history and consciousness ... This is not an expression of sociological determinism (in the sense that all action is pre-programmed) nor political conservatism: there is no logical reason why the reflexively constituted process of social reproduction demands an acceptance of *particular* habitual practices. Rather, Giddens's grounding of ontological security in routinization suggests that both existing and alternative (or oppositional) forms of life demand some sort of motivational commitment to the integration of habitual practices across space and time. (p. 94)

Therefore, routines are associated with practical consciousness. Routines are ways of ensuring security when acting in the world. Since the forms of reflection are limited to the flux of activity, it is not possible for every action in the world to go through the realm of discursive consciousness. People do not usually formulate questions regarding intentions or reasons behind certain forms of conduct, but consciousness can occur if there is a "breach" on the predictability of a routine, adding a deeper level of reflection that can be converted into words. The breaching experiments of Harold Garfinkel and Erving Goffman (Goffman, 1971) are good examples of how people can be taken by surprise when there is a breach of predictability of social conduct.

It is important to clarify that Giddens (1984) does not believe that in routine activities, action is performed without thought. The routine of those activities is something continuously "worked" by those who sustain them in everyday life. This way, in practical consciousness, reflexivity would be some kind of continuous monitoring of action, from the individual and of others, or from the continuous flow of social life. This would ensure the ability to "progress" in the scope of routines in the different contexts of which people are part.

ROUTINIZATION AND EDUCATION

Routinization helps to understand one of the difficulties young teachers encounter upon entry into schools as professionals. A large part of the work in the classroom is made up of repetitive activities. In everyday life, teachers deal with the students' behavior, with the scheduling of teaching activities, the calendar of tests and administrative duties. These activities comprise the foundation of a teacher's school life and repeat themselves for days, weeks and months. This school routine is complemented by unpredictable events, such as a new student's arrival mid-year, a curricular change dictated by legal processes, or a class with learning levels below expectations.

The repetitiveness of activities we are subjected to day-to-day is responsible for creating "social grounding." To Giddens, "(t)he term 'day-to-day' encapsulates exactly the routinized character which social life has as it stretches across time-space" (Giddens, 1984, p. xxiii). Teachers are involved in a practical level of consciousness, which means that the flow of day-to-day activities in a classroom has to be considered a pedagogic routine done with tacit knowledge. This means the things teachers really do inside the classroom are only partially captured at the discursive level.

The process of socialization of the profession occurs when young teachers assume their responsibilities in a school. Being with the school's faculty members, the students and their relatives define the courses of action to be employed. To avoid being put in jeopardy, beginning teachers normally recall a period in their own histories as students when they are learning to be around more experienced teachers. This "in school" experience is normally a much longer and much more effective process in practical examples than any other period in courses to learn to teach. Anna Carvalho and Gil Pérez (2003) call this "*formação ambiente*" or in English, "*environmental preparation.*" A person's experiences as a student have a strong power to frame the way s/he conceives her/his own action as a future teacher. For that reason, most new teaching and learning strategies from education courses succumb to the force of practical consciousness developed from the teachers' previous experience as students. The routinization act keeps the mind free for the contingency facts of school's life. As Giddens (1984) states:

> Because, I think, the apparently minor conventions of daily social life are of essential significance in curbing the sources of unconscious tension that would otherwise preoccupy most of our waking lives. (pp. xxiii–xxiv)

The embodied nature of this tacit knowledge makes it difficult to transfer knowledge received from pre-service or in-service education courses directly to a classroom situation, in which the knowledge teachers create is often and perhaps usually at a discursive level.

Recently, studies on the history of education present school, and particularly the classroom, as a place where practices and resources are well adapted to support teaching and learning in a strict sense. Jeanne-Claude Fourquin (2001) argues for a "school culture" with its way of functioning that is different from the other cultures in which education institutions are immersed. The so-called "traditional pedagogy model" results from historical processes in which a system of symbols, practices, schemas, and resources was built to self-replicate and define a way of acting inside school. Bourdieu developed the concept of *habitus* to explain the process of social reproduction (Bourdieu & Passeron, 1964, 1970). Changing the way schemas and resources are normally enacted in a classroom assumes some risk. After all, energy must be expended to disrupt the routines, patterns, and meanings inside the classroom. This means that changing routines is costly!

Most teachers are unaware of the routines in which they are involved since they used to be students in the same educational system. This is a routinization process that occurs in the course of action in the field of teaching of learning. The eventual adjustments in practices and attitudes in the classroom are made by the teachers themselves, the students, the faculty members, the family, etc. Students are usually critical of the methods used to teach them. They often prefer to be taught the "old-fashioned way" because these activities and practices in class are more predictable. Therefore, any changes in the schemas and resources employed in class could be rejected as disturbances of an established social order. This situation creates a conflicting scenario for teachers as they are unprepared to interrupt the reproduction of this system. Risk becomes an important ingredient in understanding what can be done and what should be done in class. In this sense, the breach of previously accepted standards in the established pedagogical culture should be understood as a *pedagogical risk*.

Entry into the profession of teaching means facing the pedagogical risk associated with uncertainties. On one hand, beginning teachers must adjust to established standards of conduct. On the other hand, they must adapt new pedagogical schemas and resources taken from education courses to the conditions of the schools where they will work. In this scenario, reflexivity is a contributing factor for professional success. Developing skills in situ (i.e., in the workplace) appeals as a productive way for science teacher educators to consider in a context of improving the quality of what they do. At the same time, science teacher educators should carefully ponder the role of theory. Acquiring theoretical knowledge in education courses does not ensure in itself a proper transposition to a school environment. It takes preparation to establish a day-to-day school routine in a reflexive way that involves a consciousness capable of being put into words, into thoughts and offering the possibility of change. Reflexivity entails the systematic exploration of the "unthought categories of thought which delimit the thinkable and predetermine the thought" (Bourdieu & Wacquant, 1992, p. 40).

REFERENCES

Bourdieu, P., & Passeron, J. C. (1964). *Les héritiers: Les* étudiants *et la culture.* Paris: Les Éditions de Minuit.

Bourdieu, P., & Passeron, J. C. (1970). *La reproduction: Éléments d'une théorie du système d'enseignement.* Paris: Les Éditions de Minuit.

Bourdieu, P., & Wacquant, L. (1992). *An invitation to reflexive sociology.* Chicago, IL: University of Chicago Press.

Brousseau, G. (1988). Le contrat didactique: Le milieu. *Recherches en Didactique des Mathématiques, 9*, 309–336.

Brouwer, N., & Korthagen, F. (2005). Can teacher education make a difference? *American Educational Research Journal, 42*(1), 153–224. doi:10.3102/00028312042001153.

Carvalho, A. M. P., & Pérez, D. G. (2003). *Formação de professores de ciências: Tendências e inovações.* Cortez.

Chevallard, Y. (1988). On didactic transposition theory: Some introductory notes. *Proceedings of the International Symposium on Selected Domains of Research and Development in Mathematics Education* (pp. 51–62). Bratislava, Slovakia.

Cole, A. L., & Knowles, J. G. (1993). Shattered images: Understanding expectations and realities of field experiences. *Teaching and Teacher Education, 9*, 457–471. doi:10.1016/0742-051x(93)90030-k.

Dann, H.-D., Müller-Fohrbrodt, G., & Cloetta, B. (1981). Sozialisation junger Lehrer im Beruf. Praxisschock drei Jahre später [Occupational socialization in young teachers: Practice shock three years later]. *Zeitschrift für Entwicklungspsychologie und Pädagogische Psychologie, 13*, 251–262.

Elliot, A. (Ed.). (2011). *The Routledge companion to social theory.* London: Routledge.

Feiman-Nemser, S., & Buchman, M. (1989). Describing teacher education: A framework and illustrative findings from a longitudinal study of six students. *Elementary School Journal, 89*, 365–377. doi:10.1086/461580.

Fourquin, J.-C. (2001). La pédagogie, la culture et la raison: Variations sur un thème d'Ernest Gellner. *Revue Française de Pédagogie, 135*, 131–144. doi:10.3406/rfp.2001.2807.

Geertz, C. (1973). *The interpretation of cultures.* New York, NY: Basic Books. doi:10.1007/978-3-658-13213-2_66.

Giddens, A. (1984). *The constitution of society: Outline of the theory of structuration.* Oakland, CA: University of California Press.

Goffman, E. (1971). *Relations in public: Microstudies of the public order.* New York, NY: Basic Books.

Lacey, C. (1995). Professional socialization of teachers. In W. Anderson (Ed.), *International encyclopedia of teaching and teacher education* (pp. 616–620). Oxford: Elsevier Science/Pergamon.

Lortie, D. C. (1975). *Schoolteacher, a sociological study.* Chicago, IL: University of Chicago Press. doi:10.1177/019263657505939422.

Merton, R. K., Reader, G., & Kendall, P. L. (1957). *The student-physician: Introductory studies in the sociology of medical education.* Cambridge, MA: Harvard University Press. doi:10.4159/harvard.9780674366831.

Sewell, W., Jr. (2005). *Logics of History: Social theory and social transformation.* Chicago: University of Chicago Press.

Shulman, L. (1987). Knowledge and teaching: Foundations of the new reform. *Harvard Educational Review, 57*(1), 1–23. doi:10.17763/haer.57.1.j463w79r56455411.

Veenman, S. (1984). Perceived problems of beginning teachers. *Review of Educational Research, 52*(2), 143–178. doi:10.2307/1170301.

Wideen, M., Mayer-Smith, J., & Moon, B. (1998). A critical analysis of the research on learning to teach: Making the case for an ecological perspective on inquiry. *Review of Educational Research, 68*, 130–178. doi:10.3102/00346543068002130.

Zeichner, K. M., & Tabachnick, B. R. (1981). Are the effects of university teacher education "washed out" by school experience? *Journal of Teacher Education, 32*(3), 7–11. doi:10.1177/002248718103200302.

CHAPTER TWELVE

Science Teacher Professional Development

The Blind Men and the Elephant

AIK-LING TAN

My discussion on what science teacher professional development is and its necessity in science education is based on the parable of the blind men and the elephant. This story originated from the Indian subcontinent and is about a group of blind men who touched different parts of an elephant and concluded differently what the elephant looked like. Based on this parable, I draw a parallel between science teacher professional development (the elephant) and how it is perceived and experienced by different stakeholders (the blind men). In this chapter, I make the point that because different people experience science teacher professional development differently, they also perceive the relative importance of science teacher professional development differently. Accordingly, it is often difficult not only to make conclusions about or argue for the need for professional development but also to identify just what it is about a program that affords teachers opportunities to move ahead professionally.

In many societies, the resource of time for teachers is highly limited and valued, since the responsibilities of teachers are demanding and varied—lesson preparation, counseling students, liaising with parents, working on various committees to organize school events, grade assignments, craft assessment tasks, and in some instances, particularly for science teachers, getting materials ready for laboratory sessions. The amount of time available for teacher professional development is also limited. Consequently, it is of utmost importance that the types and quality of professional development are impactful and of top-notch quality and relevance.

Based on this argument, engaging in professional development that does not directly help to improve or refine the practices of teachers will be seen as simply a waste of teachers' time, draining away funds that can potentially be used for hiring more manpower to reduce teachers' workload and freeing up time for teachers' planning and reflection. Viewing professional development as a necessary evil is tantamount to touching the rear end of an elephant and concluding that the elephant is a smelly animal, without acknowledging that defecation is necessary for the survival of the animal. But what if a person touches the trunk of an elephant? He/she probably will conclude that the elephant is likely to be tubular in shape, strong, and flexible. Analogously, professional development providers often market their programs using a structured, systematic and flexible approach to professional development. So here, we get a sense that although the "real elephant" is visible to all those who can see, the different blind men merely experience different parts of the elephant to form their conclusion. Similarly, while policy makers and administrators know the intended purposes and value of professional development, actual experiences of professional development are very much dependent on the needs of different stakeholders as they are all "touching different parts of that same elephant."

In order to make sense of teacher professional development, it is important for us to first examine the purposes of teacher professional development. Generally, it is widely understood that teacher professional development aims to: 1) increase teachers' knowledge of specific disciplines, 2) improve teachers' teaching skills (for example questioning skills, facilitation skills, etc.) in the classrooms, and 3) inform and refine teachers' practices in the classrooms according to various educational innovations and reforms (Fraser, Kennedy, Reid, & McKinney, 2007). All these purposes for professional development are based on an assumption that teachers' knowledge of their craft can be improved through engaging them in professional development.

This brings us to yet another question—what is teacher knowledge? Yew-Jin Lee (2016) argues that the construct of teacher knowledge is elusive and the tenets that make up teacher knowledge are hard to pinpoint. In fact, Lee highlighted the range of theories that attempted to help scholars make sense of teacher learning, and it ranged from:

> … pinpointing necessary certifications or personal psychological traits to a host of competencies or bodies of knowledge that enable one to be recognized as a successful teacher. Teacher effectiveness as a broad field has therefore evolved from searching from more atomistic, within-person attributes to examining excellence in professionalism from more holistic, person-in-context theories. (Lee, 2016, p. 71)

What is suggested here is the issue of how we evaluate or measure teacher learning or professional development. Do we focus on the change and what it means for an individual teacher or should we focus on positive gains systematically and collectively?

The effectiveness or impact of teacher professional development ultimately depends on its purposes. Given different situations and contexts, one can either zoom in to examine the transformation in teachers' ideas and practices, or zoom out to measure shifts in practices across the education system as a whole. In this chapter, I discuss the issue of "need" for science teacher professional development from the point of view of a science teacher, a student, and a parent. Key questions discussed include: (1) What are the perspectives about the need for science teacher professional development from different stakeholders? (2) What functions do science teacher professional development serve? and (3) What are some concerns that teachers have regarding their own professional development?

SCIENCE TEACHER PROFESSIONAL DEVELOPMENT FROM DIFFERENT STAKEHOLDERS

In this section I present three short narratives to illustrate that different stakeholders embrace multiple perspectives on science teachers' professional development. Pseudonyms are used for the stakeholders.

As a Science Teacher. Neil has been teaching elementary school science for the past seven years. He is a popular teacher who teaches in a school located in a neighborhood that is populated by residents from upper-middle class socioeconomic status.

> As an elementary school science teacher, I feel the need to engage in professional development so that I am able to use new strategies and methods to teach the same concept differently. With the infusion of technology and new ideas about how to make science teaching more effective and meaningful, it is important that as a teacher, I am cognizant of these innovations and adapt or adopt them in my classrooms. It will be ironic if I continue to teach the same way as I did ten years ago while the world outside the classroom changes. Science teachers need to move in tandem with the societal changes and demands. Nowadays, students have multiple platforms to obtain knowledge and they no longer merely rely on their textbooks and the teacher. For those reasons, I think professional development for me is important to keep me current with the latest teaching methods and also the extensive network of information out there. (Neil, October 2015)

As a Student Studying Science. Joyce is a Grade 4 student enrolled in a popular elementary school that is located in an upper-middle class neighborhood. She performs well in school examinations and is interested in school.

> I am in Grade 4 this year and have been studying science for the past two years. I think a knowledgeable science teacher is someone who is able to answer my questions immediately. I have had some inexperienced science teachers who will need a few days to come back to

> me with a response to my questions. As a student, the most important skill and knowledge that a science teacher needs to have is to be funny—the science teacher needs to be able to make the lesson fun so that the science concepts can be learnt well in an enjoyable manner. I do not even know that science teachers have to upgrade themselves professionally! My science teacher is the Head-of-Department for science and I think she knows a lot already. She does not need to go for professional development. For my friends and I, if our science teacher needs to be absent from class for some reasons, she will set some work for us. We are not bothered by her absence. (Joyce, October 2015)

As a Parent. Treia is a trained lawyer and a mother of two children who are attending elementary school. She is an involved parent who takes time to help her children with their schoolwork daily.

> As a parent of two children in elementary school, while I am concerned about the knowledge and skills of the science teacher who are teaching my children, I also assumed that as trained and certified teachers, they would be adequately equipped to teach my children. To be a good science teacher, I think the teachers need to have the ability to teach in an organized manner and also to realize how little a child actually knows. As a parent, I am not aware if the science teachers teaching my children upgrade themselves professionally. I am not very bothered by this because I feel that at the elementary school level, the change in basic knowledge of science (discoveries, etc.) should not be too drastic and thus frequent upgrading may not be necessary. Further, if the teaching skills and methods worked for previous cohorts of students, it should logically be sound and work for my children as well. I understand if teachers need to be away from school for their professional development, but the absence should not be prolonged. (Treia, October 2015)

SAME ISSUE, DIFFERENT PERSPECTIVES

I have chosen to use narratives to highlight my stance that professional development is perceived differently by different stakeholders in education because narratives are stories we tell from our personal experiences. Stories that we tell reveal how we make sense of the world and when we present our stories, it also helps others understand the existence of different interpretations of social life (Clandinin & Connelly, 2000). As Nicole Grimes (2013) wrote in her personal narrative of her identity as a Caribbean female teaching science in school, she justified her choice of using narratives by arguing that "the stories we craft describe our perceptions and our experiences, and in themselves are highly significant, as when examined closely, they provide us with information about our human culture" (p. 334). While the three short narratives from three different groups of stakeholders presented represent a narrow range of perspectives, I use them to frame my thesis that science teacher professional development is interpreted and perceived differently by stakeholders based on their lived experiences.

A Science Teacher's Perspective. From a science teacher's perspective, professional development is necessary because it helps them to stay current in their practices so that school science is relevant to students' lives. As reflected in Neil's narrative, it does not make sense for school science to lag behind the advancement in knowledge and technology that are the norms in the lives of his students. This leads to one of the key issues with which some science teachers (particularly elementary science teachers) constantly struggle: content knowledge. Keeping their content knowledge current and accurate is a struggle for elementary school science teachers, many of whom are generalist and not formally trained in any field of science (e.g. Kim & Tan, 2011). Further, teachers are generally altruistic in their profession, and they attend professional development activities to improve their professional skills and knowledge (Chang et al., 2014). However, when science teachers are engaged in professional development activities, they struggle with some tensions and dilemmas, similar to those faced by teachers of other subject domains. For example, they are torn between (1) attending professional development activities that they personally enjoy and find meaningful as opposed to attending professional development activities that are dictated by school management, and (2) deciding if their students' learning will be affected by their absence in the classroom when they are attending professional development activities (Tan, Chang, & Teng, 2015). Hence, from the perspective of teachers, professional development is a necessary endeavor within the teaching community for them to stay current in their practices but engagement in professional development activities requires them to make decisions between their personal needs as well as the needs of the school and their students. If science teachers' professional development is really necessary and a worthy activity, school leaders need to provide ongoing support for teacher change (Whitworth & Chiu, 2015). Support that schools can provide could include hiring substitute teachers for classes while teachers are away from school and explaining to students and parents about the need for teachers to be engaged in professional development activities.

A Student's Perspective. The reflection by Joyce represents blind men touching a different part of the elephant as compared to the reflection of Neil. Joyce presents a different and somewhat naive perspective of a student's understanding of teacher professional development based purely on her interaction with her teacher in her classroom. It is hence not surprising that Joyce is unaware of the fact that her science teachers engage in professional development activities to upgrade themselves. I would guess that her science teacher must be highly competent for her to comment that she does not require further professional enhancement. Unlike Neil, who emphasizes that the role of professional development is to enable science teachers to stay current in their knowledge and teaching methods of science, Joyce's idea of a good science teacher is one who is funny and is able to present concepts in an enjoyable and engaging manner. She is unaware (and hence probably it did not matter to her) of whether the strategies she experiences are progressive or traditional. Comparing the narratives of Neil and Joyce, the difficulties in assessing the impact of professional development is evident. Coming from a teacher's perspective, professional development ensures that teachers can use the latest teaching strategies to teach current and updated science concepts. Thus, for a teacher, the ability to incorporate current and progressive teaching strategies in class is one of the tenets of a good teacher and an outcome of participating in professional development activities. From a student's perspective, however, these strategies would be meaningful only if a teacher is: able to teach in a funny and lively manner; responsive to their questions; and engaging. Students' assessments of a good science lesson (and a good science teacher) may not always align with the intentions of teachers engaging in professional development activities. Although there are differences in how a student and teacher perceive the purpose of professional development, it is undeniable that students expect teachers to be competent in motivating and engaging them. These competencies can only be developed when teachers are continuously refreshing and renewing their knowledge and skill set through professional development.

A Parent's Perspective. From the perspective of parents, it is assumed that science teachers should have the necessary skills and knowledge to teach students. Treia's idea that elementary science teachers will not require as much professional development in science as compared with their counterparts teaching science at higher levels (i.e., higher than elementary). This perspective suggests that while parents are aware of different professional development needs of teachers, they embrace the fallacy that science teaching and learning in elementary school is less demanding than science teaching and learning in high school. Treia's narrative suggests that she views elementary science teachers' knowledge of how children learn to be more important than the content mastery of science concepts. Similar to Joyce's perspectives, Treia is not bothered if teachers are engaged in professional development activities as long as they are able to motivate students in their learning. Contrary to Neil's belief that acquiring progressive teaching strategies are important reasons for science teachers to engage in professional development activities, Treia is of the opinion that tried and tested strategies that have worked in the past should also work for her children. The relative merits of these two seemingly contradictory perspectives—(1) innovation in teaching strategies and (2) tried-and-tested methods for science teaching and learning—are worthy of further exploration.

DOES IT MATTER?

The question raised in this chapter relates to the differing perceptions of science teacher professional development among different stakeholders and consequently whether it matters if science teachers are engaged in professional development activities. This question challenges and problematizes the default mode of operation in many systems that place emphasis on teachers engaging in professional development activities to enable educational change and reform. In reality, we need to consider if professional development activities really matter and if they do, we need to identify the parties that they matter to (Zhang, Parker, Koehler, & Eberhardt, 2015). From the narratives and discussion above, engagement in professional development activities matters to teachers and probably school leaders as well since it affects their professional identity and self-efficacy that ultimately will have some impact on classroom practices (Palmer, 2006). However, teachers' engagement in professional development does not seem to be a priority for students and parents who are more concerned with the direct and positive interactions in the classrooms.

Currently, in educational systems across the world, many science teacher professional development activities are planned each year. In the Ministry of Education, educational bureaus and schools, funds are set aside each year to support professional development activities of teachers. In the National Institute of Education (Singapore) alone, more than 500 professional development courses are organized and attended by teachers each year. The flurry of activities suggests that professional development activities indeed make a difference to improving and changing the practices of science teachers. Since the interpretation and importance of science teacher professional development is different as viewed by different stakeholders, the way we measure impact of professional development should be multi-pronged. In this age of information technology, even in the absence of formal professional development programs, teachers have access to information on different platforms and learning can take different modes and forms. It is timely to examine the construct of professional development and define what counts as professional development activities. As researchers seeking to understand teacher professional development, we need to develop a different and tighter way to measure and describe teachers' engagement in professional development so that we can overcome the limitations and difficulties in evaluation of teachers' learning from professional development. What professional development is and means depends very much on how it is defined, who the stakeholders are, and the way that evaluation is carried out.

REFERENCES

Chang, C. H., Teng, P., Tan, A.-L., Joiko, S., Chan, Y. W., & Poh, M. S. (2014). Uncovering Singapore teachers' motivation to participate in professional development activities. *Literacy Information and Computer Education Journal, 5*(1), 1369–1375.

Clandinin, D. J., & Connelly, F. M. (2000). *Narrative inquiry: Experience and story in qualitative research.* San Francisco, CA: Jossey-Bass.

Fraser, C., Kennedy, A., Reid, L., & McKinney, S. (2007). Teachers' continuing professional development: Contested concepts, understandings and models. *Professional Development in Education, 33*, 153–169. doi:10.1080/13674580701292913.

Grimes, N. K. (2013). The nanny in the schoolhouse: The role of femme-Caribbean identity in attaining success in urban science classrooms. *Cultural Studies of Science Education, 8*, 333–353. doi:10.1007/s11422-012-9476-1.

Kim, M., & Tan, A.-L. (2011). Rethinking difficulties of teaching inquiry-based practical work: Stories from elementary pre-service teachers. *International Journal of Science Education, 33*, 465–486. doi:10.1080/09500691003639913.

Lee, Y.-J. (2016). Teacher practical knowledge: An ethnomethodological critique. *Curriculum Perspectives, 36*(1), 71–75.

Palmer, D. (2006). Sources of self-efficacy in a science methods course for primary teacher education students. *Research in Science Education, 36*, 395–409. doi:10.1007/s11165-005-9007-0.

Tan, A.-L., Chang, C.-H., & Teng, P. (2015). Tensions and dilemmas in teacher professional development. *Procedia – Social and Behavioral Sciences, 174*, 1583–1591. doi: 10.1016/j.sbspro.2015.01.808.

Whitworth, B. A., & Chiu, J. L. (2015). Professional development and teacher change: The missing leadership link. *Journal of Science Teacher Education, 26*, 121–137. doi:10.1007/s10972-014-9411-2.

Zhang, M., Parker, J., Koehler, M. J., & Eberhardt, J. (2015). Understanding inservice science teachers' needs for professional development. *Journal of Science Teacher Education, 26*, 471–496. doi:10.1007/s10972-015-9433-4.

SECTION SIX

(In)Equity and Science Education

In What Ways Does (In)equity Affect the Process of Science Education?

CHAPTER THIRTEEN

Science Education and Social Inequality

Reproduction or Confrontation?

FLAVIA REZENDE AND FERNANDA OSTERMANN

Social inequality in contemporary capitalist systems has recently gained more attention with the publication of Thomas Piketty's (2014) work, "*O Capital no Século XXI* (*Capital in the Twenty-First Century*)." Looking at the growth rate of the world's biggest patrimonies from 1987 to 2013, Piketty has shown that the world's largest wealth grew 6.7%, against 2.1% a year for the average global wealth and 1.4% a year for the average world income. Drawing on internationally reported data, Piketty points out that the unequal distribution of wealth in the first decade of this millennium is comparable in magnitude to that observed in European societies in the first decade of the 20th century. In France, for example 10% of the people had 90% of the country's wealth.

The situation is similar today, as Piketty shows that the richest 10% hold between 80% and 90% of the equity owned by the global population. He also says that it is an illusion to think that there are, in modern growth structure or in the laws of market economy, forces of convergence naturally leading to a reduction in inequality of wealth or to harmonious stabilization. Such a reduction would implicate the production and distribution of wealth by the government and building of a social state adapted to the 21st century, which would guarantee the right to education, health, and retirement.

Zygmunt Bauman (2015) has recently given attention to social inequality as well and reflected on some recurring ideologies in consumer societies around the world. One is the idea that "the chances of a decent, fulfilling and dignified life [...] is measured ostensibly by official data related to 'economic growth'" (p. 42). He shows that despite being widely publicized, this belief is not innate or "natural" for humans. In addition, Bauman brings ideas from economics experts to whom this process of constant economic growth would be only provisory until a stable stage is reached and when the quantity of produced goods could satisfy human needs. But what is observed, according to Bauman, is that after the last 60 years of capitalism in many countries, social inequality has been continually increasing at an incredible rate.

Social inequality is also supported by an ideology transmitted by the system, which advocates that the way to happiness is directly related to consumption and competitiveness (Bauman, 2015). Bauman shows that in practice, this message divides society into groups of winners and losers: winners are satisfied with their efforts and tend to regard good results as a fair reward for their innate or hard-won advantages. Losers may feel humiliated and ashamed of their poor performance, which they attribute to lack of talent and persistence. Losers consider the best way to survive is to take advantage of others, making higher grades than their neighbors and colleagues. Bauman believes that the consequence of this ideology is harmful to society because when the measurement of one's own success is guaranteed by the failure of the other, the process of repairing damage caused by inequality comes to result in more inequality.

Since that neoliberalism has spread all over the world in the last decades of the 20th century, social inequality has been accentuated. The effects of neoliberal and global capitalist agendas, the two arms of globalization (Bazzul, 2012), have been more harmful in the poorer countries. Kenneth Tobin (2011) has investigated the implications of applying neoliberalism to science education showing how global referents reproduce failure in a school in the U.S. But with some exceptions, the discussion on the impact of neoliberal reforms of education on teaching and learning has been "almost entirely absent from science education scholarship" (Carter, 2011, p. 2). Science education researchers have dedicated more attention to sociocultural changes in local adaptations of globalization affects, most of the times proposing science curricula to address issues as for example, diversity and cosmopolitanism. In the context of ecojustice and a framework of sustainability, research also has tackled the need of enhancing public understanding of science (Tobin, 2015) to educate people about science related challenges like global warming and wellness to reverse the damage to the stability in ecosystems and the "privileging of human lifestyles within frameworks that reify capitalism, neoliberalism, and competition" (Tobin, 2016, p. 31).

Our intention is not to approach social inequality as a background of curricular discussions, but to bring it to the foreground as an object of reflection, trying to present different stances from which it has been related to science education. We propose that one of the important aspects of this debate is the culturally constructed separation between natural sciences and society. We take this hypothesis to discuss why science education has remained uncritical in relation to social reality and inequality, submitted to ideologies that permeate the global consumer society, and therefore playing a political role that has mostly served the economic interests of governments. Though not always explicit, science education has been participating with this kind of position in political and educational contexts.

However, a sociocultural perspective that considers science and science education as human social activities and integrates the humanities as analytical tools and subject matter to be taught, opens a way for the issue of inequality to be addressed by research and by science education. In order to disseminate the appropriation of a sociocultural perspective as a framework to approach science education, Jay Lemke (2001) describes its main features and explains what it would mean to visualize our research objects as social activities. Lemke explains that taking this perspective as a theoretical framework first means asking questions about the role of social interaction in learning, seeing it as central and necessary and not as merely auxiliary. From the neo-Vygotskian perspectives of cognitive psychology and anthropological theory, science education and education in general are seen as a specialist socialization within a subcommunity, overcoming the opposite view of learning as an autonomous cognitive development in interaction with the real world.

Similarly, it means seeing the scientific study of the world as inseparable from the social organization of the scientists' activities, taking into account socio-interactional, organizational, sociological, biographical, historical, linguistic, semiotic and cultural as well as political, legal and economic aspects. James Wertsch (1991) also advocates an interdisciplinary perspective that expands to social sciences and humanities to establish fruitful analytical constructs to go beyond psychological scenarios.

In this chapter, we employ a sociocultural perspective when we bring together the contribution of education, sociology, epistemology and economy to reflect on the relationships between social inequality and science education. We also utilize the contribution of post-critical theories in our reflections, operating with hybridism between these theoretical frames. From post-critical studies based on Ernesto Laclau's (2011) discourse theory, politics is the exercise of decision that constitutes us as subjects, and this decision is always contingent and an option in an unforeseen set of possibilities. From this view, it is not possible to rely on fixed foundations; that is, meanings are constantly being contested, and we are continually trying to define certain concepts, such as justice and inequality.

Thus, we consider that the election of social inequality among many other possible issues as a concern for research in science education is only one singular way of constructing meaning in this research field. In the same way, an epistemological argument that we believe could be responsible for keeping social matters outside of science education is an attempt to understand ideologies that permeate education.

ISOLATION OF THE NATURAL SCIENCES AND SOME SOCIAL CONSEQUENCES

When we look at much of the work routine of scientists in the natural sciences, it is not difficult to see how they work in isolation from social complexity. This isolation was indirectly approached in a conference held in 1959, in which Charles Percy Snow (1995) coined the expression "two cultures" to point out differences between scientific and humanistic cultures. According to Snow, humanists do not know basic concepts of science, and scientists underestimate the psychological, sociological, and ethical variables of scientific production. Such cultural dichotomy is reflected in education at all levels, generating specialized professionals in one or at most two subcultures within one of the two mentioned cultures. Snow considered this separation "a serious threat to our creative, intellectual and, above all, for our everyday life" (p. 83), since educated people with the highest grades can no longer communicate with each other.

Such separation may have its origins in epistemological aspects of the natural sciences. When Thomas Kuhn (2005) describes the behavior that a scientific community assumes for choosing a problem to be investigated, he makes it clear that only the problems that have a possible solution in the paradigm will be considered as legitimate. At the moment the community elects these problems, the community relegates all the others to the categories of rejected or metaphysical; or begins to regard them as part of another discipline. Problems can also be rejected simply because they are deemed too problematic by the community. Following these laws, a paradigm can end up pushing the community away from "those relevant social problems that are not reducible to the form of a puzzle, as they cannot be stated in accordance with such instruments and concepts provided by the paradigm" (Kuhn, 2005, p. 60).

Kuhn (2005) thinks that social problems can constitute a distraction for scientists. If social problems are put aside, then normal science seems to progress rapidly because its practitioners can focus on problems that only their lack of competence could prevent them to solve. Kuhn points out that unlike engineers, physicians, and theologians, the natural scientist is not obliged to choose a problem because of its social importance or its need for urgent solution. That is exactly why they can be successful in solving the problems that they decide to investigate.

Gérard Fourez (1997) shows the isolation of the natural sciences when he portrays the school of scientific thinking that gives rise to the so-called "pure" science, in the early 19th century. According to the understanding of the time, pure science would be the one that was cleared of practical, cultural or social contexts and made use of a classical language, different from the one of the craftsmen. This science originated bodies of disciplinary knowledge structured around a well-defined rigor, strict teaching and laboratories practices inside their paradigms.

Norbert Elias (1994) draws attention to some of the consequences generated by this epistemological approach that separates science from society. Elias reminds us that the most studied features of human beings have been related to biological structure and function. The focus, then, is put on the individual organism, and little attention has been given to the fact that human beings are social beings. According to Elias, the study of the organism in isolation, the object of biological and medical sciences, has produced a linguistic and intellectual confusion that turns the singular human organism (the body) into a model of what we understand by an individual. Meanwhile, "the life in community, the society and its structures and processes are, in contrast, seen as not really natural and therefore not being actually real" (Elias, 1994, p. 159). Thus, the academic specialization has finally built a conceptual framework that postulates the opposition between nature and society.

One of the consequences of this view was its impact on the balance between personal-identity and social-identity, which has been changing since the European Middle Age. Elias (1994) points out that the overturn caused by Descartes on the human self-image from the moment he declared "Cogito, ergo sum" inverted the superposition of the social-identity over the personal-identity, which was common at the time, and shifted the social structure towards individualization. The perception of the self, devoid of the social identity, has been largely widespread since then.

This new process of autonomous thinking involved a transition between an "external" consciousness dependent on ecclesiastical authorities for an autonomous and "individual" consciousness. Elias (1994) shows how this new form of self-consciousness was linked to increasing commercialization, the formation of states, and the rise of the aristocratic and urban upper classes in Europe in the fifteenth, sixteenth and seventeenth centuries. For Elias, the Cartesian question turned out to modify not only the image of the physical world, in the light of the new methods created to investigate natural phenomena, but also the expression of human self-image.

Originating in this epistemological approach, which puts the individual as central in the act of knowledge, individualism as a way of living in society seems to be an important background to the discussion of social inequality and the way science education relates to or may come to relate with this issue. Individualism, an immediate consequence of the lack of a social identity, sounds like an alert against problems that threaten all of humanity, such as climate changes and social inequality among countries and regions of the planet.

Elias (1994) also describes two characteristics of human beings that can make a significant difference when facing global threats: our capacity for transmitting knowledge from one generation to another, this knowledge being changeable and capable of growing; and social life being able to be modified along with learning processes. If these same characteristics brought us to this stage, they may be malleable enough to produce other civilizing processes, providing us with hope to continue this discussion.

SCIENCE EDUCATION AS REPRODUCTION OF THE SYSTEM

Scientific traditions in the 19th century isolated the natural sciences from social issues and gave science a higher status that dominated secondary school and science undergraduate courses. Disciplinary sciences were considered "the most aristocratic and the ones that would give basis to all other kinds of knowledge" (Fourez, 1997, p. 52). Ignoring social complexity, pure scientists emerged like "pure technicians" who limited themselves to work within the frame of established subjects, as discussed by Kuhn (2005).

Science education inherits the isolation of natural sciences from society and other sciences. Lately, as long as science has been co-opted to legitimate policy decisions in our contemporary society, scientific knowledge has gained more authority (Lemke, 2011) as well as science education. This circumstance leads science education not to assume critical positions in relation to social and cultural processes and to be subject to conservative educational ideologies and government interests. Lemke (2006) points out that political interests focused on the formation of a scientifically prepared workforce have driven science education along the last decades. Lemke considers that this process has left most of the population aside, as it gives much emphasis to the acquisition of general and abstract principles, contributing to the alienation of science education from the issues related to social reality and from the worries faced everyday by students of different ages. Years earlier, Miguel Arroyo (1988) had already alerted us to the fact that the real victims of such a process would be the students. When students are building concepts about social life and beginning to understand the relationships between man and nature, they are often engaged in a way of thinking that is merely technical and formal, being deprived from a global comprehension of social, historical, political and cultural relations involved in the complexity of society.

Matthew Weinstein (2015) highlights neoliberalism as a damaging ideology that underpins much of science education in North America and its emphasis on STEM (Science, Technology, Engineering and Mathematics) Education. Neoliberalism, which can lead to policies that privilege markets and valorize entrepreneurial subjectivities, has driven education to a state of total reproduction of the system. Weinstein considers that neoliberal reform of education has accentuated competition and privatization and has reduced educational content to career considerations.

Jesse Bazzul (2012) also points out effects of neoliberal reforms on education, such as the significantly increased focus on standards, individualism and private funding options. Through neoliberal practices and discourses, schools that don't "perform" are being punished, as in the case of No Child Left Behind (NCLB) legislation and Race to the Top (RTTT) in the United States. At the same time, Bazzul calls attention to additional effects of neoliberalism ideology such as "underemphasizing, depoliticizing, and naturalizing the role of corporations, market logic, and private interests in science" (p. 1004).

Annette Gough (2015) discusses the relationship between STEM qualifications, industry demand and employment, highlighting an instrumentalist approach of science education in the STEM policy documents. Gouch observes that science education had already performed an instrumentalist role in the large-scale projects of the 1960s and 1970s in the U.S. and Australia. Within Brazilian context, science education is being used by the government to overcome an economic crisis, as can be interpreted from a news report published in 2015 in a major Brazilian newspaper. The text reported that a forum in which the Minister of Education was one of the speakers was carried out to discuss relations between education and the promotion of international competitiveness. One of the subjects discussed was "changing education so that education can change Brazil," which served as a basis to discuss the theme, "An impulse to modernize the infrastructure and strengthen Brazilian international competitiveness." The need for a specialized workforce was central to these discussions. Such ideas were reinforced by the Minister in an interview for the same newspaper when he stated that if schools wanted to face current challenges, they would have to make use of advanced technology and computerized systems and take part in the advance of what is called STEM: science, technology, engineering and mathematics, being all of those important to increase the economy's productiveness. The appropriation of the STEM initials by the Brazilian Minister fulfilled the purpose of making explicit the new relationship between science education and engineering that he envisaged as an important curricular component to be incorporated in the Brazilian curriculum. This proposal was likely seen as appropriate to catalyze the formation of a specialized workforce to accomplish Brazil's goals.

The need for professionals who can meet the demands of development of a more competitive and globalized market has already had an impact on teacher preparation in Brazil. Lately continuing education has been seen as a cheaper and more efficient way of educating teachers; relegating preservice teacher preparation to a lower government priority. On the other hand, improvement in the quality of education, in general measured by large-scale tests, has been attributed exclusively to teachers. Such an agenda, influenced by educational policies proposed by the Organization for Economic Cooperation and Development (OECD) and the World Bank (WB) to peripheral and central countries, has materialized in the form of increasing numbers of professional master's degree courses in education.

In Brazil, the number of professional master's courses in science education has surpassed the number of academic master's degree programs. These professional degrees emphasize teachers' mastery of scientific knowledge, which is considered more important to student learning than pedagogical knowledge. Scientific content as the core of the teacher education curriculum and the requirement that students develop instructional materials as a final dissertation suggest the intention of preparing science teachers as technical specialists. Such an orientation to continuing education provides few opportunities for teachers to consider critical issues such as relationships between education and society, curriculum policies, social purposes of high school and of science education, and social inequalities.

SCIENCE EDUCATION AS SPACE OF CONFRONTATION

By critically examining ideologies that are considered reproductive of the system, we describe some ways in which science education can play a confrontational role in the struggle against the unequal situation of humanity. Paulo Freire (1987) in the early 1970s, conceived a pedagogical method to bring emancipation to socially oppressed people in Brazil. At that time, social inequality and the large number of non-educated adults were already urgent issues in our country. Of course the dictatorial government did not invest on his revolutionary method. In the 1980s, Freire's idea of generative themes was used to develop science curriculum emphasizing its pedagogical aspects rather than the original political ones. In the 21st century, Lemke (2006) believes that to face conservative ideologies, science education should address the social issues faced by humanity, such as social justice. Lemke believes that science education should address moral and political attitudes, lest we be judged by the students, or even by history, as socially irresponsible.

The introduction of engineering into science education through STEM curriculum can be seen as a neoliberal curricular orientation when it takes a form of entrepreneurial problems for which students have to propose answers involving both global and local issues. However, these same activities can resist neoliberal discourses, when students for personal reasons, propose answers that demand social and racial justice (Weinstein, 2015). This counter position is in accordance with Fourez (1997), as he proposes two poles to classify science teaching: on one hand, we have humanist assumptions, such as the autonomy of the individuals and their possibilities of communicating and emancipating actions; and on the other hand, we have economic reasons that reside in the production of engineers and a qualified workforce. According to Fourez, science teachers will have to find their way between these two poles. He suggests that teachers analyze the ideological assumptions of the curriculum to help them reflect on the discourses they are ready to take on. Both resistance and critical ideas are aligned to what Henry Giroux (1994) had proposed in the original *Thirteen Questions* book, of which the present edition has been written in honor. Giroux defended that:

> Teachers and other cultural workers dedicated to reforming all spheres of educating as part of a wider revitalization of public life need to raise important questions regarding the relationship among knowledge and power, learning and possibility, social criticism and human dignity and how these might be understood in relation to rather than in isolation from those practices of domination, privilege, and resistance at work in the larger society. (p. 302)

Bazzul (2012) proposes that science educators can watch for all neoliberal practices and discourses that are given as natural in science education programs or science research agendas. Besides, Bazzul advocates "a broad and in-depth discussion concerning globalization, global capitalism, neoliberalism, and consumerism, where dialogue amongst science educators rises to a level of complexity that aids thoughtful praxis or academic inquiry" (p. 1005). He believes that this conceptual preparation is decisive to understand neoliberalism as an ideology that permeates science education.

But how do we prepare teachers for such a task, considering that science teacher education courses usually neglect the contribution of human sciences and the discussion of social reality? We believe that if science teachers had such academic experiences, they might come to:

> [...] understand the processes of human formation in their articulations with the social and productive life, the pedagogical theories and processes in a way they could be able to produce knowledge on education and interfere effectively in the broad or specific, institutional or not institutional pedagogical processes, based on a determined conception of society. (Kuenzer, 2011, free version, p. 684)

The ideas brought to this section so far are aligned with those of U.S. teachers and researchers who defend an agenda for social justice. According to Kenneth Zeichner (in Diniz-Pereira, 2008), this agenda emphasizes collective values, solidarity, and the power of transformation of society produced in the classrooms and teacher education programs. This group is engaged in helping future teachers understand the social and political implications of their actions and the contexts in which they work, as well as to understand how their everyday choices and/or decisions are connected to continuity axes (reproduction of the status quo) or to change axes (transformation and resistance).

We believe that the relationships between natural sciences and human sciences are a crucial feature of science teacher education that has received little attention from the community of teachers and researchers. The appropriation of human sciences discourses should not be seen as a minor detail in science teacher preparation, but as something material, that shapes the perspectives of the teachers regarding education and society in general. We had the opportunity to observe the impact of the absence of such an orientation in a meeting recently organized by the Brazilian Physical Society for a large number of secondary physics teachers. A member of the scientific committee disagreed with the suggestion that the event could examine the contributions of physics education to a less unequal society, stating that he could not see how physics education could be socially useful. Such an opinion seems to be constructed on an epistemological view of the natural sciences in total isolation from society. This example reinforces the idea that the analytical tools of human sciences can enrich science teacher education. In particular, these tools could help teachers to understand the relationships between science education and democracy and social inequality, and to analyze, for example, similarities between meritocracy in the educational processes and competition in social life.

Resisting these neoliberal ideologies would be in line with conceiving science education as equally distributed and used as a professional means not just for earning more and consuming more goods. This view is in accordance to François Flahault (in Bauman, 2015), for whom the society precedes the individuals and the search for happiness "should be focused on the promotion of experiences, institutions and other cultural and natural realities of life in community, and not on wealthy indexes that tend to reinforce the idea of human relationships as a place of competitiveness, rivalry and internal conflicts" (p. 70). With a larger knowledge base, not restricted to the knowledge of natural sciences, teachers could, for example, investigate how curriculum conceptualizes the individuals' roles in the structure of society, understanding how the curriculum addresses social issues and performs an ideological representation of science, of society, and of the world. In this investigation, it will be pivotal to understand how the issue of social inequality is being treated, if it is being accepted as a natural feature of the social system or as something that should be overcome.

A POST-CRITICAL META-ANALYSIS

In this paper, we first sought to discursively characterize a science education separated from the human sciences and from social problems, non-contesting, in defense of the neoliberal system and reproductive of social inequality. This approach also gave visibility to the flow of discourses between different countries, when we showed that STEM curriculum policy is being used in Brazil with the intention of reinforcing neoliberal purposes. This characterization of science education has moved us to focus on anti-establishment discourses and to describe an alternative close to a critical proposal of education, which we consider to be imperative in the fight for social justice.

The worldviews discursively constructed here that could be identified with the opposition between Market Principles/Enlightenment remain legitimate, like many others, even considering the frame of uncertainties and absence of ground, given by post-critical theories. In this framework, policy conceived as an intervention project aimed at a common goal to all contexts is eliminated, which means that there is no possibility of a planned and defined action that could represent the end of politics. However, political action as a struggle for the meaning of the different contextual projects carried out by each of us becomes even more important and is "able to bring a possibility of hope" (Lopes, 2013, p. 21). Thus, our singular perhaps unique way of characterizing these antagonistic projects represent opposite pathways that science education might assume in particular contexts and opposite ways of facing the problems of social inequality in the arena of constant and endless political struggle.

REFERENCES

Arroyo, M. (1988). A função social do ensino de ciências. *Em Aberto*, 7(40), 3–11.

Bauman, Z. (2015). *A riqueza de poucos beneficia todos nós?* Rio de Janeiro: Zahar.

Bazzul, J. (2012). Neoliberal ideology, global capitalism, and science education: Engaging the question of subjectivity. *Cultural Studies of Science Education, 7*, 1001–1020. doi:10.1007/s11422-012-9413-3.

Carter, L. (2011). Gathering in threads in the insensible global world: The wicked problem of globalisation and science education. *Cultural Studies of Science Education, 6*, 1–12. doi:10.1007/s11422-010-9307-1.

Diniz-Pereira, J. E. (2008). Formação docente nos Estados Unidos: Aliança conservadora e seus conflitos na atual reforma educacional norte-americana. *Educação & Sociedade, 29*, 233–252. doi:10.1590/S0101-73302008000100012.

Elias, N. (1994). *A sociedade dos indivíduos*. Rio de Janeiro: Zahar.

Fourez, G. (1997). *Alfabetización científica y tecnológica – Acerca de las finalidades de la enseñanza de las ciencias*. Buenos Aires: Ediciones Colihue.

Freire, P. (1987). *Pedagogia do oprimido*. Rio de Janeiro: Paz e Terra.

Giroux, H. A. (1994). Educational visions: What are schools for and what should we be doing in the name of education? In Joe Kincheloe & Shirley Steinberg (Eds.), *Thirteen questions: Reframing education's conversation* (pp. 295-303). New York, NY: Peter Lang.

Gough, A. (2015). STEM policy and science education: Scientistic curriculum and sociopolitical silences. *Cultural Studies of Science Education, 10*, 445–458. doi:10.1007/s11422-014-9590-3.

Kuenzer, A. Z. (2011). A formação de professores para o ensino médio: Velhos problemas, novos desafios. *Educação & Sociedade, 32*, 667–688.

Kuhn, T. (2005). *A estrutura das revoluções científicas*. São Paulo: Perspectiva.

Laclau, E. (2011). *Emancipação e diferença*. Rio de Janeiro: EDUERJ.

Lemke, J. L. (2001). Articulating communities: Sociocultural perspectives on science education. *Journal of Research on Science Teaching, 38*, 296–316. doi:10.1002/1098-2736(200103)38:3<296.

Lemke, J. L. (2006). Investigar para el futuro de la educación científica: Nuevas formas de aprender, nuevas formas de vivir. *Enseñanza de Las Ciencias, 24*(1), 5–12.

Lemke, J. L. (2011). The secret identity of science education: Masculine and politically conservative? *Cultural Studies of Science Education, 6*, 287–292. doi:10.1007/s11422-011-9326-6.

Lopes, A. C. (2013). Teorias pós-críticas, política e currículo. *Educação, Sociedade & Culturas, 39*, 7–23.

Piketty, T. (2014). *O capital no século XXI*. Rio de Janeiro: Intrínseca.

Snow, C. P. (1995). *As duas culturas e uma segunda leitura*. São Paulo: EDUSP.

Tobin, K. (2011). Global reproduction and transformation of science education. *Cultural Studies of Science Education, 6*, 127–142. doi:10.1007/s11422-010-9293-3.

Tobin, K. (2015). Science education in times of challenge | opportunity. In M. Mueller & D. J. Tippins (Eds.), *Ecojustice, citizen science and youth activism* (pp. 297–310). Dordrecht: Springer. doi:10.1007/978-3-319-11608-2_18.

Tobin, K. (2016) Collaborating on global priorities: Science education for everyone-any time and everywhere. *Cultural Studies of Science Education, 11*, 27–40. doi:10.1007/s11422-015-9708-2.

Weinstein, M. (2015, November). *NGSS, disposability, and the ambivalence of science in/under Neoliberalism*. Paper presented at the annual meeting of the 4S Conference, Denver, CO.

Wertsch, J. V. (1991). *Voces de la mente – Un enfoque sociocultural para el estudio de la acción mediada*. Madrid: Visor.

CHAPTER FOURTEEN

Dialectical Relationships and How They Shape (In)Equitable Science Learning Spaces and Places

MARIA VARELAS

> Students showed what they were learning by the questions they were asking. I didn't realize they were learning some of the things they were learning because I thought they were just fooling around in their small groups, but they were questioning each other and building on each other's ideas because this is what scientists do.
>
> —NEVEEN

> A scientist might be like if you can make something, you can make it good, then maybe you can make friends who tell you things and then you have more scientists to help you and work with their machines.
>
> —ELLIS

> [Before the Toxic Tour led by a community organizer] I didn't realize science could help facilitate student involvement in social justice.
>
> —BRITTNEY

These quotes—the first by a teacher of predominately Latino/a students in a neighborhood public school in Chicago where almost all of the students receive free or reduced lunch, the second by an African American student in a predominately African American neighborhood public school in Chicago where also almost all of the students experience poverty, and the third by an African American preservice teacher learning to teach in Chicago public schools—capture in intricate ways various dialectical relationships that are played out every day in science classrooms across the US, and the world, and shape the degree to which science classrooms are (in)equitable spaces and places for students, and especially for children and youth whose race, ethnicity, language, and class have historically marginalized them. Neeven's reflection on her classroom, her students, and whether they were learning science brings to light the dialectical relationships between wondering and knowing, between structured and unstructured conversations, between science students and scientists, between seeing students in old and new lenses. Ellis's reflection on himself as scientist makes us think about the dialectical relationships between individual and collective production of knowledge, between cognitive and social dimensions of learning, between following and leading, between product and process. Brittney's reflection on an experience she had as part of her teacher preparation program hints at the dialectical relationships between science as a field of study and social justice, between academic learning and community involvement, between abstract knowledge and real-life problems/inequities, between academic institutions and community/grassroots organizations.

SPACES AND PLACES

Every act of learning science, and any other subject, happens within spaces and places that are socioculturally and socio-politically produced with multiple dimensions—material, symbolic, cognitive, social, and emotional. Yi-Fu Tuan (1977) offers a distinction between space and place that I find compellingly relevant to the construct of equity regarding educational spaces in general, and science classrooms in particular, which are the focus of this chapter. Space is more abstract than place. In space, people move and act, feel free and unbounded. Time coexists with space defining each other; human beings experience the space and at the same time undergo "phases of tension and ease" (p. 118) that give them the sense of time. According to Tuan, "space is transformed into place as it acquires definition and meaning" (p. 136); when space feels familiar to people it becomes place. Place brings stability and organization, and a sense of place is developed as experiences in space acquire a unique sense of harmony, a rhythm that makes sense to the people living in that space, a feeling of safety. When classrooms become both places *and* spaces where children and youth experience "a dialectical movement between shelter and venture, attachment and freedom" (p. 54), learning becomes possible.

This is a critical aspect of the equity challenge that many students, and especially students from historically excluded-from-science groups, experience in science classrooms today. Being allowed to be free, being given the room to be creative, being granted the license to move in a science space beyond the confines of a meticulously designed lesson and the standards that it addresses as, at the same time, being helped to feel a part of a community, a place where relationships, ideas, feelings, habits of mind, and ways of doing things define a safe environment to live, are *both* needed if science classes are to become equitable learning hubs. Ellis in the quote at the beginning of the chapter needed "friends" around him who could add to his knowledge, and facilitate his involvement with the "machines" that he imagined he needed to do science. Such needed friendships could be formed not because Ellis likes other people or others like him, but because he has common goals with them and because there is a level of competence that others see in him—a Black boy, competence that he self-assessed as high because he has already made something "good." As spaces are transformed into places, issues of competence and how it is perceived and conceived, interactions among people and things, and a sense of progress and accomplishment play an important role in how science is experienced, especially for students whose race, ethnolinguistic background, and class have shut them out of meaningful, relevant, and just opportunities to learn.

DIALECTICAL PEDAGOGY

Being free and sheltered, along with many other dialectical relationships, shape what opportunities individual students have to learn in classrooms in the midst of other classroom members, their teacher and their peers. The dialectical pedagogy accepts contradiction as part of reality. According to Moacir Gadotti (1996):

> Contradiction is the essence of the fundamental law of dialectics ... The contradictory elements coexist in a structured reality, one not being able to exist without the other. The existence of the opposites is not a logical absurdity. It is founded on reality. (pp. 19–20)

Opposition of elements is at the center of every educational setting, including the science education classroom. At the crux of (in)equity lies how well the teacher handles authority and freedom, reproduction and transformation/creation, stability and change, order and disorder, in order to capitalize on both contradictory elements to develop learning spaces and places. As Henri Lefèbvre (1975) noted, "The dialectic method attempts to capture the connection, unity and movement that engender, oppose and make the contradictory elements clash, that break them or surpass them" (p. 238). In science, as in other fields of study, a vast range of knowledge already exists, knowledge that has been constructed by people who have been part of the professional communities that generated this knowledge. The authoritarian characteristics of the existing knowledge are undeniable; however, knowledge does not have much meaning without being juxtaposed to the thoughts that have yet to be created and the knowledge that has yet to be developed by the kinds of people who have been excluded from these professional communities. Students presently in science classrooms need to be given the chance to not only create meaningful understandings of what is known, but also exercise their freedom, and the responsibility, to imbue these understandings with their own experiences, values, beliefs, and ideologies in search of new science knowledge. As Paulo Freire (2000) eloquently captured this dialectic: "There is no freedom without authority, but there is also no authority without freedom" (p. 178).

For dialectical relationships—relationships between seemingly contradictory elements—to be embraced and balanced in order to achieve equity in science classrooms, the deficit paradigm of difference need to be relinquished. Difference needs to be constructed as an asset so that we "learn from the difference. [This] means not to consider ourselves better than others precisely because they are different from us" (Freire, 1995, p. 21). As Gadotti (1996) affirmed, "Only equality in difference and partnership are able to create the new" (p. 144), the knowledge, and the necessary social relationships that nurture the development of knowledge. Difference is implicated in science learning that I define in an expansive way as both science content learning and identity construction (Varelas, Martin, & Kane, 2012). Students who do not fit teachers' and peers' expectations and norms are perceived as troublemakers, or academically low, or as students who need to be remediated and/or closely watched. For such cases, the dialectic between individual and collective is evoked only to privilege the collective, denying, consciously or unconsciously, individual students opportunities to participate and learn so that the class's learning may not be inhibited. The irony, of course, is that the collective is constituted not only by the students who have similar and conforming ways of being relative to the existing structures, but also by those individuals whose race, ethnicity, language, gender, sexual orientation, class, (dis)ability, and other social markers accentuate their difference and compromise their belonging to the classroom community. Such students do not have either the space or the place to perceive, conceive, and live in the science world, making the science class an inequitable, oppressive setting. In the first quote at the beginning of the paper, Neveen had to consciously work against the tendency she and other teachers in the school had to consider students' (some students') attempts to actually embody science practices as "fooling around," and work towards honoring both their ways of engagement with school and with science and the importance of the less mainstream conceptions of what this engagement ought to look like.

STRUCTURE AND AGENCY

Difference is also implicated in another critical dialectic, the dialectic of structure and agency that shapes science teaching, learning, leadership, and teacher education as several science education scholars have explicated in a recent special issue of the *Journal of Research in Science Teaching* (Varelas, Settlage, & Mensah, 2015). Material, symbolic, social, cultural, and political structures permeate every space and place where students interact with science. Using Anthony Giddens' (1979, 1984) framework, William Sewell (1992) articulated that structures are themselves dialectical relationships between schemas (e.g., rules and principles), which are virtual, and resources, which are actual. Structures are the structuring properties of social systems, the schemas and resources that are implicated in the reproduction of social systems. Structures have power as they regulate people's actions within settings, but they continue to exist only if people sustain them by their actions. At the same time, structures are the result of human and non-human agency. As Giddens (1984) noted, "human beings are purposive actors, who virtually all the time know what they are doing (under some description) and why" (p. 258). But human agency emerges within sociocultural practices and is constrained by their characteristics. Agency "is not a matter of 'pure will' or absolute freedom" (Hays, 1994, p. 64), but rather the capacity to engage with structures in ways that did not exist before, influenced by the nature of the structures. And although agency is inherent in all people, it is shaped by one's sociocultural and socio-political milieu.

(In)equity in the science classroom lies within the intricate relationship between structure and agency. When all sorts of structures, from textbooks, to lab space, to racial, ethno-linguistic, and gender diversity in scientific communities, to interactional patterns, to instructional approaches, to assessment rituals, are stacked against students who have been left out of science, an intentional effort to create spaces and places where these students' agency is not only allowed to surface but also explicitly nurtured is imperative, if equity is to be restored. As Sewell (1992) wrote: "The same resourceful agency that sustains the reproduction of structures also makes possible their transformation by means of transpositions of schemas and remobilizations of resources that make the new structures recognizable as transformations of the old" (p. 27). In an inequitable system, it takes intentional work towards mobilizing students' agency to not only transform structures which have been denying them access to science, but also to explore ways in which science may contribute to the transformation of their lives outside of school.

PEDAGOGY OF LOVE

Acknowledging the dialectic of structure and agency and recognizing the ways in which this dialectic may be played out in science classrooms is paramount if the goal is to construct equitable and just science classrooms. To support students' agency so that they not just learn to function within existing structures, but also make these structures their own, changing them to fit their families' and their communities' forms of meaning making, creativity, curiosity, imagination, and aspirations, Freire's pedagogy of love (Darder, 2002) is a necessary mindset and a possibility. According to Antonia Darder (2015), the pedagogy of love includes "faith and dignity in our relationships with others, social responsibility for our world, participation in the construction of knowledge, and solidarity across our differences" (p. 51).

Understanding difference in a way that moves teacher and students away from opposition and closer to embracing a multiplicity of perspectives, experiences, ideas, and solutions can be achieved when there is mutual love not for each other but for the very *process of being with each other* in the classroom. For Freire, love is closely associated with respect, trust, hope, faith, patience, and humility—ideas that are scattered in several of his writings. Love cannot exist without caring and respect for others, and it is necessary for dialogue to take place. Dialogue, according to Freire—so critical for equity—is an act of love, where respect for each participant's ideas, gratitude for the opportunity to gain new insights from others, and fairness towards all people's positions in the world necessitate love but also create it. For Freire, who explored and denounced oppression of all sorts, one of education's goals is "the creation of a world in which it will be easier to love" (1970, p. 24), and education itself "is an act of love" (1973, p. 38).

SCIENCE AND SOCIAL JUSTICE

As Darder reminds us, love for Freire is associated with social responsibility for our world, the call to advance humanity. For science education, this involves attention to yet another dialectic, the dialectic between inducting students into science as a practice—a practice constructed by those who have been given and/or seized opportunities to become a part of the scientific community, and identify with its goals, norms, processes, products, and ways of being—and inducting students into science as a tool for social justice. There are varied interpretations of social justice that are used or assumed among educators and education researchers, and of what it means relative to science, but the one I would like to focus on here has to do with understanding and changing some of the conditions of students' lives and the lives of their communities. Freire's (1970) problem-posing approach, where teachers and students co-investigate reality through explorations of "generative themes" that are based on students' immediate concerns, realities, and lives, allows for attending to that construal of social justice. Such an approach is built on the premise of cultivating strong relationships between schools and communities, valuing local knowledge, along with global understandings, and manifesting yet another dialectic that Freire (1970) identified, the dialectic of objectivity and subjectivity: "one cannot conceive of objectivity without subjectivity" (p. 50).

For the institution of Western science, social justice is not one of the primary goals for which science is done. Moreover, given that science is implicated in eco-apartheid, environmental racism, and health injustices to name a few, many see attention to disciplinary core ideas, cross-cutting concepts, and scientific practices, specified in the new framework for K-12 science education (National Research Council, 2012), as antithetical to challenging such injustices and promoting equity and social change. Alberto Rodriguez (2015) articulated how the new framework and associated standards fall short of directly addressing issues of equity and justice beyond the rhetoric of science for all. For me, one of the complex dialectics that science educators have to attend to is the tension between helping students understand and being able to use science concepts and practices to solve problems motivated by the state of the field, and at the same time helping students interrogate the very nature of science, construct its political nature, and use science to transform communities that have been historically marginalized. Ignoring either is irresponsible and may perpetuate inequitable science classrooms. Examples of how to handle the dialectic are few, yet powerful, both in the science realm and in science classrooms. Jason Corburn (2005) showed how local insights combined with professional scientists' knowledge and techniques can improve both scientific inquiry and the health of communities. Daniel Morales-Doyle (2015; in press) offered an approach, "justice-centered science pedagogy," which makes "issues that reside at the intersection of community concerns and canonical science understandings" (p. 278) the catalyst for teaching and learning science, grounding it on evidence from an advanced placement chemistry class at an urban, neighborhood high school of working class students of color. Bringing such approaches to science teacher education may be one way to continue to push forward a science education equity agenda. For Brittney in the third quote at the beginning of the chapter, it took a structure (with particular physical, semiotic, cognitive, and socio-political dimensions)—a walking tour of a neighborhood suffering from, but also resisting, environmental racism and hearing from a community activist the history of the struggle to restore and protect the environment and the people's health—to reconsider the antithetical relationship she had formulated between science education and social justice (Varelas et al., in press).

DIALECTICS AND EQUITY

Moving towards polarities in dialectics, which exist in, and shape, educational settings including classrooms and the work that is done there, reinforces rather than weakens inequities and injustices. What is needed is acknowledging that polarities are necessary rather than inevitable and intentionally working towards their intertwinement. It is necessary that science classrooms are both spaces and places for students to experience science and use science to understand and interrogate the world (physical, sociocultural, and socio-political), not necessarily becoming scientists unless they choose so from among other possibilities. It is necessary that students engage with and know both what science has been achieved and what science is needed for the future of healthy and just science communities. It is necessary that teacher and student agency take into account structures that already exist, and use them in ways that do not unquestionably reproduce or strengthen them but in ways that examine them and transform them via an equity lens. It is necessary to strengthen individual and collective trajectories of success and growth. It is necessary to attend to both polarities in dialectical relationships and benefit from their tension, not eliminate it. Espousing a pedagogy of dialectics, difference, and love may enable science educators, along with their students, to balance centripetal and centrifugal forces, both centralizing and decentralizing, unifying and disunifying science as a practice in the world, in a way similar to that of Mikhail Bakhtin's (1981) conception of language as a heteroglot conception of the world, viewing one perspective from the perspective of another. Such balance may allow students from historically marginalized groups whose underrepresentation in science and challenges experienced in science education have been severe and longstanding to not only be offered opportunities for learning and participation but to also be accepted and recognized for who they are and who they want to become.

Author's Note: Ideas and data presented in this chapter have emerged in research supported by two US National Science Foundation grants (REC-0411593 and DUE-1439761) and a Polk Bros. Foundation grant to the author and colleagues.

REFERENCES

Bakhtin, M. M. (1981). Discourse in the novel. In M. Holquist (Ed.), *The dialogic imagination: Four essays by M. M. Bakhtin* (C. Emerson & M. Holquist, Trans.). Austin, TX: University of Texas Press.

Corburn, J. (2005). *Street science: Community knowledge and environmental health justice*. Cambridge, MA: The MIT Press.

Darder, A. (2002). *Reinventing Paulo Freire: A pedagogy of love*. Boulder, CO: Westview.

Darder, A. (2015). *Freire and education*. New York, NY: Routledge.

Freire, P. (1970). *Pedagogy of the oppressed*. New York, NY: Continuum.

Freire, P. (1973). *Education for critical consciousness*. New York, NY: Seabury Press.

Freire, P. (1995). The progressive teacher. In M. de Figueiredo-Cowen & D. Gastaldo (Eds.), *Paulo Freire at the Institute* (pp. 17–24). London: Institute of Education.

Freire, P. (2000). *Pedagogy of the oppressed* (30th anniversary ed.). New York, NY: Continuum.

Gadotti, M. (1996). *Pedagogy of praxis: A dialectical philosophy of education*. Albany, NY: State University of New York Press.

Giddens, A. (1979). *Central problems in social theory: Action, structure and contradiction in social analysis.* Berkeley & Los Angeles, CA: University of California Press.

Giddens, A. (1984). *The constitution of society: Outline of the theory of structuration.* Berkeley & Los Angeles, CA: University of California Press.

Hays, S. (1994). Structure and agency and the sticky problem of culture. *Sociological Theory, 12*(1), 57–72. doi:10.2307/202035.

Lefèbvre, H. (1975). *Formal logic, dialogical logic.* Rio Janeiro: Civilização Brasileira.

Morales-Doyle, D. (2015). *Science education as a catalyst for social change? Justice-centered pedagogy in high school chemistry* (Unpublished doctoral dissertation). University of Illinois at Chicago, Chicago, IL.

Morales-Doyle, D. (in press). Justice-centered science pedagogy: A catalyst for academic achievement and social transformation. *Science Education.*

National Research Council. (2012). *A framework for K-12 science education: Practices, crosscutting concepts, and core ideas.* Washington, DC: National Academy Press.

Rodriguez, A. J. (2015). What about a dimension of engagement, equity, and diversity practices? A critique of the next generation science standards. *Journal of Research in Science Teaching, 52*, 1031–1051. doi:10.1002/tea.21232.

Sewell, Jr., W. H. (1992). A theory of structure: Duality, agency, and transformation. *American Journal of Sociology, 98*, 1–29. doi:10.1086/229967.

Tuan, Y.-F. (1977). *Space and place: The perspective of experience.* Minneapolis, MN: University of Minnesota Press.

Varelas, M., Martin, D. B., & Kane, J. M. (2012). Content learning and identity construction: A framework to strengthen African American students' mathematics and science learning in urban elementary schools. *Human Development, 35*, 319–339. doi:10.1159/000345324.

Varelas, M., Morales-Doyle, D., Raza, S., Segura, D., Canales, K., & Mitchener, C. (in press). Community organizations' programming and the development of community science teachers. *Science Education.*

Varelas, M., Settlage, J., & Mensah, F. M. (2015). Explorations of the structure–agency dialectic as a tool for framing equity in science education. *Journal of Research in Science Teaching, 52*, 439–447. doi:10.1002/tea.21230.

CHAPTER FIFTEEN

Race and Poverty in Science Education

Questions and Tensions for the Field

MAXINE McKINNEY DE ROYSTON AND ABIOLA A. FARINDE

Despite decades of reform focused on scientific literacy *for all* (e.g., American Association for the Advancement of Science [AAAS], 1989; National Research Council [NRC], 2012) the U.S. continues to struggle with racial and economic disparities in science education and the STEM fields (Barton, 2001; Morgan, Farkas, Hillemeier, & Maczuga, 2016). In fact, while United States students taken as a whole demonstrate poor performance, disaggregated TIMSS data actually reveals a bimodal distribution of "both super and dismal achievement" (Barton, 2001, p. 904) where White students as a group outperform most nations while Black and Latino students are outperformed by nearly every nation (Berliner, 2001). Such racialized disparities legitimate critiques of "science for all" reforms as effectively "science for some" because reforms continue to rely upon student-level change rather than socio-political issues of access to science education, the narrow presentation of science as an objective and culture-free discipline, and reliance upon students continue to persist (Emdin & Lee, 2012; Varelas, Kane, Tucker-Raymond, & Pappas, 2012). Indeed, what we have come to call "science" in K-12 schools can more precisely be defined as the science that originated in the dominant "Western" world (Lee & Buxton, 2010).

We use the terms "dominant" and "non-dominant" to acknowledge the dynamics of power that operate to differentially structure and privilege certain individuals and groups' values, lived realities, and access to material, economic, social and political resources. For example, at present science education in the U.S. reinforces and privileges Western, White, male, middle-class views, practices, and identities relative to science and marginalizes those of others (Buck, Clark, Leslie-Pelecky, Lu, & Cerda-Lizarraga, 2008; Lee & Buxton, 2010). Not surprisingly, those same demographics of students continue to benefit from science education reforms and perform well in science as opposed to the racially, economically, and gender non-dominant students whose access to and performance in science initially prompted reforms (Barton, 1998).

The pervasive nature of racial disparities are evermore problematic as performance gaps in science for K-12 White girls and White boys has nearly closed, yet the scores of racially non-dominant female and male students remains depressed Aud, Hussar, Kena, Bianco, Frohlich, Kemp, & Tahan (2011). Such disparities highlight the perniciousness of race, but also highlight its intersections with poverty in ways that extend beyond categorical distinctions or static variables to considerations of how race and poverty operate as interconnected sociopolitical realities (Milner, 2013, 2015) that influence how students are positioned relative to and negotiate scientific ways of knowing, doing, and being within K-12 science education learning settings (Mensah, 2013; Warren & Roseberry, 2011). Indeed, such a focus is necessary to challenge the science for all rhetoric to live up to its inclusionary potential.

Nonetheless, science education research and teacher preparation efforts frequently focus upon best teaching practices, innovative curriculum and standards, pedagogical content knowledge and conceptualizations of what students are able to do. While important, these foci do not necessarily respond to the shifting racial and economic demographics and realities of current students nor do they disrupt inequalities relative to who has access to high quality science, who is positioned to envision themselves as scientists, and who ultimately succeeds in science (Bryan & Atwater, 2002; Kelly-Jackson & Jackson, 2011).

In this chapter, we focus on two socio-political dynamics that are increasingly salient for thinking about non-dominant students' participation, interest, and success within science: race and socioeconomic status (SES). We examine the questions: *How do race and socioeconomic status affect the field of science education? In what ways do they intersect and overlap?* And, finally, *how do these intersections affect the teaching and learning of science?* We explore a set of lingering tensions within the field and argue that across these tensions there is an undertheorization of race and SES such that they are oft treated as static, disembodied demographic variables (i.e., Black students, lower-income students, urban students) and indicators that do not reflect the lived realities of students. Specifically, we shift the discourse away from SES towards considering poverty and towards considering the intersections of race and poverty (Milner, 2015). We then offer some tools for conceptualizing race and poverty as overlapping sociopolitical and lived realities that fundamentally influence the teaching and learning of science We end with proposing politically relevant science instruction—or a pedagogy that is enacted through a teacher's political clarity about systemic forces of oppression and their student's experiences within that system—as a way to rethink teaching and learning in science, especially for racially and economically non-dominant students.

LINGERING TENSIONS IN SCIENCE EDUCATION

Below we outline three broad tensions in science education that reflect the continued salience for considering how race and socioeconomic status influence science teaching and learning, namely: lack of resources to learn science in urban areas; underrepresentation of racially and economically non-dominant persons in science education; and failure to prepare teachers for 21st century students. These tensions were distilled through reviewing science education literature over the last 30 years that reference concerns or developments in addressing issues of equity, race, and/or socioeconomic status in science education research and teaching. To be clear, this review is not intended to be exhaustive, but to highlight on-going tensions that remain unresolved despite reforms. Within each tension we point out specific issues of concern.

Lack of Resources to Learn Science in "Urban" Areas. The lack of resources with which to meaningfully engage in science continues to affect which students experience, become interested in, and are able to identify with and further pursue science. This is because the major vehicle through which many students, particularly those who can not afford out-of-school and private science learning resources (such as museums, camps, etc.), have access to science learning is through resegregated schools that are highly inequitable in their science course offerings and related resources (Tan, Barton, Turner, & Gutiérrez, 2012).

Over the last three decades, schools have resegregated by race and socioeconomic status (Darling-Hammond, 2010). At present, Black and Latino students largely attend schools with other racially and economically non-dominant students, and Latino/a students also often attend linguistically segregated schools (Orfield & Gordon, 2001). (Re)segregation tends to focus on where non-White people attend school, yet White, middle and upper class students are the most likely to attend racially homogeneous schools and to have the most access to high quality science education.

By contrast, schools serving racially and economically non-dominant students have fewer and less extensive STEM courses and resources and typically offer more low-track courses (Tan et al., 2012). In these schools, non-dominant students often encounter outdated science textbooks and resources, procedural or less rigorous science curricula and activities, and inexperienced or under-qualified science teachers. For elementary students, particularly the early grades, this may mean no access to science at all (Rodriguez, 2015). Local scenarios exist in which non-dominant students with high-interest in science may have less access to science than dominant students with low interest in science just by virtue of where each of these students is likely to go to school.

Patterns of resegregation also result in racially and economically non-dominant students being more likely to attend schools in "urban" areas that have higher rates of family and neighborhood poverty (Frankenberg & Lee, 2002). Angela Calabrese Barton's (2007) review of the research around achievement patterns in science points out that the terms "urban," "urban poverty," and "urban minority" are used synonymously and fail to appreciate the complexities and distinctions between the lived realities of the individuals encompassed within each of those groupings.

Similarly, Richard Milner (2012) argues that the term "urban" does not simply describe the density or reality within cities and around schools, but is used to convey *who* is inside of schools and how outside of school factors—such as poverty, neighborhood violence—and the like impact what happens within schools. This use of "urban" as a proxy for racially and economically non-dominant students and their experiences is rife with assumptions about who they and their communities are and what they value and are able to do. Instead, we must consider the racialized and classed geographies of opportunity in urban areas and the associated implications for non-dominant students' science learning experiences, and subsequently their attitudes and identities relative to science (Barton, 2007).

Underrepresentation of People of Color in Science. Given on-going disparities in opportunities to learn science it not surprising, albeit still alarming, that racially and economically non-dominant people are either absent from or disproportionately underrepresented in undergraduate and graduate science courses and science careers (Riegle-Crumb & King, 2010). Xueli Wang's (2013) study linked such a lack of opportunity to engage in science to many undergraduate students' intentions to major in STEM. Wang found that undergraduate students' major and career intentions are shaped by their high school achievement patterns in STEM as well as by their STEM self-efficacy beliefs and interests that are developed from their earlier experiences with math and science. In fact, early science experiences influence STEM participation and interest just as decisions to participate in STEM are often made early, frequently prior to high school (Young, 2005). Overall, racially and economically non-dominant students' participation, interest, and broader science learning pathways (Bricker & Bell, 2014) are largely shaped by the quantity, quality, and duration of non-dominant youth's exposure to science learning opportunities and resources.

The quality and nature of racially and economically non-dominant students' science learning experiences is a critical issue. On the whole, science learning environments in and out of school are ideologically and demographically dominated by White, male learners and mentors or educators (Riegle Crumb, Moore, & Ramos-Wada, 2011). For example, most science learning environments offer few to no opportunities for racially, economically, and gender non-dominant students to see or interact with adult counterparts, be they scientists, engineers, or science teachers, whose racial, economic, and/or gender life experiences and struggles mirror their own (Buck et al., 2008). This can make it difficult for such students to "see" themselves as scientists, develop a sense of belonging within science, or manage microaggressions they might experience in science learning contexts (Brown et al., 2015). Without older peers and particularly adults who can serve as culturally reflective authority figures and developmental partners, science learning environments and the domain of science will continue to be a space and context that can be isolating and disheartening for racially and economically non-dominant students who may often find themselves to be "the only one."

The demographics within science learning environments may also be reflective of how science is viewed and presented in U.S. schools as an objective discipline that is free of the racial, cultural, and gender biases of other fields (Carter, Larke, Singleton-Taylor, & Santos, 2003; Emdin & Lee, 2012). This façade of neutrality masks the historicity of Western science as a discipline concerned with asking certain questions and publicizing certain findings rather than others (Kumashiro, 2001). This plays out through how teachers privilege "Western" or dominant ways of being, knowing, and doing science (Harding, 1994) that do not align with racially and economically non-dominant students' prior experiences or ideas about natural phenomena. This privileging may alienate non-dominant students from participating and developing interests in science (Atwater, 2000).

Learning-in practice perspectives argue that learning consists of shifts in one's participation (i.e., practices) and that these processes are intertwined with how one's continuing participation in a practice frequently becomes integral to who one is (Nasir & Hand, 2008). Issues of identity, such as who is perceived as able to do science (Buck et al., 2008) and what it means to be a legitimate participant in a science learning community (Barton, 2007), thus remains a tension for racially and economically non-dominant students. For example, Nancy Brickhouse and Jennifer Potter's (2001) study suggests that racially non-dominant girls' marginalized experiences in their science classrooms made their identification with and participation in school science learning communities either not possible or not welcomed.

Likewise, Tammie Visintainer's (2015) examination of racially, and often economically, non-dominant high school students' perceptions of fit and identification with science indicates that barriers to their participation and interest are the narrow racial and practice-level representations that get conveyed to and are taken up by students around what counts as science and who can do or get to be a science person. As one of her participants shares:

> I think of them [scientists] in white gowns with clipboards and really high tech equipment and computers, and like people with glasses on and goggles and stuff… they're like labeled *scientist* and it makes them smarter then everyone because they're scientists … I've never heard of a professional Filipino scientist. Maybe it's just a lot of the other races don't feel comfortable. Here, it's a lot of low-income or working class people and they probably don't feel like they're good enough to become a scientist, cause it's like high. (Melanie)

Visintainer demonstrates how Melanie perceives the power and status she associates with scientists and how this student grapples with a visual, racialized image of who scientists are that contrasts with who students from racially and economically non-dominant communities might see themselves to be.

Part of non-dominant students' struggles to participate and become or remain interested in science may be the lack of opportunities to develop a sense of belonging within science learning communities and to identify with the practices of scientific communities. Studies by Visintainer and others document how many non-dominant students of color are socialized into a limited or a stereotypical understanding of STEM professions and professionals (Riegle-Crumb et al., 2010) and consequently perceive STEM skills and characteristics as misaligning with their own identities (Varelas, Kane, & Wylie, 2011). This conflict with the other identities they hold may also make it challenging for non-dominant students to participate and remain interested in science (Brown et al., 2015). These difficulties reflect the "thickened" (Wortham, 2004, p. 175) identities and stereotypes that are conveyed to racially and economically non-dominant youth in and outside of science classrooms that influence how they perceive themselves and are perceived by one another in relation to science.

The lack of opportunities for non-dominant students to see themselves within and identify with science is both influenced by a non-diverse science teacher workforce and has implications for the future development of a racially and economically diverse science teacher workforce. While racially non-dominant teachers are underrepresented in U.S. public schools (Ingersoll & May, 2011; Rogers-Ard, Knaus, Epstein, & Mayfield, 2013), there is also a shortage of STEM teachers (Hutchison, 2012) particularly racially non-dominant STEM teachers. There is a relatively high attrition rate among STEM teachers in general and this is exacerbated in hard-to-staff, lower-income urban schools (Ingersoll & Perda, 2010). As with other STEM careers, the lack of diversity in the teacher workforce is partly predicated by the limited number of racially and economically non-dominant students that have access to early STEM learning opportunities and become interested in and motivated to pursue further study or career options in STEM.

The lack of a diverse pool of science teachers is also a consequence of past desegregation mandates. We use the case of Black teachers as an example. In the aftermath of the landmark Supreme Court Case *Brown v. Board of Education* (1954), a substantial number of Black teachers lost employment (Tillman, 2004) as Black students were bussed to distant school districts and integrated into majority White schools (Walker, 1996). The court ordered integration of Black students into predominately White schools resulted in the merging of schools. On the whole, the decision was to transfer and bus Black students to White schools because White schools were considered superior. Such efforts led not only to the firing and displacement of countless Black educators and the closing of Black operated schools, but also to the pervasive, and lingering stereotypes about the inferiority of Black schools and Black educators, in particular (Madkins, 2011).

Today, most students are taught by a predominately White, female, middle-class, monolinguistic teaching workforce (Landsman & Lewis, 2011) despite research suggesting that a more diverse teaching workforce has the potential "to improve the academic outcomes and school experiences of students of color" (Villegas & Irvine, 2010, p. 185). One way in which this diversification may benefit racially and economically non-dominant students is through racial and cultural synchronicity between non-dominant students and teachers (Dee, 2004). Racially and economically non-dominant teachers are well-positioned to teach non-dominant students because of shared cultural and lived understanding of their students' home and community life as well as experiences as marginalized persons in society (Milner, 2006). Interestingly, the highly resegregated educational landscape suggests that racially and economically dominant students already frequently experience racial and cultural synchronicity with their similarly dominant teachers at their schools and in their science learning environments.

The disproportionate representation of racially and economically diverse teachers in U.S. public schools is also troubling when contrasted with the rapidly increasing racial, ethnic, linguistic, and economic diversity of this nation's public schools. Recent efforts to recruit racially and economically non-dominant teacher candidates continue to diversify the teaching workforce. These increases are far outpaced by the diversification of the student population and are stymied by the profession's inability to retain non-dominant teachers who are dissatisfied because of the lack of autonomy and work climate (Ingersoll & May, 2011). The lack of retention of racially and economically non-dominant teachers is exacerbated in science education, where teachers of all racial and economic backgrounds are hard to retain because of job dissatisfaction and the attractiveness of other job opportunities outside of the classroom (Ingersoll & Perda, 2010).

Failure to Prepare Science Teachers for 21st Century Students. In the sections prior we highlighted the underrepresentation of non-dominant students and teachers in STEM. This underrepresentation in STEM is occurring in direct contrast to the widespread racial and economic demographic shifts within our society and our schools. Schools serving non-dominant students are now more diverse than ever and the economic realities of many students are more severe. Darling-Hammond (2006) notes:

> In the classrooms most beginning teachers will enter, at least 25% of students live in poverty and many of them lack basic food, shelter, and health care; from 10% to 20% have identified learning differences; 15% speak a language other than English as their primary language (many more in urban settings); and about 40% are members of racial/ethnic "minority" groups, many of them recent immigrants from countries with different educational systems and cultural traditions. (p. 301)

The profiles of today's students are strikingly different than those of dominant students who were previously typified as the "traditional" student. These profiles are also more than static demographics or variables; as Darling-Hammond points out, they represent distinct sociocultural realities and likely different views of how the world operates. Within this context, we must consider how our teachers are being prepared to teach this "new" set of students, particularly when they come from different racial, cultural, and economic backgrounds than their teachers or from that which the teacher is familiar (Rodriguez, 2015). Within science education this question also requires that we rethink how science content should be taught and how teachers should be prepared to teach it (Rodriguez, 1998).

Yet, many science teacher education programs do not prepare science teachers to be comfortable and confident educators in urban environments or prepared to enact science instruction via multicultural, culturally competent, or social justice pedagogical models (Rivera Maulucci, 2013). There is also a dearth of research about the preparation of science teachers relative to multicultural issues (Seiler, 2013) even though we know that the preparation of science teachers to work in racially, culturally, linguistically, and economically non-dominant environments is critical for science teacher's efficaciousness (Fraser-Abder, Atwater, & Lee, 2006; Tal, Krajcik, & Blumenfeld, 2006). For example, Maria Rivera Maulucci (2013) examined the experience of an African American-Caribbean teacher candidate and found that despite the teacher's verbalized commitment to social justice, she lacked clarity in her feelings and her own political awareness. She thus struggled to enact the racial justice principles in her science teaching. Studies such as Rivera Maulucci's indicate that the preparation of teacher candidates cannot be limited to exploring theoretical or personal perspectives, but also has to include rich opportunities for them to enact and refine their pedagogical ideas and practices in classroom contexts similar to those in which they will teach.

When such preparation does not occur and teachers do not have other resources, such as other teachers, to support their efforts, a common response is for teachers to teach the way that they were taught or to buckle down and deepen their engagement with a traditional model with which they and others are more familiar. These forms of teaching often focus "on the ways students acquire knowledge" (Brown, Reveles, & Kelly, 2005, p. 780) rather than students' own funds of knowledge, community participations, and scientific practices and interests. These approaches have the potential to be deficit-oriented or rely upon culture of poverty perspectives that devalue students' and their communities' existing ways of participating in, thinking about, and doing science. Such science learning environments reproduce lowered expectations, cognitive passivity, and rule following rather than problem solving or conceptual learning (Seiler, Tobin, & Sokolic, 2001; Songer, Lee, & Kam, 2002). If students remain unsuccessful or unmotivated with these de-contextualized and alienating ways of teaching science, their lack of engagement and achievement in turn reifies the deficit notions (Bryan & Atwater, 2002) and lack of participation and interest in science that teachers may have initially intended to disrupt.

Culture of poverty perspectives and practices problematically position the motivation, abilities, and identities of racially and economically non-dominant students and their communities in ways that can refy deficit notions or students' self-perceptions of "incompetence in mathematics and science" (Powell, 1990, p. 292). One example of such positioning includes racial and other microaggressions (Huber & Solórzano, 2015) that can create isolating, disempowering, and symbolically violent climates within science learning environments. Microaggressions (racial, economic, gender, linguistic, etc.) are brief and commonplace verbal, nonverbal, and/or visual discriminatory insults (Pierce, 1970) that often go unnoticed by those not affected by the stereotype because of their subtle and seemingly unintentional quality that allows for plausible deniability. They manifest in a variety of forms, including a teacher's astonishment when a non-dominant student answers a question correctly or is academically successful, a quick glance at a lower income student when something is missing, the mispronouncing or altering of a student's name under the pretense that it is "hard" or in some way abnormal, and the like. These acts, whether conscious or unconscious, convey or reaffirm that someone or a group of people is inferior or occupies the status of "other." Because they are subtle, microaggressions often remain uncontested and therefore can produce cumulative physiological and psychological trauma that can weigh heavily on the consciousness and daily actions of non-dominant people (Solorzano, Ceja, & Yosso, 2000).

Microaggressions can impact non-dominant students' academic performance and development of academic interests, in part, because they are one key trigger of stereotype threat (Sue, Capodilupo, & Holder, 2008). The notion of "stereotype threat" exemplifies how deficit perspectives operate and are perpetuated in STEM classrooms in ways that connect with students' domain-specific identities and academic performance. The theory suggests that people, including K-12 students, are cognizant of stereotypes associated with their social (racial, economic, linguistic, gender, etc.) group. This awareness, in turn, creates additional stress and pressure and desires to disprove the existence of a stereotype that actually impairs one's performance and achievement, the result of which confirms the assigned stereotype that the individual was trying to disprove (Steele, 1997). The performance of students who identify with a particular domain *and* the group targeted by the stereotype, are especially hindered by stereotype threat (Steele, Spencer, & Aronson, 2002).

Given the changing face of the K-12 students in the U.S., science teachers need to be prepared to think about science differently and be prepared to teach science differently. This includes science teacher's being trained and able to reflect on their own biases, and to understand the variety of issues that might be underlying non-dominant student participation, interest, and success. Science teachers need to understand that student interest and performance are not just about that student's individual work habits and values, instead racially and economically non-dominant students' interests and participation may also be fundamentally linked to how their communities and personhood are positioned in relation to science. Without this preparation, teachers may reproduce racial and economic inequalities and also be unable to realize their own agency to affect change in their pedagogy and in the learning and achievement patterns of their students.

RETHINKING SCIENCE EDUCATION FOR THE 21ST CENTURY STUDENT

Thus far in this chapter we have examined how several tensions limit progress towards "science for all" for racially and economically non-dominant students. A lingering issue within this conversation that must be addressed is about how the terms "race" and "socioeconomic status" (SES) are used in science education, and how these usages do not effectively represent the lived realities and expressed needs of today's students. In our examination of the science education literature race and SES were generally 1) treated as demographic variables (i.e., Black students, lower-income students, urban students) and outcome indicators; and 2) used interchangeably or incorrectly as proxies for other words. As Parsons (2014) points out, there is a general undertheorization of race and SES and their intersections within science education. In the section that follows, we offer some conceptual tools for rethinking SES in terms of poverty, and for considering race and poverty as overlapping sociopolitical realities.

Conceptualizing Race and Poverty as Sociopolitical Realities. The United States as an economic and legislative entity has always been formally and informally racially and economically stratified (Omi & Winant, 2014). Events over the last decade, such as the election of President Obama and the 2008 financial crisis, have led to national conversations about whether the United States is finally "post-race" and if concerns about social inequalities should focus more explicitly on socioeconomic issues. Similar conversations are echoed within science education research and teaching communities and our determinations about how to address on-going racialized and socio-economic disparities in terms of science education access and outcomes. However, this dichotomizing of the salience of race versus that of socioeconomics is a false binary that does not acknowledge how the two constructs are historically and contemporarily intertwined in the daily lives of non-dominant students. This does not mean that one can be used as a proxy for the other, but that the tangible and intangible psycho-social and physical realities of many non-dominant students reflects the consequences of race and poverty combined.

Socioeconomic status (SES) is a socially constructed, but experientially real notion. Its meaning has shifted over time and context, yet it is a widely used term, especially in education wherein low-SES often refers to students who are eligible for free-and-reduced lunch (FRL). Questioning the undertheorization of SES and its relationship to FRL, Michael Harwell and Brandon LeBeau (2010) review existing research to offer three definitional categories of SES: 1) the social and economic life chances of individuals; 2) differential access to desired resources, and 3) an aggregate variable that enables people and communities to be hierarchically ordered based upon their ability to access, produce, and consume social resources. Using these definitions, a "low" SES student is presumed to come from a household that has less income and less educated adult household members who engage in more unskilled labor and have less social, cultural, and economic resources overall to offer their student. Reflecting upon such a characterization, Harwell and LeBeau (2010) query if SES is meant to highlight "differential access to economic resources, to social resources in the community, to social power (Rose & Pevalin, 2003), to desired education resources, or something else?" (p. 121). They argue that in using SES we know little more than which household incomes have been deemed eligible for support to buy lunch, yet our evaluations of low-SES students are clearly not solely numeric. Harwell and LeBeau urge us for better descriptions of what SES represents as connected to the strategies being used to measure SES.

Aside from FRL, socioeconomic status is commonly conceptualized as the social standing or class of an individual or group and is measured as a combination of education, income and occupation (National Center for Education Statistics, 2013). Examinations of SES expose inequities in access to resources and the related layers of privilege and social power. These layers of social power are usually referred to via the notion of class. The social power of each class refers to the ways in which an individual is subject to stereotypical presumptions about the intelligence and capacity of individuals within that group or of the group as a whole.

The status of these individuals or of this group is determined by their proximity or distance to the status quo (i.e., the middle or upper class) and to the power afforded those dominant groups.

Pushing the field of education to move away from an SES or class lens toward more explicit conversations around what it means to live in "poverty," Richard Milner (2013) points out that many racially non-dominant students' lived realities are not captured in discussions of SES, class, or even notions of poverty that simply delineate an income to person ratio or threshold. Milner articulates a definition of poverty as a sociopolitical reality that reflects the "particular characteristics and situations people find themselves in because of the amount of money and related material capital that they have or do not have" (Milner, 2013, p. 9) at a given moment in time. The linking of monetary finances with other material resources that a student or family has available makes their living conditions and constraints much clearer. It also pushes the definition of poverty away from moral suppositions about individuals or groups towards a socio-political understanding of poverty.

Like SES, race is often presented as a static term or demographic variable. For example, race and ethnicity are often used interchangeably to categorize people without appreciating how each team reflects distinct social, political, legislative, and historical constructs that index particular notions and experiences of power and privilege (Parsons, 2014). In this chapter, our focus is on race rather than ethnicity. Drawing upon Michael Omi and Howard Winant's theory of racialization (2014), we view race as "a concept, a representation or signification of identity that refers to different types of human bodies, to the perceived corporal and phenotypic markers of difference and the meanings and social practices that are ascribed to these differences" (p. 112) that has a way of "making up" people (p. 105). Race therefore operates as a *master category* that constantly organizes large and small-scale activities and interactions in explicit and implicit ways in American society and its institutions and has implications for individuals' pyscho-social and lived realities. One way that race operates as a master category is through racial "storylines" (e.g., "Science is for White people") or social discourses that mediate how individuals make sense of race and use race to position one another through their everyday actions and interactions (Nasir, Snyder, Shah, & Ross, 2013).

Na'ilah Nasir and colleagues argue that racial storylines racially and academically socialize students' ways of being and prime teachers and students into stereotyping and positioning these ways of being. These storylines or stereotypical narratives are assumed to be deterministic of individuals' or groups' ways of knowing, being, and of their ability levels thus suggesting that anyone outside of these essentialized parameters is abnormal. Storylines thus have implications for learners' perceptions of themselves and their abilities, as learners broadly and as domain-specific learners, and for the types of identities youth are afforded access to learning environments. Racial storylines reflect how the limited notion of race in U.S. society often fails to capture intra-racial complexities and diverse ways in which race is lived and performed.

To be clear, racial storylines are not limited to historically marginalized communities or those communities that remain educationally underserved such as African Americans, Latinos, and Native Americans and Pacific Islanders. Storylines called *dominant narratives* reflect how U.S. society functions to position White Americans, i.e., those persons in the U.S. that present as being of Western European ancestry, as the default or natural racial standard. These storylines are the unspoken, and at times unconscious, commonsensical norm or status quo that dictates which ways of knowing and being, which physical aesthetics and embodiments, and which values and perspectives are privileged. Under these deterministic constraints, interactions are *always* racialized and classed irrespective of who is interacting or which topic or activity they are interacting with. This includes interactions involving solely racially and economically dominant communities. The presumption that such groups or individuals are not racialized or classed reifies the dominant myth of Whiteness and economic privilege as normal and other racial and class realities as deviant.

Given the racialized and classed terrain of our society and our schools that we have outlined, the challenge for science education is to consider not "if" science education as a field is racialized or classed, but "how" it operates as a racialized and classed domain. The lingering tensions we highlighted earlier detail how science education continues to reproduce racialized and classed inequalities. With a clearer view of what we mean by race and poverty, we now turn to considering how social inequities and related dynamics can be disrupted toward affording greater equity in science education and possibly continuing to make progress towards ensuring "science for all." In the next section, we introduce politically relevant science instruction as a way to rethink science teaching and learning in ways that attend to the sociopolitical realities of race and poverty that affect non-dominant students' participation, interests, and success in science.

Politically Relevant Science Instruction. Nearly two decades ago culturally relevant pedagogy (CRP) was introduced as a form of teaching that encouraged educators to be cognizant of the types of knowledge and identities that one implicitly values, produces, and cultivates (Gay, 2000; Ladson-Billings, 1995). Initial scholarship on CRP focused on the teaching of Black children, yet it has been identified as an effective strategy for teaching students from diverse racially, ethnically, and linguistically backgrounds. It has also been examined as an equitable instruction and learning model for students from racially and economically non-dominant communities within the domain of science (Johnson, 2011; Young, 2005). The main tenets of CRP are: a) all students need to have academic success, which includes the development of high expectations for all students; b) teachers support students' development and maintenance of cultural competence by utilizing students' culture for learning and helping them develop cultural capital to succeed in the dominant culture; and c) teachers plan and implement learning experiences to develop students' critical consciousness, which will support students in challenging the status quo and the inequities in schools and communities (Ladson-Billings, 2006).

Recently it has been argued that the third tenet–developing students' sociopolitical consciousness—is the most overlooked and the most difficult for teachers to interpret and implement (Tsurusaki, Calabrese Barton, Tan, Koch, & Contento, 2013; Young, 2010). Science teachers, in particular, have expressed that it is difficult to enact a sociopolitical consciousness, much less develop that of their students (Rivera Maulucci, 2013). One issue is that in order for teachers to develop their students' sociopolitical consciousness, theirs must already be relatively well developed and their pedagogy should include a strategy that breaks down what is being learned and why it is being learned (Milner, 2011; Sleeter, 2012). Another is that some teachers feel that students, particularly younger students, cannot engage in conversations about structural inequities (Young, 2010). Finally, science educators may struggle with CRP if their school contexts are wrought with pressures to increase students' test scores or achieve grade-level content standard (Rivera Maulucci, 2013). This may be the case in schools with high populations of racially and economically non-dominant students that have a high teacher turnover rate or are training grounds for new teachers. In such cases, science teachers may focus on a narrow conception of science content rather than developing their students' critical consciousness around what is taught and helping students to understand that the learning of science is a political activity (Rivera Maulucci, 2013).

We argue for science teaching and learning to be understood as intertwined political activities, the former being teacher's awareness about the sociopolitical realities of their students and the latter being the development of students' consciousness about their sociopolitical realities and the potential power they have to alter it through using science. We offer *politically relevant science instruction* as an approach that takes up this politicized reconceptualization. It emphasizes the sociopolitical calculus—or political clarity (Beauboeuf-LaFontant, 1999)—that may characterize science teachers' thoughts and actions. Politically relevant science instruction shifts the pedagogical focus from cultural awareness and practice onto the politically relevant mindsets of science teachers as they are enacted *through* instruction (McKinney de Royston, Madkins, & Nasir, under review).

Politically relevant science instruction, like CRP, has three main tenets (McKinney de Royston et al., under review). These tenets build on those of CRP (Ladson-Billings, 1995) but focus more pointedly on the politicized nature of each tenet through acts of positioning and how that tenet connects with science education. The first is *academic success*—the key to this tenet is that the teacher rejects deficit thinking by positioning students as capable of doing science and of having the capacity to be scientists. This includes the repositioning of students' habits of mind as scientific in nature and reframing what "counts" as scientific thinking or inquiry. This also includes the explicit debunking of stereotypes about students from non-dominant communities' performance or capacity in the sciences. Finally, the teacher explicitly acknowledges the value of each student in his/her class and supports their access to scientific learning opportunities by having high expectations for their academic success in science (both achievement and learning)—adapted from.

The second tenet is *cultural competence*, which refers not only to the cultural competence of the teacher but also to how the teacher develops and maintains students' cultural competence. This is done through positioning students' racial and cultural practices and everyday ways of knowing and being as consonant with the discipline of science and scientific endeavors. Examples of this include building off of students' funds of knowledge to develop their cultural capital within the discipline of science. For example, the idea and practice of innovation is an important aspect of science that may also exist in many families, communities, and cultures. Culturally competent science teachers may highlight the spirit or practices of innovation in communities or families and connect them to innovation in science.

The third tenet of politically relevant science instruction is *sociopolitical consciousness,* which both refers to the political clarity of the teacher and the development of the critical consciousness of the student. Political clarity refers to the teacher's non-essentializing, caring, and perhaps embodied understanding about the sociopolitical realities of oppression and their students' experiences with it in and out of schools. For science teachers, this means recognizing the ways in which students from non-dominant communities' forms of knowledge and ways of being are positioned in society and within schools and classrooms generally and with respect to science and their ability to achieve in science. This understanding shapes how sociopolitically conscious science teachers care, advocate, and protect their students in and beyond their classrooms, and how they verbally and interactionally position their students as capable science learners and scientists. The development of the political consciousness of the student relies upon the political clarity of the teacher that plans and implements scientific learning experiences that develop and break down what is learned, why it is being learned, and encourages students to develop their own critique of the status quo of science as culture-free, neutral, and infallible. The politically relevant science instructor thus helps to develop students' ability to scientifically explore, critique, and test possible solutions to inequities in their school, community, or larger society. This is done with the aim of supporting students' ability to create their own critiques of societal and structural inequalities in and beyond science. This includes cultivating students' awareness of science achievement within schools as also situated within systems and hierarchies of oppression and power (Milner, 2011).

By placing a teacher's political clarity at the core of the politically relevant science teaching model, teachers engaging in this approach both have to accept the political impetus of the approach and attend to the sociopolitical realities—including oppression—that affect racially and economically non-dominant students' opportunities to learn science and pursue careers in science. As Mensah (2009) points out, this type of clarity can be developed in education programs by requiring teachers to question and challenge their own assumptions, stereotypes, beliefs and practices relative to social discourses and norms and relative to the lives of their students. This clarity may support dominant teachers in developing authentic conceptual and interactional tools to work with racially and economically non-dominant students that do not enact microaggressions or trigger stereotype threats. For this approach to be effective it would have to be infused into every aspect of the science teacher preparation program rather than be limited to one course or the equity goal for one lesson.

FINAL THOUGHTS

Over the course of this chapter, we have shown how racially and economically non-dominant students' participation, interest, and success in science is related to the availability of resources, their physical proximity to science learning environments, the presence of other non-dominant students and teachers within science learning settings. We also show how achieving "science for all" requires explicit attention to how race and poverty operate as overlapping sociopolitical realties for many non-dominant students and for the science teachers working alongside these students as they navigate issues of power, identity, and learning within science education. We argue that such work requires both looking back at issues we continue to struggle with in science education as well as moving forward with greater political clarity. The politically relevant science approach is offered as a conceptual and pedagogical tool that can support the development of teacher-student relationships and inclusive science learning environments that non-dominant students need to feel psychologically and ideologically safe to participate, become interested in, and succeed in school science and beyond.

To close, we want to highlight some opportunities and to offer some cautions. First, the adoption of the Next Generation Science Standards (NGSS) (NGSS Lead States, 2013) may represent an opportunity for science education and educators to reframe notions of science and move towards better supporting the science learning and achievement of racially and economically non-dominant students. These possibilities exist given how equity is made visible within NGSS and how NGSS emphasizes understanding science as a way of knowing, as a human endeavor, and as a field concerned about the world. NGSS standards, if taken literally, can reframe what it means to do science, towards what ends we're learning and engaging in science, and ultimately influence who is perceived as knowing and doing science. For example, among the supported practices are the asking of questions about how the world works and using the classroom or school resources or local environment to delve into these question. These shifts in perspectives about science inquiry and the types of resources that are privileged in such investigations is a more expansive understanding of science that may offer new opportunities to attend to the sociopolitical realities of non-dominant students and their communities. This framing has the potential to reclaim science as the provenance of any inquisitive student despite resources, existing stereotypes, etc.

At the same time, NGSS runs the risk of falling into the same trap of good intentions as prior reforms. One way this can happen is if we fail to understand NGSS as a "political document" (Rodriguez, 2015) that articulates certain political stances and agendas about how to support science teaching and learning. Failing to recognize NGSS as such has grave consequences for our own civic participation as educators and scholars who must hold such documents—and their creators—accountable to the socio-political realities (like "science for all") that they envision. Another way is if we don't intentionally and consistently keep questions of equity central to our everyday work as we teach science and research science education. Finally, we have to acknowledge the resistance of many teachers to changing their own ideologies and pedagogies and emphasize that such shifts are necessarily part of the work of teaching (Rodriguez, 2015).

To avoid these good intentions traps, we—as science educators and science education researchers—need to take up political stances in relation to science. Indeed, prior work in science education around CRP, social justice, and critical reflection, talks about the need for such political clarity on the part of educators (Tal et al., 2006). This means that science teachers must better understand their positionality and attitudes relative to racially and economically non-dominant students (and their lives) and consider how these understandings are continually refined and reflected in their teaching. Ultimately, we argue that in order to achieve "science for all" it is not enough to simply offer science to any student but to reclaim science, and success in science, as the provenance and civic right of any student despite where s/he lives, what s/he looks like, and where s/he comes from.

REFERENCES

American Association for the Advancement of Science (AAAS). (1989). *Science for all Americans.* New York, NY: Oxford University Press.

Atwater, M. M. (2000). Equity for Black Americans in precollege science. *Science Education, 84,* 154–179. doi:10.1002/(SICI)1098-237X(200003)84:2.

Aud, S., Hussar, W., Kena, G., Bianco, K., Frohlich, L., Kemp, J., & Tahan, K. (2011). *The Condition of Education 2011 (NCES 2011-033).* U.S. Department of Education, National Center for Education Statistics. Washington, DC: Government Printing Office.

Barton, A. C. (1998). Reframing "science for all" through the politics of poverty. *Educational Policy, 12*, 525–541. doi:10.1177/0895904898012005004.

Barton, A. C. (2001). Science education in urban settings: Seeking new ways of praxis through critical ethnography. *Journal of Research in Science Teaching, 38*, 899–917. doi:10.1002/tea.1038.

Barton, A. C. (2007). Science learning in urban settings. In S. K. Abell & N. G. Lederman (Eds.), *Handbook of research on science education* (pp. 319–343). Mahwah, NJ: Lawrence Erlbaum.

Beauboeuf-Lafontant, T. (1999). A movement against and beyond boundaries: "Politically relevant teaching" among African American teachers. *Teachers College Record, 100*, 702–723.

Berliner, D. (2001, January 28). Our schools vs. theirs: Averages that hide the true extremes. *Washington Post*, p. B3. Retrieved from http://nepc.colorado.edu/files/cerai-01-02.htm

Bricker, L. A., & Bell, P. (2014). "What comes to mind when you think of science? The perfumery!": Documenting science-related cultural learning pathways across contexts and timescales. *Journal of Research in Science Teaching, 51*, 260–285. doi:10.1002/tea.21134.

Brickhouse, N. W., & Potter, J. T. (2001). Young women's scientific identity formation in an urban context. *Journal of Research in Science Teaching, 38*, 965–980. doi:10.1002/tea.1041.

Brown, B. A., Henderson, J. B., Gray, S., Donovan, B., Sullivan, S., Patterson, A., & Waggstaff, W. (2015). From description to explanation: An empirical exploration of the African, American pipeline problem in STEM. *Journal of Research in Science Teaching, 53*, 146–177. doi:10.1002/tea.21249.

Brown, B. A., Reveles, J. M., & Kelly, G. J. (2005). Scientific literacy and discursive identity: A theoretical framework for understanding science learning. *Science Education, 89*, 779–802. doi:10.1002/sce.20069.

Bryan, L. A., & Atwater, M. M. (2002). Teacher beliefs and cultural models: A challenge for science teacher preparation programs. *Science Education, 86*, 821–839. doi:10.1002/sce.10043.

Buck, G. A., Clark, V. L. P., Leslie-Pelecky, D., Lu, Y., & Cerda-Lizarraga, P. (2008). Examining the cognitive processes used by adolescent girls and women scientists in identifying science role models: A feminist approach. *Science Education, 92*, 688–707. doi:10.1002/sce.20257.

Buck, G. A., Cook, K. L., Quigley, C. F., Prince, P., & Lucas, Y. (2014). Seeking to improve African American girls' attitudes toward science. *The Elementary School Journal, 114*, 431–453. doi:10.1086/674419.

Carter, N. P., Larke, P. J., Singleton-Taylor, G., & Santos, E. (2003). Multicultural science education: Moving beyond tradition. In S. M. Hines (Ed.), *Multicultural science education: Theory, practice, promise* (pp. 1–19). New York, NY: Peter Lang.

Darling-Hammond, L. (2006). Constructing 21st-century teacher education. *Journal of Teacher Education, 57*(3), 300–314. doi:10.1177/0022487105285962.

Darling-Hammond, L. (2010). The flat world and education: How America's commitment to equity will determine our future. *Teachers College Press*. Retrieved from http://dx.doi.org/10.5860/choice.47-6991

Dee, T. S. (2004). The race connection: Are teachers more effective with students who share their ethnicity? *Education Next, 4*(2), 52–59.

Emdin, C., & Lee, O. (2012). Hip-Hop, the Obama effect, and urban science education. *Teachers College Record, 114*(2), 1–24.

Frankenberg, E., & Lee, C. (2002). *Race in American public schools: Rapidly resegregating School districts.* Cambridge, MA: The Civil Rights Project, Harvard University.

Fraser-Abder, P., Atwater, M., & Lee, O. (2006). Research in urban science education: An essential journey. *Journal of Research in Science Teaching, 43*, 599–606. doi:10.1002/tea.20156.

Gay, G. (2000). *Culturally responsive teaching: Theory, research & practice.* New York, NY: Teachers College Press.

Harding, S. (1994). Is science multicultural? Challenges, resources, opportunities, uncertainties. In D. T. Goldberg (Ed.), *Multiculturalism: A critical reader* (pp. 344–370). Oxford: Blackwell.

Harwell, M., & LeBeau, B. (2010). Student eligibility for a free lunch as an SES measure in education research. *Educational Researcher, 39*(2), 120–131. doi:10.3102/0013189x10362578.

Huber, L. P., & Solórzano, D. G. (2015). *Racial microaggressions: What they are, what they are not, and why they matter.* Los Angeles, CA: Chicano Studies Research Center, University of California.

Hutchison, L. F. (2012). Addressing the STEM teacher shortage in American schools: Ways to recruit and retain effective STEM teachers. *Action in Teacher Education, 34*, 541–550. doi:10.1080/01626620.2012.729483.

Ingersoll, R. M., & May, H. (2011). The minority teacher shortage: Fact or fable? *Phi Delta Kappan, 93*(1), 62–65. doi:10.1177/003172171109300111.

Ingersoll, R. M., & Perda, D. (2010). Is the supply of mathematics and science teachers sufficient? *American Educational Research Journal, 47*(3), 563–594. doi:10.3102/0002831210370711.

Johnson, C. C. (2011). The road to culturally relevant pedagogy in a diverse science classroom. *Journal of Research in Science Teaching, 48*, 170–198. doi:10.1002/tea.20405.

Kelly-Jackson, C. P., & Jackson, T. O. (2011). Meeting their fullest potential: The beliefs and teaching of a culturally relevant science teacher. *Creative Education, 2*, 408. doi:10.4236/ce.2011.24059.

Kumashiro, K. K. (2001). "Posts" perspectives on anti-oppressive education in social studies, English, mathematics, and science classrooms. *Educational Researcher, 30*(3), 3–12. doi:10.3102/0013189x030003003.

Ladson-Billings, G. (1995). Toward a theory of culturally relevant pedagogy. *American Educational Research Journal, 32*(3), 465–491. doi:10.3102/00028312032003465.

Ladson-Billings, G. (2006). From the achievement gap to the education debt: Understanding achievement in US schools. *Educational Researcher, 35*(7), 3–12. doi:10.3102/0013189x035007003.

Landsman, J., & Lewis, C. (Eds.). (2011). *White teachers/diverse classrooms: Creating inclusive schools, Building on students' diversity and providing true educational equity* (2nd ed.). Sterling, VA: Stylus.

Lee, O., & Buxton, C. A. (2010). *Diversity and equity in science education: Research, policy, and practice.* Multicultural Education Series. New York, NY: Teachers College Press.

Madkins, T. C. (2011). The Black teacher shortage: A literature review of historical and contemporary trends. *Journal of Negro Education, 80*(3), 417–427.

McKinney de Royston, M., Madkins, T.C., & Nasir, N. (under review). "Mama! Do We Have Cotton Balls?": Understanding Politically Relevant Instruction in Science.

Mensah, F. M. (2009). Confronting assumptions, biases, and stereotypes in preservice teachers' conceptualizations of science teaching through the use of book club. *Journal of Research in Science Teaching, 46*, 1041-1066.

Mensah, F. M. (2013). Retrospective accounts in the formation of an agenda for diversity, equity, and social justice in science education. In J. A. Bianchini, V. L. Akerson, A. C. Barton, O. Lee, & A. J. Rodriguez (Eds.), *Moving the equity agenda forward: Equity research, practice, and policy in science education* (pp. 317–335). New York, NY: Springer.

Milner, H. R. (2006). The promise of Black teachers' success with Black students. *Educational Foundations, 20*(3–4), 89–104.

Milner, H. R. (2012). But what is urban education? *Urban Education, 47*(3), 556–561. doi:10.1177/0042085912447516.

Milner, H. R. (2013). Analyzing poverty, learning, and teaching through a critical race theory lens. *Review of Research in Education, 37*, 1–53. doi:10.3102/0091732X12459720.

Milner, H. R. (2015). *Rac(e)ing to class: Confronting poverty and race in schools and classrooms.* Cambridge, MA: Harvard.

Morgan, P. L., Farkas, G., Hillemeier, M. M., & Maczuga, S. (2016). Science achievement gaps begin very early, persist, and are largely explained by modifiable factors. *Educational Researcher, 45*(1), 18–35. doi:10.3102/0013189x16633182.

Nasir, N., & Hand, V. (2008). From the court to the classroom: Opportunities for engagement, learning, and identity in basketball and classroom mathematics. *The Journal of the Learning Sciences, 17*, 143–179. doi:10.1080/10508400801986082.

Nasir, N., Snyder, C., Shah, N., & Ross, K. (2013). Racial storylines and implications for learning. *Human Development, 55*, 285–301. doi:10.1159/000345318.

National Center for Education Statistics. (2013). *Improving the measurement of socioeconomic status for the National Assessment of Educational Progress: A theoretical foundation.* Retrieved from http://nces.ed.gov/nationsreportcard/researchcenter/socioeconomic_factors.aspx

National Research Council (NRC). (2012). *A framework for K-12 education: Practices, crosscutting concepts, and ideas.* Washington, D.C.: National Academies Press.

NGSS Lead States. (2013). *Next generation science standards: For states, by states.* Washington, DC: National Academies. doi:10.17226/18290.

Omi, M., & Winant, H. (2014). *Racial formation in the United States.* New York, NY: Routledge.

Orfield, G., & Gordon, N. (2001). *Schools more separate: Consequences of a decade of resegregation.The Civil Rights Project.* Cambridge, MA: Harvard.

Parsons, E. C. (2014). Unpacking and critically synthesizing the literature on race and ethnicity in science education. In Norman G. Lederman & Sandra K. Abell (Eds.), *Handbook of research in science education* (pp. 687–767). New York, NY: Routledge.

Pierce, C. (1970). Offensive mechanisms. In F. Barbour (Ed.), *The black seventies* (pp. 265–282). Boston, MA: Porter Sargent.

Powell, L. (1990). Factors associated with the underrepresentation of African Americans in mathematics and science. *The Journal of Negro Education, 59*, 292–298. doi:10.2307/2295564.

Riegle-Crumb, C., & King, B. (2010). Questioning a white male advantage in STEM: Examining disparities in college major by gender and race/ethnicity. *Educational Researcher, 39*, 656–664. doi:10.3102/0013189X10391657.

Riegle-Crumb, C., Moore, C., & Ramos-Wada, A. (2011). Who wants to have a career in science or math? Exploring adolescents' future aspirations by gender and race/ethnicity. *Science Education, 95*, 458–476. doi:10.1002/sce.20431.

Rivera Maulucci, M. S. (2013). Emotions and positional identity in becoming a social justice science teacher: Nicole's story. *Journal of Research in Science Teaching, 50*, 453–478. doi:10.1002/tea.21081.

Rodriguez, A. J. (1998). Strategies for counterresistance: Toward sociotransformative constructivism and learning to teach science for diversity and for understanding. *Journal of Research in Science Teaching, 35*, 589–622. doi:10.1002/(sici)1098-2736(199808)35:6<589::aid-tea2>3.0.co;2-i.

Rodriguez, A. J. (2015). What about a dimension of engagement, equity, and diversity practices? A critique of the next generation science standards. *Journal of Research in Science Teaching, 52*, 1031–1051. doi:10.1002/tea.21232.

Rogers-Ard, R., Knaus, C. B., Epstein, K., & Mayfield, K. (2013). Racial diversity sounds nice; System transformation? Not so much: Developing urban teachers of color. *Urban Education, 48*, 451–479. doi:10.1177/0042085912454441.

Rose, D., & Pevalin, D. J. (2003). The NS-SEC explained. In D. Rose & D.J. Pevalin (Eds.), *A researcher's guide to the national statistics socio-economic classification* (p. 28-43). London: Sage.

Seiler, G. (2013). New metaphors about culture: Implications for research in science teacher preparation. *Journal of Research in Science Teaching, 50*, 104–121. doi:10.1002/tea.21067.

Seiler, G., Tobin, K., & Sokolic, J. (2001). Design, technology, and science: Sites for learning, resistance, and social reproduction in urban schools. *Journal of Research in Science Teaching, 38*, 746–767. doi:10.1002/tea.1030.

Sleeter, C. E. (2012). Confronting the marginalization of culturally responsive pedagogy. *Urban Education, 47*, 562-584.

Solorzano, D., Ceja, M., & Yosso, T. (2000). Critical race theory, racial microaggressions, and campus racial climate: The experiences of African American college students. *Journal of Negro Education, 69*, 60–73. doi:10.17763/haer.79.4.m6867014157m707l.

Songer, N. B., Lee, H. S., & Kam, R. (2002). Technology rich inquiry science in urban classrooms: What are the barriers to inquiry pedagogy? *Journal of Research in Science Teaching, 39*, 128–150. doi:10.1002/tea.10013.

Steele, C. M. (1997). A threat in the air: How stereotypes shape intellectual identity and performance. *American Psychologist, 52*, 613. doi:10.1037/0003-066x.52.6.613.

Steele, C. M., Spencer, S. J., & Aronson, J. (2002). Contending with group image: The psychology of stereotype and social identity threat. *Advances in Experimental Social Psychology, 34*, 379–440. doi:10.1016/s0065-2601(02)80009-0.

Sue, D. W., Capodilupo, C. M., & Holder, A. (2008). Racial microaggressions in the life experience of Black Americans. *Professional Psychology: Research and Practice, 39*, 329. doi:10.1037/0735-7028.39.3.329.

Tal, T., Krajcik, J. S., & Blumenfeld, P. C. (2006). Urban schools' teachers enacting project based science. *Journal of Research in Science Teaching, 43*, 722–745. doi:10.1002/tea.20102.

Tan, E., Barton, A. C., Turner, E., & Gutiérrez, M. V. (2012). *Empowering science and mathematics education in urban schools.* Chicago, IL: University of Chicago Press.

Tillman, L. C. (2004). (Un)intended consequences? The impact of Brown v. Board of Education decision on the employment status of black educators. *Education and Urban Society, 36*, 280–303. doi:10.1177/0013124504264360.

Tsurusaki, B. K., Calabrese Barton, A., Tan, E., Koch, P., & Contento, I. (2013). Using transformative boundary objects to create critical engagement in science: A case study. *Science Education, 97*, 1–31. doi:10.1002/sce.21037.

Varelas, M., Kane, J. M., Tucker-Raymond, E., & Pappas, C. C. (2012). Science learning in urban elementary school classrooms: Liberatory education and issues of access, participation and achievement. In B. Fraser, K. Tobin, & C. J. McRobbie (Eds.), *Second international handbook of science education* (pp. 91–103). New York, NY: Springer.

Varelas, M., Kane, J. M., & Wylie, C. D. (2011). Young African American children's representations of self, science, and school: Making sense of difference. *Science Education, 95*, 824–851. doi:10.1002/sce.20447.

Villegas, A., & Irvine, J. (2010). Diversifying the teaching force: An examination of major arguments. *Urban Review, 42*, 175–192. doi:10.1007/s11256-010-0150-1.

Visintainer, T. A. (2015). *What is and who can do science? Supporting youth of colors' identities as learners, doers, and change agents in science* (Unpublished doctoral dissertation). University of California, Berkeley, CA.

Walker, V. S. (1996). *Their highest potential: An African American school community in the segregated South.* Chapel Hill, NC: University of North Carolina.

Wang, X. (2013). Why students choose STEM majors: Motivation, high school learning, and postsecondary context of support. *American Educational Research Journal, 50,* 1081–1121. doi:10.3102/0002831213488622.

Warren, B., & Rosebery, A. (2011). Navigating interculturality: African American male students and the science classroom. *Journal of African American Males in Education, 2,* 98–115.

Wortham, S. (2004). From good student to outcast: The emergence of a classroom identity. *Ethos,* 32, 164–187. doi:10.1525/eth.2004.32.2.164.

Young, H. (2005). Secondary education systemic issues: Addressing possible contributors to a leak in the science education pipeline and potential solutions. *Journal of Science Education and Technology, 14,* 205–216.

Young, E. (2010). Challenges to conceptualizing and actualizing culturally relevant pedagogy: How viable is the theory in classroom practice? *Journal of Teacher Education, 61*(3), 248–260. doi:10.1177/0022487109359775.

CHAPTER SIXTEEN

Dismantling Racism as a Strategy for Academic Success

ANA M. BECERRA

Before we can answer the question about the ways race affects and is affected by the educational process, we need to clarify the question itself. Asking if *race* is the cause allows us to skirt the real issue. This is a common occurrence when the conversation focuses on the academic data disaggregated by race that consistently show gaps in achievement levels between White students and students of color, leaving some to believe race is the cause. Keeping the talk limited to the notion of race allows people to disconnect from thinking about their role in maintaining systemic racism. Race is not the cause of the achievement gaps. Racism is the cause. The real question should be: In what ways does *racism* affect and is affected by the science educational process? Talking about racism requires looking inside ourselves and the ways that we unconsciously or consciously maintain racism in the institutional systems in which we live and work.

I learned about racism at an age when I could not articulate it, but I knew it. I was very young when the everyday messages I heard and actions I saw made me realize that people were valued in our society based on their skin color. I observed how the color of skin of my family was somehow not as valued as the color of White people's skin. My first big awakening occurred when I was 12 years old and lived in a town north of Sacramento, California. The year was 1966. It was Easter Sunday and like other Sundays, my parents took my sisters, brother, and me for a ride in our car. Soon we were driving away from the city onto a long stretch of highway south of Sacramento. Up ahead I could see a large group of people walking toward us carrying signs with the image of a black eagle with his wings spread out against a red background. My dad pulled the car over and parked. We all got out of the car. My dad put my baby brother on his shoulders and my mom held the hands of my little sisters as we joined the march of people. While on the walk back to Sacramento with the many people, my mom told me that they had been walking for days from the Central Valley of California and were on their way to the Capitol to tell Governor Edmund Brown that farmworkers needed improved working conditions. The march was led by leaders, including César Chávez and Dolores Huerta, of the United Farm Workers. When we finally arrived to the Capitol lawn, I saw people in the crowd become upset. My parents were also upset. My mom explained to me that although Governor Brown knew these people had walked for days to come to meet with him, he had left the city in the morning to fly down to Palm Springs to play golf with Frank Sinatra. I looked out at the sea of brown faces, and I knew we were not the people who mattered.

As young people of color, we sometimes learn about our place within a racist society subtly through micro-aggressions—the brief and commonplace daily verbal, behavioral, or environmental indignities that communicate hostile, derogatory, or negative racial slights and insults toward people of color that tell us how good or worthy of good we are. Other times it comes to us as like a slap in the face. Usually both happen.

While race is an arbitrary way of dividing humankind into different groups, racism is a system of oppression based on the concept of race that privileges White people and targets people of color. It is carried out by societal institutions or by people who have been conditioned by the society to act, consciously or unconsciously, in harmful ways toward people based on the color of their skin. Its invention has very clear historical roots in colonialism and serves the purpose of keeping some people on top and some people at the bottom; some in and some out. It is manifest through attitudes, behaviors, social structures, and institutional power. It is a system of structured dis-equality where the goods, services, rewards, privileges, and benefits of the society are available to individuals according to their presumed membership in a particular racial group (Gender Women's Studies 101, n.d.). A person of any race can have prejudices about people of other races, but only members of the privileged or dominant social group can exhibit racism because racism is prejudice plus the institutional power to enforce it.

Educators have been aware for several decades about the "pipeline" data that show higher levels by far of White recipients of advanced degrees in science and engineering in comparison to all other races. Data from the past 10 years show this to be relatively unchanged: the higher the level of academic attainment, the wider the gap (Fiegener, 2015). By providing a framework for having a discussion about the real question, we are better able to talk about why and how these gaps continue, and finally discuss ideas for what we can do about it.

History shows us that racism as a form of institutionalized oppression has had and continues to have negative consequences for students of color (Historical Timeline of Public Education in the U.S., 2006). This occurrence can have deep psychological effects on the few people of color who make it through the pipeline. A prime example is the well-known astrophysicist Neil deGrasse Tyson who questions why he ascended to the top, while so many other young people of color did not (Neil deGrasse Tyson on Racism and Sexism in Science, n.d.). When asked, "Are African Americans afraid of science?" deGrasse Tyson discusses the lack of minority representation in his field. As a student wishing to be a scientist in a White male dominated field, he encountered resistance. The young deGrasse Tyson was a curious student eager to learn more about science and become a scientist. Instead of encouraging him to pursue his goal, his teachers suggested he might want to think about becoming an athlete. He describes the struggles he had to overcome to become a scientist. It has left him wondering what it was that helped him to make it through the "forces of society" that he encountered "at every turn," which other students like him were not able to overcome. His life experiences tell him that these forces are real and before we can talk about why students of color are not represented in the sciences, we must examine these "forces of society" that leave them out of the science fields. We need to examine the ways that we, both privileged and targeted people, keep these systems, or forces, in place.

To understand the school experiences of students of color requires knowing the history and the centuries of racism upon which the U.S. education system was founded. Forty years ago, the textbooks I was assigned to read in my teacher preparation courses described children of color from backgrounds and cultures that were not from the mainstream White culture as "culturally disadvantaged" and "culturally deprived." A couple of decades later, the common term used was "students-at-risk." I believe these terms were used by well-intentioned educators speaking from their positions of dominance who assumed, unwarily perhaps, that non-White people who sometimes spoke languages other than English were somehow deprived or disadvantaged by virtue of not being White mono-lingual English speakers. These deficit theories served as the impetus for the development of Head Start and other compensatory education programs, some of which are still in existence today.

Although the terminology has changed, the dilemma of education today continues to emulate the dilemma of our society. The notion of race is socially constructed, based on our ability to identify by sight people whose ancestral origins were on different parts of the globe. As a social construct, race has served to reinforce extreme opinions and biases held by some educators and to condition significant segments of the population to accept passively inequality based on skin color as an unquestionable way of living and learning in the U.S. When this thinking becomes systemic in our institutions, such as schools, it acts as a system of oppression, or racism.

Racism requires two roles: a privileged group and a targeted group, whereby one group is systemically advantaged to the disadvantage of another group. This is not often talked about in educational circles, primarily because it tends to bring up feelings of anxiety and fear of a power shift with which many people are not comfortable. For White educators to understand how this phenomenon operates, they must be willing to do the work of examining and learning about White privilege and the ways it has targeted people of color. Likewise, people of color must understand the ways that they also hold racism as an oppressive system in place.

People often equate racism with individual prejudice, but it goes far beyond that. Institutional racism comes about when a group of people has the power to enforce their prejudices by creating social institutions to buttress their privileges. Individual thoughts, feelings, and beliefs about a group of people are the stereotype or prejudice one might hold, which remains a personal opinion. Interpersonal or group behaviors that act upon those stereotypes and prejudices lead to discrimination. When those group behaviors and actions become institutionalized and systemic in a society, then it is called oppression. Oppression around race, or racism, is ingrained in institutional and societal policies and practices. Racism is often subtle and difficult to detect. Beverly Tatum (1997) compares racism to smog: "sometimes it is so thick it is visible, other times it is less apparent, but always, day in and day out, we are breathing it in."

Over the course of U.S. history until recently, we have had a dual system of institutions: those for people of European ancestry, and those for everyone else. The segregated school system, which still exists today regardless of laws enacted to eliminate them, is an example of systemic oppression, as are hospitals, transportation, and other societal institutions. European colonization destroyed as best it could the cultures and peoples of Africa and the Americas. School curriculum has been constructed around European and Euro-American experiences. Students are regularly tested for their mastery of Euro-centric material. They are then rank-ordered and sorted into ability groups and tracks for different instruction. Those who enter school most familiar and comfortable with a Euro-American perspective about the world tend to leave school well prepared for the best college and life opportunities.

Becoming culturally White is actually a prerequisite for schooling in the U.S. Many students of color understand early on that they must navigate their home culture and language, as well as the school culture and language, sometimes believing that they have to choose one to the exclusion of the other. It was like that for me when I was discouraged from speaking our home language in the belief that it would be the best way for me to know English, the language of the classroom. I was given a name that my parents believed would be manageable by my great-grandparents and grandparents and, according to my mother, my teachers wouldn't find too difficult and try to change. That had been her experience when her teachers told her that her parents didn't know how to spell her name correctly and changed it from its Spanish spelling to English spelling. As she explained, although she knew it wasn't right, "it's just the way things were and we accepted it."

These actions and attitudes can contribute to the lack of roots or identity that can impede the academic development of a child. The risk is even greater when coupled with low economic status. Many minority parents, faced with racism and low paying jobs, have compromised their ethnic identity for a chance at the American dream. As a result, their children have lost their connection to their home language and culture. Instead they are often more familiar with negative and limiting stereotypes projected through the media than they are with poets, thinkers, and leaders who share their ancestry. To lose one's language is to sacrifice one's identity and roots. One must know where one came from before one can know where one is going.

In classrooms where varied ways of knowing, believing, assessing and behaving are considered to be innate deficiencies rather than cultural differences, the tendency for teachers to rely heavily on indoctrinated systems of racism is abundant. One result is the statistically improbable numbers of school suspensions issued to boys of color, and the number of students of color placed in special education and remedial programs. Counter to that is the disproportionally high number of White students in advanced and gifted classes.

During my years as a classroom teacher, I had the opportunity to design and teach a bilingual Gifted and Talented Education (GATE) class of students. Of course, the expectation was that I provide an enriching, rigorous, and differentiated curriculum for my students. During that same time, the director of the district's Migrant Education program asked me to teach a Saturday mathematics and science program for upper elementary and junior high migrant students. Each Saturday morning the migrant students would come to my classroom. Rather than use a different curriculum for them, I decided to teach them the curriculum I had used for my GATE students. Within a short amount of time, I was excited to see how the migrant students thrived with the "gifted" curriculum. These were students who spent most of the school days in remedial and English Language Development classes. I was also very upset knowing that during their regular school days they were deprived of an enriching and rigorous curriculum, because of the assumptions that were being made about their abilities. Of course, I needed to differentiate the curriculum according to their strengths and needs, but this was what I was also expected to do for the "gifted" students I taught during the week. At that time, I was piloting the nutrition module of the *Full Option Science System* (FOSS) curriculum (Lawrence Hall of Science, 2000), so the migrant students also were able to experience hands-on science. It was a first-time experience for them. They were so enthusiastic and came back each week eager to share about the experiments and investigations they had conducted on their own. Four years later, I was very pleased to learn that one of the students had been awarded first place at the county science fair. I was totally delighted when I learned that the theme of his science fair project was nutrition! I doubt that would have happened if the Saturday migrant program had consisted of more remedial mathematics and science lessons.

HOW DO WE TALK ABOUT RACISM

To be able to have the conversation about racism, we need to have clarity around a common language: "You can't deal with a problem if you don't name it; once you name it, you can think, talk and write about it. You can make sense of it by seeing how it is connected to other things that explain it and point towards solutions" (Johnson, 2006). There are multiple dimensions or ways of looking at the reasons why students of color are not, generally speaking, achieving at the same high academic levels as their White or Asian counterparts and this should be examined and discussed within a more holistic and historically specific sociocultural context. To have this conversation, we need a common language. In addition to the definitions of race and racism at the beginning of this chapter, it is helpful to share a common understanding of the meanings of a few basic terms:

> *PREJUDICE:* "A judgment or opinion that is formed on insufficient grounds before facts are known or in disregard of facts that contradict it. Prejudices are learned and can be replaced by new learning and understanding." (Women of Color Network, 2007, p. 1)
>
> *STEREOTYPE:* "An exaggerated or distorted belief that attributes characteristics to members of a particular group, simplistically lump them together and refuse to acknowledge differences among members of the group." (Women of Color Network, 2007, p. 1)
>
> *PRIVILEGE:* "Operates on personal, interpersonal, cultural, and institutional levels and gives advantages, favors, and benefits to members of the privileged group at the expense of members of target groups. In the United States, when talking about race, privilege is granted to White people. Privilege results in unearned access to resources that enhance one's chances of getting what one needs or influencing others in order to lead a safe, productive, and fulfilling life. It is characteristically invisible to people who have it. People in privilege groups often believe that they have earned the privileges they enjoy or that all people could have access to these privileges if only they worked to earn them (frequently articulated as, 'Pull themselves up by their bootstraps'). In fact, privileges are unearned and they are granted to people in the dominant groups whether they want those privileges or not, and regardless of their stated intent. Unlike targets of oppression, people in privilege

groups are frequently unaware that they are members of the privilege group due to the privilege of being able to see themselves as individual persons rather than stereotypes." (Women of Color Network, 2007, p. 2)

TARGETS OF OPPRESSION: "Members of social identity groups that are disenfranchised, exploited, and victimized in a variety of ways by agents of oppression (privilege group members) and the agents' systems or institutions. Targets of oppression are subject to containment, having their choices and movements restricted and limited, are seen and treated as expendable and replaceable, without an individual identity apart from their group, and are compartmentalized into narrowly defined roles" (Women of Color Network, 2007, p. 3). When talking about race, the targets of oppression are people of color. Targeting results in the denial of access, resources, and opportunities that might enhance chances of getting what one wants and influencing others. They "are subjected to exploitation, marginalization, powerlessness, cultural imperialism, and violence. Targets of oppression are kept in their place by the agent of oppression's ideology, which supports oppression by denying that it exists and blames the conditions of oppression on actions of the members of the targeted group(s)" (Women of Color Network, 2007, p. 3). "They have fewer 'life chances' or benefits as a result of their membership in a particular social group" (Women of Color Network, 2007, p. 3). For example, in addition to finding themselves on the lower levels of the academic achievement gaps, there is a higher likelihood that people of color will be arrested and incarcerated in higher numbers, earn less income, disproportionally disciplined in school, will and die at younger ages than White people. (Rudd, 2014; Whisner, 2015; Wise, 2010)

COLLUSION occurs when privilege group members and target group members work to keep racism in place. This is done intentionally or unintentionally, consciously or unconsciously, and by action, inaction or silence. (Language Justice Network, 2015, para. 29)

PRIVILEGE GROUP COLLUSION occurs when privilege groups and privilege group members do things to perpetuate systems of oppression. Also called Internalized Superiority, this is when members of privilege groups either consciously or unconsciously learn to look at themselves, each other, and society through a distorted lens such that the structural privileges they enjoy and the cultural practices and values of their group are represented as, seen as, and felt and believed to be normal and universal. (Language Justice Network, 2015, para. 69)

TARGET GROUP COLLUSION occurs when target groups and target group members internalize and transfer the systematic mistreatment to perpetuate their own oppression. Also called Internalized Oppression, this is "the destructive patterns of feelings and behaviors experienced by the targets of oppression, turned inward upon themselves and directed at each other" (Language Justice Network, 2015), whereby "people in the target group make oppression internal and personal by coming to believe that the lies, prejudices, and stereotypes about them are true. Members of target groups exhibit internalized oppression when they alter their attitudes, behaviors, speech, and self-confidence to reflect the stereotypes and norms of the privilege group. Internalized oppression can create low self-esteem, self-doubt, and even self-loathing. It can also be projected outward as fear, criticism, and distrust of members of one's own and other target groups." (Women of Color Network, 2007, p. 4)

Conversations about racism that include blaming, denial, and expressions of guilt are unproductive. Progress toward educational equity requires that educators recognize racism's effects on children, heal from how they themselves have been hurt by racism, and identify and change harmful practices and policies. Until racism is eliminated, we must give students the skills to combat it. When starting a conversation about racism with White people and people of color, I find it helpful to assure them that I hold the perspective that human beings are born without prejudice and all forms of bias, from extreme bigotry to unaware cultural biases, are acquired and actually imposed on the young person (Becerra & Weissglass, 2004). Privilege group members seem to find this particularly supportive to hear, enabling them to move onto the more challenging conversation. As a facilitator of color, it also helps to keep me hopeful in this work.

WHY WE HAVE TO TALK ABOUT RACISM

Racism, classism, and sexism and other forms of bias are serious issues facing U.S. society and education that are usually not discussed. Talking about them is necessary, not to lay blame, but to figure out better ways of raising our children and educating our students (Becerra & Weissglass, 2004). We have to talk about racism, especially in the context of collusion, because that is the glue that keeps it in place. Students of color are immersed in micro-aggressions. Privilege group collusion makes it so well-meaning educators often do not see the impact of negative messaging that students receive. Perpetrators of micro-aggressions are often unaware that they engage in such communications when they interact with racial/ethnic minorities (Sue et al., 2007).

When students of color integrate these micro-aggressions into the core of their self-identities, there are a multitude of consequences. One consequence is the tension of the burden of "acting White" felt by many students of color. This hypothesis explains that students of color limit their effort and engagement in school because they believe that doing well in school means "acting White" (Fordham & Ogbu, 1986). I recall this feeling being shared among my non-White classmates when I was in high school in the late 1960s and early 1970s. I heard it again in the early part of this century when a Native American middle school student explained to me that he did not want to do well in school because, as he said it, "I am not a White boy."

This phenomenon sometimes manifests as willful not-learning, or refusal to learn, which happens when a student's intelligence, dignity, or integrity is compromised by a teacher, an institution, or a larger social mindset. This creates a "culture of resistance" that is the collective autonomy exercised by students of color who experience marginalization in schools and society (Kohl, 1995).

STARTING WITH SELF

Educators are an important force in helping many people overcome the effects of societal bias and discrimination but they can also act as perpetrators of the inequalities and prejudices in the society. The success of an intervention depends on the interior condition of the intervener. The work calls for growing our souls while we seek to transform society (Karoff, 2006). Educator operating from an unrecognized position of a culture of power unknowingly fail to reveal the rules of the culture of power to students since they are "frequently least aware of—or least willing to acknowledge" the cultural power they hold (Delpit, 1995, p. 24).

The ways in which racism affects and is affected by the educational process requires examination and work at multiple levels. Self is the first element we must examine. It is ourselves that we take into the classroom. The self is also the level that people sometimes want to skip over, because it can be painful and, by its very nature, very personal. Yet, we have all been impacted by racism, whether as a privileged or targeted group member. We must start here.

The effects of growing up in a racist society include the planting of deep roots of internalized racism in teachers of color, as illustrated by the story of Alice, an African American mathematics and science teacher at a junior high school (Becerra, 2001). Growing up, Alice struggled to survive in her high-poverty neighborhood in South-central Los Angeles. Although she dropped out of high school at 15, she believed she was smart. Alice held onto the dream that despite the obstacles, one day she would attend college. She saw this as her only way out of the struggles she lived daily. Hers was a world of inequities and human tragedy that few in the mainstream culture will ever know. Even through her serious involvement with drugs and the birth of her son before she was 18 years old, Alice held on to her plan for finding a way to a better life. Alice had friends she knew were so smart, but were "rejected by the system." This makes her passionate about telling her students about the things that I wasn't told:

> Sometimes I think I get too passionate, and I get my feelings hurt when I don't get the response I want. I hurt a lot about that. I'm burning with this thing that I care about. If I could wake up one morning and not care about it, my life might be easier. I wouldn't have to do any of the things I do other than teach, go home, and get my check on the first. But this is the thing that I care about. It causes me to cry at night and it causes me to dance on the tables, too. (Becerra, 2001)

As Alice gains confidence in her expertise and knowledge as a teacher, she seeks ways of working with others to bring about the changes that she believes need to happen. Unfortunately, Alice also continues to be saddened and increasingly embittered by the lack of support at her school site. Alice believes racism is at work:

> I think that color has a lot to do with it. If we look at people who know mathematics, you wouldn't look at an African American, because we tend to score low or lowest ... So here is a person who traditionally hasn't scored high, wasn't in the AP classes, wasn't in the A-F classes, and here she emerges. [I think they wonder] "I never saw you in any of my classes." I think it has some backlash to it. So people don't see me as a mathematician ... my color stands in my way. (Becerra, 2001)

Alice knows she has to do a lot of personal work to overcome the effects of the racism, both the overt aggressions and the micro-aggressions that she experienced in her life, so that she does not internalize and act upon them. She works diligently to contradict the racism that she knows her students experience.

Likewise, White teachers need to reflect on the messages they have received about people of color and about themselves as White people. Students come to our classrooms with radar. They can detect the level of expectations teachers hold for them. They will rise, or sink, to that level. It can be hard work for a teacher to contradict the negative messages students have received throughout their lives about themselves, their intelligence, and what they can or cannot do. Students' internalized oppression is like a heavy load of bricks. It takes strategies and collaboration, working in partnership with the students to lift.

WHAT WE CAN DO

Educators need to move from the paradigm of talking about teaching as the subjects they teach: "I teach mathematics" or "I teach science", etc., to a mindset focused on who they teach: "I teach students"—each with his or her needs, interests, and culture. Parker Palmer (2007) said it well:

> The question we most commonly ask is the "what" question—what subjects shall we teach? When the conversation goes a bit deeper, we ask the "how" question—what methods and techniques are required to teach well? Occasionally, when it goes deeper still, we ask the "why" question—for what purpose and to what ends do we teach? But seldom, if ever, do we ask the "who" question—who is the self that teaches? How does the quality of my selfhood form—or deform—the way I relate to my students, my subject, my colleagues, and my world? How can educational institutions sustain and deepen the selfhood from which good teaching comes? (p. 4)

Educators are an important force in helping many people overcome the effects of societal bias and discrimination but schools also serve to perpetuate the inequalities and prejudices in society (Becerra & Weissglass, 2004). Racism affects the educational process when educators of color and White educators carry their collusion into the learning process. This is why we must look inside ourselves to heal from the ways we have bought into systemic oppression and perpetuate it in today's classrooms. This requires listening to and learning from others who are different from us (Davis, 2012). Deep transitional change occurs when we integrate new knowledge and take action to not do business as usual. The journey getting to this place can be filled with feelings of uncertainty, disequilibrium, and doubt. Changing our policies and practices often requires dismantling our own assumptions about what we thought we already knew.

An important way educators can turn this around is by focusing on "3 R's": Relationship, Relevancy, and Rigor. Building relationships with students means building strong learning partnerships. Through these relationships we are enabled to provide move relevant instruction for our students—relationships plus relevance enables us to increase rigor. Students are most open to learning from a teacher they know cares about them and will make the effort to know them. Having this type of access to the students requires trust. Yet, teachers and students report feeling disconnected from each other and distrustful at times. Learning partnerships are rooted in the concept of personalization that many schools have embraced, but go one step further to build relationships for the specific purpose of directly improving students' academic skills. These partnerships set the stage for the teacher's effective use of formative assessment and the students' conscious movement toward independent, self-directed learning (Hammond, 2015).

There are three components of a learning partnership: rapport, alliance, and cognitive insight. Establishing a level of rapport with individual students and the classroom community as a whole creates a positive learning environment and builds trust so that students are not so likely to feel guarded and defensive. This is especially important for struggling students who try to cover up their low achievement by acting out or trying to be invisible during class.

To make progress on this very complex problem, it will be necessary to improve alliances between people from different backgrounds, experiences, and identities (Becerra & Weissglass, 2004). To be an ally requires a commitment to dismantling oppression. This means a willingness to educate oneself about oppression, to learn from and listen to people who are targets of oppression, and to examine and challenge one's own prejudices, stereotypes, and assumptions. It means working through feelings of guilt, shame, and defensiveness in order to understand what is beneath them and what needs to be healed, to learn and practice the skills of challenging oppressive remarks, behaviors, policies, and institutional structures, and to act collaboratively with members of the target group to dismantle oppression.

Once rapport has been established, it is important to move toward building an alliance with the students—explicitly identifying a goal or "learning target" and making a pact to work together in the service of improving that identified skill or concept understanding. The alliance stage of learning partnerships is critical. Positive rapport and a strong alliance pave the way for the teacher to gain an understanding of how students are learning, called cognitive insight. It is at this stage that the deep work of interventions takes place. The teacher is able to help the students address misconceptions, as well as learn to use new strategies effectively so that they are able to self-correct and learn more independently, hence setting the stage to accelerate their own learning.

Students of color often come from cultures that stress community or the group over individuality. For example, the African world-view of Ubuntu, which essentially means "I am, because we are" fosters a concept of identity that is based on relationships with others as opposed to individuality. Similarly, people from indigenous American cultural backgrounds that include Latinos or Chicanos, frequently grow up within communities that stress the similar concept of Inla'kech—"You are my other me," which also emphasizes community rather than the European individualistic worldview of "I think, therefore I am." Both concepts, Ubuntu and Inla'kech, rely on relationships that are typically based on mutual feelings of belonging, attraction, or some other feeling of connection.

A useful framework for examining and talking about relevancy is called *Mirrors and Windows* (Schwartz, 2007). This framework provides a tool for understanding students' experiences in school. Mirrors are what students experience when they see themselves and their culture reflected back to them in their daily experiences in school and in society, which generally provide information about the dominant or privilege groups. Students from the privilege group receive many more reflections of themselves, their history, and their culture through the curriculum and the images in which they are immersed. Because privilege groups come to see themselves as the norm, it is often not obvious to them the ways that these positive ideas and images are reinforced for them on a daily basis. On the other hand, students from targeted groups experience an absence of information about the histories of people that look like them and they do not see themselves in the curriculum they are provided in school. It is as if students who are members of targeted groups are looking out a window, learning about the reflected group's history and culture through images and information that reflect the privilege group. Instead, targeted group students often receive information about themselves that is not positive, particularly through the media and other information they receive about people who look like them. Whether they are members of the privilege group that see themselves reflected through the school and society (mirrors) or are members of the targeted groups that spend their time learning a lot about the privilege group (windows), the effect on students materializes at many levels.

When considering orientation to self, students who experience mirrors most of the time tend to have a strong knowledge of and belief in self. They tend to demonstrate more self-confidence and a belief of self as "normal." On the contrary, students who experience life and school as if they were looking out a window know a lot about the mirrored group, but tend to have low knowledge and belief in self. They are apt to demonstrate lower self-confidence and a belief in self as "abnormal."

Regarding their orientation to others, for students who mostly experience mirrors there is a higher tendency for low knowledge of non-reflected groups or targeted groups, a belief that non-reflected groups are "less than," and a low trust of people from targeted races, cultures, or ethnicity. Quite the reverse, students who experience school as though they were looking through a window tend to know a lot about the reflected group's culture and history. They often internalize the belief that people who are members of the reflected group are seen as "better" and this sometimes leads to feelings of resentment or they may hold anger towards reflected group members.

With regard to students' orientations to school and academic success, those who have experienced mirrors for the majority of their schooling have a higher sense of school culture mirroring their own culture. They tend to have a stronger feeling of connection to school, and they tend to see academic success as normal, and even expected, which tips their potential toward higher academic achievement. Students who experience a school culture that does not reflect their culture tend to have a lower connection to school and feel academic success is the domain of the reflected group. They sometimes come to believe that succeeding academically equals *selling out* by trying to be the reflected culture (e.g., "trying to be White"), which leads to a rejection of school and an academic identity. When teachers provide mirrors for students who might otherwise experience only windows, they bring relevance to all students' academic experiences. Students who only experience mirrors benefit from windows into the histories, lives, experiences, and ways of knowing held by people of color.

Many years ago when I was a classroom teacher, I taught my students the science curriculum of the *Second Voyage of the Mimi* (Nelson, 1988), a curriculum with video depicting a sailing crew on a ship named the *Mimi*. The basis of the series was the exploration of the Yucatán Peninsula of Mexico. Each episode started with a fictional story that linked it to a topic of natural science, which was followed by a non-fictional lesson about the science brought out in the story. In one of the episodes, a character on the ship encountered a sting from a scorpion. The other crew members rushed her to a Mayan curandera (medicine person) who used medicinal plants to help cure her. This fictional part was followed by a lesson on medicinal botany, which taught about the research on the traditional origins of medicine plants, the history of their use, and the link to modern science. As I was watching the video, I was taken back to memories of my maternal grandmother who was a curandera and used plants to help cure illness and help people maintain wellbeing. When the video was over, I shared these memories with my students, the majority of whom were of Mexican ancestry. I asked them if they knew anyone with knowledge about plants used for medicine. The majority of my students indicated that they did. I then asked them to find someone who could tell them about how they use the medicine plants and to bring in a plant sample, if they could. I was amazed and very pleased when nearly all my students brought in the plant samples along with the written instructions given to them about their use. We proceeded to write a bilingual (Spanish and English) medicinal plants book, complete with the writings and pressed plant samples. The students were eager, because this was relevant to their lives and connected them to the elders in their communities. I can only wonder how connected I would have felt as a student to the instruction in school if I had experienced lessons like this that were relevant to my life. Even more, I am reasonably sure I would have valued what my grandma knew so much more if I'd had a teacher who told me that the knowledge of the people I come from was important.

While race benignly defines the differences in our skin colors, ethnicity and cultural backgrounds, racism is the poison that is more than the sum of individual prejudices. Racism maintains a vicious loop that affects the science educational process, and for that matter, the educational process as a whole. Oppression breathes life into itself when it becomes embedded in the educational process that institutionalizes oppressive practices and helps to maintain its place within the system. We need to reach into the past to examine the historical, institutional, and societal foundations of the practices of schooling and the way we educate our youth today. It requires that we first look within ourselves. This first step can be the most challenging. However, it is necessary if we are to make the necessary changes in the ways we educate young people in order to level the field for every student's access to quality science educational opportunities and achievement without barriers or limits.

REFERENCES

Becerra, A. M. (2001). Alice and Rachel: Teacher leaders of color. In J. E. Jacobs, J. Rossi-Becker, & G. F. Gilmer (Eds.), *Changing faces of mathematics: Perspectives on gender.* Reston, VA: National Council of Teachers ofMathematics.

Becerra, A. M., & Weissglass, J. (2004). *Take it up: Leading for educational equity.* Santa Barbara: The National Coalition for Equity and Education.

Davis, B. (2012). *How to teach students who don't look like you: Culturally responsive teaching strategies* (2nd ed.). Thousand Oaks, CA: Corwin.

Delpit, L. (1995). *Other people's children: Cultural conflict in the classroom.* New York, NY: New Press.

Fiegener, M. K. (2015). *Science and engineering degrees, by race/ethnicity of recipients: 2002–2012.* Arlington, VA: National Science Foundation.

Fordham, S., & Ogbu, J. U. (*1986*). Black students' school success: Coping with the "burden of 'acting White.'" *The Urban Review, 18*, 176–206.

Gender Women's Studies 101. (n.d.). Retrieved from https://quizlet.com/82941634/gender-womens-studies-101-flash-cards/

Hammond, Z. (2015). *Culturally responsive teaching and the brain: Promoting authentic engagement and rigor among culturally and linguistically diverse students.* Thousand Oaks, CA: Corwin.

Historical Timeline of Public Education in the U.S. (2006). Retrieved from https://www.raceforward.org/research/reports/historical-timeline-public-education-us

Johnson, A. G. (2006). *Privilege, power, and difference.* Boston, MA: McGraw-Hill.

Karoff, P. (2006). *The world we want: New dimensions in philanthropy and social change.* Lanham, MD: Alta Mira Press.

Kohl, H. (1995). *I won't learn from you: And other thoughts on creative maladjustment* (2nd ed.). New York, NY: The New Press.

Language Justice Network. (2015). *Glosario de términos de justicia social – Social justice glossary of terms.* Retrieved from http://media.wix.com/ugd/53f282_e48ea44e7a144cbdab6831961616fffa.pdf

Lawrence Hall of Science, University of California at Berkeley. (2000). *FOSS: Full option science system.* Nashua, NH: Delta Education Inc.

Neil deGrasse Tyson on Racism and Sexism in Science. (n.d.). Retrieved from http://rhrealitycheck.org/video/2014/05/01/neil-degrasse-tyson-racism-sexism-science-field/

Nelson, J. (Producer). (1988). *The Second Voyage of the Mimi* [Television series]. Public Broadcasting System.

Palmer, P. (2007). *The courage to teach.* San Francisco, CA: Jossey-Bass.

Rudd, T. (2014). *Racial disproportionality in school discipline: Implicit bias is heavily implicated.* Kirwan Institute for the Study of Race and Ethnicity, The Ohio State University. Retrieved from http://kirwaninstitute.osu.edu/racial-disproportionality-in-school-discipline-implicit-bias-is-heavily-implicated/

Schwartz, J. (2007). *Mirrors & windows: A tool for cultural relevance, understanding students' experience in our schools.* Santa Barbara, CA: Just Communities.

Sue, D. W., Capodilupo, C. M., Torino, G. C., Bucceri, J. M., Holder, A., Nadal, K. L., & Esquilin, M. (2007). Racial microaggressions in everyday life: Implications for clinical practice. *American Psychologist, 62*, 271–286.

Tatum, B. D. (1997). *Why are all the Black kids sitting together in the cafeteria? And other conversations about race.* New York, NY: Basic Books.

Whisner, M. (2015). *Race in the criminal justice system.* Retrieved from https://lib.law.washington.edu/content/guides/racecrim

Wise, T. (2010). *Colorblind: The rise of post-racial politics and the retreat from racial equity.* San Francisco, CA: City Lights Books.

Women of Color Network. (2007). *WOCN power and privilege packet.* Retrieved from http://www.wocninc.org/wp-content/uploads/2013/10/WOCN-Power-and-Privilege-Packet.pdf

CHAPTER SEVENTEEN

In What Ways Does Race Affect the Educational Process?

Challenging the Homogenization of Academia

CAROLINA CASTANO RODRIGUEZ AND LAURA BARRAZA

So you are now an Aussie!
Well, I have an Australian passport now.
So, are you no longer Mexican/Colombian?
I was born Colombian/Mexican, I will die as a Colombian/Mexican.
But didn't you have to give up your Colombian/Mexican citizenship?
If I would have asked to do that, I would not be holding an Australian Passport.
—(Personal communication with our Australian born friends, March 2015)

Critical race theorists argue that to address social inequalities associated with race, educative institutions need to move beyond the mere increase in numbers of students, academics and professional staff from diverse cultural backgrounds. What is needed is to understand all the challenges and opportunities that arise from a different culture than a white-Eurocentric culture provides (Ladson-Billings, 2012). Much has been explored regarding race in schools and the inclusion of issues of race in policy documents. However, we, the authors of this chapter, consider that issues of race among academics in tertiary education have been understudied. As female academics that have been educated in South and Meso-America and have worked in our countries of origin, Mexico and Colombia for several years before moving to Australia, we believe our experiences and narratives offer different lenses to explore race in education. Our narratives provide a counter-story to that of conceiving race from a deficit perspective. We argue that our South and

Meso-American background provides us with opportunities and visions of education that enrich, broaden and progress social justice agendas.

We start this chapter stating our view of the term "Latin America." We reject the term "Latin America" and "Latinas/os" to represent our cultural backgrounds as we consider they ignore our Indigenous and African heritage and privilege European heritage. Moreover, we consider that the low number of academic authors from South and Meso-America contributing in international peer-review journals and academic conferences in which English is the main spoken language is sometimes due to choice rather than lack of opportunities. We do not conceive of ourselves, or our cultural background, as "disadvantaged." Instead we consider that our vision and purpose as academics go beyond conferences and publications in international journals.

We argue that the discourse of opening spaces for participation of academics from non-English speaking or non-European nationalities needs to shift. If the international academic community is interested in hearing their/our voices and include them in their vision and structure of education, the conversation needs to change. It needs to move towards a vision of Meso- and South America as distinctive and strong resilient academic communities that have much to offer to the international community and whose academics in some instances choose not to participate in the international structures as a form of resistance.

In an era of globalization when there is an increase of flow of ideas, ideologies and humans across national borders, it is important for the international academic community to consider questions such as: *What does it mean for someone from another culture to move to an English speaking or Eurocentric culture? How can the international community incorporate the visions and practices of academics that come from non-Eurocentric communities without falling into colonialist or paternalistic approaches? How can the structures and vision of success better reflect the diversity of views that different cultures provide?* As Gloria Ladson-Billings (2012) argues: "Until we begin to carefully examine the way race and racialized thinking influence our work, we will continue to perpetuate destructive thinking about the capabilities of learners based on race" (p. 115). Her research focuses on learners. However, we argue that the same concerns are relevant to every structure of academia, including the vision that is applied to academics from non-English speaking or European cultures.

RETHINKING CRITICAL RACE THEORY

Critical race theory (CRT) emerged from the field of critical legal studies to analyze social inequalities between dominant and marginalized racial groups. CRT has been used as an analytical framework in education (Ladson-Billings, 1995).

CRT "challenges dominant ideology, which supports deficit notions about students of color" (Solorzano & Yosso, 2001, p. 3).

CRT serves to explore and understand how the way research functions, serve to support ongoing narratives of superiority/inferiority (Ladson-Billings, 2012). CRT discourses in education have tended to focus on a deficit perspective, which sees achievement gap as central to the CRT agenda. Achievement gap discourses frame dominant racial groups as the norm to which other ethnic groups are compared and should aspire (Foster, 1999). Similar to the achievement gap discourse, it could be argued that academia has established similar discourses that frame academics from non-English speaking or non-European countries from a deficit perspective. For instance, the international university rankings often provide a hierarchy of those "quality" universities. Within the top universities are often excluded universities from South and Meso-America (for example, the QS World University Rankings [http://www.topuniversities.com/qs-world-university-rankings] and the *Times Higher Education* World University Rankings [https://www.timeshighereducation.com/world-university-ranking]). International testing and rankings promote discourses that conceptualize academics from other cultures and ethnicity from a deficit perspective. This positioning in an international context raises serious issues of promoting sameness in academia; of providing structures and measures that all academics, academic research, and academic institutions should aspire to. This internationalization of what is constituted as quality in academia has the risk of missing the opportunity to hear and learn from those academics in non-English speaking and non-European countries. Moreover, we consider that the issue of race in academia is still understudied. For instance, is the low number of academics from non-English speaking or European countries in international conferences and research viewed by international academics from a deficit perspective or as a form of resistance? How do academics from non-English speaking or non-European countries perceive themselves and their participation in an international context? Our personal experiences in the international context of academia provide a counter-story for the deficit perspective in which academics working or coming from South and Meso-America are portrayed.

DECOLONIZING THE "SOUTH"

To decolonize *el conocimiento* (knowledge) produced in Meso- and South America in our current global neoliberal state requires epistemic decolonization and re-centering analyses of power and the cultural context in which collective identities are formed (Hernández Castillo & Suárez Navaz, 2008). Perhaps one of the greatest challenges of the 21st century in introducing a discussion of Meso- and South American women into the international community is moving beyond the stereotypes and assumptions that they may have about the region and the roles that women play in those developing nations (Zimmerman, 2006).

Throughout the past five centuries, there have been a number of prominent female figures who have played a significant role in shaping the course of Meso- and South American women's identities, e.g., La Malinche, Sor Juana Ines de la Cruz, Eva Peron, La comandante Ramona, Rigoberta Menchú, and Benedita de Silva just to mention a few (Lanyon, 1999). These women have been agents of change in the region through their intensive work in areas such as social justice, indigenous rights, equity, multiculturalism, environmental policies, and empowerment. Our identities as academics have been shaped no doubt by many of the main ideas from these women. For example, through Rigoberta Menchú (Indigenous, Nobel Peace Prize, 1992) we understand multiculturalism as an introspective journey in which we first see and recognize the diversity within ourselves and within our own existence as a means of transcendence to a peaceful coexistence of many cultures. Menchú deals not only with the political system but also with the fight for indigenous peoples to demand recognition of their values and cultural systems. As academics we apply this vision by encouraging our students to think and reflect about the notion of a diverse community and work towards respect and value of all cultures, even within our nations of origin.

In terms of shaping our theoretical, ontological and methodological practices in the field of pedagogy, not only female social leaders such as Rigoberta Menchú have provided as with role models but also education leaders and activists such as Paulo Freire have strongly influenced our work. Following those leaders, we have identified the following principles as a foundation for our work:

- *It is not enough for people to come together in dialogue in order to gain knowledge of their social reality.* They must act together upon their environment in order to critically reflect upon their reality and transform it through further action.
- *To enter into dialogue presupposes equality amongst participants.* Each must trust the others and be willing to value all voices; there must be mutual respect and love, care and commitment. Each one must question what he or she knows and realize that through dialogue existing thoughts will change and new knowledge will be created.
- *The process of developing a critical awareness of one's social reality through reflection and action.* Action is fundamental because it is the process of changing the reality. Paulo Freire (1973) argues that we all acquire social myths that have a dominant tendency, and so learning is a critical process, which depends upon uncovering real problems and actual needs.

These three main principles have had a strong impact in our experience as academics teaching in our countries of origin. Interestingly these same principles have created some of the tensions we faced as academics in Australia. We provide examples from our own experience to illustrate and provide counter-examples to colonialist and paternalistic views of South and Meso-America.

TRESPASSING THE BOUNDARIES OF ACADEMIA: GROWING AND WORKING IN A "DEVELOPING" NATION

Carolina's Experience, Growing and Working Under Ideals of Freedom and Equality in Colombia. In the center of the "plazoleta" of one of the most prestigious universities of Colombia, the Universidad Nacional de Colombia, lies a full wall size image of the Che Guevara. In fact "Plaza Che Guevara" is how students and staff call it. Although I did not study or work at Universidad Nacional de Colombia, I have learned about Che Guevara's revolutionary ideals since secondary school. At both the private university that I attended and later at the university that I worked at as an academic, ideas of freedom, social justice and peace were often included in conversations among colleagues and friends. While working as an academic, I was often participating in discussions with my colleagues regarding how to develop educative projects that contributed to addressing social justice and violence issues. Moreover, I participated in elective units that focused on peace education and critical thinking which were highly popular among students.

As an academic, I participated in projects to revise curricula from different faculties at the university and adapt them towards social justice aspects, for example, for the faculty of Medicine. I also participated in projects in other South American countries that were aimed at improving the educative programs in socio-economically disadvantaged communities. This was the nature of the goals for scholars, which went beyond publications to projects aimed at having social impact. In no moment did I feel pressure to publish in international journals or generate income from research grants as I currently constantly feel in Australia.

As an early career researcher I was mentored to create change. I felt that the quality of my work was measured in terms of how much I could empower students to contribute to help others and the impact that my research could have in communities. I grew up and worked immersed in a culture that sees education as a tool for emancipation.

Even though I did not grow up in a socio-economically disadvantaged family, my sense of social justice and interest to contribute towards it came not only from the education I experienced but also the social context in which I was immersed in Colombia while growing up. In the 1990s, Colombia was going through one of the most difficult times in history as part of the internal conflict of the nation with rebel, terrorist groups and drug cartels. Most Colombians leaving throughout the

1990s experienced situations of violence directly affecting us, and close members of our families. Those experiences shaped my perceptions of what it means to grow up among war and poverty, which provides the foundation for my passion and interest in contributing to a more just and peaceful world.

My own cultural background and the stories that make up my identity as a Colombian are what give me the strength and inspiration to be a role model as a lecturer. I do not see identity, ethnicity and cultural background as disadvantaged. Rather my experiences as a Colombian provide me with skills to be able to support others in similar situations and to work with communities which are positioned as less successful or as disadvantaged, such as poor communities with little access to basic resources or communities highly affected by violence.

Thus, rather than identifying myself from a deficit perspective, I actually see my background and experiences as an advantage to the construction of my identity as an academic working towards social justice and emancipation.

Laura's Experience Working Towards Social Justice and Multicultural Environmental Education in Mexico. Following activist and emancipatory ideologies and growing up in a nation with more than 60 indigenous groups (Comision Nacional para el Desarrollo de los Pueblos Indigenas [CDI], 2014), I offer a strong link of the social and political issues with the educative process to engage my students in transformative processes. Through my teaching practice I encourage an active, critical, autonomous and participatory approach. I support students to make cognitive links between their scientific knowledge and their real life experience. I promote a critical approach to science and environmental education, stimulating active participation, promoting pro-environmental attitudes in society and a responsible and caring commitment to solving local problems.

In all that I teach and investigate, my background as a woman from Meso-America is reflected. The educational model followed in my teaching practice contemplates two pedagogical principles: 1) creative learning through participatory teaching and 2) applicability of knowledge through participatory actions. These two principles are based in respect of ideas and creative thinking of students as individuals and active members of a community. I have developed educational programs, together with the community with participatory methodologies based on the interests and necessities of the local people.

I often provide research examples from the studies I carried in Mexico in order to show a different view and perspective of how students and participants can analyse and reflect about environmental and sustainability problems happening in their community and different parts of the world. My central focus has been in promoting and educating adults who work towards the construction of a more inclusive, responsible and just society. For instance, I have generated a participatory process of active learning with high school students, teachers and parents in rural indigenous communities of Mexico and Alaska, U.S.A. With Barbara Bodenhorn (University of Cambridge) and the support of the National Science Foundation (NSF), and other governmental and non-governmental organizations, I have organized an international exchange program where Zapotec (Oaxaca, Mexico), Purhepecha (Michoacán, Mexico) and Iñupiak (Alaska, USA) teenage students have enriched and widened their scientific and cultural experience. Ten years after this first exchange program, one the indigenous girls from Mexico who participated is now a biologist with a master's degree in Genetics and recently became an academic member of the University of La Sierra de Juarez in Oaxaca, Mexico. I have also been responsible for establishing the foundation of a rural school in San Juan Nuevo Parangaricutiro, an indigenous community in Mexico. It was created in response to the needs and interest of the community. Thanks to my negotiations and the work I promoted, 16 women of the indigenous community participated for one year in this program to become teachers who focus in more participatory approaches to education. The school has been working in the community for twelve years now.

TRESPASSING THE BOUNDARIES OF ACADEMIA BETWEEN DIFFERING SOCIETIES—WORKING IN A "DEVELOPED" NATION

Carolina's Journey to Encourage Care and Compassion in the Australian Society. During these four years as an academic in Australia, I have had little encouragement to provide an education that transforms lives, to empower students to critically analyse different social aspects of science and science education. Conversations between colleagues and with my supervisors regarding how education can contribute to change communities are often absent. Instead they often focus on the number of publications in international journals and money generated through research funding. Thus, the focus of academia in Australia has presented me with a limitation rather than opportunities to pursue the transformative goals I have as an academic. Despite this, my background and identity as an academic from Colombia have helped me to navigate those limitations. For instance, for the final unit of science education that I offer to pre-service teachers in their final year, I have included a strong focus on social aspects of science education. Students learn about diverse theories and frameworks that encourage the consideration of social aspects of science such as issues of access, ethics and social/eco justice. They design a science unit of work for one class in a school located in an area of Melbourne with large communities that include political and religious refugees. These units regard students' histories as positive and powerful components of cultural capital and incorporate their experiences as foundations for learning, acting, and positioning.

The school experience is a highlight for most students. I have been in contact with these schools since my first year of working in Australia. The rationale for this was that for many of the students, despite studying in a Catholic university whose main mission is social justice, there is little opportunity to learn and interact with socio-economically disadvantaged communities. Most of the children in the schools with which I have formed partnerships come from refugee backgrounds and are English learners, representing an experience that students otherwise might not have. In all the units I teach I follow critical pedagogies such as transformative learning theory. Transformative learning theory, first suggested by Jack Mezirow (1991), is based on the notion of education as transformational and emancipatory (see Carter, Castano Rodriguez, & Jones, 2014).

Similarly, regarding research, I have found that the conversations in Australia are focused more towards income generation and generalization of results. As my identity as an academic has been influenced by theories of emancipation and transformation, my interest in research lies in working with and for local communities. Following action research theories, the nature of the research I carry out contributes to construction of theory but is not meant to be replicated in the same in diverse social contexts. Moreover, in some cases the communities I work with have such particular needs that rarely do they receive any support from national and international organizations. Those are the communities in which I am most interested in working, as they have been disempowered and almost forgotten by society.

In fact, in the last three years I have carried out research projects with activist leaders in the field of human and animal rights. In working with activist leaders, I have found a space to work with local communities on projects that aim at transforming lives. Their work is inspiring and empowering, and highly relevant for my classes to educate future teachers who are committed to social and ecological justice. Moreover, the work with such communities and leaders has contributed to strengthening my identity as an academic committed to transforming lives. Thus, my experiences growing up and working in Colombia, and my cultural background, have provided me with the motivation and passion to work with local communities and engage in pedagogies for transformation, despite the limited support I have encountered.

Laura's Journey to Encourage Pro-environmental Attitudes in the Australian Society. My teaching at University in Australia has challenged me to continue following the participatory pedagogical and research approaches that I have adopted—giving students opportunities to reflect on their practice not only from a school based perspective but to link it to research and community projects.

Teaching environmental and sustainability issues at the tertiary level of education in Australia is for the most part too different from my experience in Mexico. I strongly believe that these differences are the result of three main things: cultural perspective; ontological and educational ideologies and the methodological approaches that I follow, in particular *dialogue* as a tool to enhance trust; and action and critical reflection. I consider that many academics in Australia are constantly more worried about assessment tasks, rubrics and standardization than the process of learning itself. This scenario, I believe, is affecting how students are responding to their own learning process, as they tend to be more focused on "what am I learning" than on "how and why am I learning it." Moreover, one of the limitations that I have encountered during my classes is a low level of critical thinking in my students, which is reflected in the difficulty of encouraging independent and reflective learning.

An example to illustrate how I am changing students' attitudes and encouraging them to become more independent and autonomous in their learning is a 3-day field excursion as part of their course. During this excursion, the main purpose is to see the individual as an active and transforming agent of learning. This means that the student as an individual has a dynamic role during the whole learning experience. This is something that the students have not been exposed to before this experience. Their learning process along the three days allows them to change their attitude from being a passive agent to being an active and reflective learner.

During the three days, we visit a farm and students support the work that is carried out there. They review the environmental study design for the Victoria Certificate of Education (VCE), contribute to building a house for the guests with a variety of materials to promote sustainability, and help plant trees as a wind barrier to protect the house. We have discussions and reflective sessions at night with all the people at the farm in order to share our ideas and learn different perspectives. It is a dynamic and rich learning experience in which students learn by doing and interacting with the local community.

My teaching plays a vital role in the promotion of environmental awareness and participation among the students. When teaching environmental issues, I also teach about values, attitudes and actions. Therefore ethical and moral issues are often included during class discussions. Developing reflective and critical thinking is an essential prerequisite for responsible citizenship. My experiences and background as a woman from Meso-America allowed me to provide students with theoretical, methodological and practical tools to understand the complexity of environmental issues.

Environmental and sustainability issues are becoming key topics to understand and reflect on the challenges that we, as a society, have to face if we are to create a more sustainable world. However, environmental issues have not been at the core of the educational agenda, and it is only in recent years that schools are incorporating these topics into the curriculum. Education for environmental sustainability involves both deep knowledge and understanding *about*, and personal and collective action *for*, creating sustainable schools, communities and societies. It also aims to have education geared towards more participative practices to provide individuals with cooperative skills in their learning processes. This means that we, academics, need to develop creative and innovative practices based on social critical models of education. As academics committed to transforming society towards sustainable ways of interacting and living with nature, we need to create spaces for transforming formal and non-formal education.

As a Mexican and a follower of Freire's (1973) dialogic thinking, I strongly believe that it is necessary to open a discussion among academics in Australia to review multiple approaches and parameters to evaluate and assess students in a more creative and empowering way that increases their level of agency and critical awareness. Only then will we be able to transform society towards sustainable and caring ways of living.

REFLECTIONS AND CONCLUDING REMARKS

In 1995, critical race theorists Gloria Ladson-Billings proposed the theory of culturally relevant pedagogy to address issues of racist ideologies. Melina Furman (2013) further argues regarding the challenges of developing culturally aware and relevant education communities:

> When seeking an international perspective, we often look to those countries where students "do well" (at least according to international exams) in science. Something apparently works in all those "successful" countries, and the challenge is to find out what works and why, in an attempt to replicate it at home. (p. 351)

Critical race theorists such as Ladson-Billings and Furman refer to issues of race regarding students. However, to address issues of race in education we need to also include discussions of race discourses in academia and among university scholars. Academia needs to create culturally relevant spaces and opportunities for those academics from non-English speaking or European countries to provide their views on education.

We were educated in nations where students have to learn to survive in a society where basic services are lacking and extreme poverty is a constant variable. Students are constantly confronted with situations in which they do not have any other option than to create, adapt, develop new ideas and commit to acting in order to transform their realities and progress the society towards a more just one. Thus, care, dialogue, reflection and action are the main drivers of our teaching and research as women from Meso- and South America. Social and political discourses are constantly present in our educative practice allowing students to become responsible for their own learning process and empowering them to take action. This requires reflexive praxis and political activism in our pedagogical approaches. We prioritize caring and transformative interactions between teachers, students, community and society in our practice. In this interaction sharing ideas, exchanging views and empowering transformative ideals are the main drivers to build knowledge. Relationships are built as a result of these interactions, and the school community acts as an intergroup and an emancipatory force within the society. These praxis and pedagogical principles have positioned us as successful academic women in South and Meso-America and have provided us with opportunities to participate in the international academic community.

As Ken Tobin argues:

> The problems associated with deficit perspectives are that they are often presented as the only game in town. For example, in arguing that poverty is detrimental to learning, without nuance that acknowledges that in fact being raised in conditions of poverty enables those concerned to learn a great deal that otherwise would not be learned can often be overlooked. What is learned in growing up in poverty often put people in good stead for success in numerous facets of social life. The cultural capital produced in the field can be used not only in that field, but because of fields not having boundaries, what is learned can be widely applicable. (Personal communication, 2014)

We agree with Tobin and consider that our own stories and identities illustrate his argument. As academics who grew up in countries where several social issues such as poverty and violence strongly affect the community, we see our upbringing and cultural background as a strength to better understand disadvantaged communities. Our cultural background and experiences also have contributed to the construction of our identities and views on education. This provides us with abilities and understandings of transformation that we feel the English-speaking and Eurocentric international academic community have overlooked.

The differences between the north and the south, to which Ruth Galván (2014) refers, have strong and complex power relationships mainly due to the distinction between the capitalist abundance (north) and the marginalized two-thirds of the population (south), regardless of geographical area. If we only consider the geopolitical aspect to start with, we will have to situate our own position as Meso- and South American women as marginalized and disadvantaged. However, our experiences and social structures have been the foundation for our ideological beliefs, construction of our identities and agency as strong female academic leaders. We do not see ourselves as marginalized and disempowered.

We position ourselves as strong leaders who sometimes opt to not participate in the structures of the international academia as a form of resistance to the homogenization we perceive in academia. We hope this chapter contributes to the understanding of academics from non-English speaking or European nations. Our cultural backgrounds and the leaders who have emerged from our nations shape the views and perspectives we hold on education. Understanding race from a position of privilege and not from a deficit perspective could help to open spaces for the construction of a more inclusive and multicultural academic community. Moreover, in going back to the start of this chapter, the quote we presented encapsulates our feelings of the diversity of perceptions others hold of our identity as South and Meso American women. It is often that we are asked about whether we had to give up our Colombian/Mexican citizenship to receive our Australian one. It is sometimes with surprise that others hear us stating that we will not consider giving up our nationality as South/Meso American as it is strongly connected with our identity as proud, strong, caring female academics. We expect more than others to understand this, we expect we can contribute to construct academic spaces of resistance where diverse identities are valued and validated.

Note: Since finishing this chapter, Laura has returned to her home country and is now working with community projects outside academia. Carolina is searching for a better work balance with academic work by exploring new non-academic spaces with better opportunities to contribute to construct peaceful communities.

REFERENCES

Carter, L., Castano Rodriguez, C., & Jones, M. (2014). Transformative learning in science education: Investigating pedagogy for action. In J. L. Bencze & S. Alsop (Eds.), *Socio-political activism and science & technology education* (pp. 531–545). Dordrecht: Springer.

Comision Nacional para el Desarrollo de los Pueblos Indigenas (CDI). (2014). *Los Pueblos Indigenas de Mexico.* Mexico, D.F.

Foster, M. (1999). Race, class, and gender in educational research: Mapping the political terrain. *Educational Policy, 13*(1), 77–85. doi:10.1177/0895904899131007.

Freire, P. (1973). *Education for critical consciousness.* New York, NY: The Seabury Press.

Furman, M. (2013). International response for part V: Equity and diversity in science education and academia: A South American perspective. In J. A. Bianchini et al. (Eds.), *Moving the equity agenda forward: Equity research: Practice and policy in science education* (pp. 351–354). Dordrecht: Springer. doi:10.1007/978-94-007-4467-7-21.

Galván, R. (2014). Chicana/Latin American feminist epistemologies of the global south (within and outside the north): Decolonizing el conocimiento and creating global alliances. *Journal of Latino/Latin American Studies, 6*(2), 135–140. doi:10.18085/llas.6.2.1160715w62582591.

Hernández Castillo, R. A., & Suárez Navaz, L. (2008). Introducción. In L. Suárez Navaz & R. A. Hernández (Eds.), *Descolonizando el feminismo: Teorías y practices desde los márgenes* (pp. 11–28). Madrid, ES: Ediciones Cátedra – Grupo Anaya.

Ladson-Billings, G. (1995). Toward a theory of culturally relevant pedagogy. *American Educational Research Journal, 32*(3), 465–491. doi:10.3102/00028312032003465.

Ladson-Billings, G. (2012). Through a glass darkly: The persistence of race in education research & scholarship. *Educational Researcher, 41*(4), 115–120. doi:10.3102/0013189X12440743.

Lanyon, A. (1999). *Malinches's conquest.* Sidney, NSW: Allen & Unwin.

Mezirow, J. (1991). *Transformative dimensions of adult learning.* San Francisco, CA: Jossey-Bass.

Solorzano, D. G., & Yosso, T. J. (2001). From racial stereotyping and deficit discourse toward a critical race theory in teacher education. *Multicultural Education, 9*(1), 2–8. Retrieved from http://search.proquest.com/docview/216501989?accountid=8194

Zimmerman, L. (2006). *Women in Latin America.* New Orleans, LA: Roger Thayer Stone Center for Latin American Studies. Retrieved from http://stonecenter.tulane.edu/uploads/Women_in_Latin_America_updated-1352754376.pdf

CHAPTER EIGHTEEN

"Where Is the ...?"

Using Intersectionality to Problematize Power and Hierarchy in Science Teaching and Learning

KATHRYN SCANTLEBURY

> When I see something that looks racist, I ask, "Where is the patriarchy in this?" When I see something that looks sexist, I ask, "Where is the heterosexism in this"? When I see something that looks homophobic, I ask, "Where is the class interest in this?"
>
> —MATSUDA (1991, P. 1189)

I became engaged with feminism during the 1970s when researchers talked about sex differences and the term was associated with student outcomes and attitudes, the comparison of girls' achievement scores on international tests to their male peers, the enrollment patterns of women and men in university courses, and women's access to occupations and careers they had traditionally been excluded from or banned from pursuing. When I moved to the USA for my doctoral studies, the word gender, rather than sex was always used, even though the data being collected, analyzed and interpreted used biological indicators to establish the categories of "female" and "male." Unfortunately, there is a lack of understanding of the distinction between the terms gender and sex among scholars and in conversations and other forms of everyday discourse gender has replaced sex (Francis & Paechter, 2015).

Science education research struggles to problematize the concept of gender and more often than not, other social categories such as race, ethnicity, socio-economic status, first language, and/or religion are positioned as a priori, ignoring gender or the nuances that come when gender is consider with|and other social categories. Mari Matsuda's series of questions that I have quoted at the beginning of this essay, highlight the importance of using intersectionality as a theoretical framework for examining the entanglement of social categories, questioning how power reinforces one's social position and the influence of society's structures on changing hierarchies. Intersectionality looks to challenge power and hierarchy, and into the 21st century the image of science remains masculine and White. What questions and the subsequent outcomes would science education research produce if we used intersectionality theory and asked, "Where is the ...?" In this essay, I re-view my ongoing education as a feminist researcher to answer the question: *"In what ways does gender affect (and is affected by) the process of science education?"* using intersectionality as a theoretical frame. This question reflects the one posed in the original "13 Questions" book.

WHAT IS INTERSECTIONALITY AND WHAT CAN IT DO FOR SCIENCE EDUCATION RESEARCH?

Within feminist research, the term intersectionality emerged from Kimberle Crenshaw's analysis of how the power axis of gender and race impacted African American women in the US's legal system (Crenshaw, 1989). Intersectionality critically examines how power and hierarchy can generate different levels of oppression that may change for various social categories and structural levels. In the two decades since Crenshaw noted the usefulness and power of intersectionality as a theory and analytical tool in feminist research, it has developed into a field of study with three perspectives. The first perspective is as a framework to examine how the overlapping of different social categories impact each other and produce different knowledge, outcomes, and inferences. Second, feminists continue to debate the definition, theoretical usefulness and methodological implications of intersectionality. Third, the strength of using intersectionality as the perspective for identifying areas of solidarity, difference, challenging dominant paradigms and producing new knowledge (May, 2015).

Looking at the Gaps in Science Education Research. For the past five years, I have had a visiting research professorship at Uppsala University's Center for Gender Research (see http://www.gender.uu.se). The center's researchers come from diverse disciplines, such as literature, systemic theology, international motherhood, childcare and kinship, literary theory, philosophy, geography, body/embodiment, museum studies, crip studies (that is, disabilities studies), law, technoscience, Sami studies (Sweden's indigenous peoples), animal studies and occupational studies. These researchers conduct their studies utilizing various and diverse theoretical frameworks, for example feminist, masculinist, queer, human-animal, post-modern, post structuralist, technosciences, new materialisms, and material feminism (see http://www.gender.uu.se). Center researchers work within and across disciplines in clusters that focus on gendered cultures which incorporate cultural and museum studies, education and science, queer, the intersect between humans and animals and family/kinship. Intersectionality is enacted daily at the center through discussions, informal presentations, and seminars. The discussions may focus on unfolding and emerging research of the scholars who visit the center for varying lengths of time, the researchers who present their work in the center's external and internal seminars, and those for whom the center is "home."

One program at the center that illustrates how intersectionality can challenge the dominant modes of thinking and isolated unconnected research ideas is the GenNa program. Through the GenNa program, "Nature/Culture and Transgressive Encounters," the Center provides researchers from the humanities, social, medical and natural sciences an intellectual and physical space to collaborate across disciplines (Bull & Fahlgren, 2016). It was within this space that I have engaged in conversations about intersectionality, material feminism, new materialisms, masculinities and lived embodiment and started to re-view science education research using intersectionality to examine the "gaps, inconsistencies, opacities, and discontinuities" (May, 2015, p. 247).

Where are the "gaps, inconsistencies, opacities, and discontinuities" in science education with regard to gender? My Swedish colleagues, Anita Hussénius, Kristina Andersson and Annica Gullberg and I posed the question "where is ...?" and examined the intersection of feminist perspectives in science education research (Hussénius, Andersson, Gullberg, & Scantlebury, 2013). This idea for this chapter emerged from a *Cultural Studies of Science Education Forum* that had focused on the reconciling sociocultural research in science education with conceptual change research (Roth, 2010). While I knew that feminist theory was rarely foregrounded in science education research, it was not until we embarked on a systematic study of what had been published in science education journals that the stark omission of gender became evident. We started with feminist/feminist theory as a search term in the ERIC database but this criterion yielded few results, so we broadened the terms to include gender, feminist, gender/feminist/equity and gender theory/feminist. Of the 11,556 peer-reviewed science education articles, 199 had at least one of those keywords and 7 had either feminist theory or gender theory: a sobering result. In our discussion we used Sandra Harding's (1986) approach of examining gender at the structural, symbolic and individual levels along and Anne Hammarström's (2005) three approaches to evaluating studies from a gender perspective to suggest how feminist theories could contribute to science education research and provided an example from our collaborative research.

One of our research projects sought to challenge the power and hierarchy in science through taking a critical approach to the cultures of science and education using a feminist approach by introducing preservice teachers to critical theories and engaging the students in using those theories to examine science and education (Hussénius, Andersson, & Gullberg, 2015). The project introduced gender perspectives into science courses taken by preservice pre-K and elementary teachers, as well as directing students to examine the un-stated gendered aspects of education in their own schooling and in the schools where they completed their field experiences. Some preservice teacher-participants became aware of "gender blindness" in schools and that science as a culture was not immune to gendered values and stereotypes.

Some preservice teachers documented examples from the schools that problematized gender as a female/male binary to include lesbian, gay, and/or transgendered perspectives. These instances provide evidence to the research group that some preservice teachers had a more sophisticated view of gender and its social construction. As Becky Francis and Carrie Paechter (2015) discuss that while gender theory has developed and evolved, research in gender and education remains "haunted" by the binary categorization of gender into female and male. They propose the simultaneous use of three different lenses to problematize gender and its binary categorization when conducting education research. Inn the first lens, the spectator view, researchers and others categorize participants as "girls, boys, women, men, other or uncategorizable" (p. 786). In the second lens, participants categorize themselves and the third lens considers the "features of the local discursive and material collage, which enable the gender" (p. 786).

But Francis and Paechter's (2015) framework may not push forward research in gender and science education because it fails to explicitly include the re-consideration of gender with other social categories and identity markers. Nor does the approach challenge the symbolic and structural levels of gender. Moreover, the theoretical constructs remain within a discursive and material collage without taking into account the recent use of material feminist theories by researchers to include the agentic nature of the material in developing new knowledge about gender in education contexts (Taylor & Ivinson, 2013). Science is at the "core" of science education, and researchers have shown that while the stereotype is that science is value-free, logical and rational, and those who study and produce science knowledge are innately talented, that image is a myth. Science education is not immune to social norms and values—ensuring that all learners have access to science is often not a priority. Nor is understanding the cultural underpinnings and norms that prevent students engaging and succeeding in science.

Staying with the Trouble: En-gendering Science Education through Critical Theories. Donna Haraway (2008) proposed that researchers need to "stay with the trouble" in order to advance feminist theorizing. In the next section I propose that science education research should stay with trouble by discussing how lived embodiment and a restructuring of intersectionality from a Baradian perspective offers the potential of interesting lines of flight for gender research in science education.

Within science education research the concept of gender is typically "bounded" into the binary of female and male. Recently, discussions of gender in the US (and other Western societies) have broadened to recognize that education institutions have to critically examine their policies and practices regarding transgender, lesbian, gay and bisexual students. Yet, heretofore, science education has rarely engaged with a broader discussion of gender and sexuality, remaining "secretly" hetereonormative (Lemke, 2011), clinging to the binary of gender as female and male. What new knowledge would our field generate in broadening its interpretation of gender and weaving through an intersectional approach how this might challenge and change science education research? There is a cadre of science education researchers embarked upon queering science/STEM education (Letts & Fifield, in press). The work of these scholars will remind science educators that gender is not binary but there are nuances to the concept, and challenge science education scholars to articulate what is meant by the term gender and how it influences their research.

Jenny Björklund, Ingvall Hellstrand, and Lisa Käll (2016) have used intersectionality to illustrate how a lived embodiment incorporates the body with discursive practices producing cultural imaginaries.

> Lived embodiment is understood as a situation where markers of identity are inscribed and expressed, reinforced, and altered. It is produced precisely as the materialization of intersecting dimensions of identity formations. (p. 100)

Cultural imaginaries are themes and images that reflect psychic and social dimensions. They discuss how a phenomenological understanding of the body takes into account the corporeal aspects of the body. While, science as masculine remains a strong cultural image, science education is placed in a paradoxical position. Science is viewed as a masculine subject, while education and its subsequent teaching as a career has long been considered suitable for females. Teaching, especially of young children, is seen as "naturally" aligned with stereotypic feminine characteristics such as nurturing, caring and working in service for others.

Their use of intersectionality in this way shows that the connection between the material and discursive practices is a different perspective to the ideas by Haraway (1998), Karen Barad (2007), Nancy Tuana (2008) and others on introducing matter into feminist theorizing, that is, material feminism, new materialism, and new material feminism. A fundamental concept within these areas of study is that matter is agentic. Barad (2007) considers that entities are not bounded but intra-act to produce phenomena that have material-discursive practices. Through intra, rather than inter, action the argument is that entities act on each other rather than remaining unchanged by those encounters. Thus, examining how humans are engaged in science education through intra-action provides researchers the possibility of producing new gendered knowledge. In the next section I provide an example of implementing intersectionality within a study on gender, science attitudes and English Language Learners.

An Intersectional Inquiry: Urban Students and/or English Language Learners and Science Attitudes. My initial foray into implementing the ideas from intersectionality was with a study of middle school students' science attitudes. I approached gender and race as structural categories rather than problematizing the definitions of those terms. The study was located in two urban, charter schools, which had significant numbers of English Language Learners (ELL). We administered a survey asking students their perceptions about experiences in their science class, friends' and parents' support for their science learning, and their views about science. Over 180 students answered the survey, with nearly equal numbers of females (n = 92) and males (n = 91). Thirty seven percent of the sample were students who self-identified as Latino/a, 33% were Asian, and 26% were African American. Within the Latin@, group students came from Mexico, Puerto Rico and the Dominican Republic and in the Asian category students had immigrated from China, Cambodia, and Vietnam. Because of the small sample size, further quantitative analysis within the categories of Latin@ and Asian was not possible.

A gender analysis showed a significant difference between girls and boys on "views about science" subscale. I then asked "Where is …?" on the scales by the three racial groups (Asian, Latin@ and African American). This showed significant differences on "what my parents/adults do" and the "views about science" scales. Another "Where is …?" and the within race and gender analysis revealed that Asian females and males had the lowest mean score on the "what my parents/adults" do scale, while ELL Latin@ students had the lowest mean scores for the "views about science scale." One survey question had asked the students what language did they speak at home, and enough students in the data sample indicated they spoke Spanish at home to warrant further analysis—to compare the responses of Latin@ students who spoke Spanish at home with their peers who reported speaking English at home. An unexpected outcome was the difference of Latin@ students' views on science compared to the other students in the sample which was attributable to whether they self-reported speaking English or Spanish at home.

A different analysis on the students' perception on whether adults at home supported their learning resulted in a significant difference between Asian and other students. The Asian students had a significantly lower mean score on the scale; further investigation using qualitative data sources (interviews with students, parents and teachers) provided the perspective that parents viewed the education of their children as the school's and teachers' responsibilities. Several of the interviewed Asian parents explained they did not help their children with homework or ask about their science learning because they expected that their children's teachers had the expertise and trusted that the teachers would ensure their children learned science.

Asking "Where is ...?" and the categories gender, race and language spoken at home resulted in unexpected outcomes. The low numbers of students in the different categories restricted the analysis; for example, Asian was a category that included students with different ethnic backgrounds such as Vietnamese, Chinese, Cambodian, and Indonesian. Another gap in these data was that only Cambodian parents agreed to be interviewed. Further, there was also an inconsistency with the "students" perceived lack of support' from the adults in their home and the assumption that Asian parents show strong support for their children's education. These outcomes suggest further study into how gender and ethnicity impacts students' attitudes towards science, the support they receive for studying science from their families, and the relationship between English language proficiency and science learning.

Intersectionality is a discursive site where different feminist positions are in critical dialogue or productive conflict. Rather than a space where different theoretical frames overlap or intersect, Nina Lykke (2011) suggests researchers use conceptual nodal points to take on lines of flight for examining data and producing new knowledge. My exploration into analyzing quantitative student data using an intersectional approach, produced a different line of flight as I considered an explanation as to why Latin@s had the lowest scores on the "views about science" scale; and Asian students perceived that the adults in their home did not support science learning. Other emergent areas for further research include students' perceptions about the teaching they experience in science class and how their friends perceive science learning.

As science education's sociocultural contexts become more complicated and diverse, intersectionality offers a theoretical framework to foreground gender as a starting context for a line of flight that would examine how gender with other social categories and contexts impact science teaching and learning. The outcomes from these studies would produce new knowledge and understandings that could challenge the continued "norm" of who is successful in science, what knowledge is scientific and the role of gender in problematizing the teaching and learning of science. Through intersectionality the concept of lived embodiment is providing feminist researchers new pathways to connect the corporeal aspects of the body with the social and psychic dimensions of gender. Intersectionality can also be an avenue for researchers to broaden "gender" analysis from a binary to multiplicative perspective that would produce new knowledge on how gender impacts science teaching and learning.

REFERENCES

Barad, K. (2007). *Meeting the universe halfway: Quantum physics and the entanglement of matter and meaning.* Durham, NC: Duke University Press.

Björklund, J., Hellstrand, I., & Käll, L. F. (2016). Marking the unmarked: Theorizing intersectionality and lived embodiment through Mammoth and Antichrist. In J. Bull & M. Fahlgren (Eds.), *Illdisciplined gender: Engaging questions of nature/culture and transgressive encounters* (pp. 99–113). Cham: Springer International Publishing.

Bull, J., & Fahlgren, M. (Eds.). (2016). *Illdisciplined gender: Nature/culture and transgressive encounters.* Cham: Springer International Publishing.

Crenshaw, K. W. (1989). Demarginalizing the intersection of race and sex: A Black feminist critique of antidiscrimination doctrine, feminist theory and antiracist politics. *University of Chicago Legal Forum, 1*(8), 139–167.

Francis, B., & Paechter, C. (2015). The problem of gender categorisation: Addressing dilemmas past and present in gender and education research. *Gender and Education, 27*, 776–790. doi:10.1080/09540253.2015.1092503.

Hammarström (2005). *Genusperspektiv på medicinen – två decenniers utveckling av medvetenheten om kön och genus inom medicinsk forskning och praktik. [Gender perspective in medicine - two decades of gender awareness development in medical research and practice].* Stockholm: Swedish National Agency for Higher Education.

Haraway, D. (1988). Situated knowledges: The science question in feminism and the privilege of partial perspective. *Feminist Studies, 14*, 575–599.

Haraway, D. (2008). *When species meet.* Minneapolis, MN: University of Minnesota Press.

Harding, S. (1986). *The science question in feminism.* Ithaca, NY: Cornell University Press.

Hussénius, A., Andersson, K., & Gullberg, A. (2015). Spotting the science culture – Integrating gender perspectives into science courses. *International Journal of Gender, Science and Technology, 7*(1), 91–103. Retrieved January 22 from http://genderandset.open.ac.uk/index.php/genderandset/issue/view/20

Hussénius, A., Andersson, K., Gullberg, A., & Scantlebury, K. (2013). Ignoring half the sky: A feminist perspective on the missing standpoints in science education research. In N. Mansour & R. Wegerif (Eds.), *Science education for diversity in knowledge society.* (pp. 301–315). New York, NY: Springer. doi:10.1007/978-94-007-4563-6.

Lemke, J. (2011). The secret identity of science education: Masculine and politically conservative? *Cultural Studies of Science Education, 6*, 287–292. doi:10.1007/s11422-011-9326-6.

Letts, W., & Fifield, S. (2018). *STEM of desire: Queer theories and science education.* Rotterdam, The Netherlands: Sense Publishing.

Lykke, N. (2011). Intersectional analysis: Black box or useful critical feminist thinking technology. In H. Lutz, M. T. Herrera Vivar, & K. Supik (Eds.), *Framing intersectionality debates on a multi-faceted concept in gender studies* (pp. 207–221). Burlington, VT: Ashgate Publishing Ltd.

Matsuda, M. (1991). Beside my sister, facing the enemy: Legal theory out of coalition. *Stanford Law Review, 43*, 1183–1192. doi:10.2307/1229035.

May, V. (2015). *Pursuing intersectionality, unsettling dominant images.* New York, NY: Routledge.

Roth, W. M. (Ed.). (2010). *Re/structuring science education: Reuniting sociological and psychological perspectives.* New York, NY: Springer.

Taylor, C. A., & Ivinson, G. (2013). Material feminisms: New directions for education. *Gender and Education, 25*, 665–670. doi:10.1080/09540253.2013.834617.

Tuana, N. (2008). Viscous porosity: Witnessing Katrina. In S. Alaimo & S. Hekman (Eds.), *Material feminisms* (pp. 188–213). Bloomington, IN: Indiana University Press.

SECTION SEVEN

Language and Science Education

In What Ways Does Language Affect the Process of Science Education?

CHAPTER NINETEEN

Reconstructing Science Education within the Language | Science Relationship

Reflections from Multilingual Contexts

SARA E. D. WILMES, CHRISTINA SIRY, ROBERTO GÓMEZ FERNÁNDEZ, AND ANNA MARIA GORGES

Our research is embedded in the multilingual national context of Luxembourg, a small diverse country in Western Europe, and as such our research participants are culturally and linguistically diverse (CLD). Luxembourg's public schools reflect the diversity of the country, with 44% of students identifying as a nationality other than Luxembourgish, and 55% speaking a first language other than Luxembourgish (Ministère de l'éducation nationale, de l'Enfance et de la Jeunesse [MENJE], 2015). Certainly, students draw on a wide variety of resources as they make meaning in science, and a key resource in this process of meaning making is language, which serves to mediate learning as well as position participants in the learning process. However, for students with proficiencies in languages other than the ones used for instruction in schools (such as the students we work with), the nuances of how language(s) can serve as resource(s) for learning are crucial for researchers and teachers to consider and understand. Science, language, and learning are interwoven, connected, and we believe, inseparable, to the processes of science education.

In this chapter we use a critical lens to deconstruct the use of language(s) in science education as we address the overarching question posted by the title of this section, "*In what ways does language affect (and is affected by) the science educational process?*" Throughout this process of deconstruction, we address several critical questions that arise from our research and lived experiences connected to the relationship between science education and language. Specifically, we address the following interrelated questions:

- *Who decides which languages are used in classrooms?*
- *How can we create classroom spaces that value diverse student resources?*
- *What is the relationship between language used in science education, power, and agency?*

We address these with the intention of drawing connections for those working with CLD students and in order to move towards reconstructing equitable science education practices. As we explore these questions, we hope that readers will draw connections to their own practices working with diverse students. We also underscore our position that these questions are only a few of the myriad of questions that we can ask ourselves as a field regarding the role of, and relationship between, language and science when examining science education.

LANGUAGE AND SCIENCE LEARNING AS INSEPARABLE

In our work with CLD students and their teachers, language is both an explicit and implicit component of teaching and learning science. Grounded in sociocultural theoretical perspectives, we consider science as a culturally embedded and contextualized enactment (e.g., Sewell, 1999). Science education then, through our theoretical lenses, is a practice, it is something that is "done" and it can emerge and evolve through interaction (e.g., Siry, Ziegler, & Max, 2012). As students "do science," they write, read, discuss, enact, negotiate, and represent their understandings in a wide variety of ways. Students draw on various forms of discursive practices (verbal, material and multimodal) as resources to engage in the practices of science and to communicate their understandings (Varelas & Pappas, 2013). These resources are at the same time structured and also structuring to the enactment of science (Tobin, 2012). In our research projects, we examine students' multimodal interactions, and explore the ways in which CLD students draw on a myriad of resources as they enact science in classroom contexts.

Our dialectical understanding of social life inspires us to look beyond binary, either-or, perspectives. Rather it supports our understandings that **both** components of a dialectical relationship exist at the same time, and evolve together within and through this relationship. Dialectical relations exist in every human situation, including "the relation between subject and object, individual and environment, self and society, outsider and community, living consciousness and phenomenal world" (Greene, 1988, p. 8). In a dialectical sense, learning is not a state of being, but a process of becoming (Freire, 1990); it is fluid and continually evolving.

We understand language as one of multiple resources for exploring the content and processes of science. In our research we seek to explore the science | language dialectic as it plays out in classroom spaces and as specifically related to teaching and learning science. As students draw on a variety of discursive multimodal meaning-making practices, their science understandings in turn are also shaped over time. Thus, language and science must be considered together as tools and resources that each can mediate students' participation in the "doing" and meaning making of science, especially when working with CLD students in multilingual contexts.

CRITICAL QUESTIONS REGARDING LANGUAGE USE AND THE SCIENCE EDUCATION PROCESS

> What disservice do we do to kids when we rigidly police the linguistic and cultural boundaries that they traverse every day with tremendous flexibility, versatility, and grace? What injuries do we do to kids from non-dominant cultural groups when we ask them to speak, read, write, and think in ways that require cutting off parts of themselves? What harm do we do to young people when we provide no space to display what they already know and can do—rather than where their "problems" lie? (Faulstich Orellana, 2016, p. 133)

This quote from Marjorie Faulstich Orellana reminds us of the dangers of not embracing "resource-rich" approaches (e.g., Siry, 2011) in working with CLD students in classrooms, and instead placing the focus on students' lack of ability to communicate in dominant, normative language settings. The risk of these types of learning settings is that they discount the ways in which students engage with "flexibility, versatility and grace" if only sanctioned representations of science practices are allowed, and allowed only in languages in which students are not proficient. Thus, we seek to problematize the role of language, particularly to meaning making in science as we consider the nuances of the relationship of language and science education. We do this by elaborating several critical questions, and drawing on descriptions of two episodes in which children are explaining science-related concepts, to illustrate the need for questioning the role of languages in mediating students' science understandings, and for problematizing the ways in which language is used by children as they engage in and with science in interactive settings.

A First Episode to Consider. The following episode took place during a lesson in a "Cours de langue maternelle intégrés" or maternal language integrated course, which makes it possible for primary-school students to attend two subject matter lessons (mainly science, but also geography and history) in their home language every week during regular school hours. The goal of cours intégrés is to support development of the home language, thus supporting the children's school success and allowing for their harmonic development (Ministère de l'Education nationale de l'Enfance et de la Jeunesse, 2015).

> Eight first-grade children are sitting on the floor of a classroom working in pairs with large sheets of paper and drawing supplies. "Draw some of the words from the poem," [about firefighters in the community, a topic in the national sciences curriculum] Maria, the teacher, instructs the students in Portuguese. Each pair of children is immersed in animated conversations about which words can be drawn on the paper. Three members of our research team are seated with the pairs, observing them work. Two girls have drawn a tree with a cat in it, and a ladder leaning on the tree. Chris, one of our researchers, asks them in Luxembourgish about what they have drawn and how it connects with the poem. Chris explains to the children that she does not speak Portuguese, and the children excitedly point to the Portuguese word for ladder [escada] written on their paper. They try to explain to Chris what this word represents. In the co-construction of the word ladder, the children draw on explanations using French with Chris and Portuguese to each other. In this collective search for the Luxembourgish word for ladder, they use their bodies to illustrate how one might move up a ladder until Chris asks them in Luxembourgish, "Oh, do you mean a ladder?" Both girls burst into giggles, and then look up at Chris as they smile, nod, and enthusiastically say in Luxembourgish, "Yes! A ladder! Exactly!"

This episode illustrates a common practice that we have seen in classrooms we have worked in, in both Luxembourg as well as in the United States, in that children are asked to draw and label what they are to be learning about. In this episode, the instructional focus is on learning about "the community," which in our context is a component of the national sciences curriculum. As children move between languages, the grasp of a term such as "ladder" is an explicit learning objective, and to explain the word, children draw, write, and act out their representations, ultimately moving from Portuguese (the language of instruction in this instance) to French (a language taught beginning in 2nd grade), to Luxembourgish (a common language of interaction in the school context) to demonstrate to Chris (an author of this chapter) what the word means.

We initially hesitated to use this example to contextualize the complex relationship between language and science education, as we believe that children should be engaged in the practices of science without an overemphasis on using isolated vocabulary words to show their understandings. Certainly being able to explain one word is not how we conceive of demonstrating science processes and understandings, however, it is a deeply embedded instructional practice in our context at the primary school level, as well as in many contexts internationally. Thus, we decided to include this illustrative example so that we can begin to recognize and learn from these normalized practices. In this episode the resources the children draw on to demonstrate their understandings; material, multimodal, and verbal, in the collective search for the word ladder in Luxembourgish, are visible. We use this first episode to situate the complexities of such an activity for children, with its explicit focus on vocabulary formulation, as they draw, write, and act out their understandings of words being used to name concepts and ideas.

Who Decides Which Languages Are Used? In our globalized world with a high rate of national as well as international mobility (United Nations Development Programme [UNDP], 2009), cultural and linguistic diversity in schools is a reality for many. The students with whom we work bring great cultural and linguistic potential to science classrooms. However, their home languages are often not valued in the classroom, their use is restricted, and they are often considered obstacles for the learning of the school language (Cummins, 2005) even "problematizing" their home languages and their families (Gómez Fernández, 2016). Therefore, CLD students are not always afforded opportunities to use all of their linguistic resources to foster their learning processes (Moore, 2014). As was illustrated in the first episode, where Maria's students could use both Luxembourgish and Portuguese to complete their task in Portuguese, allowing for these opportunities to deploy the students' linguistic resources has positive effects for students' representations.

Students come to school with a wide degree of cultural and social capital (Bourdieu, 1986) but often the capital they bring might need to be adapted to the expectations of the teachers, the school, the school system, or even the nation, in order to be valued. These expectations are linked to how societies value different languages; one language might be perceived as more valuable than others depending on the context. In Luxembourg the most valued languages in school are the languages of instruction (Luxembourgish, German and French). The use of these more valued language(s) becomes normalized or taken as *the way it is* by those within these contexts without questioning whose language is used for teaching and learning (Creese & Martin, 2008). This normalization process is reinforced by teachers who are members of the dominant linguistic group, and who have little knowledge about their students' languages (García, 2010). As a consequence, the students who are not members of the dominant linguistic group (in Luxembourg for instance, the greatest number of students speaking the non-majority languages are students who have a Portuguese linguistic and cultural background), are underrepresented in the secondary school tracks that enable access to university studies (MENJE, 2015). They are additionally underrepresented, and actually almost invisible, in the national pool of teachers. As a result of this cyclical process that favors teachers who are from the linguistic and cultural majority, most primary school teachers do not understand students' home languages, like for instance, Portuguese. Thus, creating structures that might serve to value and support CLD students' linguistic competences and to teach in ways that value the relationship between science education and language is not always at the forefront of teaching science.

As we reflect on the contradiction between our understanding that language use in interaction mediates science understandings in contrast to language policies that require students to learn in languages they are not proficient in, we ask, *who decides which languages are used in classrooms*? One way to approach this question is to consider the curricular policies and overall frameworks guiding everyday practices carried out by teachers. These policies structure the agency that teachers can take, however, according to our dialectically grounded perspectives, as teachers take agency they also can change the structures at play. Thus, teachers are the ultimate language policymakers in their classrooms, with the power to "changing and redefining policies, or even creating entirely new ones" (Menken & García, 2010, p. 4).

Unfortunately, the compartmentalization of curriculum by languages (Weber, 2014), as is often the case in many schools, does not allow for practices that value the diverse and varied ways students draw on multiple communicative resources. In Luxembourg this is the case: students are taught first in Luxembourgish in preschool, then they are taught to read and write in German beginning in first grade, and reading and writing in French is additionally added in second grade. As Jean-Jacques Weber explains, the educational system in Luxembourg envisions a "triple (or triple plus) monolingual competence" (2009, p. 90), understanding the three languages of the curriculum as separate entities; meaning, each language is taught as a bound, closed subject, qualifying for trilingualism, but a "step by step" one (Gómez Fernández, 2015). This political requirement represses students' ability to learn using the resources they bring, including their linguistic repertoires, for (co-) constructing understanding in the science classroom as the students we described in Maria's class above had the freedom to do. In asking *who decides which language is spoken*?, we turn to the meso-level interpretation of the role of the teacher in structuring the classroom activities and language expectations. We note that the question of "who decides" is also inter-related with the macro-level policy expectations as well as the more micro-, in-the-moment decisions by teachers as well as students. Recognizing however the power that the teacher holds in structuring the day-to-day unfolding of instruction and interaction in the classroom we focus herein on the role of the teacher, especially teacher as the "policy maker" as introduced above.

An example of this language policy interpretation is illustrated by the episode from Maria's class. The language of instruction in this 2-hour weekly lesson was Portuguese. However, when working in small groups, most of the students used a mix of Luxembourgish and Portuguese. We, the researchers who were present, observed how plurilingual students connected the science practice of the classroom with their everyday life strategies of meaning making in their interactions with peers. (Note: We use the terms "plurilingualism" and "multilingualism" according to the Council of Europe (2001) definition, which delimits plurilingualism at the level of the individual and multilingualism at the level of the society.) Maria created spaces to teach the language-based objectives set for this science curricular theme of "our community" (i.e., learning Portuguese words for members of the community and their roles) as well as to allow children to interact with these words and demonstrate their understandings in drawing, writing, and talking. In these peer-to-peer interactions children were able to use whichever language they needed to first make meaning around the lesson's objectives, and then demonstrate their understandings by writing in Portuguese. Thus, Maria acted as an interpreter of the rigid language policy of the school, and in her classroom, her students were able to flexibly access more of their meaning making and communicative resources during science. In this sense, it is Maria's decision that provided the structure or space, and in turn the children were able to draw on the language(s) and other modes of communication in small groups as needed before transitioning to Portuguese to complete the task at hand.

But why are some teachers more keen to create structures in which students use all their languages and transition between other communicative modes in science while others are not? Some studies have pointed out that teachers might feel uncomfortable or even insecure with students using languages the teacher does not speak and teachers might think that this is off-task talk (Martín Rojo, 2003). The critical theoretical perspectives we adopt require us to understand social, cultural, and historical processes of production and reproduction, in order to ideally work towards transformation of these processes. Thus the questions we ask in this chapter help us to gain deeper understandings of the processes that produce and reproduce a fairly consistent expectation that children should be able to name concepts in one particular language. Hopefully this will allow us to be able to find the spaces for transforming this common practice into one that better facilitates the language and science processes of CLD students.

How Can We Create Classroom Spaces That Value Diverse Student Resources? Language minority students in classrooms in general, and science classrooms in particular, can be perceived as possessing deficits in the languages of school rather than possessing competences in their home language(s) (Snell, 2013). This positioning of students as being deficient represents a common perspective in education that focuses on languages that the students "should" have, rather than recognizing the resources that students already possess and bring to the classroom.

We feel it is important at this point to pause and consider the terms used to discuss students' language abilities. Bi-lingual, tri-lingual, multi-lingual, metro-lingual, pluri-lingual—these terms have been, and continue to be used to refer to students who speak more than one "language." The point we make is that each of these terms continues to refer to named languages as singular bounded entities (Luxembourgish, Spanish, Chinese). In this sense, no matter which of these terms is used, the student is measured against the yardstick of a bounded national language. Research has shown that these bounded languages do not reflect the interwoven, connected communicative realities that people use in everyday communication. To honor these non-bounded practices, we have made a conscious choice to use the term idiolect. We suggest that by focusing on the idea of idiolect we can better understand that the sum of all languages and varieties students talk are a resource in interaction around science, and not a deficit.

Idiolect comprises the unique personal word choice (from within one named language or across named languages), grammatical language structure, and way of pronunciation that distinguishes us, like a fingerprint, from one another (Otheguy, García, & Reid, 2015). At school, however, students are required to speak only one version of a named language called "standard," which prevents students from using their full linguistic repertoire. It leads to monolingual students suppressing those features of their repertoire that are not appropriate in the particular setting or with particular people (e.g., people of authority, or children). Multilingual students have to suppress not only the same inappropriate features of the named language the monolingual students are forced to, but also all linguistic features in their repertoire from languages other than the named language required by the school. Thus, the much greater deprivation on side of the multilingual students places them at clear disadvantage. For instance, several 8-year-old students in Maria's class had Portuguese as a home language and Luxembourgish as their main language of communication at school. Other languages of everyday communication and learning at school were German and French. Their northern regional variety of Portuguese (from their family) was now enriched with some words from these other languages, creating an altogether unique idiolect. In the episode from Maria's class, these students were deploying words and grammatical structures from different languages and hence freely using their idiolect, thus increasing the chances for meaning making in the science class.

Next, we offer a second episode to illustrate that the consideration of only spoken word or written products does not create space for the diversity of resources CLD students bring to interactions in science class. This episode presents Julia, the 5-year old child of one of the authors. She has been raised multilingually, and speaks English and Luxembourgish at home. At daycare and in other day-to-day interactions outside of home, she is immersed in interactions that draw upon Luxembourgish, French, and Portuguese. The episode below is typical of the interactions Julia has with her parents and other caregivers, including teachers, and we include this to help illustrate the critical views of language | science education that we make in this chapter, and the importance of non-spoken communicative resources.

A Second Episode to Consider.

Julia is standing at the kitchen table. She is bent down, looking into a glass container. In the container there are four snails that her mother, Sara, will use for a teacher workshop. As Sara is making dinner at the stove, she has her back turned to Julia. She hears Julia

> say, "*I see it Mama. Maybe she likes it like that.*" Sara is not sure what Julia is referring to by saying "*Maybe she likes it like that.*" "*The snails?*" Sara thinks to herself. "*Can you tell me again?*" Sara asks over her shoulder, eventually turning so that she sees what Julia sees with the hopes of understanding to what she is referring. This happens often. Julia tells detailed stories and Sara is not always sure she understands what they are about. She is often left to wonder, "*Do families who all speak the same language experience the same thing when they communicate?*" Julia continues, "*The mama snail, maybe she likes it like that,*" she says pointing over her shoulder to her back with one finger and making small circles. Her gaze shifts back to the glass container holding the snails, she bends down and looks in. Sara looks with her and asks, "*What do you mean?*" Julia clarifies emphatically "*look, the mama snail,*" as she points to one of the snails and repeats her initial point, "*maybe she likes it like that,*" while pointing once again to her own back and then to the snail. Sara and Julia both peer into the container. It is then that Sara sees, slowly a row of small white bumps are being placed onto the snail's back. Julia asks, "*Does it get mama snails?*" This Sara understands. Julia uses the grammatical construction "*Does it get …?*" to mean, "*Are there?*" This is a direct translation from Luxembourgish. Julia is asking if there is such a thing as "mama" snails. Later Sara searches on the Internet for what they had seen ("snail white bumps eggs?"), and comes to understand what she thinks Julia was observing and trying to explain. The snail was laying eggs and placing them on its back. Julia had observed this, theorized about it, and was wondering about her observation.

As is the case with many children growing up with more than one language at home, Julia has a spoken vocabulary, if counted, that would number less than that of her monolingual peers. Her statement of "*maybe she likes it like that*" was her trying to explain to Sara what she observed and was wondering about the snail. Without a shared language that allows Julia to say what she is thinking, and in order for Sara to understand, Julia draws on a variety of resources in this interaction—she turns her body to the snails, uses her facial expressions to illustrate where her attention is focused, and gestures to show the snail's action and connect them to her own body as she points to her back and makes small circular gestures. She has a theory about the snail that she is wondering about, but with words alone Sara does not understand what she means in the moment. Only when Sara takes the entire set of resources into consideration (gesture, gaze, verbal, etc.) and explicitly focuses on understanding what specifically Julia is talking about does she notice the snail's shell with eggs. This episode shows an additional way that CLD children can be disadvantaged. Not only do teachers have to give students space to use different languages, but they also need to take into account students' nonverbal ways of expression.

Unfortunately, such resource-rich perspectives that honor diverse idiolects, as well as spoken and non-spoken forms of communication, often interfere with the language ideologies in society, the linguistic policies of the school, and even in some cases the language ideas of families. Due to the dialectical nature of science learning and language, language policies (explicit as well as implicit) become a barrier for the enhancement of science practices, rather than a resource for scaffolding the development of science understandings as well as language proficiencies. This contradiction leads us to question the deeply contextual role of power and agency, which we explore next.

What Is the Relationship Between Language Used in Science Education, Power and Agency? Our different research projects in multilingual classrooms have underscored the importance of teachers creating the possibilities for children to engage in the practices of science using language(s) of the children's choosing, and to engage in the flexible use of semiotic and other resources in interaction with their peers. Further, we have come to recognize the power that teacher language choice for explaining science, holds. For instance, the greater the gap between the language being used by teachers at school and the one by students, the greater the chances school success is compromised; if a teacher uses a more formal way of talking for explaining science in the classroom, most likely those students whose families speak with a more formal language at home will have more chances of success (e.g., Broeder & Kistemaker, 2015). Thus, the higher the commonalities of language(s), language varieties or even word choices used between the students and the teachers, the smoother the teaching and learning processes in the science classroom can be. However, this also means that for students who do not share the same proficiency in the prioritized way of communication in the science classroom, being required to learn in that language carries far-reaching consequences that exceed their time in school.

> The voices of the children, so prevalent and strong outside of the school, become whispers, as they acquire other ways of speaking and suppress those from home, at least during school time. As these children turn into young adults, ready for an education beyond elementary school, the silence turns to resistance, and they often not only fail in school, but also leave school. The cost to all societies of ignoring their children's differences is enormous. (García, 2010, p. 1)

The use, and indeed, the sanctioning, of diverse resources (i.e., languages, modalities, materials) in the science classroom have an important role for students and their science learning. When teachers permit, and incorporate, the possibility of a range of resources within science investigations it can serve to facilitate students taking agency (Siry, Wilmes, & Haus, 2016).

Space and its organization in the classroom is also another form of power exercised by the teacher; the way the teacher organizes the classroom space and time determines the students' interactions (Hadi-Tabassum, 2006). When students feel safer, they can deploy more freely all their resources, especially their languages, thus fostering the meaning-making process in the science class. This space in the literature is named "îles de liberté" (Bourdieu, 1982) or freedom islands, "back stage," using the theatrical metaphor from Goffman (1959) or more language-specific "translanguaging space" (Wei, 2011). It was in this space that Maria's students were able to use their home and school languages interchangeably. It was also in this space that Julia was able to draw on multiple modalities to communicate throughout the process of sense-making.

In these spaces we often find students utilizing their home languages as a tool for participating in doing science (Wilmes, 2016). Students appropriate the knowledges they construct through this participation in their own terms, through the use of their own idiolects, and through multiple nonverbal modalities, as the episode of Julia and Sara demonstrates. The use of practices that draw on diverse resources such as these in the science classroom have an important role for students and their science learning and thus should not be limited, especially given the dialectical nature of language and science.

RECONSTRUCTING SCIENCE EDUCATION WITHIN THE LANGUAGE | SCIENCE RELATIONSHIP

> All one can do for a learner en route to her forming a view of her own is to aid and abet her on her own voyage.
>
> —Bruner (1996, p. 115)

We conclude this chapter by emphasizing the need to reconstruct the notion of "science" within science education as situated within the language | science relationship, in order to "aid and abet" children on their voyages of learning in school contexts. The two episodes we have selected to illustrate our points have been chosen to represent the diverse ways in which CLD children engage in the voyage that Jerome Bruner (1996) writes about in the above quote. Our research highlights the importance of providing spaces for children to engage in dialogical interactions that draw on their many resources. However, we stress that there is no room for true dialogue in institutional structures that are built upon control, power, and hierarchy. "Anti-dialogue" (Freire, 1990, p. 46) needs to be pushed back against and vertical relationships need to be deconstructed so that classroom interactions in science can be reconstructed to become dialogical. In creating classroom spaces for dialogue in general, and within science education contexts specifically, it is important for educators and researchers to recognize that dialogical relations flow in multiple directions, and that the classroom ought to be a place where "knowledge is sought and not where it is transmitted" (Freire, 1990, p. 150). In dialogical classroom contexts, curriculum can be seen as being "created emergently in the many modes of conversation through which teachers and students dialogically make sense of topics of individual and social significance, through action, knowledge building, and reflection" (Wells, 1999, p. 98). Adopting approaches that focus on the process, and not only the product, of science education enables a focus on the ways in which students interact with each other and with the communicative and meaning-making resources available to them.

Knowledge is always in process, and part of other processes (Kincheloe, 2001). We have written elsewhere about our positioning of education as a *process of being and becoming*, in which dialogue is central (Siry, Brendel, & Frisch, in press). The value of students "talking their way into science" (e.g., Gallas, 1995) is well-documented in the science education research literature, as researchers have explored the role that students' everyday language has on their meaning making in science (e.g., Varelas et al., 2008). Dialogue, in the theoretical perspectives that we adopt, is grounded on a mutual relationship of reciprocity—one that requires listening to, as well as speaking with, the other. Speakers as well as listeners then create dialogical encounters. As we listen to our research participants, we work towards adopting critical approaches of radically listening (Winchell, Kress, & Tobin, 2016)—that is, trying to suspend our own perspectives in the moment to be able to hear and understand where the other is coming from. This requires responsively listening as well as speaking in order to provide spaces for multiple perspectives, discourses, modalities, and languages to be drawn on as resources by speakers, and be positioned that way by listeners.

> To recognize the role of perspective and vantage point, to recognize at the same time that there are always multiple perspectives and multiple vantage points, is to recognize that no accounting, disciplinary or otherwise, can ever be finished or complete. There is always more. There is always possibility. (Greene, 1988, p. 128)

In the critical theoretical sense, dialogue is loving, humble, hopeful, trusting and critical (Freire, 1990), and it moves us towards understanding the multiple, co-existing perspectives at hand within interactions. A dialogical perspective on social life grounded in critical perspectives focuses on the processes, and indeed, on the possibilities for change and transformation that is alluded to in Maxine Greene's above quote. When we reflect on Greene's quote relative to CLD students specifically, the value of the diverse resources that students bring to science is critical to consider in the process of reconstructing a science education grounded upon resources, diverse communicative repertoires, and an understanding of the centrality of the science | language relationship. In closing, we call upon educators, researchers, and those involved in all ways in the science education process to question and be critical of the ways languages are used to include and to exclude, particularly in educational contexts, and to be aware of how science and language are intertwined as one informs the construction of the other.

REFERENCES

Bourdieu, P. (1982). *Ce que parler veut dire.* L'économie des échanges linguistiques. Paris: Fayard.

Bourdieu, P. (1986). The forms of capital. In J. Richardson (Ed.), *Handbook of theory and research for the sociology of education* (pp. 241–258). New York, NY: Greenwood.

Broeder, P., & Kistemaker, M. (2015). More willingly to school: Tools for teachers to cope with linguistically diverse classrooms. *Intercultural Education, 26*, 218–234. doi:10.1080/14675986.2015.1048095.

Bruner, J. (1996). *The culture of education.* Cambridge: Harvard University Press.

Council of Europe. (2001). *Common European framework of reference for languages: Learning, teaching, assessment.* Retrieved from http://www.coe.int/t/dg4/linguistic/source/framework_en.pdf

Creese, A., & Martin, P. (2008). Introduction to volume 9. Ecology of language. In A. Creese, P. Martin, & N. Hornberger (Eds.), *Encyclopedia of language and education* (pp. xiii–xviii). Dordrecht: Springer.

Cummins, J. (2005). A proposal for action: Strategies for recognizing heritage language competence as a learning resource within the mainstream classroom. *The Modern Language Journal, 89*, 585–592.

Faulstich Orellana, M. (2016). *Immigrant children in transcultural spaces: Language, learning and love.* New York, NY: Routledge.

Freire, P. (1990). *Education for critical consciousness.* New York, NY: Continuum Press.

Gallas, K. (1995). *Talking their way into Science: Hearing children's questions and theories, responding with curricula.* New York, NY: Teachers College Press.

García, O. (2010). Foreword. In S. Ehrhart, C. Hélot, & A. Le Nevez (Eds.), *Plurilingualism and teacher education: A critical approach* (pp. 1–4). Frankfurt am Main: Peter Lang.

Goffman, E. (1959). *The presentation of self in everyday life.* New York, NY: Doubleday.

Gómez Fernández, R. (2015). El trilingüismo idealizado y el uso de lenguas no oficiales en la escuela luxemburguesa. *Lengua y Migración, 7*(2), 75–108.

Gómez Fernández, R. (2016). "Ils ne cherchent pas à apprendre le français": NAMS and multilingual primary schools in France and Luxembourg. *Carnets d'Atelier de Sociolinguistique, 11*, 133–168.

Greene, M. (1988). *The dialectic of freedom.* New York, NY: Teachers College Press.

Hadi-Tabassum, S. (2006). *Language, space and power: A critical look at bilingual education* (Vol. 55). Clevedon: Multilingual Matters.

Kincheloe, J. (2001). *Getting beyond the facts: Teaching social studies/social sciences in the twenty-first century* (2nd ed.). New York, NY: Peter Lang.

Martín Rojo, L. (Ed.). (2003). *¿Asimilar o integrar? Dilemas ante el multilingüismo en las aulas.* Madrid: Ministerio de Educación, Cultura y Deporte.

Menken, K., & García, O. (2010). Introduction. In K. Menken & O. García (Eds.), *Negotiating language policies in schools: Educators as policymakers* (pp. 1–10). New York, NY & London: Routledge.

Ministère de l'Éducation Nationale, de l'Enfance et de la Jeunesse [MENJE]. (2015). *Bildungsbericht Luxembourg 2015.* Band 1: Sonderausgabe der Chiffres Clés de L'éducation Nationale 2013/2014. Retrieved from http://www.men.public.lu/catalogue-publications/systeme-educatif/statistiques-analyses/bildungsbericht/2015/band-1.pdf

Ministère de l'Éducation Nationale, de l'Enfance et de la Jeunesse [MENJE]. (2015). *Cours intégrés en langue et culture portugaise: Vade-mecum à l'intention des communes et des enseignants.* Retrieved from http://www.men.public.lu/fr/themes-transversaux/scolarisation-eleves-etrangers/enseignement-fondamental/151006-vade-cours-portugais.pdf

Moore, S. C. K. (2014). *Language policy processes and consequences*. New York, NY: Multilingual Matters.

Otheguy, R., García, O., & Reid, W. (2015). Clarifying translanguaging and deconstructing named languages: A perspective from linguistics. *Applied Linguistics Review, 6*, 281–307. doi:10.1515/applirev-2015-0014.

Sewell, W. H. (1999). The concept(s) of culture. In V. E. Bonnell & L. Hunt (Eds.), *Beyond the cultural turn: New directions in the study of society and culture*. Berkeley & Los Angeles, CA: University of California Press.

Siry, C. (2011). Exploring the significance of resource-rich views in science education. *Cultural Studies of Science Education, 6*, 1019–1029. doi:10.1007/s11422-011-9353-3.

Siry, C., Brendel, M., & Frisch, R. (2016). Radical listening and dialogue in educational research. *International Journal of Critical Pedagogy, 7(3)*, 119-135.

Siry, C., Wilmes, S. E. D., & Haus, J. (2016). Examining children's agency within participatory structures in primary science investigations. *Learning, Culture and Social Interaction*. doi:10.1016/j.lcsi.2016.01.001.

Siry, C., Ziegler, G., & Max, C. (2012). "Doing science" through discourse-in-interaction: Young children's science investigations at the early childhood level. *Science Education, 96*, 311–326. doi:10.1002/sce.20481.

Snell, J. (2013). Dialect, interaction and class positioning at school: From deficit to difference to repertoire. *Language and Education, 27*, 110–128. doi:10.1080/09500782.2012.760584.

Tobin, K. (2012). Sociocultural perspectives on science education. In B. Fraser, K. Tobin, & C. McRobbie (Eds.), *Second international handbook of science education* (pp. 3–17). Dordrecht: Springer.

United Nations Development Programme [UNDP]. (2009). *Human Development Report 2009. Overcoming barriers: Human mobility and development*. New York, NY: Palgrave Macmillan. Retrieved from http://hdr.undp.org/sites/default/files/reports/269/hdr_2009_en_complete.pdf

Varelas, M., & Pappas, C. C. (2013). *Children's ways with science and literacy: Integrated multimodal enactments in urban elementary classrooms*. New York, NY: Routledge.

Varelas, M., Pappas, C. C., Kane, J. M., Arsenault, A., Hankes, J., & Cowan, B. M. (2008). Urban primary grade children think and talk science: Curricular and instructional practices that nurture participation and argumentation. *Science Education, 92*, 65–95. doi:10.1002/sce.2023.

Weber, J.-J. (2009). *Multilingualism, education and change*. Frankfurt am Main: Peter Lang.

Weber, J.-J. (2014). *Flexible multilingual education: Putting children's needs first*. Bristol: Multilingual Matters.

Wei, L. (2011). Moment analysis and translanguaging space: Discursive construction of identities by multilingual Chinese youth in Britain. *Journal of Pragmatics, 43*, 1222–1235. doi:10.1016/j.pragma.2010.07.035.

Wells, G. (1999). *Dialogic inquiry: Toward a sociocultural practice and theory of education*. Cambridge: Cambridge University Press.

Wilmes, S. E. D. (2016). Science workshop: Let their voices lead the way. In A. Oliveira & M. Weinburgh (Eds.), *Science teacher preparation in content-based second language acquisition* (pp. 323-340). Dordrecht: Springer.

Winchell, M., Kress, T., & Tobin, K. (2016). Teaching/learning radical listening: Joe's legacy among three generations of practitioners. In M. F. Agnello & B. Reynolds (Eds.), *Practicing critical pedagogy: The influences of Joe L. Kincheloe* (pp. 99–112). Rotterdam: Springer.

CHAPTER TWENTY

The Tension of Maintaining Language Diversity versus Extinction

Mother Tongue as a Filipino Response

PURITA P. BILBAO, DEBORAH J. TIPPINS,
AND SOPHIA (SUN KYUNG) JEONG

In recent years, as international concern about the degradation of the planetary environment has risen to the forefront, so too has concern for the world's languages which comprise the intergenerational knowledge that sustains life on Earth as we know it. Language, as part of the intergenerational knowledge essential in the creation of cultural commons, enables people to "understand relationships, make moral decisions, acquire a sense of self-identity and engage in activities that are the source of personal meaning" (Bowers, 2014, p. vii). It is widely acknowledged that loss of language can have a negative impact on biodiversity conservation. Christopher Moseley (2011) explains how indigenous names, practices, oral traditions and taxonomies are lost when a community shifts to another language, particularly a language used by more powerful groups. Today's rapid increase in migration and urbanization brings with it strong pressure to speak dominant languages that are perceived necessary for full participation in global economic systems. Gregory Anderson and K. David Harrison (2011) note that over half of the world's 6,000 languages are spoken by only a very few people. The United Nations Educational, Scientific and Cultural Organization (UNESCO) provides a useful framework for classifying these languages in terms of their status with respect to the transmission of intergenerational knowledge and the extent to which they are endangered. According to Moseley (2011):

> A language is *safe* if spoken by all generations, intergenerational transmission is uninterrupted; *vulnerable* when most children speak the language, but it may be restricted to

> certain domains (e.g., home); *definitely endangered* when children no longer learn the language as mother tongue in the home; *severely endangered* when a language is spoken by grandparents and older generations and while the parents may understand it, they do not speak it to children or among themselves; *critically endangered* when the youngest speakers are grandparents and older who speak the language partially and infrequently, and *extinct* when there are no speakers left. (pp. 14–15)

The Philippines, like many countries, has grappled with issues of policy and language instruction, particularly in relation to science and mathematics instruction in the early years. During this recent quarter century, Mother-Tongue Multilingual Education has found its niche in Filipino education. Norman Thomson and Deborah Tippins (2013) note that nations such as the Philippines are faced with different options with respect to the languages they use; these options take into account past, present and future legacies embedded in culture and mediated by power relationships. For example, nations may choose from among the following options:

> a) remain uncommitted with respect to language policy and allow things to change without interference; b) use an ex-colonial language as the official (and national language) as it is perceived to be neutral at the expense of losing cultural heritages; c) adopt the majority language where such a language is predominant; d) allocate to some of the major languages certain public roles at the regional or district level, or e) give nominal public roles or none to the smaller languages. (Thomson & Tippins, 2013, p. 239)

In the Philippines, a country comprised of more than 7,000 islands (depending on the tides), more than 170 distinct dialects are spoken, although the exact number depends on how linguists classify them. In general, the number of spoken languages is estimated based on linguistic codes, which are not intelligible or understandable. Curtis McFarland (1994), in his attempt to classify Philippine languages, identified approximately 110 distinct dialects; however, Chabacano, a Spanish Creole with two known dialects (Ternateno and Zamboangueno) constitute an additional two dialects. This rich linguistic history is not surprising in light of patterns of migration and the long history of colonization by Spain, Japan, and the United States. For example, the Philippines was under the rule of Spain for 327 years, and prior to WWII official government documents were written in Spanish. During this time, Spanish was the medium of teaching and learning, with the intent to propagate religion (Christianity), beginning with the cartella or alphabet. When the U.S. established a colonial government in the Philippines in 1900, the English language was brought to the Philippines by American teachers called Thomasites. The Thomasites established normal schools to prepare teachers and organized the curriculum into separate subjects, including science. Although the Philippines became a commonwealth in 1935 and was granted independence in 1946, its language policy in the educational context remained unclear between the years of 1935 and 1973. President John F. Kennedy sent Peace Corp volunteers to the Philippines in 1961, often referred to as a second wave of Thomasites, to teach English, science, and mathematics throughout the country. More than ten years later, in 1973, the National Board of Education declared a bilingualism policy in schools which became the basis for the Department of Education, Culture and Sports language policy (No. 25, s. 1974) entitled "Implementing Guidelines for the Policy on Bilingual Education," which mandated English as the medium of instruction for science and mathematics with Filipino designated as the language to be used for social studies, music, art, physical education, home economics, practical arts, and character education. With most school children speaking and writing in Filipino, English and their own local dialect, the Philippines has long been considered a multilingual country. While this linguistic heritage has benefitted Filipino/a youth in many ways, recent education policy as it pertains to language use in primary science and mathematics reflects a shift from multilingual instruction to Mother-Tongue Multilingual as a foundation for the teaching and learning of science and mathematics in the early years. The use of mother tongue (or the L1 language) in primary grades is supported by both empirical and anecdotal research. Else Roberts Auerbach (1993) found that the use of the first language, or mother tongue, in the primary grades provides learners with a sense of security and validates their lived experiences. He also documented how mother tongue instruction in primary grades encourages learners to take risks when they later transition to English. Aceme Nyika (2015) described the how the use of mother tongue fosters the development of innovation and creativity.

In the Philippines, the movement towards the use of mother tongue or the first language as the primary medium of instruction from pre-school until at least grade 3 has occurred simultaneously with the more recent emphasis on curriculum localization and the preservation of traditional ecological knowledge. The fragile island ecosystems of the Philippines are experiencing cultural and biodiversity losses at rapid rates as a consequence of urbanization, globalized commercialization, changing patterns of weather and climate, increased migration, war, and other factors. These threats are manifested in many ways including over-fishing, deforestation, erosion, and dynamite blasting of coral reefs to name a few. With the recognition that a national policy would be required to protect vulnerable first languages as well as cultural and ecological knowledge, the Philippines Department of Education issued an order in 2009 to institutionalize Mother Tongue-Based Multilingual Education (order No. 74) for public school primary education throughout the country (Alcudia et al., 2016). This policy was strengthened by the declaration of then President Beningno Aquino, Jr. when he said: "Learn English well and connect to the world. Learn Filipino well and connect to our country. Retain your dialect and connect to your heritage" (Aquino, 2010). UNESCO's advocacy through Education for All (EFA) and Education for Sustainable

Development (ESD) further emphasize the need for a shift to the use of mother tongue for science and mathematics in the primary grades (including Arabic for Madaris schools), with February 21st officially designated as Mother Tongue Day (UNESCO, 2008).

As a whole, primary science in the Philippines is taught in interdisciplinary ways in grade K-2 and as a separate subject beginning in grade 3. The spiraling curriculum begins in PreK with the use of the mother tongue and slowly transitions to English after grade 3. The Philippines Department of Education selected and approved nineteen dialects for mother tongue instruction: Cebuano, Itiligaynon, Ilokano, Kepampangan, Pangasinan, Tagalog, Waray, Pangasenense, Tausug, Maguindanao, Maranao, Chabacano, Ybang, Ivatan, Sambal, Akeanon, Kinaraya, Yakan and Surigaonon. The criteria for the selection of these languages are not clear and may call to question whether some majority dialects are privileged. Guiding principles were developed to facilitate the implementation of Mother-Tongue Multilingual Education. Of particular relevance to primary science is the basic principle, which maintains that all regular school-based and system-wide examinations, together with international benchmarking and assessments, should be conducted in the mother tongue.

Although the shift to a Mother-Tongue Multilingual Education policy is part of current science education reform efforts in the Philippines, its roots can be traced back to the influence of Jose V. Aguilar, a School Division Superintendent during the 1930s. At a time when public education was rapidly spreading to rural areas of the Philippines, Aguilar drew attention to the problematic role of language in separating schools from the community, and in particular rural community life. From his perspective, the use of the second language (English) in school was a world apart from the vernacular dialects spoken by students in their homes and contributed to disparities in terms of social and economic stratification. In the book, *Mother Tongue: For Teaching and Learning* (Alcudia et al., 2016), Bilbao explains how Aguilar, to address this concern, engaged a force of 1,200 teachers to reach out to rural communities by introducing farmers to the two-crop agricultural system of rice production twice a year instead of the traditional single annual planting of this crop. In order to communicate with local farmers, teachers had to speak in the vernacular dialect of the community. As part of Aguilar's vision of transforming public schools into community schools, he encouraged teachers and schools to discover the "personality" of the community and work alongside its members in the struggle to build human capacity from the bottom up. For a while the use of the mother tongue vernacular dialect was at the heart of Philippine basic education. Aguilar's early advocacy for the use of the mother tongue dialect as the medium of instruction is closely linked today not only to the revival of mother tongue use in the primary grades but also more broadly to the emergence of community centered science education and the revitalization of the commons.

Many policy and pedagogical issues surround the use of mother tongue as the medium of instruction for primary science, not only in the Philippines, but in many other countries facing similar challenges. In some cases, parents may consider an international language more important to the extent that they would rather have their children learn this language than to speak the first language at home. Care must be taken to ensure that the dialects selected for instruction do not undermine the educational participation of indigenous groups or communities whose members speak non-dominant languages. Typhoons, floods and other natural disasters, which can contribute to the relocation or resettlement of families, may increase the diversity of communities, strengthening the imperative to make primary science inclusive of all students' languages. The use of mother tongue for science instruction in the primary grades may also give primacy to academic language at the risk of ignoring theatre, drama, music, photography, art and other ways of preserving language, culture, and traditional ecological knowledge. Alternatively, the use of mother tongue may foster a broader conception of literacy, which includes many ways of "reading the world" in the manner that educators Paulo Freire and Donaldo Macedo (1987) described it. Teacher preparation issues are also important considerations in the transition to Mother-Tongue Multilingual Education in the primary grades. The recruitment of teachers who are not only prepared and committed to teaching science in the rural communities (barrios) of the Philippines, but can do so using the mother tongue as the language of instruction, is an ongoing challenge. Alongside this challenge is the need to produce locally relevant and non-colonizing curriculum resources appropriate for science teaching and learning in the mother tongue. To this end, a holistic, interdisciplinary approach will be needed to successfully attend to these issues and many more.

Meaningful science learning cannot take place in isolation from a student's first language. Alejandro Gallard and Deborah Tippins (1994) argue that, "if we are to engage students and teachers in meaningful learning we must include the voices and understandings of those who do not speak English as a first language" (p. 286). We must also listen to the perspectives that teachers, students, and other community members hold with respect to the use of mother tongue. In light of the postmodern condition we would argue that the story of Mother-Tongue Multilingual Education in the Philippines is just the beginning. It is a story that requires a deep level of analysis, from an ecojustice perspective, of the root metaphors that are carried forward through language. It is also a story that calls us to consider the commodification of language and ways that messages are reduced to information that can be stored, retrieved, packaged, articulated, and transmitted. Today's youth inhabit complex cultural and linguistic spaces that require more expansive views of the roles of both language and agency in science teaching and learning to replace the oppressive narratives of the past. Finally, it is a story that is an essential part of how humans will ultimately consider the complex issues surrounding species, language, and cultural extinctions and respond to them in productive ways.

REFERENCES

Alcudia, F. F., Bilbao, P., DeQuilla, A. C., Germinal, A., Rosano, D., & Violeta, M. (2016). *Mother tongue: For teaching and learning*. Quezon City: Lorimar Publishing Inc.

Anderson, G., & Harrison, K. D. (2011). *Global language hotspots*. Retrieved from http://www.swarthmore.edu/SocSci/langhotspots/globaltrends.html

Aquino, N. (2010). *Noynoy Aquino's 10-point basic education agenda*. Retrieved from https://www.ammado.com/nonprofit/46130/articles/14263

Auerbach, E. R. (1993). Rexamining English only in the ESL classroom. *TESOL Quarterly, 27*(1), 9–32. doi:10.2307/3586949.

Bowers, C. A. (2014). *The false promises of the digital revolution: How computers transform education, work and international development in ways that are ecologically unsustainable.* New York, NY: Peter Lang.

Freire, P., & Macedo, D. (1987). *Literacy: Reading the word and the world.* Westport, CT: Bergin & Garvey.

Gallard, A., & Tippins, D. (1994). Language diversity and science meaning: The need for a critical system of meaning. In B. McLeod (Ed.), *Language and learning: Educating linguistically diverse students* (pp. 269–289). New York, NY: State University of New York.

McFarland, C. D. (1994). Subgrouping and number of the Philippine languages-or-How many Philippine languages are there? *Philippine Journal of Linguistics, 25*(1–2), 75–84.

Moseley, C. (Ed.). (2011). *Atlas of the world's languages in danger* (3rd ed.). Paris: UNESCO Publishing.

Nyika, A. (2015). Mother tongue as a medium of instruction in developing country universities in a global context. *South African Journal of Science, 111*(1/2), 1-5. doi:10.17159/sajs.2015/20140005.

Thomson, N., & Tippins, D. J. (2013). Envisioning science teacher preparation for twenty–first century classrooms for diversity: Some tensions. In N. Mansour & R. Wegerif (Eds.), *Science education for diversity* (pp. 321–250). Dordrecht: Springer.

UNESCO. (2008). *Mother tongue instruction in early childhood education: A selected bibliography*. Paris: UNESCO.

CHAPTER TWENTY-ONE

How Perspectives from Linguistically Diverse Classrooms Can Help All Students Unlock the Language of Science

CORY BUXTON, LOURDES CARDOZO-GAIBISSO, YU XIA, AND JIONG LI

In this chapter, we discuss the specialized language that has evolved over centuries to communicate scientific ideas, and the role that this language plays in modern science education. Scientific language can be seen as a code that is used to unlock scientific thinking and communicating, but that can also appear mysterious and inaccessible to those who are first trying to make sense of it. Scientific language can also be viewed as an exclusive discourse that may intentionally or unintentionally alienate many people who then come to believe that they cannot or do not wish to engage with science. Whichever the case, it has been an ongoing national concern that too many students leave school and live their lives without basic scientific literacy; it is clear that the language of science is both part of this problem and must also be part of any solution. In this chapter, we set out to explore the nature of the language of science, its role in science education, and strategies that we might use as science educators to prevent the language of science from being a barrier to students learning (and wishing to learn) science.

We begin the chapter with a discussion of how and why the language of science evolved to function the way it does. Second, we discuss the language of science in the science classroom and some ideas about how students encounter and begin to make sense of this language, or else fail to do so, deciding that science is "not for them." Third, we consider the role that everyday language can play in understanding and explaining science concepts and how it may (or may not) serve as a bridge to adopting the academic language of science. Fourth, we consider the unique aspects of students learning science in a second (or additional) language and the roles that home language, school language, and various combinations (translanguaging) can play in supporting science learning. More specifically, we explore how a translanguaging orientation can assist teachers in understanding the dynamic role of students' multiple languages for enhancing the learning of science concepts while sustaining of students' diverse linguistic repertoires.

Fifth, we consider several practical strategies for supporting all students in learning to use the language of science by providing examples from our current research. We conclude the chapter by summarizing recommendations for language use in the science classroom and we raise some additional questions that must be answered to more fully understand how language affects and is affected by the science education process. We note that the majority of the examples we provide come from work done in U.S. contexts. While we believe that our claims and recommendations largely hold true across diverse linguistic, cultural and geographic contexts, we refer the reader to other chapters in this volume for perspectives that draw more centrally on work that does not privilege English as the dominant language of science instruction.

EVOLUTION OF THE LANGUAGE OF SCIENCE: WHY DOES SCIENCE SOUND THAT WAY?

The language of science has evolved over time in specific ways to accomplish certain purposes. It is a language that is used to construct theories, to posit explanations, and to build arguments from evidence. Thus, its features have developed to make theoretical discourse about the workings of the natural world possible. To this end, the language of science makes primary use of two types of linguistic resources, namely lexical resources and grammatical resources. Lexical resources include the technical terms that are continuously being updated as science disciplines develop, merge, splinter, and create new technologies and new ideas (e.g., bioinformatics, neuroparasitology), as well as everyday words that are given special and specific meanings in science (e.g., matter, work, fault). This aspect of the language of science can be referred to as technicality (Fang, 2006). Grammatical resources include the ways in which ideas and categories are constructed and organized so that they can be assembled to convey the particular forms of reasoning and argument that are construed to be scientific. This second aspect of the language of science can be referred to as rationality (Martin & Veel, 2005). Technicality and rationality have worked together over time as the driving forces that have shaped the language of science, and by extension, have shaped scientific thinking and understanding.

The linguists, M. A. K. Halliday and Christian Matthiessen (2004), have shown how the linguistic evolution that led to modern scientific language has its roots in early Greek philosophy, and those philosophers' desires to discuss and describe abstract objects and forms in ways that were more persistent, and therefore, easier to categorize than the transient and idiosyncratic happenings that occupy most of human perception. This can be clearly seen, for example, in Plato's notion of ideal forms (Lovejoy, 2011). For Halliday (2004), the value and the power of grammar come from its ability to turn one linguistic class of words into another class, with the resulting ideas becoming more or less stable in time and space, using what Halliday refers to as grammatical metaphor.

Grammar serves to shape and transform how we make meaning through language. The grammatical transformations that began with Greek philosophy became increasingly important to the goals of developing the natural sciences during the European Enlightenment, when Galileo, Newton, and others embraced the power of grammatical metaphor to create our modern discipline-based organization of technical knowledge, and along with it, the foundation of our modern language of science (Chomsky, 2002). Like the language of philosophy, the blossoming language of science required more stable and persistent linguistic classes for the purpose of creating technical taxonomies. To this end, the language of science began to employ what Halliday (1999) refers to as *grammatical drift* from circumstances (which are unstable and ephemeral) through processes and qualities (which are more generalizable and therefore, more permanent), toward abstract entities (e.g., conductor, osmosis, tectonic motion) that are conceptual or theoretical, but stable and persistent through time. Unlike the everyday language that we use to make sense of our idiosyncratic and concrete, but rapidly changing, life experiences, scientific language theorizes these life experiences into abstract entities, which can then be further examined, generalized, categorized, and critiqued.

Modern scientific knowledge is continually extended through the use of technicality and rationality, embodied in specialized language, grammatical metaphor, and grammatical drift. Specifically, this grammatical drift in language function often occurs through *nominalization*, the process through which circumstances and processes are linguistically modified and then expressed as qualities or entities (Halliday & Martin, 2003). This most often occurs through verbs and adjectives being reconstructed as abstract nouns, such that *move* becomes *motion* and *eroding surfaces* becomes *surface erosion*.

Nominalization is an indispensable part of how the language of science is constructed, in that it supports theorizing of everyday experience into scientific terms. Indeed, grammatical metaphor has made not only the communication of science, but even the conceptualization of scientific theory possible through its use of technicality and rationality. On the one hand, grammatical metaphor creates new technical meanings that describe and classify natural phenomena through nominalization (e.g., electromagnetic field, bioluminescence). On the other hand, grammatical metaphor supports scientific rationality by constructing arguments out of sequenced reasoning, using what Halliday and Matthiessen (2004) refer to as rheme to theme structure. In this way, technicality is packed into the object of one sentence (the rheme) which is then used as the subject of subsequent sentences (the theme), carrying with it all of the embedded technicality (e.g., The percentage of salt content in water is referred to as salinity. The salinity of water is proportional to the temperature at which it freezes. The freezing point of seawater will be slightly higher near points of fresh water outflow. Thus, part of the explanation of high rates of sea ice in Norwegian fjords is a function of their decreased salinity.). Indeed, the conceptual features that are often associated with scientific communication, such as abstraction, objectivity, dense information, and skepticism are all made possible through the use of grammatical metaphor. While we may not routinely think about the language we hear in terms of its grammatical function, we can easily distinguish language we encounter as sounding more or less scientific to the ear. That is, most of us would say that we recognize scientific language when we hear it, and this has implications for students' use of language in the science classroom, as we will discuss in detail later.

For example, the lexical density, or the ratio of words that make meaning [functional words] to words that support meaning [grammatical words], is typically much higher in scientific writing than in everyday language. This is because of the technical taxonomies in scientific language that are defined once and then reused repeatedly with the assumption that the reader has absorbed and internalized the nuances of those taxonomies. Another distinctive feature of the language of science is that the qualities of grammatical drift and nominalization are applied nearly equally to spoken and written scientific language. Our everyday spoken language is grammatically quite distinct from our standard written language, but the language of science remains largely consistent across spoken and written forms (Halliday & Martin, 2003). Whether in written or oral mode, the language of science is densely packed with objective and abstract observational meaning, while minimizing interpersonal language (such as personal pronouns), thus presenting information in a depersonalized way. Scientific writing thus serves to objectify language, making it easy for processes to be expressed without reference to the human actors involved.

In summary, the language of science has evolved over time to suit the needs of scientific communities engaged in the process of categorizing and classifying abstract constructs that describe the functioning of the natural world. These linguistic forms have developed to enable effective and efficient communication among scientists in a given field of study. However, the same linguistic qualities that support the goals of the scientific community can serve as obstacles to those outside of this community who wish to participate in the conversation.

LANGUAGE OF SCIENCE IN THE SCIENCE CLASSROOM: STUDENTS ENCOUNTER NEW WAYS OF THINKING AND COMMUNICATING

As we have already alluded, language both shapes and is shaped by our experience with the world around us. Young children begin with oral language and the immediate world of their direct experience. As they move toward school age, children's language skills continue to develop, and they begin to interact with language in new ways, such as through reading and multimedia. Reading text and engaging with multimedia open children's minds to experiences that extend beyond those that they have directly observed. At the same time, those broader experiences also broaden children's use of language, adding not only new vocabulary, but new linguistic styles, and syntax as well. As a side note, this linguistic and conceptual expansion is a primary reason why it is so important that young children be read to long before they are able to read on their own. Reading to children gives them a head start on this process of expanding both experience and linguistic forms. For example, Kathryn Fletcher and Elaine Reese (2005) found that when compared to similar children who had not been regularly read to, those who had been read to were better able to engage in conversations about a wider variety of topics and could do so through a greater variety of language.

In the specific case of science, this argument about how language and experience interact in mutually supportive ways provides a strong rationale for why science needs to be a robust part of the elementary school curriculum. Students who don't regularly study science during elementary school begin middle school lacking not only scientific experiences and concepts, but also the corresponding scientific language through which they can make sense of new science experiences. Middle school students are asked to contextualize, interpret, and explain their experiences of the natural world through the grammatical structures of science that we have discussed above, and without prior preparation, many students struggle with this task (Moje, Collazo, Carrillo, & Marx, 2001). Indeed, gaining early competence with the language of science will be increasingly critical for success in academic science because this language is central to the learning goals of the Next Generation Science Standards (NGSS Lead States, 2013). A basic ability to interpret and produce the language of science is prerequisite to the development of sustained scientific discourse, which, in turn, is needed to engage in the science and engineering practices advocated in the NGSS.

In a seminal study of language in the science classroom, Jay Lemke (1990) argued that one of the greatest obstacles to students' meaningful learning of science concepts is the overwhelming amount of time that students spend listening to and reading for information in science class. Lemke contrasts this to the very limited time that students spend talking and writing about their own science ideas and experiences. He further points out that there are powerful social class dynamics at work in these findings, such that more socioeconomically privileged students gain richer science and language experiences outside of school, which then appears as readiness for more self-directed and experiential science learning in school with correspondingly greater opportunities to practice producing the language of science through speaking and writing. In contrast, socioeconomically disadvantaged students come to the science classroom with less mastery of the language of science and then get correspondingly less opportunity to practice producing that language, as their time is spent on receptive language skills of listening and reading.

Various projects have subsequently built on Lemke's findings to design models that scaffold students' opportunities to engage in productive science speaking, writing, and visually representing. Perhaps the most systematic example has been the work of Brian Hand and colleagues to develop, test, and refine science writing heuristics (see Nam, Choi, & Hand, 2011 for a summary of this work). These heuristics have been designed to support students in making richer scientific meaning from their participation in science investigations through scaffolding students' writing of lab reports in ways that highlight personal connections and reasoning from evidence.

In this section, we have looked at how both the language and the science experiences that students bring to the classroom affect the nature of the science learning opportunities they receive, as well as how classroom science learning, in turn, influences students' language development. In the following section, we will consider how everyday language can be used to explain scientific ideas, even though students get more recognition for their science ideas when these ideas are expressed using academic language.

USING EVERYDAY LANGUAGE TO EXPRESS SCIENTIFIC THINKING WHILE LEARNING ACADEMIC LANGUAGE

Thus far, we have discussed how the language of science evolved by making use of certain linguistic features that enhance its technicality and rationality, as well as how those features hinder many students from making sense of science and from adequately expressing their thinking about science ideas. So does this mean that without mastering the language of academic science that students lack linguistic resources for making sense of and communicating scientific thinking? On the contrary, it is through everyday language that most of us first experience science learning in a variety of informal contexts as young children, and these linguistic and conceptual resources can be built upon later in the science classroom.

To begin, we should consider what we mean by everyday language. Grounded in Halliday and Matthiessen's (2004) framework of systemic functional linguistics (SFL) that we have been using to describe the language of science, we can consider everyday language in terms of its lexical resources (vocabulary) and its grammatical resources (structures). At the level of vocabulary, everyday language can be categorized as relying on words that are neither technical vocabulary specific to a given discipline (e.g., photosynthesis, electromagnetic field) nor general academic vocabulary that appears across disciplines but that is more common in written academic text and less common in general interpersonal conversation (e.g., function, primary). That is, we can say that one feature of everyday language is its low degree of technicality.

While everyday language may lack the specificity of academic language that technicality provides, it can still be used to make meaning of natural phenomena, and of course, has been used to do so for as long as humans have used language. Put another way, it is possible for everyday language to express a high degree of rationality even while exhibiting a low degree of technicality. An SFL perspective helps us to consider how everyday language is used to make meaning, by highlighting how all communication can be considered to have three purposes or functions: an ideational function, an interpersonal function, and a textual function. The ideational function of language is what allows us to make meaning of the world around us and build our own theory of experience. That is, it is through language that we make meaning of our world and ourselves (Halliday, 1999). Second, the interpersonal function of language is the way in which we use language to communicate our ideas to others. All human language, regardless of its simplicity or complexity, is well suited to this function. Finally, the textual function of language is the way in which we order and structure our ideas to create coherent language and therefore, coherent meaning. Everyday language is every bit as capable of using these three functions to create, organize, and communicate meaning as is more sophisticated academic language.

Applying this understanding, we can now consider the role that everyday language might play in students' science learning. On the one hand we can consider how learning science through everyday language might serve as a temporary state of affairs while students gradually gain mastery of academic language. From this perspective everyday language is a bridge of limited capacity that will eventually be replaced by a stronger, more permanent bridge of academic language that will be capable of supporting more advanced scientific understanding (August, Branum-Martin, Cardenas-Hagan, & Francis, 2009). On the other hand, we might consider our everyday language as fully capable of allowing us to make meaning of our experiences, including our experiences with science.

For example, the longstanding research program by members of the Chèche Konnen team (e.g., Warren, Ballenger, Ogonowski, Rosebery, & Hudicourt-Barnes, 2001) has involved case studies of students from African American, Haitian, and Latino backgrounds in both monolingual and multilingual classrooms. This body of work explicitly considers the role of everyday language in scientific sense making by investigating how students use everyday language and discourse styles as resources to understand and gradually take ownership of the language patterns used in scientific communities. For example, Josiane Hudicourt-Barnes (2003) demonstrated how argumentative discussion is a major feature of social interaction among Haitian adults and how this culturally familiar discourse pattern could then be leveraged as a resource for students as they practiced scientific argumentation in school. More recently, Warren and Rosebery (2011) have considered the value of viewing science learning as an intercultural process in which students and teachers negotiate the boundaries of language, as well as of race and culture, in relation to subject matter in order to overcome the traditional inequalities that often persist in science classrooms for students of color.

In related work, Bryan Brown and colleagues (e.g., Brown & Spang, 2008) have studied communication and interaction patterns among students and teachers of various cultural and linguistic backgrounds while learning science. Focusing on students' discursive identities as a tool for understanding student learning, Brown (2006) analyzed the interactions in a high school biology class he taught. He detailed the cultural and linguistic moves that students made in response to the classroom culture and discourse structure he was attempting to construct. He found that while marginalized students were comfortable in adopting both a scientific worldview (epistemology) and science inquiry practices, many students expressed both difficulty and resistance toward appropriating the academic language norms of science. Brown concluded that the language of science serves as a potential gatekeeper that prevents some students from adopting the culture of science, and that teachers should support students in using everyday language to make science meaning while simultaneously taking steps to support students' attempts to recast their discursive identities in ways that incorporate the academic language of science.

In this section we have argued that in addition to sociolinguistic features that distinguish everyday language from academic language, there are also sociocultural factors that can both support and limit students' desire to engage in academic language while also reducing the credit that students do get when using everyday language to express academic ideas. Everyday language may, however, help students to produce clearer and more complete science meaning, when explored in relation to the relevant academic language of science. Thus, we view everyday language as an undervalued resource for meaning making and communicating in science. While it is true that acquiring the academic language of science can serve to support and deepen scientific sense-making, it is also true that everyday language has evolved to serve the same basic communicative functions as academic language and is, therefore, capable of communicating sophisticated science ideas.

FIRST LANGUAGE, SECOND LANGUAGE AND TRANSLANGUAGING IN SCIENCE: WHAT ROLE SHOULD HOME LANGUAGE PLAY IN SCIENCE LEARNING?

A similar (but not identical) argument to the one we have made regarding the value of everyday as well as academic language for science meaning making can be made for the role of home language as well as second language in science learning. The vast majority of students in the U.S. go to schools where English monolingualism is the instructional norm in spite of the increasingly diverse multilingual student population. In these classrooms, instructional practices continue to reproduce a model that sustains a monolingual view of academic language and literacy (Hornberger & Link, 2012). There are, however, currently more than 460 home languages spoken by students in U.S. public schools, with Spanish remaining the most frequent after English, accounting for over 75% of students who speak a home language other than English (Ryan, 2013). More often than not, these emergent bilingual students are placed in segregated English as a Second Language (ESOL) classes for at least part of their school day, despite conflicting research on the efficacy of such classes in supporting students' long-term academic learning. For example, a recent study by Dafney Dabach (2015) investigated the ways in which differentiated instruction for emergent bilingual learners (such as pull-out ESOL) may hinder students' opportunities to access more advanced content courses, such as science. Similarly, a recent study by Yasuko Kanno and Sara Kangas (2014) documented how a school district in Pennsylvania placed all students classified as English language learners in low-level sheltered science and math classes as high school freshmen regardless of their past academic performance in middle school. From those sheltered classes during freshman year, there was no viable course taking path that would allow these emergent bilingual students to take advanced placement science or math classes by their senior year, regardless of their subsequent grades or academic performance. In effect, this language-based policy limited emergent bilingual students' college options by prohibiting them from taking the most rigorous science and math courses available, an admissions criterion at most selective universities.

While pull-out ESOL courses raise one set of problems for emergent bilingual learners, the most common alternative, the sink or swim approach of placing emergent bilinguals in general education content classes with little or no language support is fraught with another set of challenges. In such settings, emergent bilinguals have access, in theory, to more rigorous content but often delivered through instruction that fails to account for the challenges of making meaning in a second language. Barring major changes in how schools prioritize staffing decisions, it will never be possible for trained ESOL teachers to "push in" to general education classes as co-teachers in sufficient numbers to provide adequate language support for all the emergent bilinguals in our schools. This leaves as the most viable option the enhancement of professional learning for general education teachers in how to support emergent bilinguals in their content area, combined with the limited support that ESOL teachers and bilingual paraprofessionals can provide. With the exception of a few states (e.g., California, Florida) where ESOL endorsement is a required component of all teacher preparation programs, the vast majority of U.S. teachers have received little or no preparation in ESOL instruction (Bunch, 2013). In the science classroom, there are many basic but effective strategies that teachers can use to support their emergent bilingual students, and these strategies begin with an assumption of students' linguistic competence.

Rather than viewing emergent bilingual students from the deficit perspective of limited English proficiency, it is also possible to view these students' emergent multilingualism as an asset for science learning. To take advantage of this linguistic asset, however, educators must embrace the value of students' home language for supporting science learning (Wilson, 2013). Such a multilingual approach can be conceptualized in several ways. First, home language can be seen as a bridge to the language of schooling (English). Science learning support provided in the home language can thus be seen as a way to prevent students from falling behind in science content while they are gaining competence in English. The shift from predominantly pull out ESOL instruction for emergent bilinguals to push-in support in general education science classes has typically been done with this rationale in mind (Haneda & Alexander, 2015).

Second, a student's home language(s) can be seen as a valuable resource for learning science in its own right. From this perspective, home language can and should continue to be used to support science learning without looking to simply replace the home language with English. This is the practice in most two-way bilingual programs and in some home language preservation programs in which students' home languages become a powerful resource to learn science and other academic content. While this model has been shown to be successful on a limited scale (Cummins & Swain, 2014), the shortage of trained bilingual science teachers makes this model difficult to scale up.

A third alternative for conceptualizing the relationship between home language and second language in science teaching is the model of translanguaging. Translanguaging refers to the "multiple discursive practices in which bilinguals engage in order to make sense of their bilingual worlds" (Garcia & Wei, 2013, p. 22). Within the classroom context, this translanguaging plays out as students use their entire linguistic repertoire and semantic resources to construct their own understandings of academic information. In a translanguaging framework, home languages are not excluded, hidden, or separated, but rather, are integrated into instruction to support understanding, reflection, and communication about academic content. In this way all of a student's linguistic resources are valued, included, and brought to bear as students make meaning not only of science but also of their own evolving academic identities. What makes a pedagogy of translanguaging unique is its transdisciplinarity, emerging from the multifaceted communications of multilingual individuals (Garcia & Wei, 2013).

Unlike a two-way bilingual approach, a translanguaging approach does not require teachers to be proficient in the students' home language(s). Teachers can work with translations generated by students or others, can make visible the use of cognates when possible, and can create word/phrase walls to provide students with the concepts in the language(s) most familiar to them as well as in the new language(s) they are learning. Multimodal resources such as visual cues and representations, videos and multilingual online tools can also be used. Most importantly, students should be encouraged to work together on meaningful and challenging tasks, such as lab investigations, freely using and combining all of their linguistic resources to make meaning and communicate ideas. In the following section we illustrate how we have blended principles and practices of translanguaging with other practices for supporting all students in learning and using the language of science as part of an ongoing research and teacher professional learning project.

PRACTICAL STRATEGIES FROM OUR CURRENT RESEARCH FOR SUPPORTING STUDENTS IN LEARNING AND USING THE LANGUAGE OF SCIENCE

In this section we use practical examples from our current work on the Language-rich Inquiry Science with English Language Learners through Biotechnology (LISELL-B) project to further clarify various ideas about the language of science that we have discussed thus far in this chapter. In the LISELL-B project we promote a number of specific strategies for supporting all students in unpacking and gradually taking ownership of both the functional and the grammatical aspects of the language of science. We work with teachers to develop resources and instructional approaches that make the technical and the rational features of the language of science explicit for students, thus helping to demystify this language. In developing these practices, we took as our starting points a number of instructional strategies that have been suggested by Lemke (1990) and by Zhihui Fang, Mary Schleppegrell, and Annabelle Lukin (2008).

Build Students' "Word Consciousness" by Breaking Down the Technical Nature of Scientific Vocabulary. The technical vocabulary of science overwhelmingly comes from Greek and Latin roots, prefixes and suffixes (e.g., arthropoda, nocturnal). Therefore, teachers can help students understand the technical features of the language of science by explicitly analyzing technical terms they encounter in the science class through their prefixes, suffixes and roots, discussing how the words function in a text, and by practicing using these terms in multiple ways. We note that speakers of the Romance languages (e.g., Spanish, French, Italian, Portuguese), which evolved from Latin, have a natural advantage with the technical vocabulary of science in that many of these technical vocabulary terms have cognates that are common everyday words in the Romance languages. For example, the Spanish words *utilizer* (utilize), *capacidad* (capacity), and *respiración* (respiration) would all be considered common everyday terms in Spanish while their English cognates would be considered technical vocabulary that have other more common everyday English language synonyms.

Understand the Abstract and Depersonalized Nature of Scientific Language Created through the Use of Grammatical Drift, Nominalization and Passive Voice. In addition to its technical vocabulary, another key feature of scientific discourse is the use of abstract nouns. Abstract nouns (e.g., significance, distillation) are created by taking processes (verbs) or qualities (adjectives) and expressing them as abstract things (nouns) through the process of nominalization. Nominalization removes or suppresses the actor (often the scientist) and promotes the use of passive voice, making the world seem more stable and making science seem more objective by focusing on what is done rather than on who or what is doing it (e.g., trees are cut down resulting in loss of soil; measurements are taken to check for increased ozone levels). Teachers can help students to understand the abstract and depersonalized nature of science discourse by explicitly pointing out abstract nouns and passive voice in science texts and by having students re-write science texts (such as for a "younger sibling") in ways that explicitly include the actors, use active voice, and make the language less abstract and more concrete.

Understand the Dense Nature of the Language of Science through Practice Deconstructing and Constructing Clauses of Science Text. Science text is dense and concise writing is valued, meaning that science writers typically pack a large amount of information into a limited number of sentences. Clauses in science texts typically contain many more processes, participants, and circumstances than do clauses in everyday language (e.g., the net force being applied to the baseball under acceleration). These dense clauses are usually formed by a noun (baseball) with multiple pre-modifiers (the net force being applied) and post-modifiers (under acceleration). Teachers can help students to understand the dense nature of science discourse by breaking down (deconstructing) complex clauses into their parts and by practicing expanding simple nouns by adding their own pre- and post-modifiers. This kind of two-way re-writing to simplify dense clauses and to add density to simple clauses helps build students' fluency with the language of science at the same time that it enhances conceptual science understanding.

Teach Students How to Express Science Ideas in Multiple Ways. Students often learn to express things in science in just one way (They learn that "a hypothesis is an educated guess" or that "density is the mass divided by the volume"). Teachers may assume that when students say these things that the students possess an underlying conceptual understanding of what these statements mean. In fact, asking for elaboration will often uncover that the student has memorized this response with only a limited conceptual understanding. If a person can only say something in one way, it is likely that they have simply memorized it. Expressing that idea in multiple ways provides evidence of a fuller understanding while also deepening that understanding. Teachers can help students practice expressing science ideas (both spoken and written) in multiple ways to reinforce this idea that to "say it another way" is evidence of true understanding.

Give Students More Practice Talking Science. The typical question-answer-evaluation (often referred to as Initiation-Response-Evaluation or IRE dialogue) discourse pattern (Mehan, 1979) between teacher and students fails to give students adequate opportunity to talk, to ask their own questions, and to discuss science while doing practical work. Teachers can help by consciously changing the discourse pattern in their classroom with the goal of giving students more frequent opportunities to ask questions, to talk in pairs and peer groups, to explain scientific concepts, and generally to use more language than just short responses to teacher questions. Prompts can be provided to scaffold students' talk about science, helping them to build confidence and fluency with the language of science. Sarah Michaels and Cathy O'Connor (2011) have highlighted a series of classroom norms that support productive discourse as well as more specific "science talk moves" that serve to guide rich, interactive science discussions by replacing shallow and formulaic IRE discourse as the classroom norm.

Make Time for Students to Discuss Their Everyday Understandings of Topics Using Everyday Language and Translanguaging. Related to the previous practice, students need to be able to compare everyday ways of talking and thinking about topics with more scientific ways of talking and thinking about those same topics and to do so using their full linguistic repertoires. Even if these everyday understandings are not scientifically correct, they serve as a valuable starting point for learning the relevant science concepts and relevant science language. In addition, moving back and forth between expressing understandings in both everyday and academic language can enhance meaning making. Teachers can help students make these connections by dedicating time to explicitly discussing the everyday ways of talking and thinking so that these can then be linked to the more scientific approaches.

Students Need Practice Writing in Scientific Genres, Including on Assessments. In addition to more opportunities to talk science, students need regular and ongoing practice writing in scientific genres as an important part of learning to communicate scientifically. As we noted earlier, the spoken and written forms of scientific language are quite similar, and thus, practice with science writing can support improved science talk and vice versa. Formative assessments such as lab reports can provide such ongoing practice, especially if students are required to write in full sentences or even full paragraphs (Hand, Wallace, & Yang, 2004). Summative assessments, such as science tests, should also require substantive written answers because in the end, tests send the clearest message about what is important to learn (Buxton, Allexsaht-Snider, Kim, & Cohen, 2014). Multiple choice tests tell students that what is important is memorization, not meaning. Written responses highlight the value of meaning. Teachers can help students learn to take ownership of the language of science by requiring them to both write and talk science on a daily basis.

Build General Academic Vocabulary in Multiple Languages. General academic vocabulary refers to words (e.g., indicate, feature, benefit) that are common in written academic texts, assessments, standards, and teacher talk across multiple content areas, but are rare in students' oral, conversational, every day language. This vocabulary is a linguistic blind spot that is often overlooked in school, because teachers, textbook authors, and test developers routinely make the incorrect assumption that most students are already proficient with this vocabulary. Catherine Snow and colleagues (e.g., Lawrence, Capotosto, Branum-Martin, White, & Snow, 2012) have identified problems with students' comprehension of general academic vocabulary as a major hurdle in science learning. For example, the glossaries and other resources in science textbooks and science websites are full of this unexplained general academic vocabulary, making it nearly impossible for many students to use these resources to get the help they need to comprehend the meanings of science (as well as other academic area) texts. To address this issue, Avril Coxhead (2011) compiled an academic word list of the most commonly used general academic vocabulary across the disciplines that can serve as a basis for supporting teachers' thoughtful development of this vocabulary in meaningful contexts. Such vocabulary support can and should take place in all content area classes, including science class to help all students, and particularly emergent bilingual learners, to better understand, engage in, and communicate about science as well as the other content areas.

When taken together, these eight practices can teach students a great deal about how to use the language of science. Many of these practices can be reinforced in only a few minutes a day. Others require teachers and students to change some of their normal classroom practices (such as moving away from the traditional IRE discourse pattern). Adopting these practices can support teachers and students in making the language of science, as expressed in science texts, discourse, standards, and tests, more accessible and more meaningful.

SUMMARIZING HOW PERSPECTIVES FROM LINGUISTICALLY DIVERSE CLASSROOMS CAN HELP ALL STUDENTS UNLOCK THE LANGUAGE OF SCIENCE

In this chapter, we have highlighted a number of features of the language of science as well as strategies to help make these features explicit for students and teachers as a way to provide a partial answer to the overarching question: In what ways does language affect (and in what ways is it affected by) the science education process? We have described the evolution of scientific language, how this language first appears and begins to develop in the science classroom, the intersections between academic language, everyday language, home language and second language, as well as an overview of practical strategies for supporting students in learning and applying the language of science. In reviewing these topics, we have highlighted ways in which the language of science often becomes an obstacle to students' science learning. Through grammatical metaphor and nominalization, processes and qualities in scientific texts are densely packed with a large amount of information in a comparatively small space. Although numerous technical terms with specialized meanings are part of the overall effect of why the language of science sounds the way it does, the unique nature of scientific language lies more with its grammatical construction than with its technical vocabulary. That is, the problems that learners face with technical terminology arise less from the technical terms themselves and more from the complex relationship these terms have with one another. Thus, while the language of science serves the needs of the scientific community well, in the science classroom this same language can act to diminish students' confidence and often their interest in science.

Despite the challenge that the linguistic features of science present, there is also good evidence that this language can be understood, mastered and put to use by all students through explicit study and practice. In general, to promote more meaningful linguistic comprehension and production from their students, science teachers should consider how to build the following features into their instruction:

- Create the authentic need for students to meaningfully interact using language: Language is not just how we communicate our thinking; it is also fundamental to how we think. Thus, without a reason to communicate their ideas, students' thinking and understanding are diminished. Put another way, the person doing the communicating is usually the person doing the most learning. When students are given science tasks that require them to do and communicate things together, science learning will follow.
- Ensure that every student's ideas are sought and valued: Every student, regardless of background or language competence, has meaningful lived experience that can be related to topics of scientific study. Teaching strategies that prompt students to communicate (using all linguistic resources they have at their disposal) the connections they see between lived experiences and current topics of study build students' motivation to learn while also building language skills.
- Support student discussion of reasons behind linguistic choices: As discussed earlier, features of the language of science can and should be taught explicitly as part of science instruction to build students' metalinguistic skills. Articulating the reasons for the language choices they make can help students think intentionally about the language that most clearly conveys the ideas they are trying to communicate.
- Support student tenacity in understanding each other's ideas rather than just the teacher trying to understand: All students struggle to clearly articulate emergent understandings. In the traditional IRE discourse structure, students are generally not expected to be responsible for understanding and evaluating the ideas of their classmates, with that task reserved for the teacher through whom all communication passes. When students are responsible for understanding their peers' ideas, this requires everyone to work harder to clarify and communicate their understandings.
- Use successive approximations that increase in detail both in building conceptual knowledge and linguistic abilities: The task of gaining competence in using the language of science is a long term goal that takes years to develop. While setting high expectations for all students, teachers should also realize that students' ownership of the language of science will progress in steps. Language teachers often use the idea of successive approximations toward a language goal as a model for thinking about how language

develops. Keeping running records of students' language use over time will allow teachers and students to recognize growing aptitude with the language of science.

- Use posters, drawings, models, and other visual resources to contextualize and focus language: Academic language use is facilitated when we have something concrete to talk about (think about the slide shows we often rely upon when giving professional presentations). Such resources can include student created, teacher created, or professionally created visuals. Models of scientific phenomena, whether conceptual, physical, mathematical, etc., can be particularly powerful tools for focusing language on describing scientific concepts and processes.
- Send home topics and discussion questions with students to prompt science conversations among students and family members and to seek out home-school connections: Family is an underutilized resource for building science communication skills. Related to the value of having an authentic audience, students should be encouraged to explain the science ideas they are learning in school to parents, siblings or any other family members who are willing to listen and engage. This activity can initially be scaffolded by sending home explicit discussion questions, but over time, students should be expected to take responsibility for structuring these science discussions at home.

To conclude, in this chapter we have argued that by building explicit language goals and linguistic support into all lessons, science teachers can ensure that the language of science comes to be a positive resource for students rather than an academic obstacle that needs to be overcome. The LISELL-B project, as well as many other research and professional learning projects, continue to develop and test resources, such as model lessons, lab notebook formats, and linguistically scaffolded resources to support all students in effectively utilizing the language of science in the contexts of science class work, assignments, and assessments. We are at a point where much is known about how the language of science is constructed and what strategies can be successful in supporting its development. The large questions that remain are how best to prepare and support teachers to guide all their students in acquiring and applying this specialized language and how to work with policy makers, publishers, and test developers to effectively integrate what we know about how students learn the language of science. Without systemic reforms that address each of these areas simultaneously, we will be forced to continue with a piecemeal approach that fails to take advantage of how the language of science might support students' science meaning making.

REFERENCES

August, D., Branum-Martin, L., Cardenas-Hagan, E., & Francis, D. J. (2009). The impact of an instructional intervention on the science and language learning of middle grade English language learners. *Journal of Research on Educational Effectiveness, 2*, 345–376. doi:10.1080/19345740903217623.

Brown, B. A. (2006). "It isn't slang that can be said about this stuff": Language, identity, and appropriating science discourse. *Journal of Research in Science Teaching, 43*, 96–126. doi: 10.1002/tea.20096.

Brown, B. A., & Spang, E. (2008). Double talk: Synthesizing everyday and science language in the classroom. *Science Education, 92*, 708–732. doi:10.1002/sce.20251.

Bunch, G. C. (2013). Pedagogical language knowledge preparing mainstream teachers for English learners in the new standards era. *Review of Research in Education, 37*, 298–341. doi:10.3102/0091732X12461772.

Buxton, C., Allexsaht-Snider, M., Kim, S., & Cohen, A. (2014). Potential benefits of bilingual constructed responses science assessments for emergent bilingual learners. *Double Helix, 1*(2), 1-21. Retrieved from http://qudoublehelixjournal.org/index.php/dh/article/view/31/156

Chomsky, N. (2002). *On nature and language.* Cambridge: Cambridge University Press.

Coxhead, A. (2011). The Academic Word List 10 years on: Research and teaching implications. *TESOL Quarterly, 45*, 355–362. doi:10.5054/tq.2011.254528

Cummins, J., & Swain, M. (2014). *Bilingualism in education: Aspects of theory, research and practice.* New York, NY: Routledge.

Dabach, D. B. (2015). Teacher placement into immigrant English learner classrooms limiting access in comprehensive high schools. *American Educational Research Journal, 52*, 243–274. doi:10.3102/0002831215574725.

Fang, Z. (2006). The language demands of science reading in middle school. *International Journal of Science Education, 28*, 491–520. doi:10.1080/09500690500339092.

Fang, Z., Schleppegrell, M. J., & Lukin, A. (2008). *Reading in secondary content areas: A language-based pedagogy.* Ann Arbor: University of Michigan Press.

Fletcher, K. L., & Reese, E. (2005). Picture book reading with young children: A conceptual framework. *Developmental Review, 25*(1), 64–103. doi:10.1016/j.dr.2004.08.009.

Garcia, O., & Wei, L. (2013). *Translanguaging: Language, bilingualism and education.* New York, NY: Palgrave Macmillan. doi:10.1080/14708477.2014.950828.

Halliday, M. A. K. (1999). Grammar and the construction of educational knowledge. In R. Berry (Ed.), *Language analysis, description and pedagogy* (pp. 70–87). Hong Kong: Lignan University.

Halliday, M. A. K. (2004). *The language of science.* London: Continuum.

Halliday, M. A. K., & Martin, J. R. (2003). *Writing science: Literacy and discursive power.* New York, NY: Taylor & Francis.

Halliday, M. A. K., & Matthiessen, C. M. I. M. (2004). *Introducing functional grammar.* New York, NY: Edward Arnold.

Hand, B., Wallace, C. W., & Yang, E. M. (2004). Using a Science Writing Heuristic to enhance learning outcomes from laboratory activities in seventh grade science: Quantitative and qualitative aspects. *International Journal of Science Education, 26*, 131–149. doi:10.1080/0950069032000070252.

Haneda, M., & Alexander, M. (2015). ESL teacher advocacy beyond the classroom. *Teaching and Teacher Education, 49*, 149–158. doi:10.1016/j.tate.2015.03.009.

Hornberger, N. H., & Link, H. (2012). Translanguaging and transnational literacies in multilingual classrooms: A biliteracy lens. *International Journal of Bilingual Education and Bilingualism, 15*(3), 261–278. doi:10.1080/00405841.2012.726051.

Hudicourt-Barnes, J. (2003). The use of argumentation in Haitian Creole science classrooms. *Harvard Educational Review, 73*(10), 73–93. doi:10.17763/haer.73.1.hnq801u57400l877.

Kanno, Y., & Kangas, S. N. (2014). "I'm not going to be, like, for the AP": English language learners' limited access to advanced college-preparatory courses in high school. *American Educational Research Journal, 51*, 848–878. doi:10.3102/0002831214544716.

Lawrence, J. F., Capotosto, L., Branum-Martin, L., White, C., & Snow, C. E. (2012). Language proficiency, home-language status, and English vocabulary development: A longitudinal follow-up of the Word Generation program. *Bilingualism: Language and Cognition, 15*, 437–451. doi:10.1017/S1366728911000393.

Lemke, J. L. (1990). *Talking science: Language, learning, and values.* Norwood, NJ: Ablex.

Lovejoy, A. O. (2011). *The great chain of being: A study of the history of an idea.* New Brunswick, NJ: Transaction Publishers.

Martin, J. R., & Veel, R. (Eds.). (2005). *Reading science: Critical and functional perspectives on discourses of science.* New York, NY: Routledge.

Mehan, H. (1979). *Learning lessons: Social organization in the classroom.* Cambridge, MA: Harvard University.

Michaels, S., & O'Connor, C. (2011). *The talk science primer.* Retrieved from http://inquiryproject.terc.edu/prof_dev/Goals_and_Moves.cfm

Moje, E. B., Collazo, T., Carrillo, R., & Marx, R. W. (2001). "Maestro, what is 'quality'?": Language, literacy, and discourse in project based science. *Journal of Research in Science Teaching, 38*, 469–498. doi:10.1002/tea.1014.

Nam, J., Choi, A., & Hand, B. (2011). Implementation of the science writing heuristic (SWH) approach in 8th grade science classrooms. *International Journal of Science and Mathematics Education, 9*(5), 1111–1133. doi:10.1007/s10763-010-9250-3.

NGSS Lead States. (2013). *Next generation science standards: For states, by states.* Washington, DC: The National Academies Press.

Ryan, C. (2013). Language use in the United States: 2011. *American Community Survey Reports, 3*, 1-16. Retrieved from https://www.census.gov/prod/2013pubs/acs-22.pdf

Warren, B., Ballenger, C., Ogonowski, M., Rosebery, A., & Hudicourt-Barnes, J. (2001). Rethinking diversity in learning science: The logic of everyday sense-making. *Journal of Research in Science Teaching, 38*, 529–552. doi:10.1002/tea.1017.

Warren, B., & Rosebery, A. (2011). Navigating interculturality: African American male students in the science classroom. *Journal of African American Males in Education, 2*(1), 98–115.

Wilson, S. M. (2013). Professional development for science teachers. *Science, 340*, 310–313. doi:10.1126/science.1230725.

SECTION EIGHT

Religion and Science Education

In What Ways Does Religion Affect the Process of Science Education?

CHAPTER TWENTY-TWO

Why a Chapter on Religion in a Book on Science Education?

MICHAEL J. REISS

Worldwide, religion remains of importance to many people, including young people. High quality international comparative data are difficult to come by for a number of reasons: the various surveys often fail to distinguish between beliefs and practice; sampling is often poor; what is understood by the questions can depend on language. Nevertheless, probably around a half of all people affirm and to some extent put into practice a religious faith, with the proportion falling in many Western countries (including the US) and increasing in a number of other countries (British Humanist Association, 2016; Pew Research Center, 2015). Islam is the fastest growing faith, but Christianity is still predicted to remain the faith with most adherents for at least the next few decades.

Furthermore, to the surprise of many science educators, the delight of some and the frustration of others, issues to do with religion seem increasingly to be of importance in school science lessons, science museums and some other educational settings. In this chapter I begin by examining the nature of religion and the nature of science before looking at possible ways in which religion might relate to science. I then consider what the aims of education might be, before considering whether or not religion has implications for science education (and *vice versa*) and, if it does, how science teaching might take account of religious belief (Reiss, 2014).

To many science educators, especially those of an agnostic/atheistic persuasion, even raising the possibility that religion might be considered in some way within science education raises suspicions that this is an attempt to find a way of getting religion into the science classroom for religious rather than scientific reasons. This is not the intention here. In terms of the nature of science, part of the argument is that considering religion can, on occasions, be useful simply for helping learners better understand why certain things come under the purview of science and others don't.

Another argument for considering religion within science education proceeds much as an argument for considering history in science education might. While science can be learnt and studied in an historical vacuum, there are reasons why it can be helpful to study science in its historical contexts. For a start, this helps one understand better why certain sorts of science were pursued at certain times. Wars, for instance, have sometimes led to advances in chemistry, physics and information science (e.g., explosives, missile trajectories, code breaking), while certain botanical disciplines, such as systematics and taxonomy, have flourished during periods of colonisation. Much biology is studied in the hope that medical advances will result. Then there is the observation that for many learners understanding science in historical contexts can aid motivation. Science courses that take contexts and applications into account are now quite widespread (cf. science-technology-society and socio-scientific issues/movements).

Similarly, while many students enjoy learning about the pure science of genetics and evolution, others are motivated *and* may come to understand the science better if they appreciate something of the diversity of religious beliefs held by such principal protagonists as Charles Darwin, Joseph Hooker, Thomas Huxley and Gregor Mendel and the religious views (including the diversity of religious views) of the cultures in which they lived and worked.

More generally, there are a number of places where religion and science interact. Consider, first, the question of "authority" and the scriptures as a source of authority. To the great majority of religious believers, the scriptures of their religion (the Tanakh, the Christian Bible, the Qur'an, the Vedas, including the Upanishads, the Guru Granth Sahib, the various collections in Buddhism, etc.) have an especial authority by very virtue of being scripture. This is completely different from the authority of science. Isaac Newton's *Principia* and Darwin's *On the Origin of Species* are wonderful books, but they do not have any permanence other than that which derives from their success in explaining observable phenomena of the material world and enabling people to see the material world through Newtonian/Darwinian eyes. Indeed, as is well known, Darwin knew almost nothing of the mechanism of inheritance despite the whole of his argument relying on inheritance, so parts of *The Origin* were completely out of date over a hundred years ago.

Then consider the possibility of miracles, where the word is used not in its everyday sense (and the sense in which it is sometimes used in scripture), namely "remarkable," "completely unexpected" or "wonderful" (as in the newspaper heading "My miracle baby"), but in its narrower meaning of "contrary to the laws of nature." Scientists who do not accept the occurrence of miracles can react to this latter notion of miracles in one of three ways: i) miracles are impossible (because they are contrary to the laws of nature); ii) miracles are outside of science (because they are contrary to the laws of nature); iii) miracles are very rare events that haven't yet been incorporated within the body of science but will be (as rare events, e.g., eclipses, and mysterious creatures, e.g., farm animals with two heads or seven legs, have been).

In the following sections, I address such issues.

THE NATURE OF RELIGION

There are many religions and it is difficult to answer the question "What is the nature of religion?" in a way that satisfies the members of all religions. Nevertheless, the following, derived from Ninian Smart (1989) and John Hinnells (1991), are generally characteristic of most religions (Reiss, 2008a):

- Religions have a *practical and ritual dimension* that encompass such elements as worship, preaching, prayer, yoga, meditation and other approaches to stilling the self.
- The *experiential and emotional dimension* of religions has at one pole the rare visions given to some of the crucial figures in a religion's history, such as that of Arjuna in the *Bhagavad Gita*, the revelation to Moses at the burning bush in *Exodus* and Saul on the road to Damascus in *Acts*. At the other pole are the experiences and emotions of many religious adherents, whether a once-in-a-lifetime discernment of the transcendent or a more frequent feeling of the presence of God either in corporate worship or in the stillness of one's heart.
- All religions hand down, whether orally or in writing, vital stories that comprise the *narrative or mythic dimension*, for example the story of the birth, life, death, resurrection and ascension of Jesus in the Christian scriptures. For some religious adherents such stories are believed literally, for others they are understood symbolically.
- The *doctrinal and philosophical dimension* arises, in part, from the narrative/mythic dimension as theologians within a religion work to integrate these stories into a more general view of the world. Thus the early Christian church came to its understanding of the doctrine of the Trinity by combining the central claim of the Jewish religion—that there is but one God—with its understanding of the life and teaching of Jesus Christ and the working of the Holy Spirit.

- If doctrine attempts to define the beliefs of a community of believers, the *ethical and legal dimension* regulates how believers act. So Sunni Islam has its Five Pillars, while Judaism has the Ten Commandments and other regulations in the Torah and Buddhism its Five Precepts.
- The *social and institutional dimension* of a religion relates to its corporate manifestation, for example the Sangha—the order of monks and nuns founded by the Buddha to carry on the teaching of the Dharma—in Buddhism, the umma'—the whole Muslim community—in Islam, and the Church—the communion of believers comprising the body of Christ—in Christianity.
- Finally, there is the *material dimension* to each religion, namely the fruits of religious belief as shown by places of worship (e.g., synagogues, temples and churches), religious artefacts (e.g., Eastern Orthodox icons and Hindu statues) and sites of special meaning (e.g., the river Ganges, Mount Fuji and Uluru (Ayers Rock)).

As will be discussed below, the relationship between science and religion has changed over the years (Al-Hayani, 2005; Szerszynski, 2005); indeed, the use of the singular, "relationship," risks giving the impression that there is only one way in which the two relate. Nevertheless, there are two key issues: one is to do with understandings of reality; the other to do with evidence and authority. Most religions hold that reality consists of more than the objective world, and many religions give weight to personal and/or (depending on the religion) institutional authority in a way that science generally strives not to.

For example, there is a very large religious and theological literature on the world to come, i.e., life after death, (e.g., Hick, 1976/1985). However, to labour the point, although some (notably Atkins, 2011) have argued that science disproves the existence of life after death, it can be maintained that science has little or nothing to say about this question because life after death exists or would exist outside of or beyond the realm to which science relates. Furthermore, many religious believers are likely to find the pronouncements on the question of life after death by even the most reputable of scientists to be of much less significance than the few recorded words of their religion's founder(s).

THE NATURE OF SCIENCE

The term "nature of science" is understood in a number of ways, but at its heart is knowledge about how, and to a lesser extent why, science is undertaken. So the nature of science includes issues about the fields of scientific enquiry and the methods used in that enquiry as well as, to a certain extent, something about the purpose of science.

A key point about the fields of scientific enquiry is that these have shifted over time. In large measure this is simply because of developments in instrumentation. We can now study events that happen at very low temperatures, at distances, at speeds and at magnifications that simply were not possible in the past. What is still unclear is the extent to which certain matters currently outside of mainstream science will one day fall within the compass of science. Take dreams, for example. It may be that these will remain too subjective for science but it may be that developments in the recording and interpretation of brain activity will mean that we can obtain a sufficiently objective record and understanding of dreams for them to be amenable to rigorous scientific study.

But the scope of science has also shifted for reasons that are more to do with theorisation than with technical advances (Reiss, 2013). Consider beauty. Aesthetics for a long time was not considered a scientific field. But there is now, within psychology and evolutionary biology, growing scientific study of beauty and desire (e.g., Buss, 2003). Indeed, a number of the social sciences are being nibbled away at by the natural sciences. Despite such movements in the fields of scientific enquiry and in the actual methods employed by scientists, the overarching methods of science (what a social scientist might term its methodology) have shifted far less, certainly for several hundreds of years, arguably for longer than that.

As is well known, Robert Merton characterised science as open-minded, universalist, disinterested and communal (Merton, 1973). For Merton, science is a group activity; even though certain scientists work alone, science, within its various sub-disciplines, is largely about bringing together into a single account the contributions of many different scientists to produce an overall coherent model of one aspect of reality. In this sense, science is (or should be) impersonal. Allied to the notion of science being open-minded, disinterested and impersonal is the notion of scientific objectivity. The data collected and perused by scientists must be objective in the sense that they should be independent of those doing the collecting (cf. Daston & Galison, 2007)—the idealised "view from nowhere."

Karl Popper emphasised the falsifiability of scientific theories (Popper, 1934/1972): unless one can imagine collecting data that would allow one to refute a theory, the theory isn't scientific. The same applies to scientific hypotheses. So, iconically, the hypothesis "All swans are white" is scientific because we can imagine finding a bird that is manifestly a swan (in terms of its anatomy, physiology and behaviour) but is not white. Indeed, this is precisely what happened when early White explorers returned from Australia with tales of black swans.

Popper's ideas easily give rise to a view of science in which knowledge accumulates over time as new theories are proposed and new data collected to distinguish between conflicting theories. Much school experimentation in science is Popperian: we see a rainbow and hypothesise that white light is split up into light of different colours as it is refracted through a transparent medium (water droplets);

we test this by attempting to refract white light through a glass prism; we find the same colours of the rainbow are produced and our hypothesis is confirmed. Until some new evidence causes it to be falsified, we accept it (Reiss, 2008a).

Thomas Kuhn made a number of seminal contributions but he is most remembered nowadays for his argument that while the Popperian account of science holds well during periods of normal science when a single paradigm holds sway, such as the Ptolemaic model of the structure of the solar system (in which the Earth is at the centre) or the Newtonian understanding of motion and gravity, it breaks down when a scientific crisis occurs (Kuhn, 1970). At the time of such a crisis, a scientific revolution happens during which a new paradigm, such as the Copernican model of the structure of the solar system or Einstein's theory of relativity, begins to replace (initially to co-exist with) the previously accepted paradigm. The central point is that the change of allegiance from scientists believing in one paradigm to their believing in another cannot, Kuhn argues, be fully explained by the Popperian account of falsifiability.

The above account of the nature of science portrays science as what John Ziman (2000) has termed "academic science." Ziman argues that such a portrayal was reasonably valid between about 1850 and 1950 in European and American universities but that since then we have entered a phase largely characterised by "post-academic science." Post-academic science is increasingly transdisciplinary and utilitarian, with a requirement to produce value for money. It is more influenced by politics, it is more industrialised, and it is more bureaucratic. The effect of these changes is to make the boundaries around the domain of science a bit fuzzier. Of course, if one accepts the contributions of the social study of science (e.g., Yearley, 2005) one finds that these boundaries become fuzzier still. The argument in this chapter does not *rely* on such a reading of science though someone who is persuaded by the "Strong Programme" (Bloor, 1991) within the sociology of scientific knowledge (i.e., the notion that even valid scientific theories are amenable to sociological investigation of their truth claims) is much more likely to accept the worth of science educators considering the importance of religion as one of many factors that influence the way science is practised and scientific knowledge produced.

UNDERSTANDINGS OF POSSIBLE RELATIONSHIPS BETWEEN SCIENCE AND RELIGION

It is clear that there can be a number of axes on which the science/religion issue can be examined. For example, the effects of the practical and ritual dimension are being investigated by scientific studies that examine such things as the efficacy of prayer and the neurological consequences of meditation (e.g., Lee & Newberg,

2005); a number of analyses of religious faith, informed by contemporary understandings of evolutionary psychology, behavioural ecology and sociobiology, examine the possibility or conclude that religious faith can be explained by science (e.g., Dennett, 2006); the narrative/mythic dimension of religion clearly connects with scientific accounts of such matters as the origins of the cosmos and the evolution of life; the doctrinal and philosophical dimension can lead to understandings that may agree or disagree with standard scientific ones (e.g., about the status of the human embryo); and the ethical and legal dimension can lead to firm views about such matters as land ownership, usury and euthanasia.

There is now a very large literature on the relationship between science and religion (e.g., Clayton & Simpson, 2006). Indeed, the journal *Zygon* specialises in this area. John Hedley Brooke (1991) provides a particularly thorough historical study of the relationship between science and religion. Brooke's aim is "to reveal something of the complexity of the relationship between science and religion as they have interacted in the past" (p. 321). He concludes:

> Popular generalizations about that relationship, whether couched in terms of war or peace, simply do not stand up to serious investigation. There is no such thing as *the* relationship between science and religion. It is what different individuals and communities have made of it in a plethora of different contexts. Not only has the problematic interface between them shifted over time, but there is also a high degree of artificiality in abstracting the science and the religion of earlier centuries to see how they were related. (Brooke, 1991, p. 321)

Perhaps the best known categorisation of the ways in which the relationship between science and religion can be understood was provided by Ian Barbour (1990). Barbour himself updated this book (Barbour, 1997) and since 1990 there has been a considerable literature about the ways in which science and religion relate (e.g., Haught, 1995; Stenmark, 2004). Nevertheless, Barbour's (1990) typology continues to dominate the literature and so is used here. Barbour, who focuses especially on epistemological assumptions of recent Western authors, identifies four main groupings.

First, there is the relationship of *conflict*; "first" simply because it is the first in Barbour's list and first, perhaps, also in the minds of many people, whether or not they have a religious faith. Barbour doesn't give a reason for the order of his listing but at least two can be suggested: comprehensibility and familiarity. It is both easy and familiar (given Barbour's declared focus on recent Western authors) to see the relationship between science and religion as one of conflict. However, Barbour sees limitations in this way of understanding the science/religion issue. As he memorably puts it:

> In a fight between a boa constrictor and a wart-hog, the victor, whichever it is, swallows the vanquished. In scientific materialism, science swallows religion. In biblical literalism,

> religion swallows science. The fight can be avoided if they occupy separate territories or if, as I will suggest, they each pursue more appropriate diets. (Barbour, 1990, p. 4)

Barbour's second grouping is *independence* (e.g., Gould, 1999). Science and religion may be seen as independent for two main reasons: because they use distinctive methods or because they function as different languages. In any event, the result is that each is seen as distinct from the other and as enjoying its own autonomy:

> Each has its own distinctive domain and its characteristic methods that can be justified on its own terms. Proponents of this view say there are two jurisdictions and each party must keep off the other's turf. Each must tend to its own business and not meddle in the affairs of the others. Each mode of inquiry is selective and has its limitations. (Barbour, 1990, p. 10)

Barbour's third grouping moves beyond conflict and independence to *dialogue* (cf. Watts, 1998). As an example of dialogue, Barbour points out how our understanding of astronomy has forced us to ask why the initial conditions were present that allowed the universe to evolve. The point is not that the findings of science require a religious faith—that would be for the wart-hog of religion to swallow the boa constrictor of science. Rather the point is that scientific advances can give rise (no claim is made that they do for all people) to religious questions, so that a dialogue ensues.

Barbour's final grouping is one in which the relationship between science and religion is seen to be one of *integration* (cf. Peacocke, 2001). For example, in natural theology it is held that the existence of God can be deduced from aspects of nature rather than from revelation or religious experience (e.g., Ray, 1691/2005). Natural theology has rather fallen out of favour (but see Polkinghorne, 2006). A more modern version is process theology, which rejects a view of the world in which purely natural events (characterised by an absence of divine activity) are interspersed with occasional gaps where God acts. Rather, for process theologians, every event is understood "to be jointly the product of the entity's past, its own action, and the action of God" (Barbour, 1990, p. 29). Furthermore, God is not the Unmoved Mover of Thomas Aquinas but instead acts reciprocally with the world.

I think it can be difficult for those who have never had a religious faith, or only have one rather tenuously, to imagine what a life is like that is lived wholly within a religious ordering. For such a person, the relationship between science and their faith may be described as "integrated" though this is to give an epistemological framing to the relationship, whereas what may be going on is that the person has little overt interest in the precise nature of the relationship between science and religion other than that there can clearly be no conflict between them.

Having examined possible relationships between science and religion, I turn now to the issue of determinism to discuss at a more fine-grained level how religion and science can co-relate.

DETERMINISM

The science-religion issue is often examined simply by recourse to certain cause célèbres—Galileo and Copernican heliocentrism, Darwin and evolution, arguments about the sanctity of life, for example. At a school level, examinations of such particular instances of the relationship between science and religion, along with a more general consideration about how science and religion can relate, are perfectly appropriate. However, there are certain "higher order" questions that teachers and curriculum developers need to consider to decide whether they can be introduced meaningfully at a school level. One such central question is about whether nature is deterministic and, if it/she is not, whether that has anything to do with divine action. Theologically speaking, this is part of the more general question as to how (for those who have a religious faith) God acts in nature (Dixon, 2008).

The post-Newtonian advent in the early 20th century of quantum theory and, later in the same century, of chaos theory has led many to wonder whether within either or both of these two frameworks might lie a space for divine action in a way that does not contradict the scientific worldview in the way that miracles seem to. For almost anyone who has not studied quantum physics to at least first degree level, it is exceptionally difficult to understand what is going on that is relevant to the science-religion issue but a core concept is that of determinism, which results from the issue of the relationship of measurement to reality (e.g., Bhaskar, 1978).

As is well known, in 1927 Werner Heisenberg argued that certain key physical variables that had previously been presumed to be independent (e.g., the position and momentum of an object) are linked. Measuring one to a very high degree of precision necessarily means that the other cannot be so precisely determined. Thus far, there is not a great deal that is of interest to the non-physicist—the issue appears to be one of epistemology. However:

> Heisenberg himself took a more radical view—he saw this limitation as a property of nature rather than an artefact of experimentalism. This radical interpretation of uncertainty as an ontological principle of indeterminism implies that quantum mechanics is inherently statistical—it deals with probabilities rather than well-defined classical trajectories. Such a view is clearly inimical to classical determinism. (Osborn, 2005, p. 132)

Put somewhat loosely, a number of people have tried to find room for divine action in this indeterminancy. No consensus yet exists as to the validity of this search though, on balance, the current views seem to be that such a search is mistaken for reasons both of theology and of physics. A particularly helpful, though demanding, analysis of both the theology and the physics is provided by Nicholas Saunders (2002). Beginning with the theology, Saunders draws on the widespread distinction between general and special forms of divine action. In the words of Michael Langford, general divine action is "the government of the universe through the universal laws that control or influence nature, man, and history, without the need for specific or ad hoc acts of divine will" (Langford, 1981, p. 11). On the other hand, special divine action is characterised by:

> Those actions of God that pertain to a *particular* time and place in creation as distinct from another. This is a broad category and includes the traditional understanding of "miracles", the notion of particular providence, responses to intercessionary prayer, God's personal actions, and some forms of religious experience. (Saunders, 2002, p. 20)

Oversimplifying considerably, all religions are comfortable with the notion of general divine action but they differ both among and within themselves considerably in their understanding of specific divine action. In particular, many leading theologians are uncomfortable with the notion of specific divine action so defined for a number of reasons including the particular problems for the occurrence of suffering that it raises (if suffering can sometimes be averted miraculously, why isn't it always or, at least, much more often?) and the apparent shortcomings, including capriciousness, suggested by a divine being who relies on occasional exercises of supernatural activity to keep things moving along (cf. Kenny, 1992).

Going onto the physics, Saunders is sceptical of attempts to locate the possibility of specific divine action in quantum or chaos theory. The argument here becomes even more technical and depends, in respect of quantum theory, on whether one accepts the standard (Copenhagen) interpretation of reality (in which Schrödinger's cat is either dead or alive before the box is opened) or the more radical interpretation (in which the cat is both dead and alive). In both cases, though, as well as in the case of chaos theory (sometimes termed "complexity theory" on the grounds that it deals with systems that are deterministic but unpredictable because of their exquisite sensitivity to small changes in their initial conditions) Saunders rejects attempts to find opportunities for specific divine activity in the science.

THE AIMS OF EDUCATION

The aims of science education have been much discussed but they need to be seen within the broader question of the aims of education (Reiss, 2015). John White and I have argued that that there are two fundamental aims of school education, namely to enable each learner to lead a life that is personally flourishing and to help others to do so too (Reiss & White, 2013). Such a school education would lead to greater social justice (e.g., Griffiths, 2003) and also the utilitarian aim of greater net happiness.

Much science education can contribute to this overall aim by helping students get a good understanding of how science builds up knowledge and of key scientific knowledge that has been built up. Some of this knowledge can be of use to just about everyone, for example that to do with health and environmental issues; other knowledge is preparatory for those who will go on to use science in their careers.

One of the ways in which the aims of human flourishing, social justice and greater net happiness are likely to be advanced is if schools help people live together. Of course, much of this is the job of citizenship / moral education (which different countries name in various ways), but both science education and religious education (in those countries that have it) can play their part. A good science education—though one with a curriculum rather different from that which currently exists in any country, so far as I am aware—might look at why we stereotype others and at why we are more likely to be dismissive of or cruel towards those who are different from us.

More modestly, a science curriculum might help students realise that scientific knowledge is not to be equated with other forms of knowledge. A mathematician, for example, does not ascertain that the sum of the interior angles of a flat triangle add up to 180 degrees by going out into the field, as a scientist might, collecting samples of triangles, bringing them back to the lab, growing them in Petri dishes and determining, from a sample of 73 triangles, that the sum of their interior angles equals 179.4 ± 0.8 degrees.

Similarly, the natural sciences are not a good vehicle for deciding whether we should sell foods made from genetically modified organisms without labelling them, what makes for a great painting or how scripture should be interpreted. These matters are better dealt with elsewhere. However, a good science education might help students to be less dismissive of other forms of knowledge. Just as a good religious education (in countries which have such a subject) might help students realise that there is a whole discipline of how to read and interpret the scriptures, that there is a range of ways of understanding how science and religion relate.

THE APPROACH OF WORLDVIEWS

Before going on to consider the pedagogical implications of all this, mention should be made of one approach to the science-religion issue that has become prominent within science education and is of considerable pedagogical value—namely the concept of worldviews. The essence of a worldview, as the word itself implies, is that it is a way of conceiving and understanding the world that one inhabits. So, someone with a traditional Christian worldview is likely to believe that the world is fundamentally good but has become corrupted as a result of human sin. However, for such a person, there is always the hope of redemption and one of the tasks of Christians is to live their lives so as to help bring about the kingdom of God.

On the other hand, someone with an atheistic worldview is likely to believe that the world is morally neutral and that there are no ultimate purposes in life beyond those that we decide for ourselves. Which of these two worldviews one finds the more convincing and conducive says much about oneself.

Creationism can profitably be seen not as a simple misconception that careful science teaching can correct, as careful science teaching might hope to persuade a student that an object continues at uniform velocity unless acted on by a net force, or that most of the dry mass of a plant comes from air as opposed to the soil. Rather, a student who believes in creationism can be seen as inhabiting a non-scientific worldview, a very different way of seeing the world compared to the scientific perspective. The pedagogical significance of this comes largely from the observation that one very rarely changes one's worldview as a result of one or two lessons, however well taught, whereas one may indeed replace a misconception with its scientifically validated alternative after such a brief teaching sequence (Reiss, 2008b).

The probable reason for this difference in the difficulty of replacing worldviews and misconceptions is two-fold. First, a student is likely to have far more of personal significance invested in a religious worldview than a scientific misconception. It is clear that the personal implications of abandoning a belief in a literal reading of the chronology of *Genesis*, including the six days of creation as six periods each of 24 hours, are far greater than of discarding a presumption that plants gain most of their dry mass from the soil. Secondly, many scientific misconceptions are relatively discrete—one can discard one without this affecting much else of one's scientific understanding. Abandoning creationism entails accepting the notion of Deep Time, the relatedness of all life and, perhaps more significantly, that God acts in very different ways to how one has previously supposed.

PEDAGOGICAL IMPLICATIONS

The question of the significance of religious issues for science education can be considered at the intended, the implemented and the attained curriculum levels. In a school setting there are therefore implications for curriculum developers, for classroom teachers and for learners. In this section I concentrate on teachers (whether in school science classrooms or informal settings) and learners.

Science teaching is demanding for teachers, particularly in a school setting. I have discussed elsewhere whether or not it is realistic to expect science teachers to deal with ethical issues in science lessons (Reiss, 1999). Although there are examples of this happening successfully (Jones, McKim, & Reiss, 2010; Reiss, 2008c), this is far from always being the case. It seems even more optimistic to expect science teachers to deal with religious issues, even when these are restricted to religious issues that relate to science.

The aim of including religion in science learning is not primarily to teach about religion but to enable richer and more effective ways to enable students to understand certain ideas within science and to help them appreciate better certain topics where science and religion interact. If science teachers, or other communicators of science, do deal with religious issues, or science issues that have religious connotations, I recommend that they be both true to science and respectful of their students and others, irrespective of such people's religious beliefs. Indeed, nothing pedagogically is to be gained by denigrating or ridiculing students.

The principle of respect for students has implications for assessment too. Well-designed examination material should be able to test student knowledge of science and its methods without expecting students to have to convert, or pretend that they have converted, to a materialistic set of beliefs. So, for example, while it is appropriate to ask students to explain how the standard neo-Darwinian theory of evolution attempts to account for today's biodiversity, it is not appropriate to ask students to explain how the geological sciences prove that the Earth is billions of years old.

Perhaps the most important implications of religion for the teaching of science come when teaching about the nature of science. It can be a useful exercise with some students for science educators to get them to consider whether such topics as astrology, ghosts, paranormal phenomena and miracles fall within the scope of science or not. The aim is not to smuggle such topics into science but to get students more rigorously to think about what science is and how it proceeds. I remember one student of mine who undertook a survey among her peers to see whether, as predicted by astrology, their astrological birth sign was related to their personality, using a validated measure of personality. It wasn't. That student learnt something about science that was of value to her that I suspect she might not have learnt had I told her not to be silly and instead research something within mainstream science.

James Skehan and Craig Nelson (2000) point out that science educators generally do not do a good job of providing students with criteria to compare the strength of great scientific ideas. They emphasise the value of enabling students to develop skills of critical thinking when considering controversial topics, such as evolution, and provide a valuable list of eight criteria for comparing major scientific theories:

1. How many lines of independent evidence support the theory?
2. How many previously unconnected areas of knowledge did a theory tie together?
3. Does the theory make precise predictions?

4. How clear are the causal mechanisms?
5. Does the theory adequately explain the ultimate origin of the systems it describes and explains?
6. Is the theory scientifically controversial, or only publicly or politically controversial?
7. Is the theory fundamental to many practical benefits embraced by our economic system?
8. Is the theory widely understood and accepted by the general public?

There would seem to be much of value in encouraging students to consider such questions when examining the validity of evolution or other major scientific theories.

Tonie Stolberg and Geoff Teece (2011) write about how to teach the science-religion issue but address their advice to specialist teachers of religion, not science, which provides a useful counterweight to much of this chapter. They point out that religious education teachers often assume that they should be neutral when teaching about controversial issues, yet this can be unrealistic and may not be the most effective way of teaching:

> Teachers may well feel that adopting a neutral stance—focusing on "the facts", giving a "balanced" picture—is most likely to be the "safest" one to adopt. In practice, this is a very difficult strategy to achieve. The choice of facts you present (or withhold), the "expert" opinions you share with your students and all the other educational judgments—in terms of the resources chosen and time devoted to the issue being explored—makes the effort of teaching religion and science issues in this way unrealistic. As with all controversial issues, however, your students need to be taught to examine critically the information they are given and the attitudes to values that have led to its production. So, rather than seeking to "not get involved", you should be explicit about the aims and objectives of any exercise so that your students are aware of the circumstances in which they are being asked for their opinions and share the basis for their thinking. (p. 71)

Specific Issues to Do with Creationism. Part of the purpose of school science lessons is to introduce students to the main conclusions of science, and the theory of evolution is one of science's main conclusions. For this reason, school biology and earth science lessons should present students with the scientific consensus about evolution, and parents, I argue, should not have the right to withdraw their children from such lessons. At the same time, science teachers should be respectful of any students who do not accept the theory of evolution for religious (or any other) reasons.

Science teachers should not get into theological discussions, for example about the interpretation of scripture. They should stick to the science. If they are fortunate enough to have one or more students who are articulate, able and willing to present any of the various creationist arguments against the scientific evidence for evolution (e.g., that the theory of evolution contradicts the second law of thermodynamics, that radioactive dating techniques make unwarranted assumptions about the constancy of decay rates, that evolution from inorganic precursors is impossible in the same way that modern science disproved theories of spontaneous generation, that humans and dinosaurs coexisted in the fossil record), they should use such contributions to prompt the rest of the class to think rigorously and critically about the arguments and standard accounts of the evidence for evolution.

My own experience of teaching the theory of evolution for some thirty years to school students, undergraduate biologists, trainee science teachers, members of the general public and others is that people who do not accept the theory of evolution for religious reasons are most unlikely to change their views as a result of one or two lessons on the topic, and others have concluded similarly (e.g., Long, 2011). However, that is no reason not to teach the theory of evolution to this population. One can gain a better understanding of something without necessarily accepting it. Furthermore, recent work suggests that careful and respectful teaching about evolution can indeed make students who initially reject evolution considerably more likely to accept at least some aspects of the theory of evolution (Winslow, Staver, & Scharmann, 2011).

For sites of informal education there are some issues that are the same for schools and some that are different. The principle of respect for students and others, irrespective of their religious beliefs, holds in the same way but the principle of being true to science manifests itself somewhat differently depending on the nature of the site of informal education. If the site is one that identifies itself as being scientific, for example a science or natural history museum or centre, then it can validly attempt to convince visitors that the standard scientific position is correct. Other informal education sites may have less of a science agenda. In any event, any museum or other site of informal learning should be able to prepare carefully and access resources in a way that may not be possible for a classroom teacher, so as to ensure that an exhibition, a display, a taught session or an outreach activity does deal with relevant religious issues.

Finally, there are an increasing number of creationist museums (e.g., http://creationmuseum.org/) and zoos (e.g., www.noahsarkzoofarm.co.uk/). Perhaps somewhat optimistically, I would ask those running such creationist places of learning to make one concession to evolution. I do not expect them to promote evolution, but it is reasonable to ask them to make it clear that the scientific consensus is that the theory of evolution and not creationism is the best available explanation for the history and diversity of life. It is perfectly acceptable for those running creationist institutions to critique evolution and to try to persuade those visiting such institutions that the standard evolutionary account is wrong. But just as science teachers with no religious faith should respect students who have creationist views, so creationists should not misrepresent creationism as being in the scientific mainstream. It is not.

REFERENCES

Al-Hayani, F. A. (2005). Islam and science: Contradiction or concordance. *Zygon, 40*, 565–576. doi:10.1111/j.1467-9744.2005.00688.x.

Atkins, P. (2011). *On being: A scientist's exploration of the great questions of existence.* Oxford: Oxford University Press.

Barbour, I. G. (1990). *Religion in an age of science: The Gifford Lectures 1989–1991* (Vol. 1). London: SCM.

Barbour, I. G. (1997). *Religion and science: Historical and contemporary issues*. London: Harper Collins.

Bhaskar, R. (1978). *A realist theory of science*. Sussex: Harvester Press.

Bloor, S. (1991). *Knowledge and social imagery* (2nd ed.). Chicago, IL: University of Chicago Press.

British Humanist Association. (2016). *Religion and belief: Some surveys and statistics*. Retrieved January 31, 2016 from https://humanism.org.uk/campaigns/religion-and-belief-some-surveys-and-statistics/

Brooke, J. H. (1991) *Science and religion: Some historical perspectives*. Cambridge: Cambridge University Press.

Buss, D. M. (2003). *The evolution of desire: Strategies of human mating* (Rev. ed.). New York, NY: Basic Books.

Clayton, P., & Simpson, Z. (2006). *The Oxford handbook of religion and science*. Oxford: Oxford University Press.

Daston, L., & Galison, P. (2007). *Objectivity*. New York, NY: Zone Books.

Dennett, D. C. (2006). *Breaking the spell: Religion as a natural phenomenon*. London: Allen Lane.

Dixon, T. (2008). *Science and religion: A very short introduction*. Oxford: Oxford University Press.

Gould, S. J. (1999). *Rocks of ages: Science and religion in the fullness of life*. New York, NY: Ballantin.

Griffiths, M. (2003). *Action for social justice in education: Fairly different*. Philadelphia, PA: Open University Press.

Haught, J. F. (1995). *Science and religion: From conflict to conversation*. Mahway, NJ: Paulist Press.

Hick, J. (1976/1985). *Death and eternal life*. Basingstoke: Macmillan.

Hinnells, J. R. (1991) *A Handbook of living religions*. London: Penguin Books.

Jones, A., McKim, A., & Reiss, M. (Eds.). (2010). *Ethics in the science and technology classroom: A new approach to teaching and learning*. Rotterdam: Sense.

Kenny, A. (1992). *What is faith? Essays in the philosophy of religion*. Oxford: Oxford University Press.

Kuhn, T. S. (1970). *The structure of scientific revolutions* (2nd ed.). Chicago, IL: University of Chicago Press.

Langford, M. (1981). *Providence*. London: SCM.

Lee, B. Y., & Newberg, A. B. (2005). Religion and health: A review and critical analysis. *Zygon, 40*, 443–468. doi:10.1111/j.1467-9744.2005.00674.x.

Long, D. E. (2011). *Evolution and religion in American education: An ethnography*. Dordrecht: Springer.

Merton, R. K. (1973). *The sociology of science: Theoretical and empirical investigations*. Chicago, IL: University of Chicago Press.

Osborn, L. (2005). Theology and the new physics. In C. Southgate (Ed.), *God, humanity and the cosmos* (2nd ed., pp. 119–153). London: T & T Clark.

Peacocke, A. (2001). *Paths from science towards God: The end of all our exploring*. Oxford: Oneworld.

Pew Research Center. (2015). *The future of world religions: Population growth projections, 2010-2050*. Retrieved January 31, 2016 from http://www.pewforum.org/2015/04/02/religious-projections-2010-2050/#projected-growth-map

Polkinghorne, J. (2006). Where is natural theology today? *Science & Christian Belief, 18*, 169–179.

Popper, K. R. (1934/1972). *The logic of scientific discovery*. London: Hutchinson.

Ray, J. (1691/2005). *The wisdom of God manifested in the works of the creation*. London: Ray Society.

Reiss, M. J. (1999). Teaching ethics in science. *Studies in Science Education, 34*, 115–140. doi:10.1080/03057269908560151.

Reiss, M. J. (2008a). Should science educators deal with the science/religion issue? *Studies in cience Education, 44*, 157–186. doi:10.1080/03057260802264214.

Reiss, M. J. (2008b). Teaching evolution in a creationist environment: An approach based on worldviews, not misconceptions. *School Science Review, 90*(331), 49–56.

Reiss, M. J. (2008c). The use of ethical frameworks by students following a new science course for 16-18 year-olds. *Science & Education, 17*, 889–902. doi:10.1007/s11191-006-9070-6.

Reiss, M. J. (2013). Religion in science education. In N. Mansour & R. Wegerif (Eds.), *Science education for diversity: Theory and practice* (pp. 317–328). Dordrecht: Springer.

Reiss, M. J. (2014). What significance does Christianity have for science education? In M. Matthews (Ed.), *Handbook of historical and philosophical research in science education* (pp. 1637–1662). Dordrecht: Springer.

Reiss, M. J. (2015). Learning for a better world: futures in science education. In D. Corrigan, C. Buntting, J. Dillon, A. Jones, & R. Gunstone (Eds.), *The future in learning science: What's in it for the learner?* (pp. 19–34). Dordrecht: Springer.

Reiss, M. J., & White, J. (2013). *An aims-based curriculum: The significance of human flourishing for schools*. London: IOE Press.

Saunders, N. (2002). *Divine action and modern science*. Cambridge: Cambridge University Press.

Skehan, J. W., & Nelson, C. E. (2000). *The creation controversy & the science classroom*. Arlington, VA: NSTA Press.

Smart, N. (1989). *The world's religions: Old traditions and modern transformations*. Cambridge: Cambridge University Press.

Stenmark, M. (2004). *How to relate science and religion: A multidimensional model*. Grand Rapids, MI: Wm B Eerdmans.

Stolberg, T., & Teece, G. (2011). *Teaching religion and science: Effective pedagogy and practical approaches for RE teachers*. Abingdon: Routledge.

Szerszynski, B. (2005). Rethinking the secular: Science, technology, and religion today. *Zygon, 40*, 813–822. doi:10.1111/j.1467-9744.2005.00709.x.

Watts, F. (Ed.). (1998). *Science meets faith*. London: SPCK.

Winslow, M. W., Staver, J. R., & Scharmann, L. C. (2011). Evolution and personal religious belief: Christian university biology-related majors' search for reconciliation. *Journal of Research in Science Teaching, 48*, 1026–1049. doi:10.1002/tea.20417.

Yearley, S. (2005). *Making sense of science: Understanding the social study of science*. London: Sage.

Ziman, J. (2000). *Real science: What it is and what it means*. Cambridge: Cambridge University Press.

CHAPTER TWENTY-THREE

Religion in Science Classes

Heresy or Constructive Pedagogy?

NIDHAL GUESSOUM

I teach introductory astronomy for non-science university students almost every semester. And each time I describe the formation and evolution of Earth as a planet, and as I reach the evolution of Earth's atmosphere, the liquid water on the surface, the habitability of the planet, and the appearance of life and its effect on the atmosphere, I invariably get explicit or implicit religious questions. Those include: "Sir, are you adopting an evolutionary (Darwinian) paradigm?"; "Sir, did God play any direct role in this, or was he confined to the (initial) creation of the universe?"; "Sir, could life appear without God's intervention?", etc. In fact, even when I describe astronomical phenomena, e.g., eclipses and meteors, some student will invariably ask me either in class or after class: "Sir, how does that compare or relate to Qur'anic verses that describe these phenomena?", or "What do you say about the many claims that a lot of what science has discovered (lately) can be found in the Qur'an?", etc.

I actually have it easy, as I only teach astronomy, not biology, although there are clearly enough religion-related topics in astronomy for difficult situations such as the above. God help those who teach biology, particularly evolution and genetics (mutations, etc.), at least in my part of the world. Rana Dajani, an associate professor of biochemistry at the Hashemite University in Jordan, has written and given many talks on her experience of teaching evolution to Muslim students (Dajani, 2012, 2015a, 2015b). Salman Hameed, an associate professor of integrated science and humanities at Hampshire College in Massachusetts, USA, has been studying Muslim attitudes with regard to biological and human evolution; he and his colleagues found many cases of biology teachers in the Muslim world who do not believe in Darwin's theory and tell their students to only learn it for the exams, should that come up (Hameed, 2008). Even in advanced seminars and workshops, I have seen participants' faces turn somber, their minds closing up, when a speaker refuses to compromise on origin-creation issues (cosmos, life, humans, etc.). In fact, in many cases, when they hear a professor insisting on natural explanations for some "sensitive" issues (origin of life, origin of humans, etc.), students see it as an attempt to indoctrinate them with some particular ideology (materialism, mostly). Even if the teacher remains neutral and injects no particular ideological slant to the material, students who see larger implications to the topics being addressed (origin of the universe, evolution of life, etc.) will likely come out with the impression that science, remaining so disconnected from God, is inherently materialistic/atheistic, or at the very least representing a cold endeavor that has no interest in the concerns and aspirations of humans, both as individuals and as societies.

The effect of students' religious or cultural mindsets (their "worldviews") on their interaction with science topics is very strong and clear in the Muslim world, but it can also be seen elsewhere, at least in societies and cultures where religious teachings are still strong. In her study of 101 first-year Arab and Jewish students in Israel, some religiously observant, others traditional, and others secular, Ester Aflalo (2013) reported that: a) neither the students' previous scientific background nor their being Arab or Jewish correlated much with their perception of the nature of science (NOS); b) their religious beliefs, either Jewish or Muslim, was "the most distinguishing factor amongst the participants' perceptions of the NOS. The more religious a student is the greater the weight he affords culture and society vis-à-vis science, and supports to a lesser extent the freedom of inquiry and the variability of science."

Sometimes it is the professor who lets his/her religious dogmas affect his/her scientific knowledge: reports of creationist teachers who distort the content of biology/science courses are not rare; large studies include Randy Moore's (2008) four-year survey of 1,465 freshmen university students who had taken high school biology in public schools in various regions of the United States; Slate magazine recently reported that thousands of publicly funded schools in the United States allow or even encourage creationist teachers to teach "alternatives" to evolution (Kirk, 2014). Elise K. Burton (2011) reviewed evolution and creationism in middle-eastern education, focusing on Iran, Israel, Saudi Arabia, and Turkey, and showed surprising situations. In Iran, contrary to what many people might expect, evolution is widely taught in both high schools and universities, while in Israel, not only is evolution merely an optional (not mandatory) topic in the curriculum, "many Israeli teachers avoid teaching the optional unit … precisely because it is perceived as a controversial topic that may cause teacher-student conflict" (p. 302). In Turkey, decades of rising religious attitudes and campaigns have affected students' acceptance and understanding of evolutionary theory (Peker, Comert, & Kence, 2010).

In all such cases, students and teachers are simply trying to construct a coherent picture of the world, one which fuses their previously acquired religious or cultural views with the newly received scientific knowledge. Sometimes this makes people close up on the science, sometimes it leads them to abruptly reverse their worldviews.

The science education community has for the past decade or so come to realize the importance of this issue, for several reasons. Wolff-Michael Roth (2010) relates how his 1997 paper (Roth, 1997), where the interaction of scientific and religious discourses in a physics class was discussed, was "deemed too sensitive" (p. 163) and was rejected by at least two journals before finally seeing the light and ultimately recognized as seminal.

Now, however, it is well understood and accepted that student learning does not occur in a vacuum or on a blank slate; students' cultural backgrounds and beliefs are an essential factor in the learning process overall and in shaping one's understanding of science (methodology, theories, results, facts), in particular. Indeed, the extent to which the nature of science will be understood depends not only on the curriculum and the teaching methods but also and perhaps more on the students' and the teachers' correct navigation of the interacting streams of reason, belief, experiment, critical thinking, etc. Secondly, the science education community (teachers and students) has become much more culturally diverse, even in the West where millions of migrants from around the world have settled in recent times. Thirdly, the science-religion debates, which have recently reignited among scholars, scientists, philosophers, and theologians, have slowly found their way to education journals (Reiss, 2010), societies, associations, conferences, and even courts (the most famous one in recent years being the *Kitzmiller v. Dover* Area School District intelligence-design trial). Indeed, there are legal issues involved in some countries (USA, France, and others) as to whether any religious theme can be brought up or discussed in science classes or sometimes in any classes (in public schools).

And last but not least, there are important competency issues involved in the topic: are science teachers well enough equipped to address religion-infused questions (such as those that I receive from my astronomy students or those that Dajani's students repeatedly raise); are religion teachers knowledgeable enough about science, its philosophy, and its theories and results, to answer such students' questions? Indeed, to whom should students address their questions; should there be special "bridge" courses, as has been proposed (e.g., Mimouni, 2015)?

Needless to say, allowing religious views to be voiced in a science classroom is hugely fraught with various dangers: creationism may sneak in and find fertile ground (Taşkın, 2014), and likewise for religious fundamentalism, pseudo-science, presumed scientific content in sacred books (the so-called "miraculous scientific content of the Qur'an", known as I'jaz, or in the Vedas, or other sacred texts—see the chapter on the topic in my book *Islam's Quantum Question: Reconciling Modern Science and Muslim Tradition* (Guessoum, 2011)). At the very least, there is a risk of inducing in the students' minds a confusion or a blurring of the lines between solid science and erroneous but widely held ideas, between beliefs and systematic methods of uncovering facts about the world, etc. Still, one must not ignore students' pre-conceptions, which have taken root in their minds from their parental or communal teachings and which often come from religious "knowledge." One must carefully address all the interrelated issues of: where various ideas come from, how they relate to physical or metaphysical realms, how much evidence (historical, scientific, etc.) one has on this or that "fact," what methodology is proper for each type of "knowledge," how to distinguish between scientific assertions and philosophical interpretations thereof, etc. Ignoring students' beliefs is, in my and many other educators' opinions, not a constructive and productive approach to the issue of the intermingling of religious ideas with scientific knowledge.

STUDENTS' BELIEFS, SOCIAL CONTEXTS, AND CULTURAL RESPONSIVENESS

People carry many beliefs that have slowly taken form and crystallized from various educational and life experiences. Those may come from religious principles (there is a creator; we humans are special creatures put here on earth for a purpose; we have a spiritual dimension and activity; there have been divine revelations and sacred books; etc.) or other philosophies (humanism, equality, care for nature, etc.). Such beliefs are often passed on (with modification and mutation) to people's children, who then use them as guiding principles in life and confront them or attempt to apply them to their social and educational environment. This then affects the learning process, as teachers and students bring their own mindsets, values, and worldviews into the classroom. Studies have shown this belief factor to be one of the most essential ones in science education (Calderhead, 1996; Mansour, 2008; Reiss, 2009; Roth, 1997; Southerland & Scharmann, 2013; Stolberg, 2007).

In March–April 2010, I conducted a survey on science, education, and religion among students at my university, the American University of Sharjah. 105 students (73 females; 32 males) were surveyed, most of them freshmen or sophomores, non-science majors. 89% of them were Muslim, and 81% Arab. Below are the results to the main questions that I asked them:

- 75% disagreed strongly or somewhat with the statement "I think there are conflicts between what my religion teaches me and what Science claims."
- On the issue of reconciliation between science and religion ("I try to reconcile the facts of Science with the truths of my religion"), 68% said yes.
- On whether religious issues should be discussed in science classes ("I think religious ideas/ viewpoints, e.g., on the creation of the world, biological evolution, etc., *should be* discussed by Science professors"), 66% of the surveyed students said yes.
- On whether Modern Science or Modern Technology has (have) made it harder for people to be religious, only 29% said Modern Science does but 56% said Modern Technology does make it harder for people to be religious.
- 90% of the students responded favorably to the question: "Do you believe that the Qur'an contains explicit statements that are now known to be scientific facts?" Only 8% preferred the statement: "The Qur'an deals with natural phenomena and alludes to scientific facts but only vaguely."

Keeping in mind that this survey was conducted at the American University of Sharjah, an English-language, liberal-education type of institution and a socially diverse and liberal campus (by regional standards), it seems clear that there is a rather strong desire among students to have those issues of religion and science addressed in their learning environments, perhaps even in the science classrooms, at least in communities where religion(s) is(are) a strong component of people's identities and cultures.

A number of authors have noted the importance of taking into consideration students' beliefs and cultural contexts when attempting to impart scientific knowledge and the appropriate kind of "cultural responsiveness" that can be adopted. Jackson, Doster, Meadows, and Wood (1995) highlighted the dissonance and even alienation that many students (in a Western context) feel when a topic like human evolution is presented to them without any attempt to address their beliefs and worldviews. The authors recommended the adoption of a pedagogy that diffuses the tension between the students' religious views and the scientific approach. Similarly, Shipman, Brickhouse, Dagher, and Letts (2002) introduced small modifications to an astronomy course to try to address philosophical or religious views in a way that is both rigorous and pedagogically sound.

Dajani (2015), teaching evolution in a Muslim environment, has adopted a pedagogy that fully acknowledges students' prior misgivings and resistance toward the subject. By openly discussing all ideas inside the classroom, she is able to at least clear up students' misconceptions about the theory and clarify the important methodological principles at play in the subject. Similarly, P. Lynne Honey, a professor of psychology, has recently argued, "Why I teach the controversy: using creationism to teach critical thinking" (Honey, 2015, p. 793). Indeed, noticing the recent increase in her students' interests in creationism and intelligent design (due to media coverage of events such as the Dover trial), she wondered, "whether ignoring supernatural views allowed them to remain as viable 'alternatives' to scientific hypotheses, in the minds of students" (p. 793). As a psychology professor, she was interested in discussing both the scientific method and the cognitive errors associated with non-science views. Responding to those in the science and science education community who fear that discussing religious and pseudo-scientific views in the science classroom might elevate their status and raise their credibility among students and the public and is a favorite ploy of creationists (Scott, 2007), she pointed out that her pedagogic approach is consistent with the "teaching as persuasion" model (Alexander, Fives, Buehl, & Mulhern, 2002), where different views are discussed and some are then rejected on the basis of demerit. She also likened her approach to a vaccination against pseudo-scientific arguments.

A number of science education specialists have recently been urging the adoption of some forms of "cultural responsiveness" toward students' backgrounds. For example, Jewell Cooper and Catherine Matthews (2005) have written: "Science teachers must become acquainted with their students, especially within the communities in which they live. By doing so, science becomes a contextualized engagement and a culturally relevant experience, one that allows students to link their daily experiences to what they do in class" (p. 52). Geneva Gay (2003) defined "culturally responsive teaching" as "using the cultural characteristics, experiences, and perspectives of ethnically diverse students as conduits for teaching them more effectively" (p. 106). Likewise, Moyra Keane (2008), reporting on a field study in a rural African setting, stated: "By understanding key aspects of an African worldview, science educators can contribute to both education and community well-being" (p. 587). Perhaps most educators will agree that some forms and levels of "cultural responsiveness" are important to adopt (e.g., Hernandez, Morales, & Schroyer, 2013); the question then is: what exactly should one do and what caveats should be put in place to ensure that the primary educational objectives are upheld and no diversions occur? I will come back to this shortly.

NATURE OF SCIENCE (NOS) AND NATURE OF RELIGION (NOR)

Science and Religion constitute perhaps the two main ways to look at the world in which we live. To put it simply and succinctly: Science describes how the (physical) world works; Religion presents a philosophy of life and existence, its meaning and purpose. If one is able to clearly define and understand the nature of each (the Nature of Science, NOS, and the Nature of Religion, NOR), then navigating the straits where the two come in contact and constructing a coherent overall picture becomes feasible and, I say, necessary, even for those who do not consider Religion to be important for themselves but should see its relevance and impact on society more widely.

This is where we science educators, and most particularly those who construct science curricula and pre-service teacher education programs, have largely failed. Unfortunately, science teachers do not always fully and clearly understand the nature of science and how and where it distinguishes itself from religion without necessarily opposing it. And likewise and even more so, religion teachers do not always understand the nature of religion, how it differs from science, and where the limits of each lie. This is particularly true in my part of the world (the Arab-Muslim world), but—I suspect—is also true, to varying degrees, in other parts of the world.

Let me briefly cite the results of a recent study (Mansour, 2008) that explored the views of 75 Egyptian science teachers on science and religion. Questions included:

> Do your religious conceptions affect the way you teach science? Who do you think should be responsible for teaching the relationship between science and religion: religious education, or science education? Do you feel confident or worried when you teach science/religion issues (for example, cloning, evolution, creation)? And how do you teach these topics? (Mansour, 2008, p. 562)

Nasser Mansour (2011) further reported interesting results that may be surprising to Westerners but not so much for educators and observers from the region. First, adopting Barbour's taxonomy of science-religion relations (Ian G. Barbour's seminal and influential *Religion in an Age of Science* (1990) had summarized the ways in which Science and Religion inter-relate as four main categories: Conflict, Independence, Dialogue, and Integration), Mansour found that 61% of the teachers subscribe to the Integration model, 19% adopt the Dialogue approach, 13% view the relation through the Independence lens, and less than 7% see Conflict between the two. More interestingly, 53% put the responsibility for addressing science-religion issues on the science teachers' shoulders, 20% assign it to the religious teachers, and 27% see it as a dual responsibility. They believe it is important to gradually and increasingly address those topics from elementary to secondary school. Interestingly, 60% of the science teachers (mostly secondary-level science educators) reported "no confidence" in addressing science-religion topics. More alarmingly, 40% of them reported a negative attitude toward "non-Muslim sciences." On this particular point, Mansour (2011) commented that "[t]he attitude of science teachers towards non-Islamic Western scientists reflects the teachers' concerns about the ethics and values of science" (p. 304). More generally on the results of his study, he commented that:

> Some teachers did not just ascribe to naïve views of the nature of science, but also to naïve views of the Islamic perspective of science and scientific investigation. They argued that science is an ever-changing phenomenon and scientists' assumptions and predictions may be wrong, whereas the teachings of Islam are eternal and not subject to human error. The purpose of doing research in science, they argued, is to validate the Qur'an or establish the truth of the Qur'an. (p. 305)

This study and its author's comments clearly show that there is a serious issue of (mis)understanding both the nature of science and the nature of religion by science teachers, at least in some cultural contexts.

On the other side of the equation, one finds students carrying several essential misconceptions about science, including:

- Viewing scientific theories as mere conjectures/speculations/propositions; this often stems from a serious misunderstanding of the term "theory" or even of "hypothesis."
- Believing that not only can "real/deep science" prove God, but a correct goal for scientific research should be to find God and show His glory. This view of science, as Mansour (2011) found, is also often carried by science teachers with strong religious (Islamic) beliefs.
- Confusing the natures of the Qur'an with that of science, Muslim students (falling prey to the current widespread populist phenomenon of "the miraculous scientific content of the Qur'an"), instead of seeing their holy book as encouraging them to contemplate and reflect on nature/creation and extract knowledge through the process of exploration and research, widely believe that much specific scientific information can be found in the Book itself (90% of the students I surveyed believed this!)

We science educators have much work to do to clarify the nature of science and its methodology, which is based on systematic research, gathering evidence, and confronting it with hypotheses and testing models. Furthermore, in religious communities, there is a strong need to distinguish the nature of science from the nature of religion. This will not occur by simply ignoring the interaction between the two in the minds of students and teachers. This will require some careful approaches, perhaps with some experimentation to produce the best methods and reach the best results.

POSSIBLE APPROACHES

If we agree that there are both needs and benefits to addressing religion-related questions when teaching some topics of science (the origins, evolutions, and ages of the universe, Earth, life, and humans; biodiversity; genetic engineering; etc.), and assuming that there are no legal obstacles to addressing such topics in the classroom, what fruitful approaches can be adopted?

There are a few principles that one must operate under in order to ensure that the process does not slip away from its intended objectives:

- First and foremost, an accurate picture of the nature of science must be presented and defended at all times. Teachers and students cannot be allowed to distort science or compromise its methodology in order to "harmonize" the (well-established) scientific knowledge with religious views.
- Secondly, teachers must make sure to clarify and underscore for students what is well established in science from what is still unproven or currently being researched and debated among scientists. The ages of the universe and of the Earth are *facts* and are not open for acceptance or rejection, unlike (so far) the multiverse *hypothesis*, for example.
- Thirdly, and very importantly, whatever worldviews the teachers may carry, they must always not only do their utmost to remain neutral but also explain and uphold the neutrality of science vis-à-vis worldviews. In other words, even though science adopts a naturalistic methodology (all physical phenomena must be explained using natural causes only), it does not imply that God does not exist or that the world has no purpose or plan behind it; this principle simply declares such philosophical questions (God's existence, purpose of existence, etc.) to be beyond the remit of science and leaves them open for various possible interpretations, theistic, agnostic, or atheistic.

As to the specific approaches that can be adopted and applied in dealing with science and religion in the educational environment, they may be divided into three general frameworks:

A. The NOMA (Non-Overlapping Magisteria) approach, which was proposed by the late paleontologist and science popularizer Stephen J. Gould (and also subscribed to by the late astronomer and science popularizer Carl Sagan and others), and which has been adopted in many quarters as an adequate framework for separation between science and religion. Michael Reiss (2008) summarizes what he sees as a number of advantages to this approach: an epistemological justification for the separation of science and religion; an emotional way for religious learners to handle topics where science says something that conflicts with their prior views; a legal framework in places where religion is not allowed in the classrooms; etc. (Reiss, however, is fully cognizant that this does not suit all individuals and groups.)

In the Islamic context, Jamal Mimouni (2015) has defended this approach on the basis of the following argumentation: "Certainly the intermingling of science with religion is a no-go endeavor, as it brings no benefit for either side. What we mean by this is … the encroachment of theological considerations to the naturalistic explanations." (p. 85) I fully agree with this general principle; however, this does not address the questions at hand, namely how to deal with topics that surface in science courses but are strongly related to religion and/or to students' cultures and worldviews. Moreover, treating the two domains in total separation does not give students an opportunity to compare and contrast the different natures of religion and science, their methodologies, and their approaches to truth claims (Reiss, 2008).

NOMA is not completely inapplicable in the educational context. It is quite useful as a preliminary principle in that it requires careful understanding of the domains of application—and the nature of—both religion and science and the necessity for everyone to distinguish between them. But more crucially, there will always be some areas of human cognition and behavior where both seem to apply, such as ethics, philosophy, and worldviews. It is agreed, of course, that certain areas will remain the sole province of "jurisdiction" of either science or religion.

B. The SOMA (Softly Overlapping Magisteria) approach, which I gingerly coined (Guessoum, 2012) in contra-distinction to the "hard separation" approach of NOMA. Indeed, as I have related above, my long teaching experience at universities in the Arab-Muslim world had shown me that students in this culture find it near-impossible to put aside their religious education and mindset when dealing with scientific ideas, facts, models, or theories. Thus, telling them to just separate the two magisteria could not be an educationally satisfactory approach; in fact, it would just leave those preconceptions unchallenged—to be happily taken up by the preachers and the *I'jazists*. The need for reconciliation and harmony has been constant and pressing. Furthermore, Muslims find it difficult to digest any "separation" of domains, widely believing (quite erroneously) that Islam is a complete system, which covers every aspect of life. That is why "secularism" is often strongly resisted in Muslim, considering it as simply an attempt to exclude or tone down religion in society's various domains of activity (judicial, educational, etc.).

One example of how religious ideas can be touched up in science courses was provided by Shipman et al. (2002), where an astronomy course raised a few religious ideas, minimally and softly, that is without advocating for any particular viewpoint, but rather simply acknowledging that a number of astronomical topics bring up some religious or worldview issues and there can be discussions and various perspectives on them.

C. The "Bridge Courses" approach (Mimouni, 2015), where topics at the intersection of science and religion (e.g., origins topics and ethics issues) are addressed in interdisciplinary or multidisciplinary ways, either by teachers/professors with competencies or by teams of teachers who bring a diverse and complementary expertise and perspectives.

In the last two approaches (SOMA and Bridge Courses), clear guidelines will have to be provided to the teachers and some training (or workshops) will be quite useful and even necessary to show teachers how to act in various situations and how to address different types of questions and attitudes from students. In particular, teachers will need to be trained to quickly recognize the methodology that is most appropriate for each type of question, i.e., empirical evidence for and robustness of a given theory, multiplicity of religious views on a given topic, critical thinking, etc. Teachers and students should also be encouraged to look at issues from multiple sociological and historical perspectives: how humans' (and scientists' and theologians') views have evolved on various issues.

WHAT TO DO: CONSTRUCTIVE APPROACHES AND DIFFICULTIES

I believe that the last few decades, particularly with the internationalization of the science education debates, have shown that religion is, in several regions of the world if not in all of it, an important factor both in education generally and in science education specifically. The way teachers and students interact with the science curriculum, indeed how they understand the nature and the methodology of science, in contra-distinction to the methodologies of other human endeavors (religion, politics, media, etc.), is often strongly—albeit at times implicitly or surreptitiously—affected by teachers' and students' religious backgrounds and mindsets.

How to address this issue is far from obvious, but the importance of addressing it cannot be overstated. We need to first realize that teachers are not always objective and neutral transmitters of scientific knowledge, and students are not blank slates and passive and fully absorbing recipients. Secondly, we need to carefully consider and devise constructive and effective ways on how to deal with students' beliefs in the education process. Simply telling students that their beliefs are wrong (as in the case of creationism) or that religion-based ideas are not to be brought into the science classroom, is certainly not a constructive educational approach. On the contrary, allowing such topics to be raised has a number of advantages: (a) it tells the students that learning is an open and tolerant process; nothing is off the table; (b) it gives the teacher a chance to clarify the nature of science and its methodology (empirical evidence, repeatability of measurements, theoretical robustness, falsifiability, etc.); (c) such discussions can be a fruitful way to develop critical thinking in students and to equip them with arguments to defend against pseudo- and anti-science views (Honey, 2015; Reiss, 2008).

Needless to say, there are risks in such endeavors and approaches. I have briefly mentioned above the "Trojan horse" of creationism, which has been documented in a number of places; likewise, probably, for climate change and other such topics. Many teachers have been guilty of bringing their ideological biases into the classroom and filling students' heads with them in lieu of proper science. For this reason, clear guidelines and training would need to be developed to ensure that the exercise does not turn into a free-for-all, where whoever believes more strongly in one view ends up getting a chance to impart his/her beliefs. Science education specialists, curriculum designers, and policy makers need to carefully draw red lines between, on the one hand, cultural responsiveness and open-minded, constructive discussion, and, on the other hand, opening the door so widely as to end up compromising scientific objectivity, methodology, and rigor.

How to deal with this issue is not limited to education specialists. I believe there are important roles to be played by other stakeholders, particularly parents through parents-teachers associations. And there are other factors that will contribute and complicate the issue, such as textbooks, educational technologies, social media, and online platforms. These factors should be explored in other research and publications.

REFERENCES

Aflalo, E. (2013). Religious belief: The main impact on the perception of the nature of science on student teachers. *Cultural Studies of Science Education, 8*, 623–641. doi:10.1007/s11422-013-9504-9.

Alexander, P. A., Fives, H., Buehl, M. M., & Mulhern, J. (2002). Teaching as persuasion. *Teaching and Teacher Education, 18*, 795–813. doi:10.1016/S0742-051X(02)00044-6.

Barbour, I. G. (1990). *Religion in an age of science: The Gifford Lectures 1989–1991* (Vol. 1). London: SCM.

Burton, E. K. (2011). Evolution and creationism in Middle Eastern education: A new perspective. *Evolution, 65*, 301–304. doi:10.1111/j.1558-5646.2010.01113.x.

Calderhead, J. (1996). Teachers: beliefs and knowledge. In D. Berliner & R. Calfee (Eds.), *Handbook of educational psychology* (pp. 708–725). New York, NY: Macmillan.

Cooper, J., & Matthews, C. (2005). A decade of concern: A review of multicultural science education issues in *The Science Teacher*. *The Science Teacher, 72*(3), 49–52.

Dajani, R. (2012). Evolution and Islam's quantum question. *Zygon, 47*, 343–353. doi:10.1111/j.1467-9744.2012.01259.x.

Dajani, R. (2015a). *Evolution and Islam – Is there a contradiction?* Retrieved from http://muslim-science.com/task-force-essay-evolution-and-islam-is-there-a-contradiction/

Dajani, R. (2015b). Why I teach evolution to Muslim students. *Nature, 520*, 409. *doi*:10.1038/520409a.

Gay, G. (2003). *Becoming multicultural educators: Personal journey toward professional agency*. San Francisco, CA: Jossey-Bass.

Guessoum, N. (2011). *Islam's quantum question: Reconciling modern science and Muslim tradition.* London: IB Tauris.

Guessoum, N. (2012). Issues and agendas of Islam and science. *Zygon, 47*, 367–387. doi:10.1111/j.1467-9744.2012.01261.x.

Hameed, S. (2008). Bracing for Islamic creationism. *Science, 322*, 1637–1638. *doi*:10.1126/*science*.1163672.

Hernandez, C. M., Morales, A. R., & Schroyer, M. G. (2013). The development of a model of culturally responsive science and mathematics teaching. *Cultural Studies of Science Education, 8*, 803–820. doi:10.1007/ s11422-013-9544-1.

Honey, P. L. (2015). Why I teach the controversy: Using creationism to teach critical thinking. *Frontiers in Psychology, 6*, 793. doi:10.3389/fpsyg.2015.00793.

Jackson, D. F., Doster, E. C., Meadows, L., & Wood, T. (1995). Hearts and minds in the science classroom: The education of a confirmed evolutionist. *Journal of Research in Science Teaching, 32*, 585–612.

Keane, M. (2008). Science education and worldview. *Cultural Studies of Science Education, 3*, 587–621. *doi*:10.1007/s11422-007-9086-5.

Kirk, C. (2014). *Map: Publicly funded schools that are allowed to teach creationism.* Retrieved from http://www.slate.com/articles/health_and_science/science/2014/01/creationism_in_public_schools_mapped_where_tax_money_supports_alternatives.html

Mansour, N. (2008). Religious beliefs: A hidden variable in the performance of science teachers in the classroom. *European Educational Research Journal, 7*, 557–576. doi:10.2304/eerj.2008.7.4.557.

Mansour, N. (2011). Science teachers' views of science and religion vs. the Islamic perspective: Conflicting or compatible? *Science Education, 95*, 281–309. doi:10.1002/sce.20418.

Mimouni, J. (2015). *Should religion be kept out of the science classroom? Report of the Task Force on Science at Universities of the Muslim World.* Retrieved from http://comstech.org/docs/Science_at_Universities_of_the_Muslim_World.pdf

Moore, R. (2008). Creationism in the biology classroom: What do teachers teach and how do they teach it? *The American Biology Teacher, 70*(2), 79–84. doi:10.1662/0002-7685(2008)70[79:CITBCW]2.0.CO;2.

Peker, D., Comert, G. G., & Kence, A. (2010). Three decades of anti-evolution campaigns and its results: Turkish undergraduates' acceptance and understanding of the biological evolution theory. *Science Education, 19*, 739–755. *doi*:10.1007/s11191-009-9199-1.

Reiss, M. (2008). Should science educators deal with the science/religion issue? *Studies in Science Education, 44*, 157–186. doi:10.1080/03057260802264214.

Reiss, M. (2009). Imagining the world: The significance of religious worldviews for science education. *Science Education, 18*, 783–796. doi:10.1007/s11191-007-9091-9.

Reiss, M. (2010). Science and religion: Implications for science educators. *Cultural Studies of Science Education, 5*, 91–101. doi:10.1007/s11422-009-9211-8.

Roth, W.-M. (1997). The interaction of students' scientific and religious discourses: Two case studies. *International Journal of Science Education, 19*, 125–146. doi:10.1080/0950069970190201.

Roth, W.-M. (2010). Science and religion in a high school physics class: Revisiting the source materials of "The interaction of scientific and religious discourses". *Cultural Studies of Science Education, 5*, 163–167. doi:10.1007/s11422-009-9224-3.

Scott, E. C. (2007). What's wrong with the "Teach the Controversy" slogan? *McGill Journal of Education, 42*, 307–315.

Shipman, H. L., Brickhouse, N. W., Dagher, Z., & Letts, William J. IV. (2002). Changes in student views of religion and science in a college astronomy course. *Science Education*, 86, 526–547. *doi*:10.1002/sce.10029.

Southerland, S. A., & Scharmann, L. C. (2013). Acknowledging the religious beliefs students bring into the science classroom: Using the bounded nature of science. *Theory into Practice, 52*(1), 59–65. doi:10.1080/07351690.2013.743778.

Stolberg, T. (2007). The religio-scientific frameworks of pre-service primary teachers: An analysis of their influence on their teaching of science. *International Journal of Science Education, 29*, 909–930. doi:10.1080/09500690600924934.

Taşkın, Ö. (2014). An exploratory examination of Islamic values in science education: Islamization of science teaching and learning via constructivism. *Cultural Studies of Science Education, 9*, 855–875. doi:10.1007/s11422-013-9553-0.

SECTION NINE

Families and Science Education

What Is the Role of Families in the Process of Science Education?

CHAPTER TWENTY-FOUR

Exploring Families' Roles in Science Learning

From El Bosque in Veracruz, México to the Southeastern U.S.

MARTHA ALLEXSAHT-SNIDER

For the past 20 years, as a teacher, teacher educator, and researcher, I have worked in a range of communities in the U.S., México, and Thailand, where I have seen the integral role that families play in children's learning of science. I have observed multigenerational, informal, family explorations of the natural and physical world occurring in backyards, fields, and neighborhoods, in garages and parks, and in ponds and gardens. I have also encountered families from a wide range of educational, ethnic, and socioeconomic backgrounds around the world who take on active roles in fostering children's and adolescents' curiosity about science and encourage their children to take advantage of opportunities for advancing academic learning and college and career pathways linked to science. In my current work with Latino/a families in a project designed to improve science teaching and learning for immigrant students in the southeastern U.S., I am acting on a fundamental commitment to developing pedagogical spaces in- and out-of-school that ensure all students and families are offered opportunities and encouragement to learn about and understand the natural world around them and to pursue aspirations for education and occupations grounded in science.

In formal school contexts, as I have worked with beginning and experienced science teachers and teacher educators, I have seen gaps in these educators' understandings of how to partner effectively with diverse families to build on family- and community-based aspects of science learning to further children's school-based science education. At the national and international levels, policy documents and organizations outline roles for families in supporting children's interests and aspirations related to science (e.g., National Science Teachers Association [NSTA], 2009; Redding [UNESCO], 2000). These organizations assert that families are critical in reaching global and national goals for expanding a science literate citizenry and a workforce that is ready to innovate to solve challenging problems in arenas such as health, the environment, and social welfare, and to foster economic development. However, in two of the most important policy documents that are currently guiding the science education community in the United States, the *Framework for K–12 Science Education* (National Research Council [NRC], 2012) and the *Next Generation Science Standards* (NGSS Lead States, 2013), the role of parents and families in influencing science education, and the outside-of-school informal science learning contexts in which families are often key participants, are barely visible.

In this chapter, I offer some perspectives for educators seeking to widen their views of the ways in which diverse families around the globe do and can participate in informal science learning with their young and adolescent children. I begin with a snapshot of informal science learning outside of school in the context of family life in rural México. I present this vignette as inspiration for science educators to use as a jumping off point for autobiographical self-reflection and biographic inquiry about the ways in which family and community experiences with informal science learning, in the settings where they grew up, played a role in fostering their own and others' interests in science. The snapshot is also offered to provoke critical reflection on the ways in which these kinds of informal science learning contexts could be linked with school-based science learning in communities encompassing families from a wide range of backgrounds. This snapshot is followed by a discussion of the potential for principles of "life-long, life-wide, and life-deep" learning (Banks et al., 2007, p. 15) to provide a framework that could support science educators in conceptualizing new ways of partnering with families and building linkages for students between their informal science learning in family and community contexts and their formal science learning at school. Finally, I relate an account of a *Steps to College through Science* project that has engaged Latino/a middle and high school students and their families and teachers in bilingual family workshops in two school districts over seven years. A brief analysis of project activities and participants' experiences illustrates how integrating perspectives on life-long, life-wide, and life-deep learning in such workshop settings can serve to link home and school learning in science.

LEVERAGING CHILDREN'S INFORMAL SCIENCE LEARNING IN THE COMMUNITY OF EL BOSQUE

From 2000 to 2009, I had the opportunity to spend time in rural schools and communities in Veracruz, México. I led an educator exchange program for teachers from the U.S. state of Georgia who were encountering an upswing in Latino/a immigrant students populating their classrooms. Each year of the educator exchange program, during a three-week study abroad experience, teachers lived with families, studied Spanish, and visited schools in the state of Veracruz, México. In addition, we spent time, along with Mexican and U.S. colleagues collaborating in the project, in rural schools outside of Xalapa, Veracruz, examining *la vida cotidiana* (everyday life) to better understand the sociocultural nature of rural school environments and the personal experiences of Mexican immigrant children before coming to the U.S. This research has been shared in several articles (e.g., Bryan & Allexsaht-Snider, 2008; Bryan & McLaughlin, 2005). One of the Mexican teachers we came to know well, Alberto, is portrayed in an episode we documented of informal science learning that he orchestrated in the small agricultural community of El Bosque where the one-room multi-grade rural school served 30 students in grades 1 through 6. The portrayal from the article is paraphrased here:

> The multigenerational families in El Bosque own their own houses and small plots of land and have worked for generations harvesting mango, papaya, and chayote, in addition to cultivating fruits and vegetables and corn and beans. A few family members have left for bigger cities and towns, and a few have migrated north. Children and families speak a regional dialect of Spanish at home and at school. Most parents have completed primary school and have developed basic literacy skills. The families and community come together for traditional Catholic religious holiday celebrations (e.g., the festival of the local patron saint), rites of passage (e.g., the Quinceañera, a party for 15-year-old daughters), and holidays recognizing family roles (e.g., Mothers' Day, Fathers' Day, Grandparents' Day), and the important role of educators in the community (e.g., Teachers' Day). The school serves as the central location of community events. Often prior to a celebration, Alberto connects the celebration to the curriculum and children will make preparations during the school day.
>
> Situated in the coastal lowlands alongside a river with nearby hills, community funds of knowledge about nature and naturally occurring events are linked to observations of seasonal and weather conditions critical to families' agricultural work. Investigating the plants, animals and the physical features of the landscape around them are a part of daily life of school and work for families in El Bosque. As part of a larger unit on plant life/diversity of life, Alberto capitalized on a cultural resource in El Bosque—the ethnobotanical knowledge of a student's grandmother—by designing a medicinal plant tour for her to lead with the students. During such tours and walks in the community, the language that flowed between teacher, students, relatives, and/or guests did not include the language of horticulture (binomial nomenclature) but rather common local names. Plants were discussed in terms of local uses (e.g., aloe plant's ointment for rashes and burns, *nopales* [prickly pear leaves] for consumption in salads or as a dietary aid for diabetes). From this and other similar experiences, children as young as first grade became adept at identifying a myriad of flora and fauna. At a young age students were able to draw, in accurate detail, parts of plants found in their local community. They knew the lifecycles and growing seasons of trees, plants, animals, and insects in the local area. They were able to explain how the wheel-and-axle system of a well worked and construct two- and three-dimensional

> geometric shapes with clay and sticks from their own backyards. The teacher's use of local contexts for learning established a common ground of local language and observations of phenomena that were meaningful to the students. Alberto was able to draw upon these rich resources to embed school knowledge about science into the everyday home and community context of his students' lives. (adapted from Bryan & Allexsaht-Snider, 2008, p. 55)

As you picture a grandmother who is leading a medicinal plant tour, a group of elementary age children, other family members, and the children's teacher ambling around the hills, fields, and arroyos of the community of El Bosque, it is not difficult to imagine that teacher moving the group into the classroom, drawing on families' cultural knowledge that he valued as a learning resource and making connections to important concepts related to a unit of study on plant life and diversity of life. As a launching point for building these kinds of home-school-community connections, teachers and teacher educators might consider their own experiences with family- and community-based informal science learning. How were these experiences similar to or different from the experiences of their students, who may hail from rural, suburban, and urban settings across the U. S. and around the world? How can science educators draw on their own experiences to better understand their students' and their families' experiences and how might they make connections to more formal science learning in the classroom?

SCIENCE LEARNING WITH FAMILIES THAT IS LIFE-LONG, LIFE-WIDE, AND LIFE-DEEP

The snapshot portrayal of children's experiences with science learning in family contexts outside of school illustrates the idea that learning is a "life-long, life-wide, and life-deep" (Banks et al., 2007) endeavor. This model of learning can serve as a vision for engaging families in the process of science education in ways that foster curiosity, creativity, problem posing, and problem solving, while promoting ownership of science language, practices, and thinking. Family-oriented science learning activities can support this development in out-of-school locales such as museums and parks, in the communities where families live, and in PreK-12 and postsecondary formal education settings, as well as through multimedia contexts for learning that now include radio, television, movies, the Internet, and apps accessed on handheld devices.

The 2007 report from the Learning in Informal and Formal Environments (LIFE) Center (Banks et al., 2007) advocates four principles that can help educators to understand and build upon the science learning that occurs in the homes, communities, and cultures of all students, regardless of their backgrounds. These principles include the idea that *learning is situated in broad socio-economic and historical contexts and is mediated by local cultural practices and perspectives.* Families' agrarian and labor practices in El Bosque of harvesting mango, papaya, and chayote, and cultivating fruits, vegetables, corn, and beans, formed the backdrop for lessons integrating science, social studies, and literacy across the first through sixth grade levels in the one-room multi-grade school. As we saw in the story of a grandmother guiding a nature walk in the neighborhood to identify medicinal plants, a second LIFE principle is that *learning takes place not only in school but also in the multiple contexts and valued practices of everyday lives across the life span.* Alberto, the teacher in El Bosque, acted on his belief in the third LIFE principle, that *all learners need multiple sources of support from a variety of institutions to promote their personal and intellectual development,* as he regularly invited family members to be a part of his students' science and other learning experiences. Finally, the last LIFE principle, that *learning is facilitated when learners are encouraged to use their home and community language resources* as a basis for expanding their linguistic repertoires, came alive in the medicinal plant walk example where children learned to link their developing knowledge of concepts such as the lifecycle of plants to the native plants they were identifying and the local names of those plants used in everyday conversation among their family members and friends (principles excerpted from Banks et al., 2007, p. 15).

Building on a funds of knowledge approach (González, Moll, & Amanti, 2005), the LIFE report argues that teachers can increase academic development, including science learning, of all their students if they build upon the knowledge, skills, and languages that students acquire in the learning environments of their homes and communities, as well as in the varied spaces of schooling (not only in classrooms, but also in clubs, sports, and other school-sponsored extracurricular activities). In this model, life-long learning refers to the ways that people acquire needed skills (e.g., how to ride a bike) and information (e.g., wearing sunscreen protects you from sunburn) or desired abilities (e.g., how to play a musical instrument). This kind of learning carries through from childhood into old age and for many people may come from formal instruction and learning that is "picked up" through observation and practice, or in other cases may be learned unconsciously. Thus, in the example of El Bosque, children on the medicinal plants tour with a grandmother learn that nopales can help with diabetes symptoms.

Life-wide learning refers to the range of experiences that guide children and adults in adapting to new situations that occur throughout the day, and help us deal with hardships and gain comfort in our lives. For example, how do people deal with unfamiliar situations: how do we figure out how to protect our family when a hurricane is predicted or how to use mass transit in a new city? Similarly, how do we make new friends, construct a healthy diet, or balance a household budget? In El Bosque, we saw a grandmother knowledgeable about medicinal plants learning to serve as an educator, at the same time as a teacher was learning to serve as a cultural broker.

The third LIFE principle, life-deep learning, refers to moral, ethical, social, and spiritual values that guide what we believe, how we act, and how we judge ourselves and others. How and where do these values develop and evolve across the lifespan and how do we put them into practice in our lives? Respect for elders in the community and appreciation of their knowledge was a value woven into the children's interactions with the grandmother in El Bosque, as they skipped ahead and eagerly pointed out and identified plants she brought to their attention.

When we consider the full gamut of life-long, life-wide, and life-deep learning, we can see that science education and scientific problem solving (both formal and informal) are relevant to much of this learning in the modern world of the 21st century. Moreover, parents and families are central to this learning as they support their children in gaining, interpreting, applying, and valuing different kinds of learning and as they themselves continue to learn throughout their lifetimes. In the next section, I outline aspects of life-long, life-wide, and life-deep learning that have developed in the ongoing program of Steps to College through Science bilingual family workshops that are one part of the Language-rich Inquiry Science with English Language Learners through Biotechnology (LISELL-B) project (Buxton et al., 2016).

STEPS TO COLLEGE THROUGH SCIENCE BILINGUAL FAMILY WORKSHOPS

The LISELL-B project is located in two school districts in the Southeastern U.S. that have experienced a rising Latino/a population. Since 2009, as part of the LISELL-B project, we have offered three bilingual family science workshops per year in each district for immigrant families, their middle and high school aged children, and their children's science and ESOL teachers. An average of approximately 50 students, 40 parents, and 7 teachers participate in each workshop. The 4-hour long workshops take place in various settings, including university campuses, technical colleges, and school district career academies.

If you attended a workshop on a typical Saturday, you would see groups of children, mothers, fathers, uncles, and grandmothers attending: a) a bilingual family conversation session with opportunities to talk with Latino/a and other college students, sharing hard-won practical knowledge about planning for and paying for post-secondary education; b) a bilingual science investigation activity session, such as Squishy Circuits, where students, their families and their teachers work together to use models to construct explanations of electrical circuits; and c) a career pathways/biotechnology session where students and families might visit a Biofuels and Fermentation lab and discuss careers and studies in STEM-related fields with university faculty and college students. Each workshop ends with a shared meal to encourage informal conversation among teachers, families, college students, and researchers.

Addressing the fourth of the life-long, life-wide, life-deep learning principles (using home and community language resources), all activities in the LISELL-B family workshops are conducted bilingually to encourage children's development of bilingual and biliterate skills, and to facilitate their communication about science with their families in both English and Spanish. Bilingual school liaisons, project staff members, and/or family members provide explanations and encourage question-posing as scientists explain their work or as university personnel and students explain admissions procedures and programs. For example, in a recent family workshop we conducted a language-rich science investigation we call *Bubbleology*. Families received a bilingual investigation guide, with the suggestion that they record the family's ideas in either English or Spanish or both languages, using everyday and academic language to express their scientific understandings and wonderings as they explored the properties of bubbles and attempted to create the biggest and longest lasting bubbles that they could, using experimental practices.

Reflecting LIFE Center Principle #3 (the need for multiple sources of support from a variety of institutions), families in the Steps to College workshops frequently state that an important reason they continue to attend the family workshops is because they have an opportunity to build relationships with their children's science and ESOL teachers, and to learn from university and college faculty and students about how to support their children's interests in science. Parents also talk about how much they value meeting and interacting with other parents who hold similar aspirations for their children's academic development in middle and high school, leading to attending college.

The ways in which families are invited to articulate local cultural practices and perspectives (LIFE Center Principle #1) in the context of exploring prior knowledge and scientific understandings in the bilingual science investigation activities is best illustrated with a vignette drawn from a book chapter written about the workshops (Buxton, Allexsaht-Snider, & Rivera, 2012). This vignette also sheds light on the ways in which families learn science in the practices of everyday lives across the lifespan (LIFE Center Principle #2), and children come to see their family's experiences as meaningful resources in scientific sense-making:

Alka-Seltzer Rocket Lab (2/27/10):

> Parents and students were working in family groups testing one possible variable that might affect the flight of their Alka-Seltzer Rockets. "So we have to test if powdered Alka-Seltzer or chunks of Alka-Seltzer works best," Jorge, one of the students, clarified. The project director reminded the family that before testing anything, he wanted them to

> think about what they already know. He asked them to formulate a hypothesis about which one might be better, and to think about what evidence would support their hypothesis. Jorge's father, Manuel, suggested that the powder would be better and would work faster. When asked why he thought that, he responded, "Because it's the same as when you grind up an asperina or when you use El Polvo Goodies ... You know ... the headache powder. It works faster because it's a powder. It works faster in your stomach. My mom always did that when I was a kid." "I never knew why you used that powder," responded Jorge. "I thought it was just one of those things from México, but I didn't know there was really a science reason." (Buxton et al., 2012, p. 241)

In the Steps to College through Science bilingual family workshops, we use activities as simple as constructing Alka-Seltzer rockets, that require only minimal materials, to teach and model, for both children and their parents, important science investigation practices such as controlling variables to design a fair test. Families explore the language and history of science as they work together on tasks such as making their own light bulb using batteries, wires, cups, and pencil lead. Children and parents begin to develop a sense of science as a discipline that is not remote and inaccessible, but instead something comprehensible that applies to their daily lives. We have found that when families participate in science together, this allows parents to become aware of expectations that are being placed upon their children in middle and high school. This model also allows students to see their parents as co-learners and resources, drawing on parents' funds of knowledge and interests in science constructed in their work, in informal science experiences in family and community settings, and in their previous schooling. With grandmothers, aunts, uncles, and older college-age siblings participating, as well as parents, children, and teachers, we have seen the idea of dynamic science learning across the lifespan come to life for families, teachers, and university educators who engage together in the workshop settings.

WHAT ARE THE POSSIBILITIES FOR FAMILIES' ROLES IN CHILDREN'S SCIENCE LEARNING?

In sharing the experiences of families in México and Georgia, framed in terms of a model of life-long, life-wide, and life-deep learning, I have attempted to show the potential for science educators in conceptualizing science learning as taking place across the lifespan, within families, in contexts that bridge home, community, and school. Such a conceptualization, with broad views of the funds of knowledge and aspirations families from diverse cultural and experiential backgrounds would bring to the task of collaborating in supporting national and international goals of deepening children's science learning and expanding a science literate citizenry, is critical. We need innovative ideas to encourage science educators in both formal and informal science learning settings, and in PreK-12 schools and in postsecondary institutions, along with scientists in communities and workplaces, to partner with families to create dynamic, lifelong science learning opportunities for people of all ages.

In closing, I pose several questions for all of us to ponder as we consider our own roles as educators in relation to the roles of families in the process of science education: What new ideas for engaging families collaboratively with schools to promote interest in science and aspirations for academic development in science across the lifespan emerged as you explored the stories related in this chapter? What aspects of your particular contexts are untapped resources to support the goal of engaging families in exploring science together? What resources possessed by the diverse families in your communities can you begin to see as assets that can be activated to foster broader scientific literacy in your community? How can local efforts to engage families in doing science together be spread regionally, nationally, and globally? How can engaging families collaboratively in science support the goal of expanding the science-linked workforce around the globe?

REFERENCES

Banks, J. A., Au, K. H., Ball, A. F., Bell, P., Gordon, E. W., Gutiérrez, K. D., … Zhou, M. (2007). *Learning in and out of school in diverse environments. Life-long, life-wide, life-deep.* Seattle, WA: The LIFE (Learning in Informal and Formal Environments) Center, University of Washington, Stanford University, and SRI International. Retrieved from http://life-slc.org/docs/Banks_etal-LIFE-Diversity-Report.pdf

Bryan, L., & Allexsaht-Snider, M. (2008). Community contexts for understanding nature and naturally occurring events in rural schools in Mexico. *L1 – Educational Studies in Language and Literature, 8*(1), 43–68.

Bryan, L., & McLaughlin, H. J. (2005). Teaching and learning in rural Mexico: A portrait of student responsibility in everyday school life. *Teaching and Teacher Education, 21*, 33–48. doi:10.1016/j.tate.2004.11.004.

Buxton, C., Allexsaht-Snider, M., Hernandez, Y., Aghasaleh, R., Cardozo-Gaibisso, L., & Kirmaci, M. (2016). A design-based model of science teacher professional learning in the LISELL-B project. In A. Oliveira & M. Weinburgh (Eds.), *Science teacher preparation in content-based second language acquisition* (pp. 179-196). New York, NY: Springer.

Buxton, C., Allexsaht-Snider, M., & Rivera, C. (2012). Science, language and families: Constructing a model of steps to college through language-rich science inquiry. In J. Bianchini, V. Atkerson, A. Calebrese Barton, O. Lee, & A. Rodriguez (Eds.), *Moving the equity agenda forward: Equity research, practice and policy in science education* (pp. 241–259). New York, NY: Springer.

González, N., Moll, L. C., & Amanti, C. (2005). Introduction: Theorizing practices. In N. Gonzalez, L. C. Moll, & C. Amanti (Eds.), *Funds of knowledge* (pp. 1–28). New York, NY: Routledge.

National Research Council (NRC). (2012). *A framework for K-12 science education: Practices, crosscutting concepts, and core ideas*. Washington, DC: National Academies Press.

National Science Teachers Association (NSTA). (2009). *NSTA position statement: Parent involvement in science learning* (Adopted by the NSTA Board of Directors, April 2009). Retrieved from http://www.nsta.org/about/positions/parents.aspx

NGSS Lead States. (2013). *Next generation science standards: For states, by states*. Washington, DC: National Academies Press.

Redding, S. (2000). *Parents and learning*. Geneva: UNESCO. Retrieved from http://www.ibe.unesco.org/publications/practices.htm

CHAPTER TWENTY-FIVE

From Opening Portals to Creating New Pathways

LEANNE M. AVERY

The collection of essays in this book is being written in an extraordinary context in which humanity is grappling with unprecedented issues such as global climate change, rampant energy extraction, food sovereignty, increasing poverty, and inequitable human rights to name just a few. Because we are faced with these unprecedented threatening complex problems, now more than ever before, we are compelled to diversify the fields of science and science education in ways that intentionally and purposefully open every portal to alternative ways of knowing in order to address and solve these wicked problems (Kassam & Avery, 2013). Simply put, we must make the semipermeable or "selectively permeable," permeable.

As this chapter illuminates, the local/familial/intergenerational knowledge is plentiful, rich in sociocultural and ecological assemblages, and well documented and researched. Journals such as the *Cultural Studies of Science Education* (CSSE) are contributing to changing the conversation from a positivist approach to a more socioculturally diverse approach to science education. However, being a practitioner at heart, I am of the opinion that *doing* makes things happen. Getting this unique knowledge base through the school door and into the classroom to be utilized and built upon in meaningful ways is the greatest challenge, particularly in this era of standardization. Thus, in answering this section's the question, I *do* the following: describe the educational context from which this question arose; present a growing body of research that clearly demonstrates that the foundations are in place to create pathways for local/familial/intergenerational knowledge to become an essential component in the teaching and learning of science; offer pedagogical strategies, programmatic structures, and alternative methodologies that will aid in shoring up the portals for synergizing sociocultural assemblages and science education; and contribute to the growing movement that recognizes the critical role that diverse knowledges will play in saving the planet.

THE ROLE OF FAMILIES AND INTERGENERATIONAL KNOWLEDGE IN SCIENCE EDUCATION

In reflecting back on my almost 30 years in science education, first as a high school science teacher in a rural school in upstate New York to the past 14 years in higher education (as a doctoral student and then professor), I have observed that families (parents, caregivers) have been allowed to play a minor part in the conversation and given set roles/designated places and spaces more so in the after school curriculum, and to a much lesser extent, within the school curriculum. These spaces range from participating in parent-teacher conferences, PTA and Booster Club organizations, science fairs, open houses, as well as making a guest appearance in the "assess prior knowledge" and/or "family home connection" section of a preservice teacher's lesson plan. Research also continues to explore the role parental involvement plays in diversity and equity in science education (Rodriguez, Collins-Parks, & Garza, 2013). Position statements by science teacher organizations such as the National Association of Science Teachers (NSTA) boldly state:

> The [NSTA] believes the involvement of parents and other caregivers in their children's learning is crucial to their children's interest in and ability to learn science. Research shows that when parents play an active role, their children achieve greater success as learners, regardless of socioeconomic status, ethnic/racial background, or the parents' own levels of education (Henderson & Mapp, 2002; PTA, 1999; Pate & Andrews, 2006). Furthermore, the more intensely parents are involved, the more confident and engaged their children are as learners and the more beneficial the effects on their achievement (Cotton & Wikelund, 2001) … Parents and other caregivers have a critical role to play in encouraging and supporting their children's science learning at home, in school, and throughout their community. Teachers also play an important role in this effort and can be valuable partners with parents in cultivating science learning confidence and skills in school-age youth. (NSTA, 2015, pp. 1–2)

Although deemed commendable by some in various instances and contexts, in my opinion and experience, these set and predetermined roles are restricted, constricted and to some extent, regulated and standardized. Access to the educational system, and to science, technology, engineering and mathematics (STEM) education in particular, continues to be an ongoing area of scrutiny, as we still observe the underrepresentation of diverse people and perspectives all along the STEM pathway. This access, or lack thereof, starts early (Archer et al., 2012), when the interplay of family habitus and capital render science aspirations more believable for some children (usually the middle class) than other children. Also according to Archer et al. (2012), "Social inequalities in the distribution of social capital and differentially classed family habitus combine to produce uneven (classed, racialized) patterns in children's science aspirations and potential future participation" (p. 881). Across almost all data sets, one theme persists—the importance of the role of family in students' science aspirations and achievement.

PARENTAL AND SCHOOL CONNECTION/ INVOLVEMENT: ONE-WAY PATHWAY

What correspondingly seems to be missing from view is both a commitment to, and subsequent action by, the educational system to better create welcoming opportunities where families and communities feel *valued* and therefore inspired and confident to walk through the school door. Alternative ways of knowing (local knowledge, Indigenous knowledge, intergenerational knowledge) and the holders of these forms of knowledge merit validation as critical cogs in the education of our youth. Position statements such as the one offered by NSTA and others seemingly communicate to local communities and families that their role (or perhaps the expectation placed on them by the larger system) lies within a narrow framework of what families and communities "should" be doing *outside* of school so that their children are ready to do what the current standards demand "they do" once they are *inside* school. It seems to be a one-way path (Corbett, 2007).

Approximately 20 years ago, Joyce Epstein (1995) identified six overlapping spheres of influence that relate to the role of parental involvement in schools: (1) parenting, (2) communication (from school to parents), (3) involvement at school (volunteering in school-related organizations), (4) learning at home (helping with homework), (5) decision-making (in external groups), and (6) community collaborations (assisting parents to get involved in community-based organizations). Although helpful, they still fall into the realm of external involvement as a one-way street. More recently, Alberto Rodriguez et al. (2013) posited three new categories to "draw attention to the importance of connecting the school curriculum to the students' everyday lives, the realities of the community in which they live, and their own sense of agency" (p. 56). Action possibilities include making teaching and the curriculum multiculturally and socially relevant. Such an approach is supported by research on teaching and learning, which distinctly reveals that when the science curriculum is more connected to the students' local knowledge, everyday lives and culture, learners are more engaged and performed better academically

(Avery & Kassam, 2011). Furthermore, Okhee Lee and Aurolyn Luykx (2007) recommend ways in which science education researchers might aim to better identify and better connect children's cultural resources and experiences gained outside of school to classroom science:

> Future research should aim to identify those cultural and linguistic experiences that can serve as intellectual resources for science learning, as well as those beliefs and practices that may be discontinuous with the specific demands of science disciplines. This will require a balanced view of nonmainstream students' intellectual resources as well as of the challenges they face in learning science. An expanded knowledge base around students' non-school experiences related to science could offer a stronger foundation for science curriculum and instruction. Students of all backgrounds should be provided with academically challenging learning opportunities that allow them to explore and construct meanings based on their own linguistic and cultural experiences. (p. 30)

Also, assigning meaningful socially and culturally relevant homework can allow students (and their parents) to appreciate the authentic benefits of science knowledge and how that knowledge interconnects with their own cultural practices. Rodriguez and colleagues explore the role of student agency has on science achievement (2013). They define student agency as "an individual's ability to affect change to improve his or her own conditions" and they postulate that a "student with a strong sense of agency would be expected to have higher educational aspirations and a stronger self-concept—two important attributes that have been associated with higher student achievement and participation at school" (p. 57).

These offerings provide schools with essential steps to help them to become a more welcoming place for families who may be less involved in school-related activities, not because they do not value their children's education, but because they are intimidated by language and/or cultural obstacles as well as by their own feelings of inadequacy in subjects such as science (Kaya & Lundeen, 2010). Kenneth Tobin (2015) elaborates on this point in stating, "Mainstream perspectives of what counts as science have marginalized many discourses, which are often regarded as nonscientific (e.g., Indigenous knowledges). In this way, the domain of science education can be impoverished and may not connect with everyday life experiences of important stakeholders" (p. 4). Others claim that the current national agenda of standardization undermines cognitive diversity of rural American society (Kassam, Avery, & Ruelle, 2017). Uncritical standardization may ignore the experiential knowledge that children and families bring to the classroom, and this ensures a *pedagogy of erasure* as classroom practice (Eppley, 2011). Some would go as far as to say that "this top-down [highly bureaucratic and political] approach also serves to separate young people from communities and is simultaneously incompatible with thinking" (Cornett, 2015, p. 11).

The aforementioned new categories have in common a purposeful commitment to connect families to schools and schools to families by mooring the science within the foundation of children's and families' everyday lives and cultural experiences rather than the other way around. This begs the question, why then do the barriers still persist? In what follows, I propose that the loophole lies in the lack of purposeful and explicit action, not in the lack of knowledge. Three overarching explicit courses of action are discussed and within each course, exemplar programs or projects are presented to illustrate the various ways in which families can play a critical role in science education.

CREATING MULTIPLE PATHWAYS: MAKING THE SEMIPERMEABLE, PERMEABLE

Construction is involved in paving a two-way street. It takes knowledge, planning, blueprints, imagination, perseverance, dedication, deconstruction, and above all, purposeful action. Three particular courses of action are presented next—that when employed in parallel—have the net effect of making semipermeable boundaries, permeable: (1) utilizing methodologies that bring alternative ways of knowing explicitly into the science curriculum, (2) employing a place-based pedagogical approach to science education, and (3) implementing innovative programmatic structures and strategies in K-16 science and science teacher education that purposefully connect families to science education.

Drawing on case studies of children's local knowledge, community elders' local knowledge, Indigenous and non-Indigenous intergenerational knowledge, I demonstrate the value, diversity, and necessity of local and intergenerational knowledge of science and engineering in rural contexts. I also discuss implications for urban and suburban environments via innovative research methodologies that honor context and transcend content. The rich case studies illuminate the many ways in which the semipermeable can be made permeable by purposefully reinvigorating deep rooted practices and knowledges to contextualize educational experiences for *all* children and enhance critical place-based pedagogies that focus on structuring the teaching and learning of STEM around local history, culture, economy and the environment. Preserving and/or reinvigorating intergenerational knowledge in school science and within teacher education programmatic structures, offers a unique way to connect local communities to school and local places to global science, thereby enriching children's science learning.

Utilizing Methodologies that Explicitly Connect Families and Communities to School Science. A growing body of research unveils several contexts where instances of situated learning can be used as portals to explicitly connect families and communities to school science. This research includes investigations into the ways in which rural children learn from families in places and spaces outside of school; alternative ways of learning and knowing that have been passed down from generation to generation through traditions of oral knowledge and practices/apprenticeships inherent in everyday rural life; and human ecological studies with Indigenous communities that reveal how knowledge is complex in connectivity between beings and their habitats. This body of research clearly shows that the foundation is in place for anchoring school science within the foundation of children's knowledge that has been gleaned from home, family, habitat and community.

Using explicit methodologies that engage families directly in the teaching and learning of science offers clear pathways for local knowledge to enter classroom science. Methodologies such as photodocumentation and Heuristic Inquiry (Hains, 2007) connect the learner with original sources of knowledge in real and meaningful ways. The learner becomes the conduit by which local knowledge enters the school curriculum. In working with rural children in upstate New York, my colleague Karim-Aly Kassam and I (Avery & Kassam, 2011) provided children with cameras and invited them to take photographs of what they perceived as science and/or engineering taking place in their everyday lives and communities. We interviewed the children about their photographs and asked them to describe why they took the pictures that they did, what they knew about what is depicted in the photos, and how they knew what they knew regarding what was shown in the photos. This methodology of photodocumentation enabled children to capture and discuss what they observed and/or learned about science and engineering taking place in their habitats. The children learned by observing or doing or both, and they learned from family members, particularly parents and grandparents. By paying close attention to the children's discussion and engaging in conversation with them, we also learned that they intuitively and intellectually recognized that cultural and social processes are connected to their habitat, that they themselves are dependent on habitat, and that they are active participants within their communities and broader ecosystems (Kassam & Avery, 2013).

What does this mean in terms of the role of families in science education? The study not only reveals how and where children learn, it also illuminates just how strong the first-hand learning connection is between parents and children. Capturing and documenting these connections provides insights for teacher educators on how, where, why and when children learn as well as providing a concrete method by which they can directly participate in and contribute to their own science learning. The teacher has the opportunity to help connect or map school science to the child's local knowledge in tangible and durable ways. For example, creating cognitive mapping structures (Bischoff, Avery, Feldt-Golden,

& French, 2010) is just one way that teachers might assist students in connecting what they have learned at home from families to what they are learning in school. In addition, cognitive mapping illuminates the diverse pathways of learning exhibited by children as they assimilate new concepts and revise and enhance existing schema. Purposefully identifying personal, cultural and community assets of alternative ways of learning and knowing as revealed by these particular methodologies provides tangible moorings by which knowledge structures can be meaningfully attached and enriched.

Recently, another colleague, Bryan Hains and I explored the complex nature of rural tenets that were shared with us by community elders, and we examined the historic scientific roots of these sayings (Avery & Hains, 2017). In this research, we used a Heuristic Inquiry approach whereby interviewers co-collaborate with interviewees in conversation around local knowledge and how it is transmitted via oral traditions and/or practices. We uncovered a wealth of information regarding the diverse rural sociocultural and ecological connections and the situated macro and micro-contexts from which these tenets arise. These tenets, many steeped in science, both augment and differ from traditional western science and science curricula by illuminating the ways in which oral traditions are embedded in place, people, and culture. The complex connectivity of the knowledge held by elders in memories, stories and places are rich in cultural and ecological principles.

How does intergenerational knowledge intersect with the role families might play in science education? Interviewing community elders is another method in which school children might engage as part of their science, social studies, or civics coursework. In so doing, children can make intergenerational connections to family or community members and trace historical roots of knowledge as well as exploring the sociocultural underpinnings of various STEM disciplines using the framework of the sociology of scientific knowledge as one approach (Meyer & Avery, 2010). This interdisciplinary approach enables the classroom teacher (and students) to interconnect various content areas around meaningful contexts that might include for example, nurturing an earth stewardship perspective within science education and the broader community. According to Cornett (2015), "young people learn much better when they are active participants in things that matter in their communities" (p. 6).

Elsewhere, other researchers have engaged in similar projects that involved interviewing community elders about local history and ecology (Cornett, 2015) as well as working with Indigenous communities in mapping their sacred sites and hunting resources (Kassam, 2009). Also, Morgan Ruelle and Karim-Aly Kassam (2013) examined the diversity of plant knowledge found among elders who live in the Standing Rock Nation in the Great Plains region. Ruelle and Kassam found that the elders' seasonal knowledge about the plants might inevitably aid their community's adaptation to climate change. These collaborations facilitated intergenerational communication and intercultural understanding. Documenting knowledge and/or resources through oral interviews and community mapping provided models for youth to write their own family histories, and resources that enhance school science. Mapping in particular brought together diverse community members around a central theme of resource preservation that provided interactions around shared knowledges that also served to strengthen ties within the community as well as between the community and school.

Photodocumentation, Heuristic Inquiry, and community mapping are exemplar methodologies that involve processes and products that ultimately illuminate the powerful connections between what is learned outside of school to what is learned inside of school. When these methodologies are holistically implemented in classrooms in ways that connect children's local and familial knowledge to school science, children's access to and success in science is enhanced (Hine, Podsiedlik, Avery, & Herman, 2015). The projects presented here created strong and vibrant synergies between families, communities, and schools. Culturally significant products are co-generated (traditional western science partnered with local and Indigenous knowledge) and utilized within communities and schools in tangible ways. Thus, the work has meaning because it often contributes to helping solve community issues that impact families and schools. Furthermore, the overall product strengthens ties and lays the groundwork for future endeavors. The next step—in some ways the most challenging—is building durable and sustainable portals throughout the educational fortress so that these types of synergies can gain traction and grow beyond a one-time event. This portal construction involves commitment across multiple levels: students and families, teachers, administrators, and policy makers.

Employing a Place-Based Pedagogical Approach to Science Education. Securing successful portal construction rests upon laying a strong theoretical foundation from which to frame subsequent infrastructures of support. Place-Based Education (PBE) offers pedagogical strategies and programmatic structures aimed at creating, sustaining and growing pathways and portals to secure a vibrant role for families and intergenerational knowledge in science education (both in K-12 schools and in teacher education).

Place-based education focuses on structuring learning around local history, culture, language, environment, and economy:

> PBE uses the local environment as a starting point to teach [numerous] subjects. Emphasizing hands-on, real-world learning experiences, this approach increases academic achievement, strengthens students' ties to their community, enhances students' appreciation for the natural world, and creates a heightened commitment to serving as contributing citizens. (Sobel, 2005, p. 7)

Gregory Smith (2002) states that PBE: a) emerges from a specific place and includes indigenous cultural and nature studies; b) is multidisciplinary and experiential; c) includes local internships/entrepreneurial opportunities; d) connects students with the community involving them in decision making and real-world problem solving; and e) reflects a much wider-ranging learning paradigm than simply learning to take a test. The legitimization of local knowledges through place-conscious pedagogies bridges the gaps between children's local/familial/intergenerational knowledge and school and global science.

Building on this work of place-based educators, David Gruenewald advocates for a "critical pedagogy of place" in educational settings (2003). According to Gruenewald:

> A critical pedagogy … explicitly makes the limits and simulations of the classroom problematic. It insists that students and teachers actually experience and interrogate the places outside of school—as part of the school curriculum—that are the local context of shared cultural politics … Acknowledging that experience as a geographical context opens the way to admitting critical social and ecological concerns into one's understanding of place, and the role of places in education. (p. 9)

A critical pedagogy of place provides a framework to anchor and empower local/familial/intergenerational knowledge by moving the discussion from teachers' attempts to activate prior knowledge through a cursory introduction to the day's lesson to anchoring the entire curriculum within the children's experience. A critical approach gives power as legitimate learning spaces to places outside of school and voice to alternative ways of knowing. The PBE framework brings theory and practice together in ways that undergird key teaching strategies that enhance access to science in rural, urban and suburban communities.

Programmatic Structures: Making Institutional Pathways and Portals. Critical place-based practices need to get traction early on within teachers' careers so that the role of families in science education is on the forefront, and practices that ingrain this role within school systems become habitualized. Several essential steps are needed in order to create an infrastructure for changing science teacher practice from a more standardized approach to a place-based approach. First, cultivate teacher place-based pre-and inservice professional development programs that at the core are committed to changing the way science is taught by valuing and utilizing local knowledge. Second, scaffold content and pedagogy across grade levels and connect local knowledge with educational standards, which can enhance students' engagement in science. Third, provide preservice teachers and students with concrete experiences that show them how their local/familial/intergenerational knowledge connects with the broader science community, which can inspire children's science interests and pursuits. (I cannot underscore enough the significance of implementing strategies that are extraordinarily explicit in nature, and in practice. I find that strategies and pedagogical practices that are vacant of explicit procedures ultimately go unrecognized and unfulfilled.) Fourth, by establishing clinically-rich preservice teacher education programs where semester to full year college education courses are situated in regional partnering schools, preservice teachers are able to become full members of the school community, they become emotionally invested in the school, they truly get to know their students and cooperating teachers, and they have multiple opportunities to make meaningful and powerful connections between families, home, school, community, content, and context. What is more, having successful teacher professional development programs that are situated in contexts where local and Indigenous knowledge is legitimized leads to stronger synergies and outcomes (Boyer, 2006).

COURSES OF ACTION *IN ACTION*

There are a variety of ongoing projects taking place both nationally and internationally in teacher education where preservice teachers are learning pedagogies and practices explicitly geared at connecting families to school science education. These projects incorporate several, if not all, of the essential steps listed above. In my own practice, I implement two interdisciplinary projects that are situated within my science methods courses that are part of clinically-rich partnerships between local rural schools and the college. In the STEM-based Learning Center Project, my science methods preservice teachers utilize photodocumentation to create interactive interdisciplinary learning centers that engage children with science and/or engineering connected to everyday life. In so doing, preservice teachers experience first-hand, that teaching science using the content standards need not be top-down, but in fact, can be born from a grassroots initiative grounded in their own local knowledge and/or interest in science. The Learning Center Project is therefore rooted in place-based education pedagogical practice and serves as a teaching tool that preservice teachers can use in their own future classrooms as well as an experiential learning activity, which their future students can also create. The Learning Center Project deliberately connects the science in everyday life (often rich in sociocultural assemblages) to global science to help school children traverse the path from concrete science concepts (contextualized) to more abstract science concepts (decontextualized or which might be currently absent in their lifeworld experiences).

In a further effort to provide additional explicit place-based experiences that deliberately unveil the role families can play in school science, my science methods preservice teachers also engage in the Role of Families in Science Education Project. Using the Heuristic Inquiry approach, preservice teachers interview a family member about an instance where they learned about something related to science and/or engineering (purposely very broad in nature) from that family member, write a reflection commentary, create a culminating creative project that captures their experience, and they then share their project with their peers and with the school children in our clinically-rich classrooms.

Some of the emergent themes that evolved from these two projects included: baking, cooking, building, hunting and fishing, learning to drive a car, learning to ride a bike, gardening, visiting parents' places of work, building/engineering something with a family member, and family farming. Several preservice teachers also have shared that their experience of looking through old photographs with family members was heartwarming as were the shared stories that emerged in so doing. One preservice teacher in particular commented on the "power of pictures." Some insightful quotes (recurring themes) that came from the presentations include:

- "Science is a tradition and our parents are our first teachers" [several preservice teachers shared this insight]
- "Science is hands-on—you need to experience it yourself" [J on learning to hunt with his father and brother]
- "Science comes from families, from generation to generation. It's creative, messy and you learn what works and what doesn't" [S on cooking with her mother]
- "Family is the first place that we learn." [D on interviewing her father about cooking Italian food for the family]
- "Hunting is in your genes—you are born into it. Our family has been hunting for five generations and over time, we notice how animal populations change—this is science"! [A reflecting on her family's history of hunting]
- "We now see what the children bring to the classroom" [A on making connections to teaching science]

As embodied in the sample quotes listed above, preservice teachers were able to concretely make connections between what they had learned from their families outside of school to what is within the science curriculum inside school. They also gained new perspectives on what they experienced at home as *being* science—they observed science in everyday things—and witnessed how this has the potential to enrich children's learning in school science (as well as their own!).

In order for preservice teachers to appreciate/value children's local knowledge and the role families and intergenerational knowledge can play in science education, it is critical that they experience it first hand and make the connections for themselves. Like children, they too, learn by seeing and/or doing. For example, in one of our clinically-rich classrooms, we used the Role of Families in Science Education Project to amplify the growing interest of participating teachers and students in engaging in the *National Public Radio's*

Story Corps Project: The Great Thanksgiving Day Listen (https://storycorps.me/about/the-great-thanksgiving-listen/). The preservice teachers taught the grades 5 and 6 children interviewing techniques that they learned from the Role of Families in Science Education Project to help the children prepare for their interviews with family members on Thanksgiving weekend. This exemplifies gaining traction—putting pedagogical strategies into practice in meaningful ways that serve to secure the connection between families and schools.

Elsewhere, the *Will of the Ancestors* (Ayuluk et al., 2015) project set in Anchorage Alaska brought together community elders, Indigenous inservice and preservice teachers from the University of Alaska, parents, and elementary students to develop a Science, Technology, Engineering, Arts, and Mathematics (STEAM)-based curriculum project that included a local atlas of plants and animals, a language revitalization children's book, and an associated community-based writing workshop. This particular program brought together key stakeholders around a community concern of preserving and revitalizing native language that resulted in the creation of an innovative interdisciplinary curriculum that became an integral component of the overall school curriculum. The pairing of these two curricula enriched the role of families and communities in the science education program.

In her Hawaiian/Indigenous Science teacher development program, Pauline Chinn (2011) "incorporates Indigenous, sustainable practices and personal, place-based activities for teachers that enable them to create a conceptual framework for transforming mainstream science curricula into meaningful place and problem-based curricula relevant to local places" (p. 75). By intentionally embedding Hawaiian Indigenous knowledge and practices directly within the science curriculum, teachers were able to readily implement culturally-rich perspectives that diversified traditional school science in ways that honored local ways of knowing and doing.

In their place-conscious preservice teacher education program in Australia, teacher educators Simone White and Jo-Anne Reid (2008) focus on rural pedagogies that center on place-conscious education and student teaching placements in rural communities that position student teachers as "co-researchers in documenting the knowledge and social practices that constitute the children's homes, lives" (p. 7) and communities. They state,

> [Place-based approaches] build a place-conscious teacher subject—and that a teacher's consciousness of place in devising and planning learning experience brings about particular sorts of curriculum. As teachers come to know, and know about, a particular rural place, and come to understand its relationships to, and with other places, they are developing knowledge, sensitivities, awareness, skills, attitudes, and abilities that will allow them to feel more at home and more powerful in a rural setting. (p. 6)

Rural economies, communities, and teacher education programs are purposefully linked enabling preservice teachers to recognize the importance of place, and to value the lives of their rural students. White and another colleague, Jodie Kline (2012) have gone further to develop a rural teacher curriculum package to support teacher educators in preparing their future teachers for the challenges and opportunities of teaching in rural communities. By availing themselves of the sociocultural richness of local rural places, communities and people, these PBE programs and resources thoroughly assimilated families in science education.

Several take-home messages emerge from these projects and studies. First and foremost, value is placed on alternative ways of knowing. In order for families to *have* a role in science education, their role must be *valued*. Value begets action. Concrete infrastructures, place-based practices, numerous resources and critical strategies of pedagogical awareness when made available and put in place, result in families becoming an integral part of science education in meaningful and powerful ways. Second, a logistical programmatic (but essential) commonality among the teacher education programs presented here is that the preservice teachers (and subsequently their students) are provided with concrete experiences of how their local/familial/intergenerational knowledge connects with the broader science community, which can inspire their own motivation in the teaching and learning of science (Avery, 2013). Both explicit and implicit instruction, and more the former than the latter, is key in elementary preservice science teacher education in order to facilitate the teachers' proclivities for connecting local science/community to global science/community (Avery & Meyer, 2012). Third, when these aforementioned structures and strategies are in situ, familial knowledge and expertise has a real portal of entry and as several of these examples demonstrate, that expertise and knowledge becomes formalized within the school curriculum. In order for the circle to be completed—*sustaining* portals so that families can have a role in science education—PBE programmatic structures must also be put in place in K-12 settings, too. Programmatic examples that focus on K-12 students and settings are presented next.

Projects and procedures that link local communities to K-12 school settings take many forms. These may vary in size and scope and can range from community-school gardens centered around a small rural school to larger urban gardening models that include multiple participating neighborhoods (Krasny & Tidball, 2016). Other examples may include school-based partnerships not only with neighboring communities but also in tandem with local and national organizations as well as state-level agencies such as State Departments of Environmental Conservation, United States Department of Agriculture, among others. Key teacher leaders often play a crucial role in spurring interest in these collaborative projects and then via strategic planning with other school and community leaders, they gain traction, momentum and are able to move these efforts forward. These efforts also require educators and administrators to expand their conceptualization of curricula beyond traditional forms that include textbooks, science kits and other standardized classroom materials to encouraging students to interrogate places, spaces, and knowledges outside of school.

For example, one unique model that encompasses multiple levels of organizational partnership is the St. Marys Area Middle School Environmental Center. Set in rural Pennsylvania, St. Marys created an Environmental Center on their campus to serve students, teachers, families, and local and regional communities. The Center was a collaborative effort among teachers, students, families, community members, master gardeners, local and state agencies, wildlife organizations, sportsmen/women clubs, and other local and regional partners. The Center has a trout nursery, outdoor classroom, greenhouse, community gardens, and nature trail. Through teachers' use of place-conscious pedagogies, in partnership with families, community members, and content specialists, students learn about their local environment, use their local and intergenerational knowledge, and connect what they have learned to state and national STEM standards (Avery, 2013).

The *Local Fisheries Knowledge Project* (Bartsch, 2008) situated in the state of Maine enables high school students to better understand required content by engaging with local resources. High school students interviewed local fishermen and then documented and preserved the fishermen's knowledge and experiences for future generations. Students explored the connections between the local fisheries, local oceanography, their local communities and their own lives. Participating teachers integrated this project into their curriculum, which is aligned with many of the state of Maine's learning standards. Engaging with their surroundings and exploring environmental issues that impact their community (Duvall & Zint, 2007), spurs both teachers' and students' learning of science content while employing their local knowledge skills. Teachers' willingness to explore curricula beyond the classroom door enables their students to investigate the role of place and community in their own science education thus widening the sphere of content knowledge to include local expertise and intergenerational knowledge.

The *Cornell Summer Snapshot Series for Educators* is a teacher professional development program that models engaging, place-based instruction and design-based experiences focused on STEM curricula both in and outside of school settings (Hine et al., 2015). The *Summer Snapshot Series* was initiated by a team of college faculty, outreach educators and rural teacher leaders who were interested in providing STEM experiences for rural youth. The program provides teams of teachers with time for collaborative lesson preparation along with the opportunity to practice implementing these lessons during the summer as well as gathering authentic feedback from students, families, community members and program partners in order to make appropriate revisions for school-year implementation. This program takes place at Cornell University during the summer and includes a teacher component, a student component and a capstone event for participants' families. Over the years, program leaders have identified two strategies that are key to providing meaningful learning opportunities by legitimizing students' values, enhancing student motivation and fostering engagement in STEM activities with rural students: Place-based Education and Engineering Design-based experiences. A third feature of the program involves connecting students to the STEM pipeline, accomplished by hosting summer learning experiences on the campus of a major research university where STEM role models and artifacts are made available to these youths. Collectively, these three programmatic strategies draw explicit connections between local knowledge and families, motivational and attitudinal constructs, and student agency in STEM.

In each of the examples presented here, families and communities ultimately had a role in the science education program because explicit portals were created and sustained that enabled alternative ways of knowing to be brought into the school science context and curriculum. The various levels of commitment across teachers, students, administrators, districts and university/college partners as well as families and communities are all equally vital to programmatic success. Each participant recognized the value in and impact of connecting local people, local knowledge and community context to school and national curricula and agendas. Actions were taken that secured a place for local and Indigenous knowledge, families, and community within the curriculum because multiple parties were willing to extend their conceptualizations (and boundaries) of what curricula is, and how it can be implemented both in and outside the classroom door.

The three courses of action presented here-valuing and utilizing alternative ways of knowing, undergirding and framing programmatic structures within a strong theoretical plan and creating, growing and sustaining programmatic structures that incorporate families and communities as integral components of science education—also honor and transcend context. Although the presentations and discussions presented in this chapter are situated in science education, they have ingredients and processes that can be ported to other contexts and content. The courses of action involve people interrelating with each other and with their habitats. The methodologies, pedagogies and programs also illuminate the complex interconnections between people, practice, places, stories, culture, knowledge and environments. New and old types of knowledge surface, subtle nuances are unveiled, powerful connections are made, and new spaces for creative activity, conversations, and collaborations emerge—all of which provide fertile possibilities and pathways for entry into school science and beyond.

MOVING FORWARD: FROM OPENING PORTALS TO CREATING NEW PATHWAYS

Very recently, in a guest editorial in the Ecological Society of America's prestigious science journal, *Frontiers in Ecology*, several renowned and diverse scholars called for the implementation of the "earth stewardship" into the practice of science in addition to change the role of science in society (Sayre et al., 2013). It is quite unusual for an editorial of this nature to appear in a historically positivistic journal. Moreover, in this editorial, the team of authors made several cutting-edge and compelling recommendations to the scientific community:

> (1) Expand the concept of stewardship to encompass all lands and waters: urban as well as rural; the open oceans as well as lakes, rivers, and coasts; and areas actively managed as well as those set aside from direct human exploitation. (2) Work alongside various stakeholders—policy makers, professionals, employers and employees, educators, artists, clergy and laypeople, young and old—to envision and realize a future that is more secure, sustainable, and just. (3) Include people not only as variables affecting ecosystems but also as participants in those ecosystems and in the practice of science itself. (4) Acknowledge that social justice and environmental health are not separate or separable concerns, but are interconnected. (5) Understand processes and practices at multiple scales, interacting in complex and often non-linear ways. (6) Value a diversity of "ways of knowing," including local and indigenous knowledge, about landscapes and natural resources, and (7) Embrace stewardship as an ethical guide to the practice of science. (p. 339)

In so *doing*, these scholars are changing both the conversation and the practice of science, and thereby, science education. Local, familial, intergenerational Indigenous and non-Indigenous knowledges are now located at the forefront, and are deemed necessary ingredients for tackling unprecedented issues of global climate change, food sovereignty, diminishing fresh water resources and the like. People as well as other beings are viewed as participants, not variables. Linearity is expanded to multiple pathways. Ethics becomes an essential priority in guiding scientific practice. The time has come for educational institutions to follow suit.

Indeed, several scholars in the field of science education are asking timely and provocative questions that are very much in concert with the work of finding and opening new pathways and changing past practices (Long & Avery, 2017). The common themes among these new efforts include: employing decolonizing methodologies in science education research (Abrams et al., 2014); exploring issues of social and environmental justice in school science (Lowen-Trudeau, 2017), and the role administrative leadership might play in debunking deficit perspectives and changing school policies and practices (Eppley & Shannon, in press); revisiting and reinvigorating critical pedagogies of place in science education (Huffling, Carlone, & Benavides, 2017); and utilizing theoretical underpinnings from the field of Appalachian Studies to advance a critical science education perspective (Kingsolver, 2011, 2017). Thus, moving forward, these upcoming lines of scholarship will continue to expand the necessary vibrant conversations and actions in ways that expand portals for alternative ways of knowing that include families, communities and Indigenous and non-Indigenous intergenerational knowledge. These scholars are undoubtedly transforming the science education into a very real promise of pluralism (Avery & Hains, 2017) and one that offers new places and spaces for innovations to take root and prosper.

REFERENCES

Abrams, E., Yore, L. D., Bang, M., Brayboy, B., Castagno, A., Kidman, J., … Yen, C.-F. (2014). Scientific literacy for all: Culturally relevant schooling for indigenous learners. In N. G. Lederman & S. K. Abell (Eds.), *Handbook of research in science education, Volume II* (pp. 671–696). New York, NY & Oxon: Routledge.

Archer, L., DeWitt, J., Osborne, J., Dillon, J., Willis, B., & Wong, B. (2012). Science aspirations, capital, and family habitus: How families shape children's engagement and identification with science. *American Educational Research Journal, 49*, 881–908. doi:10.3102/0002831211433290.

Avery, L. M. (2013). Rural science education: Valuing local knowledge. *Theory into Practice, 52*(1), 28–35. doi:10.1080/07351690.2013.743769.

Avery, L. M., & Hains, B. J. (2017). Oral traditions: A contextual framework for complex science concepts. *Cultural Studies of Science Education, 12*, 129–166. doi: 10.1007/s11422-016-9761-5.

Avery, L. M., & Kassam, K.-A. (2011). Phronesis: Children's local rural knowledge of science and engineering. *Journal of Research in Rural Education, 26*(2), 1–18.

Avery, L. M., & Meyer, D. Z. (2012). Teaching science as science is practiced: Opportunities and limits for enhancing preservice elementary teachers' self-efficacy for science and science teaching. *School Science & Mathematics, 112*, 395–409. doi:10.1111/j.19498594.2012.00159.x.

Ayuluk, F., Ayuluk, J., Friday-Tall, S., Matchian, M. T., Tulim, P., Olson, L., Unin, L., Shields, A. J., Coulter, C., Ohle, K., & Ortega, I. (2015). Sustaining place, language, and culture together. *Science Education & Civic Engagement: An International Journal,* 7(2), 69-77. Retrieved from http://sencer.net/wp content/uploads/2016/12/seceij_summer_2015_issue.v4.pdf

Bartsch, J. (2008). Youth as resources in revitalizing communities. In D. A. Gruenewald and G. A. Smith (Eds.), *Place-based education in the global age: Local diversity*, pp. 65-83. New York: Lawrence Erlbaum.

Bischoff, P. J., Avery, L. M., Feldt-Golden, C., & French, P. (2010). An analysis of knowledge structure, diversity and diagnostic capabilities among pre-service science teachers within the domain of oxidation and reduction chemistry. *Journal of Science Teacher Education, 21*, 411–429. doi:10.1007/s10972-010-9188-x.

Boyer, P. (2006). *Building community: Reforming math and science education in rural schools.* Retrieved June 11, 2011 from www.ankn.uaf.edu/publications/building_community.pdf

Chinn, P. (2011). *Developing a sense of place and an environmental ethic: A transformative role for Hawaiian/Indigenous science in teacher education?* Retrieved July 10, 2010 from jan.ucc.nau.edu/~jar/HOH/HOH-5.pdf

Corbett, M. (2007). *Learning to leave: The irony of schooling in a coastal community.* Black Point, NS: Fernwood Publishing.

Cornett, B. (2015). *Reclaiming public education: Common sense approaches.* Dayton, OH: Kettering Foundation.

Duvall, J., & Zint, M. (2007). A review of research on the effectiveness of environmental education in promoting intergenerational learning. *The Journal of Environmental Education, 38*(4), 14–24. doi:10.3200/JOEE.38.4.14-24.

Eppley, K. (2011). Reading mastery as pedagogy of erasure. *Journal of Research in Rural Education, 26*(13), 1–5.

Eppley, K., & Shannon, P. (in press). Literacy education for the lumps and divots of smart cities and rural places. In S. Williams & A. Grooms (Eds.), *The politics of educational opportunity in rural contexts.* Scottsdale, AZ: Information Age Press.

Epstein, J. (1995). School/family/community partnerships: Caring for the children we share. *Phi Delta Kappan, 76*, 701–712. doi: 10.1177/003172171009200326.

Gruenewald, D. (2003). The best of both worlds: A critical pedagogy of place. *Educational Researcher, 32*(4), 3–12. doi:10.3102/0013189X032004003.

Hains, B. J. (2007). *The dilemma of student downshifting: Pedagogical practices which influence downshifting in high school agricultural science programs* (Unpublished doctoral dissertation). Purdue University, West Lafayette, IN.

Hine, L. K., Podsiedlik, M., Avery, L. M., & Herman, E. (2015). Summer science snapshot: A developing partnership model to spark interest in STEM among rural learners. *Excelsior, 10*(1), 4-20.

Huffling, L. D., Carlone, H., & Benavides, A. (2017). Re-inhabiting place in contemporary rural communities: Moving toward a critical pedagogy of place. *Cultural Studies of Science Education, 12, 33–43.* doi: 10.1007/s11422-016-9756 2.

Kassam, K.-A. (2009). *Biocultural diversity and indigenous ways of knowing: Human ecology in the Arctic.* Calgary, AB: University of Calgary Press/Arctic Institute of North America.

Kassam, K.-A., & Avery, L. M. (2013). The Oikos of rural children: A lesson for the adults in experiential education. *Journal of Sustainability Education, 5*, 1–17.

Kassam, K.-A., Avery, L. M., & Ruelle, M. (2017). The cognitive relevance of Indigenous and rural: Why is it critical to survival? *Cultural Studies of Science Education, 12*, 97-118. doi: 10.1007/s11422-016-9745-5.

Kaya, S., & Lundeen, C. (2010). Capturing parents' individual and institutional interest toward involvement in science education. *Journal of Science Teacher Education, 21*, 825–841. doi:10.1007/s10972-009-9173-4.

Kingsolver, A. (2011). Crossing county lines in Appalachia. *Anthropologies.* Retrieved July 8, 2015 from http://www.anthropologiesproject.org/2011/11/crossing-county-lines-in- appalachia.html

Kingsolver, A. (2017). Practical resources for critical science education in rural Appalachia. *Cultural Studies of Science Education, 12,* 219-225. doi: 10.1007/s11422-016-9755-3

Krasny, M. E., & Tidball, K. G. (2016). *Garden Mosaics Project.* Retrieved May 22, 2016 from http://communitygardennews.org/gardenmosaics/

Lee, O., & Luykx, A. (2007). Science education and student diversity: Race/ethnicity, language, culture, and socioeconomic status. In S. K. Abell & N. G. Lederman (Eds.), *Handbook of research in science education* (2nd ed., pp. 171–197). Mahwah, NJ: Lawrence Erlbaum Associates.

Long, D., & Avery, L. M. (Eds.) (2017). Cultural studies of rural science education. *Cultural Studies of Science Education, 12*, 1-5. doi: 10.1007/s11422-016-9744-6.

Lowen-Trudeau, G. (2017). Gateway to understanding: Indigenous ecological activism & education in urban, rural, & remote contexts. *Cultural Studies of Science Education, 12*, 119-128. doi: 10.1007/s11422-016-9746-4.

Meyer, D. Z., & Avery, L. A. (2010). A third use of sociology of scientific knowledge: A lens for studying teacher practice. *Studies in Science Education, 46*(2), 153–178. doi:10.1080/03057267.2010.504546.

National Science Teachers Association (NSTA). (2015). *Parent involvement in science learning.* Retrieved October 13, 2015 from http://www.nsta.org/docs/PositionStatement_ParentInvolvement.pdf

Rodriguez, A. J., Collins-Parks, T., & Garza, J. (2013). Interpreting research on parent involvement and connecting it to the science classroom. *Theory into Practice, 52*(1), 51–58. doi:10.1080/07351690.2013.743775.

Ruelle, M. L., & Kassam, K. (2013). Foodways transmission in the Standing Rock Nation. *Food and Foodways, 21*, 315–339. doi:10.1080/07409710.2013.850007.

Sayre, N., Kelty, R., Simmons, M., Clayton, S., Kassam, K.-A., Pickett, S., & Chapin, F. (2013). Invitation to earth stewardship. *Frontiers in Ecology, 11*, 339–339. doi:10.1890/1540-9295-11.7.339.

Smith, G. (2002). Place-based education: Learning to be where we are. *Phi Delta Kappan, 83*, 584–594. doi:10.1177/003172170208300806.

Sobel, D. (2005). *Place-based education: Connecting classrooms & communities.* Great Barrington, MA: Orion Society.

Tobin, K. (2015). The sociocultural turn: Beyond theoretical imperialism and the imperative of learning from difference. In C. Milne, K. Tobin, & D. DeGennaro (Eds.), *Sociocultural studies and implications for science education* (pp. 3–31). Dordrecht: Springer.

White, S., & Kline, J. (2012). Developing a rural teacher education curriculum package. *Rural Educator, 33*, 36–43.

White, S., & Reid, J. (2008). Placing teachers? Sustaining rural schooling through place-consciousness in teacher education. *Journal of Research in Rural Education, 23*, 1–11.

SECTION TEN

Culture of Science and Science Education

In What Way Does the Culture of Science and Images of Science Affect the Process of Science Education?

CHAPTER TWENTY-SIX

Forming Bonds—Breaking Bonds

ANITA HUSSÉNIUS

POINT OF DEPARTURE—FORMING BONDS

I had my first experience in studying chemistry in 8th grade at the age of 14. It was love at first sight. I had a strong feeling that this was "my subject," this was what I wanted to dedicate my life to. Afterwards, I have often wondered what it was that gave me that feeling, why I felt such a strong "identification" with, and to, chemistry. There is no simple answer, but I can see some important factors. My childhood was not an uncomplicated road towards adult life and by the age of ten, I had assumed several adult responsibilities. No one in my family or among my relatives had a higher education—we were working class. My desire to attain a different and higher status than the one my family held became a major driving force. The high status of science, the image of science as being abstract and difficult attracted me. And I have to admit, although it might be a retrospective rationalization, that the fact that most of my contemporaries had difficulties in understanding the concepts, the formulas, the reactions and did not like chemistry at all, contributed to making me special. A subject with an aura of demanding high intelligence, a subject for smart students, gave it attractiveness and empowered me. Thus, I saw this knowledge domain as a pathway to a better life. And yes, in that regard it was a fortunate choice.

I have been fascinated and dazzled by the molecular world. Being able to understand, explain and predict phenomena in the visible macro level, based on knowledge of how atoms and molecules interact with each other, have given me a sense of self, a firm foundation to rest on, an identity. How many times have I introduced myself as "I am a chemist", with strong emphasis on "am"? And my love for chemistry remains and is still very strong, albeit different today than about 45 years ago when I was 14 years old. In this text, I draw partly from my own experience in being part of and a devoted co-actor and co-creator of the academic chemistry culture, and partly from how I look upon the same culture when I distanced myself and consider it from an outsider's perspective.

Open Up Eyes. During my years as a university senior lecturer in chemistry, as a researcher in physical organic chemistry and later in science education, I have devoted parallel to my regular work much time to popularizing activities. I have held popular science lectures and workshops, presented chemistry on the basis of everyday perspectives and tried to find interesting inputs. I could not understand that not everyone else thought that science in general and chemistry in particular was interesting, exciting, and fascinating. Evidently, I am not alone in having dedicated myself to such a mission; there are many others who with great commitment and seriousness engage in trying to convert those who *do not understand.* Peter Atkins (2013) writes in the preface to the book, *What Is Chemistry?,* that:

> I hope to open your eyes and show you a fascinating, intellectually and economically important world, that of chemistry. Chemistry, I have to admit, has an unhappy reputation. People remember it from their schooldays as a subject that was largely incomprehensible, fact-rich but understanding-poor, smelly, and so far removed from the real world of events and pleasures … I want to change all that. I want to encourage you to look anew at chemistry, through modern unprejudiced eyes, with those memories and attitudes swept away and replaced by comprehension and appreciation. I want to show you the world through a chemist's eyes … (p. *v*)

I think this mission is important. Just like Atkins, I wanted to make others see and realize what I myself experienced as so self-evident: that all is chemistry; that everything that happens and we can see and experience in the macroscopic world is also expressed at the molecular level. At the same time, there is something chafing, and this chafing links to what is expressed at the end of the quote from Atkins above: "I want to show you the world through a chemist's eyes" (p. v). And that in turn is linked both to science education and to the campaigns and investments made in many parts of the world to come to terms with the declining interest in science, to get young people to choose studies and careers in these subject areas; campaigns and initiatives designed to arouse interest, to show how exciting, fascinating and important physics, chemistry, engineering, and to some extent biology are. We want them to look upon these subjects with our eyes. But how often do we look at our subjects with their eyes? What will we see if we do? And what will be the implications for science education? These questions I want to address and discuss in this chapter.

BREAKING BONDS THROUGH IMAGES

There are some (problematic) images of science that are conveyed in society by, for example, media and educational institutions, as well as by scientists themselves. These images, with historical roots, are reproduced and reconstructed over and over again. A main component is the message that natural science subjects are especially difficult and require special talents, something that has been critiqued and problematized by feminist philosophers of science like Donna Haraway (1997), Sandra Harding (1986), and Evelyn Fox Keller (1985). According to such elitism, not everyone can pursue these disciplines, and this excluding practice affects mainly women, but also men. Other aspects put to the fore by feminists' critique of science are dominant positivistic views and for being a male preserve, where the latter aspect is reflected in, for example, the skewed gender distribution in certain subjects such as physics. A recently published study in *Science* by Sarah-Jane Leslie, Andrei Cimpian, Meredith Meyer, and Edward Freeland (2015) highlights this problem. In a large-scale U.S. nationwide study of academics from 30 disciplines, the researchers found that women are underrepresented in fields whose practitioners believe that raw, innate talent is the main requirement for success, because women are stereotyped as not possessing such talent. Moreover, disciplines that valued giftedness over dedication rated themselves as less welcoming to women (Leslie et al., 2015).

The Image of a Genius. To illustrate the elitist image, I will give some examples of what scientists, two Nobel laureates, convey about the science disciplines, science activities and what it means to be a scientist. In Sweden we have a radio program called "Vetenskapsradion" (the Science radio). At the end of the year, in connection to the ceremony for the award of the Nobel Prize, the Science Radio broadcast a number of programs where the Nobel laureates are interviewed and get an opportunity to talk about their work, but also about their growing up and what led them to become a scientist. So, what is the image they choose to forefront? All programs since 2009 are available on Internet, and below is an excerpt from an interview with Adam Riess, who was awarded the Nobel Prize in physics 2011, for the discovery of the accelerating expansion of the Universe through observations of distant supernovae. He talks about dark energy and at the end of the program he says:

> What we need I think, for the next Einstein to come along and to sort of synthesize these clues into a deeper theory. I, I imagine someday having a conversation where this theorist saying okay this is what we know about dark energy: We know it accelerates the universe, we know it fills in the missing energy budget in the universe, we know it seems to have repulsive gravity, we know it seems to been around for, you know, more than twelve billion years. All this, all this sorts of facts, the same way maybe a detective would, would collect facts about a suspect. But at some point I imagine, this theorist taking all these and going off to a quiet room and coming back and say, okay I have a new idea about gravity and it fits all that. (Riess, 2011)

This quote, coming from a man who has received the Nobel Prize, is quite remarkable. Although he is one of the most successful physicists in the world, he is kind of longing for a Salvatore, a single theorist, a genius who will enter a quiet room and then come back with the solution. This image conveyed about what it means to be a successful physicist is hardly something that an average person can identify with. He doesn't even identify himself with the picture he draws. That is one thing. The other thing is whether he conveys a realistic picture of physics practice, which in other contexts (also by Nobel Prize winners) is described as a team work, a team effort, often with a lot of people involved; something that is also shown in the number of authors on scientific publications.

Another laureate is James Rothman, who was awarded the Nobel Prize in medicine or physiology 2013, for the explanation of how the body's transport machinery works in and between cells. James Rothman initially decided to become a physicist and started to study physics. In the interview he explains how he reluctantly allowed himself to be persuaded by his father to broaden his horizons and visit a biology lecture, despite his negative beliefs:

> My father said, you know, Jim you might consider taking just a little biology. And I didn't like the little biology I had in high school. It just seemed like a lot of memorizing to me, and physics, you know, if you're smart at it, really you don't have to go to the classes, you can figure it out on your own … So, I said I'll go to one lecture. One, one lecture, because every physicist knew biologists were stupid. That's the first thing they teach you, actually the first thing they teach you in physics is that physicist are really smart, maybe the mathematicians are little smarter but they're, they are not quite smart enough to be physicists. (Rothman, 2013)

Again this illustrates an elitist view of science and also shows the hierarchical relationships or status order between the scientific disciplines, with physics at the top. But, what both of these Nobel Prize awarded scientists do is to reproduce the image of a genius, and although not explicitly stated, preferentially a male.

In Sweden, the Nobel Prize, the announcement of the laureates, the ceremony etc. is a big yearly event that attracts a lot of attention, in the media and in a variety of contexts. During these couple of weeks, science is recognized more than at any other time of the year for its achievements and its importance to people and society. The event is used to pay attention to science's developments, to engage and evoke interest among the general public and specifically among students at all levels of education. And the laureates are role models in this endeavor. Thus, what they say and the image of science they mediate will impact in a way that is not easily erased afterwards, if problematic.

Let us pause for a moment and consider the image of the genius. Why is it so strong and so tenacious? Why is it reproduced by scientists, even by Nobel laureates who hardly need to doubt about their own geniality? I would claim that the reason lies in our egos. Scientists have a position, a power on the basis of the genius aura. We do not want to lose an identity that makes us smart, talented and special in our own and other people's eyes.

The Image of Students. At a meeting, those responsible for chemistry education at a university discussed the problem of diminishing recruitment of students, that the students' knowledge is inadequate and that most of them do not complete their studies. I was present as an observer within the framework of a research project and listened to their discussion. Each year approximately 25–30 students started to study at the bachelor program in chemistry, after one year only half of them were left. At the third year, 5–10 students remained and of those only a few completed the education and took their exams. At this meeting, one of the attendants said:

> They [the students] are not prepared for education in the same way as before. All can start, there is no competition to the study places. As they fail, they must be less prepared than before. We teach the same way as before. Maybe we have lowered some requirements. (Field notes, October 2015)

The quotation shows a perception that neither teaching nor the subject content is to be blamed, but the students. Since all students that apply can start the program, there is implicit complaining about no longer attracting the brightest students. Concerning the teaching carried through in their chemistry education nothing has changed, apart from possibly some minor changes in terms of lowering the level of requirements, but that is all. Accordingly, the problem is placed outside of the science discourse, the teachers and those responsible for the education and instead, the problem is attached to the students. This reasoning is of the same kind as the reasoning behind a majority of recruitment campaigns addressing those potential students who do not realize how exciting science studies and activities can be and therefore have to be convinced. The problem rests with them.

If I presuppose the hypothesis that we want to be viewed as smart and special, and that we want this image to survive, then it seems likely that we want to recruit the brightest students, e.g., the same kind of students that we were. If so, one could argue that there is no reason to adapt and change courses and teaching in such a way that more students pass and fewer students drop-out. Of course, that kind of argument is never explicit and I doubt anyone would admit such thoughts.

But still, the way recruitment problems and drop-out is articulated and the lack of activities and initiatives that really means a change can be interpreted as caused by such underlying ideas. Such unconscious thoughts may take the form of a resistance and a refusal to see beyond "the students are the problem."

BREAKING BONDS THROUGH THE "RIGHT ANSWER" CULTURE

Natural science's results, its descriptions of phenomena and the concepts used as explanations are depicted within sciences as objective and universally valid. There is a truth claim. In school and higher education, these truth claims represents the foundation for what I would call, the "right answer" culture. One reason is probably that the knowledge conveyed is far from the research frontier, and in many respects has been and is static. Sometimes the packaging is new—modern age adapted, new teaching aids—but the content is the same. Concepts, laws, theories are commonly presented as rigid truths, compared to the way they are communicated and debated within the scientific community itself, and thereby convey a stereotyped positivistic view of science. Moreover, while the curriculum in school addresses the scientific phenomena and concepts explicitly, it also mediates an implicit message of a hierarchy of science practices and who can access and participate in that practice, as expressed by Douglas Roberts and Leif Östman (1998):

> Science textbooks, teachers, and classrooms teach a lot more than the scientific meanings of concepts, principles, laws, and theories. Most of the extras are taught implicitly, often by what is not stated. Students are taught about power and authority, for example. They are taught what knowledge and what kind of knowledge, is worth knowing and whether they can master it. (p. ix)

I shall give two examples of how the message of the "right" kind of knowledge is conveyed through assessments in chemistry. Although, over the last decade and more, there have been many laudable and to a large extent successful programs that have introduced context-based learning, problem solving approaches, and holistic perspectives (Ram, 1999), these have not reflected in the assessments. And, assessment is a driver for students not only in higher education but also in compulsory school, though strength of motivation may depend on the national assessment culture. If a skill, phenomenon, concept or knowledge segment is deemed to be assessable in a way that contributes to a student's goal (e.g., high score, degree), then it is likely to be given a high priority by the students. In his study, Stuart Bennett (2004) analyzed 82 first-year university chemistry examination papers from UK, USA and Australia. In total the tests consisted of 627 questions and 95% of the problems were of the type with given data, a familiar method, and one correct answer. Another common feature of the examinations was a mismatch between outcomes claimed and outcomes tested (Bennett, 2004). Like a red thread through these assessments, the message communicated is that you need to learn the "right" facts, you need to give the right answer, and there is one and only one correct answer.

In another study, Marie Ståhl and I analyzed the Swedish national assessments in chemistry for grade 9, looking for norms and values explicitly and implicitly transferred through the test items (Ståhl & Hussénius, 2017). A majority (60–70%) of the questions were to be found within what we chose to call a "disciplinary discourse", focusing on disciplinary issues as concepts and phenomena, as well as how to act and speak in the scientific discourse. These questions have only one correct answer, and they define the boundaries of chemistry (Ståhl & Hussénius, 2017). We found three sub-categories to the disciplinary discourse: naked, backdrop and border discourses. Test items belonging to the naked discourse mediates the chemistry discipline in its most naked form, seemingly value-free, logical and rational. The backdrop discourse test items convey the same companion meanings (Östman, 1998), although they try to establish some sort of interaction through contextualized backdrops and use pronouns to address the reader. The border discourse is slightly different; the test items are not primarily about chemistry itself but about how to act and argue in a scientific context, and to distinguish a scientific assertion and method from a non-scientific (Ståhl & Hussénius, 2017). Not only do the tests portray what is the right kind of knowledge, they also portray who is the "real" scientist and an elitist view emerges from the words, pictures and content selection: natural scientists are preferably men/boys, who use the right scientific approach. This exclusion is not only implied in the portrayal of girls'/women's lack of knowledge but also by the choices of traditionally male-dominated subject areas that the tests focus on: cars, the petroleum industry and metal extraction (Ståhl & Hussénius, 2017).

Through the national tests, policy makers and curriculum committees have an influence on the design of teaching in school, something confirmed by Malena Lidar and Eva Lundqvist (2013). They found that teachers thought the tests were helpful in concretizing the goals in the syllabi and gaining confirmation of their teaching from the tests. Again, we see the same kind of mediated image of what science is, this time not only to the students but also to their teachers and thereby having substantial impact of what is regarded to be the "right science".

BELONGING AND IMITATION

If one has been attracted to science (as I have been) and has chosen such a career (as I have done), then one has been schooled into the science culture from the first contact with anyone of the science subjects and continuously during training at more advanced levels. I longed to belong to this context, I imitated and reproduced what was around me to fit in. I learned the science university culture, acclimated myself to the jargon, made an effort to be part of the culture, and the culture became a part of me. To fit in, I pressed down and forgot parts of me that were within me; I made my working class background invisible, I was unable to see that I was treated differently because I was a woman. I kept within the boundaries that defined this environment, implicitly as well as explicitly—I did nothing and thought nothing that went beyond what was expected. Identity is produced in specific knowledge contexts through repeated performances, and I went through an identity-(re)forming process and became a chemist. As such, I had learned the norm, unaware that it was a norm; I was an integrated part of the norm, and I conveyed the norm to others.

The norm becomes the obvious, the right. The science teaching environment, the teaching laboratory, is a policed space. Protocols are followed and the variety of possible identity enactments is limited. If the norm is that there is "one right answer", then a context in which there is not so purely a right or wrong is perceived as blurred or fuzzed. In Andreas Ottemo's (2015) dissertation *Gender, Body, Desire, and Technology: Passion and Instrumentality in Two Technical University Programs*, it appears that students perceive an environmental course in exactly this way, as something that deviates from the norm or even as an anomaly. A quote from one of Ottemo's informants, who perceived the environment of the course as "very vague", reads as follows:

> It was not quite like that "this is right and this is wrong", but it was more like that "if you can justify it, it's right" as well as [...] In all cases within the science program at high school and even as such person I am, I like that "this is right—this is wrong". But in this course, it was more like "it need not be so". For as long as you can prove or explain it yourself, justifying it as well, it could be right anyway. Even if someone else thinks this is wrong. (p. 211, translated from Swedish by the author)

In the thesis there are several similar examples of the kind seen in the quote above: they show a conformation in and affirmation of a subject culture, which is manifested in a distancing from what is not part of the (mediated and read-off) culture. As Staffan Bergwik (2014) stated, "Students learn to become scientists by way of imitating the proper role and self-image, as well as knowledge, of their forebears" (p. 499). Although I myself once acted in a similar way, expressed similar views and have an understanding of the underlying mechanism, I feel today a certain utmost concern that we still shape science identities of this kind. For those students who do not adopt and feel at home within a community of restricted and rigid norms of science activities, these norms may constitute the struggle per se; the students don't fit in and therefore choose another path. Attrition from a bachelor program in science to which I referred earlier in this chapter may be caused by such non-fitting-into-the norm reaction.

EYES WILLING BUT UNABLE TO SEE

Coming from another university in another city, I participated as an outsider in an inter-department workshop for university chemistry teachers. The general purpose of the meeting was to identify and discuss challenges that the department faced in its teaching activities. Prior to the meeting, a literature list of articles within chemistry education research and development had been provided to read in advance of the meeting. Among other things, the meeting focused on topics such as students' retention (instead of drop-out), varied forms of examinations, learning in laboratory exercises, assessment, and pedagogical development. The head of department opened the workshop by stressing the importance of teaching, that it always can be improved and that we can learn from each other. Thereafter the director of studies continued:

> We need to understand what is going on. Chemistry students, where are they and what has happened? [...] We have to stop talking about the students. These are the students we have. [...] We have not changed the way of teaching chemistry in the last 100 years. Can we change something? This is the focus of the discussion today. (Field notes, October 2015)

During part of the day, I sat with a group who discussed learning in laboratory exercises and listened to their conversations. It was an ethnically diverse group of teachers, several coming from non-European continents. After a while, comments made by various participants, with some variation of content, formed a pattern of similar narrative stories. The kind of narrative was the same whether the teacher came from Mexico, Russia, China, Sweden or some other country. These stories were based on their own experiences of chemistry studies; they talked from experiences of laboratory exercises that had been part of the courses they had taken as students. All emphasized the advantages in how the labs had been organized and carried through. They emphasized what they experienced as positive aspects, and because it worked for them, it should work (the same way) for others. Thus, the solution that emerged was to continue to do laboratory teaching in more or less the same way as before, because it worked (for those who completed their studies in chemistry and have remained in academia). I do not think these teachers were aware that they, with their proposals for how to "change" lab teaching, reproduced laboratory work as work as usual.

Although the ambition level was high and the awareness of a need to do something was evident from how the workshop was planned and introduced, what remained in my mind when I left the workshop at the end of the day was the experience of a chemistry education community trapped in its own conception of how to teach the subject. They struggled to find new ways to address a problematic situation, but repeatedly returned to the starting point: science as usual. And although summoned at the beginning of the day that "We have to stop talking about the students," this was the main theme and explanatory model that permeated the workshop: the problem rests with the students.

TAKING ANOTHER PERSPECTIVE

One way of interpreting culture is that our social world is constructed through cultural connections or relations that includes or excludes. As members of a culture, its processes and habits are incorporated into our minds and bodies, making them largely invisible to us. As science teachers we are insiders, we are part of the science education culture, we are part of its norms and likely also accommodators of cultural values and norms. As teachers we need to be able to take a step outside the borders that define the culture we belong to and look at the present norms from an outsider's perspective. Only then can we understand those who are not attracted to and interested in the same way and by the same things as us who are part of the culture. A feminist perspective means being able to see and problematize norms; a feminist perspective therefore provides a tool to take an outsider position, which is necessary if we want to change science teaching and education.

Kevin Kumashiro (2000) uses the notion "the other" for marginalized groups in the society, i.e., that are other than the norm, when discussing anti-oppressive education. Through a critical examination of how various educational institutions work to change oppression within education and teaching, he categorized different approaches to such a change from four problem formulations: education for the other, education about the other, education that is critical to privileging and othering, and education that changes the students and society (Kumashiro, 2000). If we apply Kumashiro's categories to science education, the formulation of the problem can be: there is a declining interest in science among young people; young people do not choose science courses/educations; a large proportion of students studying natural sciences drop-out from their education. The numerous and extensive investment in recruitment campaigns, activities to stimulate interest, the use of successful (female) role models, etc. carried out in many parts of the world, can be subsumed in the category of education for the other, or reformulated to interest-evoking investments for the other. It is eminently urgent to instead focus on the last two categories and work for a scientific education that is critical of privileging and elitism. And instead of shaping students into a strict science norm, strive to change the education and thereby also scientists and society. My point is similar to the arguments once fore-fronted concerning women's participation in science education and research: new groups of students, students who today do not fit-in to the present science education norm may contribute with new ways to look upon and carry through science activities; they may contribute with another kind of creativity and resourcefulness, finding other ways to approach some of the big challenges for society.

In a sense I, throughout this chapter, have helped to maintain chemistry as it is and has been, although I at the same time have been revealing this. To challenge prevailing norms I think the use of dual strategies is necessary, although such a strategy at the same time demands an awareness of risks associated with the duality. On one hand teachers must make visible and problematize the norms, and thereby risk a result of consolidating the same. On the other hand teachers must strive to challenge, deconstruct and undermine these norms. Each one of us, science teachers, within schools and universities must strive to adopt an outsider position and observe the area and the activities of which we are part. We need to scrutinize what we say, do, and embody as if we looked through someone else's eyes instead of through our own.

FORMING NEW BONDS—QUEERING SCIENCE EDUCATION

I want to conclude this chapter by suggesting how a feminist theory can be interpreted and applied on science teaching. The word queer is a central concept within queer theory and although a broad term that can include several different things, it basically means a questioning of the heterosexual norm. Renita Sörensdotter (2010) highlights how queer theories, ones emerged in gender and sexuality studies, also are relevant in other areas and usages. In a way queer theory has appropriated the word queer and transformed its original meaning. As an adjective queer means odd, strange, peculiar, weird, atypical or out of the ordinary, to mention a few of a long list of synonyms. If the word instead is used as a verb it means foil, threaten and undermine, among others. Thus queering science education can mean *doing something out of the ordinary*, to *foil* and *undermine* what is considered to be an ordinary, taking for granted, subject content when teaching science subjects. By using the "queer" thinking logics (questioning, critical thinking, challenging norms, etc.) within the science educational field, it can be an important tool in efforts to identify and deconstruct companion meanings such us being value-free, objective, high demanding, male coded and requiring brilliance, giftedness, etc. In her article Sörensdotter (2010) gives a number of examples from her own teaching. For example, how she used pictures and movies to evoke emotions and engage students intellectually, which is then taken as the starting point for a process of shifting and changing understandings of norm and deviation. Analogously, queering science education can integrate new content and alternative forms of teaching depending on which the norms are that constitute the target, which the norms are that need to be challenged. My assumption is that in-depth knowledge of the science culture, that is beliefs and norms, which are linked to the natural sciences, can provide students access to new ways of relating to and working with science. By problematizing and visualizing the science culture while teaching the content of the subject courses, students may get tools to make informed choices about how they want to relate to this culture (Hussénius, Andersson, & Gullberg, 2015). Moreover, students can become our companions and catalysts towards an increased awareness of stereotypic norms that need to be challenged, because students' cultural blindness is generally not as deep rooted as our own. If so, science education will be transformed, not totally changed but widened, and new groups of students might form bonds to these subjects.

END OF JOURNEY

My professional journey as a chemist and academic started with a strong desire and ambition to *become* a chemist, to be a part of and an equal co-actor in such a community. I was successful and managed to a large extent in this endeavor. I have been a dedicated teacher who received teaching awards, and with great fervor tried to convey the same feeling for the subject as I felt myself. It took decades before I realized that my field of vision was limited, that there were problematic norms embedded in the science community, in science culture, and that I conveyed and replicated these norms. It was only in connection with my feminist awakening that I came in contact with theories and tools that helped me pay attention to, make visible and challenge existing norms and hierarchies. I am still fascinated and dazzled by the molecular world. I still feel a great satisfaction of being able to understand, explain and predict phenomena in the visible macro level, based on knowledge of how atoms and molecules interact with each other. I want others to experience the same.

ACKNOWLEDGEMENTS

The idea of using Vetenskapsradions interviews with Nobel laureates is originally Kristina Andersson's, and I thank her for letting me borrow this idea and use it in this chapter. Financial support from the Swedish Research Council through grant 2014-939 is gratefully acknowledged.

REFERENCES

Atkins, P. (2013). *What is chemistry?* Oxford: Oxford University Press.

Bennett, S. W. (2004). Assessment in chemistry and the role of examinations. *University Chemistry Education, 8*, 52–57.

Bergwik, S. (2014). The historicity of the physics class: Enactments, mimes and imitation. *Cultural Studies of Science Education, 9*, 495–501. doi:10.1007/s11422-013-9497-4.

Haraway, D. (1997). *Modest_Witness@Second_Millenium.FemaleMan©_Meets_OncoMouse™*. London: Routledge.

Harding, S. (1986). *The science question in feminism*. Ithaca, NY: Cornell University Press.

Hussénius, A., Andersson, K., & Gullberg, A. (2015). Spotting the science culture – Integrating gender perspectives into science courses. *International Journal of Gender, Science and Technology, 7*(1), 91–103.

Keller, E. F. (1985). *Reflections on gender and science*. New Haven, CT: Yale University Press.

Kumashiro, K. K. (2000). Toward a theory of anti-oppressive education. *Review of Educational Research, 70*(1), 25–53. doi:10.3102/00346543070001025.

Leslie, S.-J., Cimpian, A., Meyer, M., & Freeland, E. (2015). Expectations of brilliance underlie gender distributions across academic disciplines. *Science, 347*(6219), 262–265. doi:10.1126/science.1261375.

Lidar, M., & Lundqvist, E. (2013). Nationella prov i NO och lärares val av undervisningsinnehåll. *Utbildning och Demokrati, 22*(3), 85–107.

Östman, L. (1998). How companion meanings are expressed by science education discourse. In D. Roberts & L. Östman (Eds.), *Problems of meanings in science curriculum* (pp. 54–70). New York, NY: Teachers College Press.

Ottemo, A. (2015). *Gender, body, desire, and technology. Passion and instrumentality in two technical university programs* (Doctoral dissertation). University of Gothenburg, Gothenburg, Sweden.

Ram, P. (1999). Problem-based learning in undergraduate education. *Journal of Chemical Education, 76*, 1122. doi:10.1021/ed076p1122.

Riess, A. (2011, December 9). In *Vetenskapsradion*. Retrieved October 14, 2015 from http://sverigesradio.se/sida/avsnitt/52467?programid=412

Roberts, D. A., & Östman, L. (Eds.). (1998). *Problems of meaning in science curriculum*. New York, NY: Teachers College Press.

Rothman, J. (2013, November 22). In *Vetenskapsradion*. Retrieved November 22, 2013 from http://sverigesradio.se/sida/avsnitt/285262?programid=412

Sörensdotter, R. (2010). En störande, utmanande och obekväm pedagogik: Om queerteoriernas relevans for en normbrytande undervisning. In J. Bromseth & Frida Darj (Eds.), *Normkritisk pedagogik. Makt, lärande och strategier för förändring* (pp. 135–154). Uppsala, Sweden: Uppsala University.

Ståhl, M., & Hussénius, A. (2017). Chemistry inside an epistemological box! Discursive exclusions and inclusions in Swedish national tests in chemistry. *Cultural Studies of Science Education, 12*, 395–423. doi: 10.1007/s11422-016-9730-z

CHAPTER TWENTY-SEVEN

On the Need to Engage in Constructive, Productive, Scientific Discourses on Highly Controversial and Emotionally Charged Topics

CASSIE F. QUIGLEY AND S. MEGAN CHE

Most of us can attest to the efficacy and prudency of a tactic of avoidance when potentially controversial topics are raised—particularly in settings such as family gatherings or social functions—the adage warning us not to discuss religion or politics reflects a widespread policy of skirting such issues. Thus, many of us are not experienced or particularly well versed in the art of engaging in constructive discourse involving highly charged topics like politics, religion, or, particularly relevant in this age: climate change, abortion, gun sense/gun control, or civil rights. However, when the stakes become high enough, constructive debate can be powerful in our efforts to forge paths forward. We might find that an ability to construct discourse that catalyzes insights rather than dictates a particular, rigid perspective is a highly valuable skill—a skill, which both because of the content and the processes of discourses, should be fostered in academic settings like schools and universities. However, in absence of productive discourse processes in educational settings, the public often leans to the media for examples of productive dialogue (Olausson, 2009). In this chapter, we describe the public's understanding of controversial scientific issues, the ways in which the field of science education has attempted to prepare a scientifically literate populace, and the challenges the field faces in attempting to do this. To elucidate this issue, we use climate change as an example case of our argument. Next, we describe what happens when the public looks to the media for examples of evidence-based dialogue and what the implications are for science education when the media truncates dialogue, thus providing an inaccurate model of productive dialogue.

MEDIA'S ROLE IN DISTANCING CLIMATE CHANGE SCIENCE AND PUBLIC

The media is often a major source of information and opinions about science for citizens. Public perceptions of science are significantly influenced by media's representations of scientific knowledge (Carvalho, 2007). As Dorothy Nelkin (1995) noted, the public often makes sense of science "less through direct experience or past education than through the filter of journalistic language and imagery" (p. 2). Like most topics that the media presents, when the media discusses science topics, science becomes more than dissemination of information, but also a reconstruction of that information—choices of content, people who will be interviewed, etc. that are all a part of the journalistic process that translates science to the public (Nelkin, 1995). As such, particular worldviews are highlighted, examined, and transformed in media discourses, while others are excluded from them. Therefore, as science educators seeking to understand the ways students and communities make sense of science topics, it is important for us to recognize the influences on their knowledge construction.

As mentioned above, we use climate change as an example case for our argument. This example is timely as climate change has been a particular focus of the media for the past few decades, and as a result scholars have contributed to a better understanding of media representation of climate change. Maxwell T. Boykoff and Jules Boykoff (2004) argued that journalistic tactics led to narrow demonstration of knowledge on climate change in the U.S. press with emphasis on those who deny human involvement, giving attention to the notion of "climate change deniers." Policy-makers often expect scientists to provide answers to problems that are debated in the media and make a variety of public uses of science to legitimize certain programs, policies, or action. This is a reasonable request, as the scientists are often those with immediate knowledge of emergent literature. Through this relationship, an important link was established between citizens, scientists, politicians and media professionals, the connection between science and politics has become increasingly public and scientific. As a result, science is more susceptible to critique, contestation, and transformation. While scientists should engage in debate that is translatable/comprehensible to a wide audience, the manner in which that debate occurs is often not a productive one; instead it often dismisses the opponent's ideas without any consideration.

In a *National Geographic* article (Achenbach, 2015), Marcia McNutt states the public should question scientists because that is what science is—questioning; however, she then hedges: "… they [the public] should use the scientific method or trust people who use the scientific method, to decide which way they fall on those questions" (p. 47). While we agree with McNutt, that using scientific processes to garner evidence is important, we also feel the notion of "trust the scientists" has landed science into the current situation where the public is at odds with science [Note: We challenge the use of "the scientific method" and instead will use the term "scientific processes" to denote an inquiry process that does not follow a rigid set of steps often associated with the term scientific method. See William McComas (1996) and Mark Windschitl, Jessica Thompson, and Melissa Braaten's (2008) pieces for a full examination of the differences.] For example, there exists a 38-point gap between scientists and the public about the public's understanding of human's roles in current climate change: 88% of scientists state that climate change is mostly due to human activity, while only 50% of the public hold this same idea (Pew Research Center [PEW], 2015). What the *National Geographic* article does not address is not only the how the public's mistrust for science is leaking into the classrooms around the U.S. through media, but also why mistrust for science is being taught by science classroom teachers.

SCIENCE TEACHERS' BELIEF SYSTEMS ABOUT CONTROVERSIAL ISSUES

Often science teachers engage in well-reasoned practices surrounding scientific issues and pose questions for further attention based on students' past and current understandings. However, in order to do this for controversial issues such as climate change, teachers need to have an opportunity to explore their own beliefs about this topic. Current evidence indicates that university students preparing to become science teachers (preservice science teachers) are, to a large degree, unconvinced about the existence of human-induced climate change (Lambert, Lindgren, & Bleicher, 2012). In fact, Julie Lambert and her colleagues found that almost half of preservice science teachers (n=149) surveyed believed that climate change was a result of natural variability, changes in the sun's energy output, or the cyclical changes in Earth's tilt and orbit. Although our chance to forestall climate change by reducing our environmental footprint may well have passed (Solomon,

Plattner, Knutti, & Friedlingstein, 2009), the importance of educating younger generations about the existence of and potential ramifications for climate change remains, because they will be the ones to cope with the consequences of increased carbon dioxide levels. Increasing younger generations' understandings of climate change is a multi-pronged effort of which K-12 science education is a crucial component. However, if the preservice teachers whom we are currently preparing to teach science are still largely unconvinced of the existence of climate change, we miss an opportunity to ask young people to think about, in a reasoned way, past and current evidence regarding climate.

Philip Bell (2000) notes productive dialogue can be powerful in helping students understand the complex problems and help them see beyond their opinions. For example, when students are able to construct an evidence-based argument, this dialogue can provide insights into the students' own value systems. However, the skill of dialogue should be fostered in academic settings such as schools and universities. This includes ensuring the content necessary to hold a productive discussion is present as well as incorporating strategies to ensure personal beliefs are not attacked.

Science education has not always been successful with controversial issues. For example, when teaching evolution, science teachers often try to circumvent religious beliefs with a dual-theory approach, but Craig Nelson (2007) discourages this practice. Because no other science content is taught with a dual-theory approach, this approach makes evolution look less scientifically accepted than other theories. Another common approach is allowing students to debate evolution vs. creationism. The problem with this approach is that students are left trying to make sense of complex processes that are connected and embedded in their identity with little support to do this (Cook, 2009).

Dan Kahan and his colleagues (2012) state that one of the reasons science education struggles with approaching controversial issues is because certain controversial science topics such as climate change are more than just scientific issues. They are also indicators of who you are and with whom you identify. For example, Bob Inglis, the former Republican South Carolina congressman, stated he believed in climate change and quickly found himself without a job. People have a strong need to fit in with people whom they feel share their value system, and that will trump science especially "when there is no downside to ignoring science" (Achenbach, 2015, p. 45). However, for topics such as climate change, there is a downside. As the 2014 Intergovernmental Panel on Climate Change (IPCC) reported, with increased warming there will be, "severe, pervasive, and irreversible impacts" including "substantial species extinction, large risks to global and regional food security" (p. 14). As more and more people distance themselves from science, the likelihood of being prepared to handle the consequences of climate change diminish.

THE DISTANCE BETWEEN SCIENCE AND THE PUBLIC

Concern regarding the distance between science and the public is nothing new (Ziman, 1991). Associations such as the Committee on the Public Understanding of Science in England, and research into this area have existed since the 19th century (Wynne, Simmons, Waterton, Hughes, & Shackley, 2001). Nevertheless, there is renewed interest, as widespread concern about environmental problems has brought a set of specific scientific issues to public attention (Olausson, 2009). Until recently, scientific topics such as global increase in temperature, increase in carbon dioxide in the atmosphere, and sea level rises would have only been reserved for science class; now global warming jokes are present in television programs, popular newspaper articles, and political debates. For this reason, an examination into the ways in which people form opinions and beliefs about climate change offers a vital, current example of the handling of science in the public sphere. Clearly, there are many ways in which people formulate knowledge about controversial environmental issues, and we provide insight into one of those ways. In such situations, one reasonable response is to turn to institutions of learning to provide information about the controversy. That is, we expect, by and large, for teachers and professors to engage in well-reasoned and well-supported practices surrounding students' past and current states of knowledge, and to pose questions for further attention based on students' past and current understandings.

Tensions between scientific understandings and popular explanations of phenomena have existed for hundreds/thousands of years, since the beginning of our construction of science, and several well-known, public clashes might inform current tensions. Historically, clashes between science and popular understandings were centered on scientific knowledge that was perceived, particularly by the religious hierarchy, as oppositional to faith and therefore a threat to that faith. Obvious instances of clashes of this nature include Galileo's forced recantation of heliocentrism and his subsequent, lengthy house arrest, and the condemnation of Darwin's theory of evolution. These dramatic clashes involved more than individual belief systems; they also challenged religious authority and control. We still grapple with the teaching of evolution in classrooms today. Clearly, if scientific evidence were all that was necessary to convince people, this debate would have died down long ago, and we would likely not be accepting the Bible as a scientific classroom text (Alters & Alters, 2001).

Many scientists contend that science and religion (here, our focus is on Christianity) are not inherently conflicting and that the apparently contentious relationship between the two has been overblown. Francis Collins asserts in a popular newspaper column by Richard Gray (2009), for instance, that "once you set aside an insistence on an ultra-literal interpretation of Genesis, you can arrive at a conclusion which is quite comfortable for me as a believer and as a scientist, that yes Darwin was right" (p. 1). This viewpoint, given the overwhelming majority of people in the U.S. who state that scientific developments make society better and who identify as Creationists (Masci, 2007), may hold potential as a way to convince Creationists of scientific understandings because this viewpoint refrains from oppositional positioning of science and religion. The crucial question may be how to set aside a literal insistence on at least selected Bible excerpts. This question is vital because nearly two-thirds of people in the U.S. would continue to hold a religious belief even if that belief had been disproven by science (Masci, 2007). For many teachers of science, negotiating religious and scientific understandings is tantamount to a professional responsibility. Below, we explore and explicate several approaches for science teachers to engage students in productive dialogue with controversial topics.

OPPORTUNITIES FOR PRODUCTIVE INQUIRY DIALOGUE

Clearly, scientific controversial issues such as climate change present a challenge for science education. Much scholarship is devoted to strategies for convincing people of the validity of science (Saylan & Blumstein, 2011). In such situations, one response is to turn to institutions of learning to provide information about the controversy. Because of the personal connections to one's understanding of climate change, perhaps emotional reactions to climate change may be connected to methods of science teaching. Findings indicate, and are supported by our own experiences with preservice science teachers, that presenting scientific observations, such as carbon dioxide trends, ice core data, glacial area change, and Keeling curve trends are often not enough to help someone understand the science behind changes in climate (Keene & Blumstein, 2010). The persistence of ideas, even in the light of scientific evidence refuting those ideas, suggests that people engage emotions in addition to their ideas when they are tightly woven into a deeply held belief system (Weber & Stern, 2011). In such instances, a threat to one's belief system is often perceived as a threat to one's identity and self-interest. Making progress in helping people to understand the validity of scientific theories like climate change must include a thoughtful consideration of the belief systems, including emotions and interests, strongly tied to the perpetuation of the prevalent issues. That is, understanding and acknowledging ways in which people conflate belief systems, identity, and academic inquiry can be productive in efforts to talk to, rather than ignoring. Clarity regarding the expectations for productive dialogue can be helpful in contrasting academic and inquiry dialogue with other forms of dialogue in which belief systems, in the relative absence of established evidence, can play a larger role.

Opportunities for engaging in reasoned dialogue are a priceless benefit of universal public education. We consider a particular form of dialogue developed to explicate and, to the extent possible, facilitate a shared understanding of the aims and procedures of verbal dialogic interactions. This shared understanding of why and how we—as teachers as students—agree to engage each other on a discourse plane is crucial to meaning construction, especially when the dialogic topic is contentious. Thus, for the purposes of explicating verbal dialogic processes, Douglas Walton's (1989) characterization of dialogue as "a sequence of exchanges of messages or speech acts between two (or more) participants" is meaningful (p. 3). For academic and scientific settings, evidence and reason are hallmarks of argumentation, and Walton points out that, "Reasoned argument presupposes an arguer who is, at least to some degree, open-minded, serious, and trustworthy in collaborating in joint dialogue" (p. 153). In addition to the often unmet criteria of open-minded, authentic participation in dialogue, common areas of confusion frequently arise from ambiguities in what the argument is, what the argument is about, what the argument may be taken to be, and what the purpose of the argument is (Walton, 1989). Because different contexts of dialogue entail different goals and different criteria for admittance of propositions, it is important to establish clarity regarding the central issue of the argument and hence the particular form of dialogue best suited to address the argument. Public academic educational spaces provide numerous opportunities to engage students in coming to know more about each of these various dialogic contexts (NGSS Lead States, 2013).

Walton (1989) characterizes several contexts, or forms of dialogue, including the personal quarrel, debate, persuasion dialogue, inquiry dialogue, negotiation dialogue, and educational dialogue. The personal quarrel is distinguished by personal attack, appeal to emotions, loss of "balanced" perspective, and a desire to "win" an argument no matter what. We view discourse as James Gee (2004) does; to mean verbal and non-verbal processes of meaning making and meaning sharing. Ubiquitous as it may be, this type of dialogue is clearly neither academic nor reasoned, frequently devolving into insults and emotional arguments that would not be deemed valid or admissible in other dialogic contexts. The form of dialogue above the level of the personal quarrel, according to Walton's (1989) classification, is the debate, or forensic debate, in which the goal is to win the debate by impressing the third-party evaluator or judge of the argument. Though the processes of forensic debate are highly structured, standards for the level of logic or reason expected in debates are often permissive, frequently allowing personal allegations and fallacious arguments. To be successful in a debate—that is, to win the vote or votes of the judge(s)—does not necessarily entail logically sound, reasonable, or coherent arguments. The goal in a persuasive dialogue, however, is to persuade the other person(s) of one's own conclusion or point of view. The methods by which one attempts to persuade another include establishing a linkage of premises that the other person agrees to, resulting in a concession, based on inference from the linkage of agreed-to premises, to one's point of view. This linkage of premises may be established internally, from the chain of reasoning, and/or externally, by introducing external scientific sources. In contrast to the reason-based nature of persuasive dialogue, the negotiation dialogue makes no pretense to ideals based on strong arguments because the goal of negotiation dialogue is self-interest, which is sought through bargaining. Whereas the goal of negotiation dialogue is to get something, the goal of educational dialogue is for one party to impart something, most often information or facts, to another party. Educational dialogue is familiar to us from the traditional, dated teaching method of lecture.

Inquiry dialogue is a particular form of persuasive dialogue. In this context of argument, each participant has a thesis (conclusion) to prove. The strategy is for each participant to prove his/her thesis from the concessions of the other participants. Argument is based on the concessions of the other participants, who can concede any proposition they care to when engaged in persuasive argument. However, in inquiry dialogue, the propositions must be premises that are known to be true, that have been established as reliable knowledge to the satisfaction of all parties to the inquiry. The basic goal of the inquiry dialogue is to acquire knowledge, as the inquiry process seeks out knowledge or evidence. Evidential priority is the key feature of the inquiry dialogue, for the inquiry dialogue is strongly directed toward deriving conclusions from premises that can be well established on solid evidence. In the inquiry dialogue, participants are supposed to be neutral investigators of an objective truth to the extent that this is possible. In this way, inquiry dialogue is a cooperative rather than an adversarial context of dialogue.

In dialogue, a participant not only has an obligation to make a serious effort to fulfill her own goal in the dialogue, but also an obligation to allow the other parties to fulfill their own obligations—these obligations engender rules of dialogue. These rules include giving other parties reasonable opportunities to reply to questions or to make a point. Specifically, rules of dialogue require that a respondent answer questions cooperatively, an expectation that gives rise to several prohibitions of dialogue, including (Walton, 1989):

- Unlicensed shifting from one type of dialogue to another (i.e., shifting from debate to inquiry or from persuasion to quarrel)
- Failure to meet a burden of proof or to defend a commitment when challenged (i.e., when asked to provide evidence supporting climate change, the response is, "It's obvious. Why don't you believe in climate change?")
- Appealing to external sources of proof without backing up your argument properly (i.e., making the claim that because a scientist refutes the existence of climate change, then climate change must not exist).

- Failing to reply to questions or unduly evasive replies (i.e. making statements such as "You just don't understand.")

One way of framing the interplay between reason and emotion in discourses of scientific debates is to focus on the processes that frequently occur during these opposing verbal exchanges when people with irreconcilable views try to persuade the other of the validity of their own stance; as they do so, they often stray from the procedures of persuasive argument. During this process, called schismogenesis, by anthropologist Gregory Bateson (1935), one person in the debate makes a claim and the other person responds with a counter-claim of his/her own, rather than respond to the strengths and weaknesses of the first claim. By essentially ignoring the first claim, the debater has set the stage for talking at rather than talking with the person—a situation, which often increases the disagreement between the two debaters. Thus, instead of a reasoned discussion of evidence-based rational arguments, the discourse frequently devolves into volleys of increasingly emotional claims and counter-claims. In an schismogenetic argument about climate change, for instance, one debater might begin with a claim that current climate changes are human-induced. The other participant might respond by claiming that the causes are historical changes in climate and that these changes have always and will always occur. In response, the first debater might claim that the changes in climate are more frequent and are more erratic. The second debater might then argue that climate change is due to variations in the sun's energy output. Neither of the debaters is willing to respond to the claims of the other or to acknowledge the other's claim as worthy of consideration. This type of discussion can masquerade as persuasive dialogue, but in these situations neither participant fulfills the obligations of dialogue.

When Inquiry Dialogue Breaks Down. Unfortunately, the rules of dialogue are often broken, particularly in the media. With controversial issues, for instance—if the public is uncertain about science content, people can begin making judgments about knowledge and scientific inquiry. The reasons that breakdowns in productive dialogue in the media are so problematic include the truncation of reasoned inquiry that occurs with breakdowns and the absence of opportunities to engage in such inquiry in K-12 academic settings. These opportunities could provide informative contrasts to the quarrels students are often exposed to in the media. Further, because of the relative inaction in public K-12 schools to engage students in productive dialogue, the main source of examples for participating in discourse becomes popular media (Carpentier, 2011).

One example of truncated dialogue through a breakdown in productive discourse in popular media is the Meet the Press debate in which Bill Nye, "The Science Guy," and Marsha Blackburn, Congresswoman (Republican from Tennessee), vice-chair of the House Energy and Commerce committee, discuss climate change (O'Brien, 2014). The host, David Gregory, begins with a clip of extreme weather in the Northeast and drought out in the West and asks Nye, "In this moment in this extreme weather moment, is it creating a new urgency to act?" Nye responds that it is time to "do everything all at once" but also calls on US innovation to "be a world leader in new technologies" to secure financial security. At this point, Gregory does not permit Blackburn to respond directly, but instead asks the same question of her "is there a new urgency to act?" Because the host does not allow Blackburn to respond to Nye's claims, the host is not encouraging debate or dialogue or even a discussion. The congresswoman responds by making her own claim:

> David, what I think it brings to mind is how we utilize the information that we have. And we all know, and I think that Bill would probably agree with this, that neither he nor I are climate scientists. He is an engineer and an actor. I am a member of congress and what we have to do is look at the information from climate scientists. And as you've said there is not agreement around the fact of what exactly is causing this. Even the President's own science and technology office head said, no one single weather event is due specifically to climate change.

At this point, the host does not allow for a direct response but rather shuts down opportunity for inquiry dialogue by flatly claiming, "There is consensus."

As the discussion continues, both participants attempt, in spite of the moderator, to respond to the other's claims. For instance, Blackburn attempts to weaken Nye's claim of evidence supporting climate change by asserting, "And when you look at the fact that we have gone 320 ppm, 0.032 to 0.040, 400 ppm you have to realize it is really very slight. There is not consensus." Blackburn has introduced two propositions at once—that the increase is small in magnitude and that there is not consensus on the existence of climate change. Nye responds to both of these claims at separate points in the remainder of the discussion, in spite of the moderator's interjection that, "You can pick out particular skeptics but you cannot really say that there are hundreds of scientists that have gotten together to conspire to manipulate data." The reason this interjection is problematic for the discourse is because the facilitator's role is now unclear—is the facilitator responsible for holding the parties to the rules of discourse or is the facilitator participating in the dialogue?

Despite this confusion, Nye and Blackburn find ways to circumvent the moderator and have the beginnings of a schism. Blackburn begins by stating:

> Let's say everything that Bill says is wrong is wrong. Then let's say that. What are you going to do about it? What would that policy be and will it have an impact? Now even Director McCarthy from the EPA stated reaching all the 26 US goals is not gonna have an impact globally. You don't make good laws, sustainable laws, when you are making them on hypotheses or theories or unproven science.

There are a few claims Nye can choose to respond to; he focuses on the last proposition that climate change theory is founded on unproven science. Nye asserts:

> Once again the congresswoman is trying to do is introduce doubt. What I would encourage everyone to do is back up and let's agree on the facts. Would you say the Antarctic has less ice than it used to? When you asserted that a change from 320 ppm to 400 ppm is insignificant. My goodness that is 30%! That is an enormous change.

Here, in this exchange, the discussants are able to engage in inquiry dialogue. Each cites evidence for their claims and Nye is able to provide counterpoints for the claims made about the atmospheric content of carbon. However, Blackburn is not given the opportunity to respond to Nye's question, "Would you say that the Antarctic has less ice than it used to?" If Blackburn agrees with this statement, then the discussants have found common ground, which could have provided the space for the pair to move forward.

When Gregory does not provide the opportunity for common understandings, he is breaking a rule of productive dialogue (unlicensed shifting from one dialogue to another). When schools or universities do not foster the space for productive dialogue, people look to public spaces, such as the media for examples of ways to engage in controversial scientific conversations; and when the media falls short of creating productive inquiry dialogue, the result could be a less scientifically literate populace.

FOSTERING PRODUCTIVE DIALOGUE

To foster an argumentative environment with the potential for generating a shared space where students can learn from one another, Bateson argues there must be a moment in the dialogue that prevents the schismogenic unit (i.e., the discourse of people or the group in the discussion) from destroying itself through excessive disagreement. Otherwise, the schismogenic debates can devolve into quarrels, which represent "argument at its worst" and can potentially harm personal and professional relationships to the point that normal social functioning becomes increasingly difficult. Walton notes that, "Reasonable standards of good argument should be designed to prevent argument from deteriorating into the personal quarrel … when an argument descends to the level of the quarrel, it is usually in deep trouble" (pp. 3–4). This is one reason the tone of the discourse is important to its success in generating shared insights and understandings.

When debates devolve into schismogenic discourse, in spite of efforts to clarify the processes of debate, there are strategies for constructive and reasoned paths forward. For example, a debate facilitator or a participant, drawing on Bateson's notions of shifting the schism to encourage communication, could achieve common ground and focus on working towards a shared goal. Raimo Tuomela (2013) would call this action, collective action, which could ease the schismogenic strain. For example, if during the dialogue between Nye and Blackburn, they had the opportunity to agree that Antarctica has less ice than before, then perhaps that could have been an opportunity to think about similar goals. However, both teachers and students need practice looking for opportunities for collective action and practice in how to ease strain during dialogue in order to understand how and when to intervene.

Another tactic is to have teachers watch for a shift in tone and language that often is an attempt to make it less confrontational and more inclusive. This often includes a shift from "you" towards "we" (Bateson, 1935). These shifts in tone are what Bateson calls "ratcheting-down" (p. x), which includes less definitive language such as "maybe" or "perhaps." These shifts in tone often indicate that the person is more open to listening to another viewpoint. To support students in these moments, teachers can look for these shifts in tone and language use, and encourage the students to continue the conversation—or to bring the dialogue back to those moments where this shift occurred.

In this chapter, we argue that when students do not have opportunities to engage in productive inquiry dialogue—the goal of which is to broaden perspectives—or worse, they consistently participate in unproductive dialogue, the type of knowledge they garner is scientifically inaccurate. When the rules of inquiry dialogue are consistently broken, the type of dialogue moves from inquiry to quarrel, which Walton (1989) notes that the method becomes a personal attack and represents that lowest level of argument and that, "most of the logical lessons to be drawn from the quarrel turn out to be pathological" (p. 4).

In academic settings such as schools and universities, one way to forestall rule breaking during dialogue can derail into quarrels, is to explicitly and clearly delineate the nature of the dialogue to ensue and to clearly state the rules and expectations of entering into such a dialogue. Simply making these boundaries, rules, and expectations clear to all students can help them to understand why some dialogues go astray and how to further attempts at dialogue. This is one way to prevent the degeneration of dialogue into destructive schismogenesis—a place that hinders learning, as neither side is willing to listen to one another. The purpose of a facilitator, or in the classroom—the teacher, in this type of discourse is to continue as necessary to explicate and clarify the rules for that particular discourse and to call participants to order when violations occur.

One challenge of providing opportunities for students to engage in inquiry dialogue is if the standards that dictate much of instruction do not encourage this type of teaching. As states in the USA move to create "less controversial" standards, opportunities for students to authentically participate in meaningful inquiry dialogue decrease. Students often encounter challenges when attempting to participate in productive inquiry dialogue (Osborne, Erduran, & Simon, 2004). Without sustained opportunities to participate in this type of discourse, students are not likely to participate in these conversations in meaningful ways. In order to deal with environmental issues, we must be able to engage in reasoned dialogue and inquiry discourse, and prevent discussions from devolving into quarrels. Hopefully, if we can keep the conversations going, we can move towards common ground and ultimately solutions.

REFERENCES

Achenbach, J. (2015, March). Why do reasonable people doubt science? *National Geographic*, pp. 30–47.

Alters, B. J., & Alters, S. (2001). *Defending evolution in the classroom: A guide to the creation/evolution controversy*. Burlington, MA: Jones & Bartlett Learning.

Bateson, G. (1935). Culture contact and schismogenesis. *Man, 35*, 178–183. doi:10.2307/2789408.

Bell, P. (2000). Scientific arguments as learning artifacts: Designing for learning from the web with KIE. *International Journal of Science Education, 22*, 797–817. doi:10.1080/095006900412284.

Boykoff, M. T., & Boykoff, J. M. (2004). Balance as bias: Global warming and the US prestige press. *Global Environmental Change, 14*, 125–136.

Carpentier, N. (2011). *Media and participation: A site of ideological-democratic struggle* (p. 408). Bristol, UK: Intellect Ltd.

Carvalho, A. (2007). Ideological cultures and media discourses on scientific knowledge: Re-reading news on climate change. *Public Understanding of Science, 16*, 223–243. doi:10.1177/0963662506066775.

Cook, K. (2009). A suggested project-based evolution unit for high school: Teaching content through application. *The American Biology Teacher, 71*(2), 95–98.

Gee, J. P. (2004). Discourse analysis: What makes it critical? In R. Rogers (Ed.), *An introduction to critical discourse analysis in education*. Mahwah, NJ: Lawrence Erlbaum Associates.

Gray, R. (2009, March 16). Science and faith: The conflict. *The Telegraph*. Retrieved from http://www.telegraph.co.uk/news/science/4999924/Science-and-faith-the-conflict.htmlWeb

Kahan, D. M., Peters, E., Wittlin, M., Slovic, P., Ouellette, L. L., Braman, D., & Mandel, G. (2012). The polarizing impact of science literacy and numeracy on perceived climate change risks. *Nature Climate Change, 2*, 732–735. doi:10.1038/nclimate1547.

Keene, M., & Blumstein, D. T. (2010). Environmental education: A time of change, a time for change. *Evaluation and Program Planning, 33*, 201–204. doi:10.1016/j.evalprogplan.2009.07.014.

Lambert, J. L., Lindgren, J., & Bleicher, R. E. (2012). Assessing elementary science methods students' understanding about global climate change. *International Journal of Science Education, 34*, 1167–1188. doi:10.1080/09500693.2011.633938.

Masci, D. (2007). *How the public resolves conflicts between faith and science*. Pew Forum on Religion and Public Life. Retrieved from http://pewforum.org/Science-and-Bioethics/How-the-Public-Resolves-Conflicts-Between-Faith-and-Science.aspx

McComas, W. F. (1996). Ten myths of science: Reexamining what we think we know about the nature of science. *School Science and Mathematics, 96*(1), 10.

Nelkin, D. (1995). *Selling science*. New York, NY: Freeman.

Nelson, C. E. (2007). Teaching evolution effectively: A central dilemma and alternative strategies. *McGill Journal of Education, 42*, 265–283.

NGSS Lead States. (2013). *Next generation science standards: For States, by States*. Washington, DC: The National Academies Press.

O'Brien, M. (2014, February 16). "Science guy" Bill Nye debates lawmaker on climate change. *NBC News*. Retrieved from www.nbcnews.com/politics

Olausson, U. (2009). Global warming—global responsibility? Media frames of collective action and scientific certainty. *Public Understanding of Science, 18*, 421–436. doi:10.1177/0963662507081242.

Osborne, J., Erduran, S., & Simon, S. (2004). Enhancing the quality of dialogue in school science. *Journal of Research in Science Teaching, 41*, 994–1020. doi:10.1002/tea.20035.

Pew Research Center. (2015). *Major gaps between the public, scientists on key issues*. Retrieved from http://www.pewinternet.org/2015/01/29/public-and-scientists-views-on-science-and-society/

Saylan, C., & Blumstein, D. T. (2011). *The failure of environmental education (and how we can fix it)*. Oakland, CA: University of California Press.

Solomon, S., Plattner, G. K., Knutti, R., & Friedlingstein, P. (2009). Irreversible climate change due to carbon dioxide emissions. *Proceedings of the National Academy of Sciences of the United States of America, 106*, 1704–1709. doi:10.1073/pnas.0812721106.

Tuomela, R. (2013). Collective acceptance, social institutions, and social reality. *American Journal of Economics and Sociology, 62*, 123–165. doi:10.1111/1536-7150.t01-1-00005.

Walton, D. N. (1989). *Informal logic: A handbook for critical dialogue*. Cambridge: Cambridge University Press.

Weber, E. U., & Stern, P. C. (2011). Public understanding of climate change in the United States. *American Psychologist, 66*, 315. doi:0.1037/a0023253.

Windschitl, M., Thompson, J., & Braaten, M. (2008). Beyond the scientific method: Model based inquiry as a new paradigm of preference for school science investigations. *Science Education, 92*, 941–967.

Wynne, B., Simmons, P., Waterton, C., Hughes, P., & Shackley, S. (2001). Institutional cultures and the management of global environmental risks in the United Kingdom. *Learning to Manage Global Environmental Risks, 1*, 93–114.

Ziman, J. (1991). Public understanding of science. *Science, Technology & Human Values, 16*, 99–105.

SECTION ELEVEN

Science Education Reform

What Have Been the Goals and the Effects of the Attempts to Improve Science Education Over the Last Decade?

CHAPTER TWENTY-EIGHT

Science Education Reform

Can Students Learn Science While Navigating Oppressive Schools Within an Oppressive Society?

ROWHEA ELMESKY

When unarmed Michael Brown was shot to death by a police officer on the neighborhood streets in Ferguson, Missouri at the young age of 18, his death sparked a massive outcry for the valuing of Black lives and for the re-evaluation of America's judicial system. "We are all Michael Brown!" became a symbolic statement signifying the rejection of a broken system that uses excessive force on marginalized groups, especially African American males. In fact, in the months that followed, inflamed by Ferguson, individuals and groups mobilized the national Black Lives Matter movement as additional tragic examples, like Freddie Gray, Eric Garner and Tamir Rice, surfaced one after another. On March 5, 2014, the extensive *Investigation of the Ferguson Police Department* report, compiled by the United States Department of Justice Civil Rights Division, illuminated disgraceful law enforcement practices. The report found clear and insurmountable evidence that policing patterns in Ferguson reflected discriminatory intent, exacerbated racial bias, and revealed racial disparities. Moreover, the report highlighted a deep mistrust between parts of the community and the police department.

It is within this volatile context of distrust, social control and aggressive policing practices that marginalized youth, in Ferguson and beyond, live and attend school. By understanding society as nested social spaces or fields with boundaries that are porous, allowing cultural practices and symbols to seep from one field to another (Sewell, 1999), African American students in Ferguson, like Michael

Brown, experience a culture of disempowerment both inside and outside of school walls. As documented by the U.S. Department of Justice, schools within Ferguson were found to replicate and reinforce punitive neighborhood practices. The report found, "FPD's [Ferguson Police Department's] approach to policing impacts how its officers interact with students, as well, leading them to treat routine discipline issues as criminal matters and to use force when communication and de-escalation techniques would likely resolve the conflict (p. 37)." Student Resource Officers were found to use unnecessary force on students, for example, pushing a female student into lockers and drive-stunning a middle school student with an electronic control weapon (ECW). Unfortunately, Ferguson is not an anomaly; rather, it is indicative of hundreds of schools serving marginalized populations where students are disciplined through practices that punish, isolate and alienate. These students face unhealthy, daily encounters with symbolically violent practices, as well as with consciously and unconsciously-enacted forms of micro-aggressions or verbal, behavioral, or environmental factors that degrade them. Moreover, racial disproportionality in the use of exclusionary discipline practices is a nationally documented phenomenon. For example, Erica Shirley and Dewey Cornell (2012) found that Africa American and Black students are suspended about three times more than their White peers. Another study found that while African Americans make up 17% of the American public school student population, they represent 32% of all students suspended (Hinojosa, 2008).

One may wonder what is the connection between these recent "political" events and science education reform, and one may question why I would open a chapter, focused upon improving science education, by highlighting neighborhood policing practices and patterns of discipline disproportionality occurring in schools. In this chapter, I argue that the goals of improving science education and the United States' constant endeavour to become a global leader in the areas of science, technology, engineering and mathematics has yet to be achieved *precisely* due to the dearth of research that addresses the ways that students' experiences inside of science classrooms replicate and reinforce the same repressive structures that they must navigate outside of school. Whereas some researchers have pointed out the critical role that external stressors like poverty play in impacting students' performance in school (Berliner, 2009, 2013), such research downplays the role of schools in helping students combat the onerous social and cultural factors affecting them. David Berliner (2013) wrote:

> What is it that keeps politicians and others now castigating teachers and public schools from acknowledging this simple social science fact, a fact that is not in dispute: Outside-of-school factors are three times more powerful in affecting student achievement than are the inside-the-school factors (Berliner, 2009)? And why wouldn't that be so? Do the math! On average, by age 18, children and youth have spent about 10 percent of their lives in what we call schools, while spending around 90 percent of their lives in family and

> neighborhood. Thus, if families and neighborhoods are dysfunctional or toxic, their chance to influence youth is nine times greater than the schools'!

While Berliner should be commended for bringing to the forefront the importance of attending to the external factors affecting students' performance in school, this argument shifts responsibility away from the school, and doesn't allow one to ask, what happens when both the school environment and the home/neighborhood environment are toxic? After fifteen years of studying the sociocultural dimensions of science education for students who are economically/racially/ethnically marginalized in society, I have learned that there is profound insight to be gained when we understand that students' experiences inside school are very similar to those experiences outside of school—especially the ways in which they are perceived and treated by those with authority. Students will continue to experience failure in science classrooms unless we address the critical role that administrators', teachers' and support staff's practices and cultural perspectives also play in the students' academic performance. Specifically, I argue that our society is compelled to begin to decipher the poor achievement of marginalized youth in science and other STEM areas as linked to the culture of social control and dehumanization facing students, both outside and within the educational institutions that are entrusted to care for children. Hence, there is a need for learning how to interrupt science teaching practices that reproduce the same deficit perspectives and unjust treatment that youth are experiencing in outside-of-school contexts. Until we stop pretending that the schools present a "safe haven" that protects children from oppressive structures and learn to address issues of social justice holistically, high quality science teaching and learning opportunities will continue to be inaccessible to a large segment of our population.

"WE PEOPLE": DEHUMANIZATION AND DEFICIT VIEWS OF MARGINALIZED YOUTH

> My best friend just passed away, and I tried to tell my teachers like, this was going on. And most of them are just like, "Well, you just gotta make up the work. And you gotta do this." And I'm like, "I'm not mentally here. I'm here because I have to be." Because if I don't, they gonna say I missed too many days of school. But I'm not really here, so I just need that extra help. And it seem like they don't care. Like, they push it off. I know they got a lot of kids—I know a lot of kids coming through the doors and have problems, too, **but we [are] people.** I know you [are] people, too, but it's hard. (Daneesha, Local High School, student focus group, Spring 2015)

Death of family members or friends, as described in Daneesha's vignette, is one of the many types of physical, social, emotional and mental health related stressors being navigated by marginalized students. They often carry these loads through school doorways only to find themselves facing adults who may not recognize or understand their challenges and who interpret their lack of engagement in the science classroom as intentional and possibly malicious. High levels of control, enforced by exclusionary policies, lead to minimally supportive school environments that label "apathetic," "troublemaker," or "unmotivated" students and respond to them by replicating hegemonic external societal structures. Consequently, powerfully stated in a two-word proclamation—"we people"—students feel compelled to remind the adults in their schools that they should be treated with compassion in order for learning to occur.

My recent work in one predominantly Black high school in the midwest, Local High School (LHS), documented the perspectives of four different stakeholder groups at the school—administrators, teachers, support staff, and students—regarding the school culture. Consistently emergent throughout the student focus group data was their request to be humanized as well as their desire to humanize their teachers. One student reflected on the relationships between teachers and students in the school: "There could have been disconnection between students and the teacher, and it kind of goes both ways with one not realizing that the other is another person. … because you don't really treat the other as a human interaction." The LHS teacher-student disconnection, referenced by the student in this quote, was repeatedly cited during conversations with all of the stakeholders. Despite the teachers' well-vocalized intentions of helping students to succeed academically and continue onto college, students expressed that they did not trust their teachers, consistently felt disempowered in school, and experienced unpleasant interactions during which they felt minimized. Furthermore, students repeatedly complained of being falsely accused of severe infractions such as cheating, plagiarizing, and skipping class, and they expressed the strong desire for learning within an environment that supported positive teacher-student relationships.

LHS is not very different from other high schools serving student populations that are economically disadvantaged and racially/ethnically marginalized. The teaching staff in these schools tends to be majority White, comes from backgrounds starkly different from their students, and has lived experiences that barely resemble the students' daily lives—all of which contributes to the development of deficit perspectives about the marginalized students. When teachers lack asset-oriented views about their students, their deficit-oriented ideas (e.g., poor students of color are dishonest) creep into the classroom and either inhibit the development of student-teacher relationships or interrupt opportunities for strengthening such relationships. During one focus group, a senior, Alisha, shared a vignette capturing the frustration she experienced when her teacher questioned her integrity and honesty.

> It was a group quiz and this boy comes up and bothers me and my partner … and I was like, "Move your paper so she [the teacher] doesn't think we're cheating." And as soon as I said that, she's like, "Give me both of your papers. You're cheating." So, I was like, "What?"

> Like, "What?" She's like, "Give me your paper." She gets up. He's trying to explain—he's like, "No, no. We weren't cheating." I'm like, "We weren't cheating." She was like, "Yeah, you were." So, this is when I'm like, "Well, on the normal, we [the teacher and I] have a good relationship—why won't you listen to me? Why won't you try to understand?" (Alisha, Local High School, student focus group, Spring 2015)

In addition to causing a strain to teacher-student relationships, deficit views about students impact their ability to successfully access the science discipline in multiple ways. When teachers see their students as capable science learners, they are more likely to implement innovative pedagogical approaches in the classroom in search of mechanisms to provide opportunities for their students to learn science. For example, Cristobal Carambo and Anita Abraham, two teachers with whom I worked at City High in a large northeast city, converted their classrooms in an under-resourced school into interactive, dynamic spaces where students could participate in hallmark chemistry laboratories as well as challenging activities such as DNA electrophoresis, growing fast plants, and comparative anatomy dissections. Even in a school where science was being taught in a book-centric and rote manner, these two teachers with asset-oriented mindsets were able to transform ordinary science classrooms into places where students were engaged in inquiry learning. Whereas both of these science teachers felt that the students deserved quality science instruction and believed in the intellectual capabilities of the school's 98% Black student population, of whom 80% were receiving free or reduced lunches, unfortunately, these two teachers do not represent the majority of those teaching in schools with high numbers of marginalized youth.

STRENGTHENING TEACHER-STUDENT RELATIONSHIPS

The disconnection experienced by many marginalized groups of students in schools like LHS is shaped by the manner in which the students' embodied ways of being or embodied forms of cultural capital (Bourdieu, 1986) are received by their teachers. Students who bring attitudes, social norms and/or discursive practices that do not align well with dominant standards, as defined by society and the school, struggle with finding acceptance in the classroom and with building connections with their teachers. Some students, like Ayesha in the quote below, believe that teachers will only become truly invested in their students' growth and development when they obtain greater insight and understanding of a student's "home life." This awareness will, she argues, foster the emergence of stronger teacher-student relationships.

> We [the school] could maybe have some type of workshops with student/teacher relationships. 'Cause I think that—I really think that is really important, because it makes such a difference. … When you really get a good relationship with your teacher, your teacher is

> gonna want to see you grow—he or she. So, they're gonna do things to try to help you out. "Hey, you need to pay attention. Hey, maybe you can stay after school so you can understand this." … So, if they get a good understanding like, of your home life—they don't have to know everything, but just how it goes—why, sometimes, you can't get this [homework] done and whatever—that might make [your teacher] … care a lot more. (Ayesha, Local High School, Focus Group, Spring 2015)

Ayesha's reflection and passionate call for teachers to understand their students' lived experiences, illuminates the extent to which the students' embodied cultural capital may be currently overlooked, misunderstood or minimized in the school environment—especially when students enact capital in non-dominant forms. That is, negative notions of marginalized students' capabilities and motivation are reinforced as students and teachers interact in classrooms—especially when students engage in practices that look and sound different from what the dominant school culture defines as appropriate or respectful. Her comments are also insightful in pointing out that students' communication with teachers about their lives does not only facilitate acceptance and positive feelings for the students, but also improves their cognitive engagement and academic performance in school.

> And by having the relationship with teachers and knowing how to keep those good relationships—that would probably help grades, 'cause I know it helped mine. It would probably help the discipline levels—or, you know, the discipline reports and things like that. And overall, like, have the student body feel good about coming to school and not feel discouraged about it or not feel as intimidated by it. So, if you could get something like that going on for students that are struggling with student/teacher relationships, then that might help out some. (Ayesha, Local High School, Focus Group, Spring 2015)

FIGHTING THE DEFICIT MINDSET: LEARNING TO VALUE STUDENTS' CULTURAL CAPITAL

Racially marginalized and economically disadvantaged children have long been denied access to quality science education experiences. A contributor to this reality is the lack of recognition of the cultural resources that marginalized children possess that can help them to be successful in science learning environments. Being perceived through deficit lenses relegates them to being ignored, misunderstood or regarded as inferior by mainstream standards, and often lead to exclusionary practices that remove a child from a science-learning environment rather than facilitate their inclusion. Sometimes entire curriculum decisions are made based upon negative views of what the children are capable of doing. For example, for over a decade, laboratory setups (i.e., a dissection) that required the use of sharp objects were replaced by rote activities and worksheets at City High because students were thought to be violent and capable of harming each other with dissection tools.

In contrast, in my research, I have found that marginalized youth are owners of a wealth of resources that can contribute to their participation in science classrooms, yet teachers—being primarily from the dominant White, middle to upper middle class population—may not understand their students' lives enough to recognize, appreciate, or make connections to their school science experiences. In light of oppressive structures that marginalized children may experience outside of school, (e.g., poverty, neighborhood isolation, power stratifications, and excessive policing practices), they have not had the same opportunities to develop ways of being in the world that "fit in" or carry symbolic value with the dominant class. What's more, there is little or no valuing of the embodied cultural capital that they do bring to school. Even when they express mainstream attitudes such as persistence and values such as the achievement ideology or utilize discursive resources that can assist in scientific inquiry and argumentation, more often than not, miscues and misunderstandings arise, due to the negativity already surrounding the student population.

MENTAL HEALTH, MINDSET AND MINDFULNESS

Students need to be in environments that are reinforcing and supportive. They need to feel that their voices are valued, have a sense of purpose and drive, and experience opportunities for exercising leadership and other forms of agency in their lives. A culture that is characterized predominantly by deficit views derails the focus from developing classrooms of learners who are engaged, responsible critical thinkers, genuinely interested or invested in the content matter. When students are subjected to being perceived negatively both inside and outside of school, their feelings of being dehumanized extend into their classrooms and lead to failed interactions with teachers. It feeds the development of negative attitudes associated not just with schools, but also with science specifically. If solidarity and relationships take precedence, however, students may be more willing to engage in classroom activities even if the required tasks are not intrinsically valued. An LHS administrator reflected on student test scores in relation to her teachers' abilities to form relationships with her students, illuminating the power of relationships:

> She gets the best results out of the kids on EOC. … so Ms. Smith is not—I'd say after you look at her teaching—she's not the best teacher in the building, but she forms really good relationships with the kids so when they go in to take an EOC or a test or something, they want to perform well for her because they don't want to disappoint her. (Administrator, Local High School, Spring 2015)

The creation of more successful learning environments will require ideological shifts amongst not just the science teachers but the school personnel at large so that under-represented students can feel humanized and respected and encouraged to bring their discursive resources and cultural practices to their science learning experiences. In addition to focusing upon shifting the adults' fixed mindsets about marginalized youth within schools, there must be an emphasis on providing science teaching that recognizes and embraces the multitude of resources embodied within populations of economically disadvantaged youth, even if such capital is considered as non-dominant in mainstream society. These students will then become more engaged in schools and in science classrooms. Thus, recognizing the capital of racially marginalized and economically disadvantaged students and embracing the ways in which those resources can help them develop school science skills is crucial for shaping a positive classroom environment. In understanding failed teacher and student interactions in school as replications of deficit perspectives and power differentiation based on capital possession from outside of school spaces, certainly, great possibilities for expanded opportunity in science classrooms arise by moving beyond identifying "shortages" to recognizing children's skills, dispositions, beliefs, attitudes, and language resources.

There are broad public health implications if we can address and minimize the micro-aggressions experienced by marginalized students, especially as they transition from adolescence to adulthood. Mental health and well-being are critical components to becoming a productive individual and fulfilling one's purpose in life. Whereas society recognizes that students' health can be impacted negatively by frequent exposure to outside-of-school stressors such as poverty, family dynamics, neighborhood isolation, drugs and crime, there is minimal attention directed to understanding the impact of inside-of-school stressors. Schools with marginalized students are infused with deficit mindsets, punitive approaches, and minimal support structures are available to help them cope with the high levels of stress they may be enduring. I close the chapter by reminding us of Daneesha's passionate cry for humanization. Schools must become safe spaces filled with adults who are mindful of the oppressive structures facing marginalized youth, dedicated to transforming rather than replicating such structures, constantly challenging themselves to embrace an asset-oriented mindset, and focused on producing new teaching practices that will communicate to students that they are truly "people"—with the embodied resources to engage critically with their school, with STEM and with the world at large.

REFERENCES

Berliner, D. C. (2009). *Poverty and potential: Out-of-school factors and school success.* Retrieved from http://epicpolicy.org/publication/poverty-and-potential

Berliner, D. C. (2013). Effects of inequality and poverty vs. teachers and schooling on America's youth. *Teachers College Record, 115*, 1–26.

Bourdieu, P. (1986). The forms of capital. In J.G. Richardson (Ed.), *Handbook of theory and research for the sociology of education* (pp. 241-258). New York, NY: Greenwood Press.

Hinojosa, M. S. (2008). Black-white differences in school suspension: Effect on student beliefs about teachers. *Sociological Spectrum, 28,* 175-193.

Sewell, W. H. (1999). The concept(s) of culture. In V. E. Bonnell & L. Hunt (Eds.), *Beyond the cultural turn: New directions in the study of society and culture* (pp. 35–61). Berkeley, CA: University of California Press.

Shirley, E. L., & Cornell, D. G. (2012). The contribution of student perceptions of school climate to understanding the disproportionate punishment of African American students in a middle school. *School Psychology International, 33,* 115–134. doi:10.1177/0143034311406815.

United States Department of Justice Civil Rights Division. (2015, March 5). *Investigation of the Ferguson police department.* Retrieved from https://www.justice.gov/sites/default/files/opa/press-releases/attachments/2015/03/04/ferguson_police_department_report.pdf

CHAPTER TWENTY-NINE

Science Education Reform

Reflecting on the Past and Raising Questions for the Future

LUCY AVRAAMIDOU AND LYNN A. BRYAN

There is no question that our rapidly changing world in the 21st century presents profound and intractable global challenges that defy simple technical and conventional solutions. Grand socio-political issues ranging from overpopulation, global warming, climate change, and loss of biodiversity to poverty, racism and prejudice, terrorism, and gender inequality—all require ambitious, principled, and ethical solutions that will have an impact beyond the boundaries of a single society. All place new demands on and raise increasingly high expectations for the role and quality of science education as an instrument for viable solutions.

As former science teachers and current science educators, we both have lived—in different countries (U.S., U.K, and Cyprus)—through reform movements in science education and have experienced the evolution of science education over many decades. We have encountered and experienced science education reform from a number of perspectives: as a high school teacher before the days of science standards, as an elementary teacher at the time when standards were being implemented, as university level instructors who work with prospective teachers, as professional development leaders, as a parent of a child in the K-12 science education system, not to mention as science students ourselves. All of these perspectives influence our observations of and positions about science education reform over the last few decades. Thus, in this chapter our aim is not to elaborate on the full history of science education reform, as literature is replete with historical descriptions that chronicle science education reform in a number of countries around the world, but to raise a few questions through critical reflection regarding some of the features of science education reform, particularly in the U.S. and Europe, over the last two decades. This time frame seems apropos, as it was 20 years ago that the second edition of *13 Questions: Reframing Education's Conversation* (Kincheloe & Steinberg, 1996) was published. In addition, a major science education reform document in the U.S., the *National Science Education Standards* (National Research Council [NRC], 1996), was released in the same year. So what has happened as a result of science education reforms over the last two decades? What goals, purposes and rhetoric have driven science education reform and what have been the consequences of the implementation of these reforms?

WHAT HAS DRIVEN SCIENCE EDUCATION REFORM IN THE LAST TWO DECADES?

In both Europe and the U.S., science education reform efforts over the last two decades typically have come about as a response to *crises*, predominantly framed in terms of economic vitality and technological competition. As Paul DeHart Hurd (1986) noted decades ago in one of his analyses of the history of science education reform in the U.S. that the enduring need to reconceive school science has been motivated by one crisis to the next for nearly a century:

> The issue has always been how to best prepare citizens to live in a culture most often described in terms of achievements in science and technology. Whether the speakers have been scientists or educators, the recommendations for reform have shown remarkable agreement. From one educational crisis to the next, we find a continuing refinement of the rationale for improving quality science education. (p. 355)

Perhaps, the most recent example of science education reform can be found in the context of the European Union (EU), which currently faces two major related issues: a multi-year debt crisis and a steady decline of interest towards science-related careers. These two crises are exemplified in various reports that have been published in recent years, for example: *Europe Needs More Scientists* (European Commission [EC], 2004), *Science Education in Europe: Critical Reflections* (Osborne & Dillon, 2008), *Science Education NOW: A Renewed Pedagogy for the Future of Europe* (EC, 2007), and *Encouraging Student Interest in Science and Technology Studies* (Organisation for Economic Co-operation and Development [OECD], 2008).

Following these reports, in August 2015, the European Commission published the report, *Science Education for Responsible Citizenship* (EC, 2015), which offers a 21st century vision for science for society within the broader European agenda. The report places emphasis on the process of aligning research and innovation to the values, needs and expectations of society, referred to as "Responsible Research and Innovation" (RRI). The main objectives of the report emphasize the following: attention to promoting RRI and enhancing public understanding of scientific findings; enhancement of the quality of teaching; collaboration between formal, non-formal, and informal educational providers; and connecting innovation and science education strategies at local, regional, national, European and international levels.

Unlike European initiatives for reform that emphasize the role of science to society, in the U.S., numerous reports published over the last several decades have called for a re-visioning of science education with the goal of making the U.S. the most competitive nation in the world. Going back to 1983, the National Commission on Excellence in Education argued in their report, *A Nation at Risk: The Imperative for Educational Reform* (National Commission on Excellence in Education, 1983) that academic standards had fallen in the U.S. and warned of a national education crisis: "Our Nation is at risk. Our once unchallenged preeminence in commerce, industry, science, and technological innovation is being overtaken by competitors throughout the world" (p. 3).

Dozens of reports issued over the subsequent years would support the commission's conclusions. In 1989, the American Association for the Advancement of Science (AAAS) published *Science for All Americans* (Rutherford & Ahlgren, 1990). This was a report to the nation on what constitutes literacy in science, mathematics, and technology and the steps necessary to achieve it were, in fact, a response to the call for standards-based reform. By 1993, Project 2061 (1993) had developed *Benchmarks for Science Literacy*, a set of specific K-12 learning goals to enable educators to help students achieve science literacy by the time they graduate high school. *Science for All Americans* and *Benchmarks for Science Literacy* served as a foundation for the subsequent *National Science Education Standards* (NRC, 1996), which focused on scientific inquiry as the new approach to science teaching and learning (NRC, 2000).

In the years that followed Project 2061 and the NSES, U.S. government and corporate leaders expressed growing concern, based on a confluence of indicators, that the future vitality of the U.S. science and technology enterprise and the growth of highly-skilled jobs were in serious jeopardy. As a result, an alarming report entitled, "Rising above the Gathering Storm: Energizing and Employing America for a Brighter Economic Future" (Institute of Medicine, National Academy of Sciences, and National Academy of Engineering [IM/NAS/NAE], 2007), decried the dismal state of math and science education and called for an "increase [in] America's talent pool by vastly improving K-12 science and mathematics education" (p. 5) and enlarging the pipeline of students who are prepared to enter college and graduate with a degree in science, engineering, or mathematics by increasing the number of students who pass AP and IB science and mathematics courses" (p. 6). By this time, syntheses of key recommendations for science and technology education in major reports from business, industry, government agencies, and associated groups were being reported. One of the main findings of this work was that the recommendations were driven largely by goals related to the role of science and technology in U.S. economic competitiveness, as opposed to addressing science and technology *education* (Bybee & Fuchs, 2006).

WHAT HAVE BEEN SOME OF THE EFFECTS OF REFORM IN SCIENCE EDUCATION?

Much has been written, especially within the U.S context, about the rationale, goals, and processes of reform in science education over the last half century (see, for example, DeBoer, 2000). In what follows, we offer our reflections on and raise questions about four intertwined effects of reform in science education and issues related to those effects: (a) the increasingly myopic and melodramatic rhetoric surrounding science education reform; (b) the de-professionalization of teaching; (c) the aftermath of standards; and, (d) a testing and accountability culture.

Science Education Reform Rhetoric. The ideology driving science education reform in the U.S. and many other countries, particularly industrialized nations, has essentially remained constant over the last several decades. Impending crises and sober outlooks have traditionally framed reform narratives that admonished those within the educational system to reform science education. However, what we have seen over the last 20 years is a mounting intensity and an increasingly myopic nature reflected in the language of reform—a building urgency communicated through hyperbolical metaphors, melodramatic pronouncements, and blunt declarations. For example, U.S. documents warn that "storms" are gathering (IM/NAS/NAE, 2007) and are "approaching Category 5" (National Academy of Sciences, National Academy of Engineering, and Institute of Medicine [NAS/NAE/IM], 2010), and STEM is experiencing a "leaky pipeline" that will compromise a "race to the top" (U.S. Department of Education, 2009). STEM education has become a Titan holding the world on its shoulders: "STEM education will determine whether the United States will remain a leader among nations and whether we will be able to solve immense challenges in such areas as energy, health, environmental protection, and national security" (President's Council of Advisors on Science and Technology, 2010, p. 1).

In addition, the rhetoric of science education reform has become much more myopic. In the 1980s and 1990s within the U.S., issues of equity were given an explicit part of the narrative with slogans such as "Science for all Americans" (Rutherford & Ahlgren, 1990), "Science for All" (American Association for the Advancement of Science, Project 206 [AAAS], 1995), and "principles of equity and excellence" (NRC, 1996). It seems that increasingly over the last two decades, science education reform rhetoric has shifted its center of mass away from equity and excellence and toward competition and standardization. Calls for reform in science education have been driven by questions such as: "What are the top 10 actions, in priority order, that federal policymakers could take to enhance the science and technology enterprise so that the United States can successfully compete, prosper, and be secure in the global community of the 21st century?" (151 Cong. Rec. 22769, 2005).

Further, in examining educational reform rhetoric, it is important to consider the institutions/agencies that drive these reform initiatives (e.g., World Bank, International Monetary Fund), the kinds of narratives and ideologies that guide these initiatives (e.g., neoliberalism, capitalism), and the intended outcomes (e.g., economical goals, political-cultural goals). The drive for economic growth in recent decades (indeed, the last half century) has led to substantial increases in wealth for large numbers of people across the globe. However, despite significant gains in global economic output, there is evidence that rather than reducing inequalities, our current social, political and economic systems are exacerbating inequalities (Dabla-Norris, Kochhar, Suphaphiphat, Ricka, & Tsounta, 2015). *Is this where science education should lead us?* When science education reform initiatives are framed in terms of global economic competitiveness and dominance, in terms of privilege and power, science education becomes a tool—*but, for whom do we teach science and to what end?*

De-professionalization of Teaching. Efforts for large-scale educational reform, particularly in the U.S., have typically been top-down, legislated, mandated by the courts, and promoted by government agencies and organizations (Kahle, 2007). In other words, the recommendations for reform within schools have primarily come from outside the public school system and have usually been disseminated through legislation, policy reports and publications that prescribe specific programs, practices, and curricula. Perhaps, most critical has been the way standardized education has affected science teachers' practices over the last two decades. George DeBoer (2000) noted that the science education reform initiatives in the U.S. have significantly reduced the amount of control that individual teachers and school districts have over their own science programs. However, the erosion of teacher decision-making and autonomy—the de-professionalization of teaching—is not limited to U.S. classrooms. As Bernie Froese-Germain (2010) described, a teacher in Ontario, Canada reflects on the changes he has observed over his career:

> Over my 20-plus years of classroom teaching, I have become convinced that the debates about education are now, more than ever, populated by an excess of experts whose purported expertise is inversely proportionate to the amount of time they have spent inside a classroom. Those of us who have taught in classrooms for many years have seen scores

> of bandwagons come and go—but the vast majority of the time we have not seen these initiatives produce the panacea of results they had professed to obtain. Short-term ideas—often politically-driven, inconsistent and incoherent rather than providing concrete information, realistic ideas or authentic suggestions—are intruding and micromanaging my classroom practice. As a result, I usually find that, more and more, my freedom to use professional judgment and, therefore, my autonomy as an educator has been and is being curtailed. (p. 25)

These words likely echo the sentiments of many other teachers around the world struggling to enact reform recommendations and to ultimately feel pressure to "teach to the test." This is the case in Cyprus, as well, where Lucy worked as a teacher educator for a decade.

Since adopting a centralized curriculum in 1960, Cyprus, a small and diverse island in the Mediterranean, has been struggling to adopt reform recommendations as proposed by the EU and to improve its TIMSS and PISA scores. In conjunction with a set of national goals, the science curriculum is overloaded with units, which are virtually impossible to teach in the number of science hours, and many of which are not in line with the specific geographical and cultural context. In her first-year of teaching, Lucy taught in a small rural elementary school (150 students), the children coming from eight different small villages where the majority of parents, economically disadvantaged, were employed in the areas of farming and agriculture. Consequently, the students at this school had a great interest in farming and agriculture as well, as they were spending most of their time after school helping their parents.

Yet, the national curriculum did not make any specific reference (nor left room for potential reference by the teachers) to these rural and isolated settings. Instead the standards proposed the same science learning goals for the students in small rural schools as for students in large urban schools—disconnected from the students' sociocultural backgrounds, knowledge, interests, and experiences. Inevitably, the students showed no interest in science and were disengaged with the classroom activities. Three months into the year, Lucy in consultation with the head-master and another science teacher, selected units that could realistically be taught in the year, and revised them to make them more context-specific and relevant to her students. As one might expect, students' interest in and engagement with science and learning were superb. However, the students' scores on the end-of-year national exams were very low, given that some units were not taught and specific goals had not been met. This dilemma is undoubtedly one that many other science teachers have experienced. Should she teach a set of disconnected, uncontextualized science facts from a national curriculum so that students can score well (or at least better) on a standardized national exam? Or should she teach less content in a manner that more deeply engages and connects to students' real lives and from which students learn more deeply but score worse on a test (because they "cover" less material that is tested)? Do teachers really have a choice? *For what goals do we teach science and whose professional expertise decides those goals?*

The Aftermath of Standards. One profound change over the last two decades in the day-to-day instruction in U.S. science classrooms began with the standards-based movement of the 1990s. Right after the release of the *Benchmarks for Science Literacy* (Project 2061, 1993), the *Improving America's Schools Act* (1994) was authorized, for the first time making standards-based education a law. By 1996, the *National Science Education Standards* (NRC, 1996) debuted, and with the No Child Left Behind (NCLB) Act of 2001 (NCLB, 2002), the standards-based movement intensified as content standards were integrally linked with test-based accountability measures.

Early in the science standards movement, William Kyle (1996) warned of the unintended ways in which science standards may be used:

> In recent years, members of various professional organizations, including members of the science education community, have been caught up in the process of creating and promulgating academic standards, but they have not attended to the social, economic, and political agenda that engaged them in this process. Nor have they analyzed the terrain in which such standards will be utilized. Science educators should not be surprised, then, in the likely event that such standards are used in unintended ways. (p. 1044)

Much has been documented since this time about the standards being used in "unintended ways." Indeed, if you have a child in the K-12 school system, you may have had direct experience with these "unintended ways." Undoubtedly, many science educators, students, parents, etc. have observed lists of standards written in the corner of backboards and whiteboards in science classrooms at nearly every grade level—standards that are mandated to be written on the board to report which ones are being addressed in instruction. On any given day of the academic year, teachers must be prepared for an administrator to walk into the room and take note of the science standards related to that day's instruction. It has become a regular feature in many of the Indiana science classrooms in which Lynn has worked with teachers (and certainly classrooms elsewhere and of other disciples, as well). What has evolved over the last two decades is an increasingly rigid adherence to standards—not as a guide or a road map—but as the curriculum. Under intense scrutiny of administrators and others who oversee accountability measures in school, teachers feel very little, if any, latitude to deviate from the standards or the textbooks that have been produced to correlate any given state's standards. This rigidity is epitomized in the following brief anecdote. A few weeks before the end of the school year, Lynn's middle school daughter brought home a study guide for a comprehensive exam over the entire year of science. The study guide consisted of a word-for-word reprint of the Indiana Academic Standards for science for her daughter's grade level. To repeat, twelve and thirteen-year-old students were given the list of state science standards for their grade level to use as the study guide for an end-of-year comprehensive exam! The state science standards had not only become *the* curriculum, they had become the study guide. Should there be any doubt that these children had "covered" all of the science standards, one need only look as far as the study guide for the end-of-year final exam. This "unintended" use of standards is but one example of the aftermath of the marriage between standards-based reform and test-based accountability policies. As John Settlage and Lee Meadows (2002) described over a decade ago:

> The effects of the 1996 standards-based reform linger in schools. The standards are used as a curriculum and as such become a set of disjointed facts that students need to know. Quality science curriculum [has become] a casualty of the standards-based reform as teachers are obligated to address lists of unrelated topics within their science courses. (p. 119)

Unfortunately, in many U.S. classrooms, the effects of the 1996 standards-based reform that Settlage and Meadows describe persists today. *For whom are the standards intended and for what goals?*

To their credit, the authors of standards documents, whether national or state level, remind their audience that standards are not the curriculum. For example, the National Research Council (1996) explicitly states in the NSES: "All users and reviewers are reminded that the content described is not a science curriculum" (NRC, 1996, p. 111). The Indiana Department of Education (2016) website repeats this caution no fewer than three times on their website: "The standards are not curriculum;" "While the standards may be used as the basis for curriculum, the Indiana Academic Standards for Science are not a curriculum;" "… the standards do not prescribe any particular curriculum and it is encouraged to seek out strategies that best meet the needs of the individual educator's classroom."

To emphasize, we do not disagree with the concept of standards; however, we do disagree with many aspects of the standards-based culture that has evolved (e.g., excessive focus in K-12 classrooms on "coverage"; narrowing and oversimplifying of the curriculum; lack of resources to support the implementation over time; educational publishing companies' "cashing in" on the standards and testing culture). We agree that standards share an improved vision of science education and have emphasized innovative approaches to science teaching (e.g., the focus on scientific inquiry stimulated by the NSES). Across the spectrum of inquiry-based practices, different researchers around the world have conducted studies that provide evidence of the benefits of implementing inquiry-based approaches to science teaching and learning (e.g., Crawford, 2000). The outcomes of these studies illustrate the positive impact of inquiry-based science to students' learning and students' understandings of the nature of science and the work of scientists (Schwarz & Gwekwerere, 2007). There is also promise in the latest U.S. science education reform effort to develop rigorous and coherent K-12 science standards, the *Next Generation Science Standards* (NGSS Lead States, 2013). Unlike previous standards, the NGSS combine three distinct and equally important dimensions to learning science (cross-cutting concepts, science and engineering practices, and disciplinary core ideas) to form each standards. Solidly based on *A Framework for K–12 Science Education: Practices, Crosscutting Concepts, and Core Ideas* (NRC, 2012), the NGSS describe learning goals that become increasing more complex from grades K to 12, and focus on the competencies in science and engineering that all students should be able to demonstrate. *Considering the aftermath of the last two decades of science standards, what will/can be different in the implementation of the NGSS to ensure rigorous and high quality science education for American students? How will the standards culture change? What will prevent the NGSS from being implemented in "unintended ways" that lead to far-reaching unintended effects?*

Testing and Accountability Policies. Entangled within an argument that the future of the nation's economic dominance and technological superiority is the use of national rankings on large-scale, external international assessments such as PISA and TIMMs as indicators of a nation's economic and technological future. Test scores have become accepted as "a valid indicator of the current state of affairs and sufficient justification for state and federal governments to exert more control over the direction the science education program should take" (DeBoer, 2000, p. 594). That is, such tests have been afforded an elevated status in determining whether or not a particular nation has the economic and technological capacity to be a serious global competitor. In turn, policy decisions essentially rest on these tests. Many countries, particularly industrialized nations, justify their educational reforms through their performance on PISA (Figazzolo, 2009): "Half of the [57] respondents report that their government has launched reforms or initiatives in education as a consequence of or with reference to PISA 2006: rankings and results have been used to justify and sustain these actions" (p. 8). For example, as a result of Australia's performance on the 2013 PISA, the Australian government declared in its Australia Education Act (AEA), that one of the central objects of the act is "for Australia to be placed, by 2025, in the top 5 highest performing countries based on the performance of school students in reading, mathematics and science" (AEA, 2013, p. 1). Any one of the hundreds of U.S. science education reform documents and reports in the last twenty years is replete with references to PISA or TIMMS or other standardized test scores.

Researchers around the world have criticized the emphasis placed on these tests and particularly the fact that reform efforts have been designed to serve these tests. As Froese-Germain (2011) argued:

> The culture of standards-based accountability and data-driven school improvement distorts the educational process, and can result in "gaming the system," leading to cynical, quick-fix strategies to appease administrative superiors and create the appearances of improvement that would keep politicians and the public at bay. (p. 40)

Should the worth of a nation's mean test score be elevated to the point of driving its education agenda? If scores on exams like PISA and TIMMS are to be used, then it is incumbent on those who make decisions based on these scores to systematically examine the results and how data are interpreted: the mean scores only reflect a small piece of a large image, and so we have to look beyond the means. As Gorur and Wu (2015) warned, if a country "insists on using PISA to inform policy, it would do well to explore the data in much greater detail. Average scores obscure far more than they reveal. Using average performance scores and rankings to inform policy is leading to damaging policy decisions" (p. 622). Particularly in light of decades of research on learning and assessment as well as volumes of critiques of the use of large-scale external measures like PISA, we have to ask, *how has a single measure (or single scores from various tests) become and continued to be so influential in science education reform and education reform in general?*

MOVING FORWARD

In this chapter so far, we have shared a few of our observations of and reflections on the rhetoric that has formed the basis of science education reform recommendations, especially in the U.S., over the past two decades. In doing so, we have identified some problematic issues, ranging from the framing of reform rhetoric to testing and accountability policies. Built upon these reflections about the past, we offer a few ideas for re-envisioning science education and concurrently raise questions for the future.

Reframing the Rhetoric. A fundamental question that was raised in this chapter is: *what rhetoric has driven science education reform efforts in the last 2 decades?* As we described earlier, the rhetoric of reform in science education has short-sightedly envisioned science serving economic and technological goals, instead of educational and societal goals. Clearly, we need to reframe the way we talk about science education. Lev Vygotsky (1986) reminds us of the "union of word and thought" (p. 212)—that the language we use reveals how we think and who we are. We need to carefully examine the rhetoric of reform, the origins of the rhetoric and the way in which this rhetoric is propagated. Too often, the ideological narratives do not involve educators in their construction; teacher voices are writ large silent. Instead, educators react to the policies and mandates of reform from above. *Are the values and goals that are communicated in the reform shared?* Shared values and goals mean that stakeholders, including educators, have to be at the table; yet too often the rhetoric of reform demonizes the very stakeholders who are called to enact the change.

In reframing the way that we talk about science education, we also need to reframe the goals and purposes of science education and swing the pendulum back to goals of science education that reflect the relational nature of science—rather than preparing a nation for global competition, science education should be preparing a nation for global citizenship. We would like to see science education reform shift from a rhetoric of *science for economy* to a rhetoric of *science for society and social justice,* adopting more sociocultural than cognitive approaches. Essentially, science education reform should have at the heart of its account concepts such as equity, language, multiculturalism, identity, and social justice. A competition framework for reform is exclusionary and has translated at the classroom level into a parallel subtext; that is, if you are not "a natural" in science or mathematics or not interested in STEM disciplines, then you are not "smart," you are not motivated, you do not work hard enough. Ask any middle school or high school student who proclaims to not like science (or mathematics) or not be "good" at science (or mathematics), and you are likely to hear some version of this subtext. Instead, we need to communicate that the goals and purposes of science education include understanding and explaining what is happening in the world, and how scientific phenomena affects our lives, the lives of others, and the planet itself. We need to develop students' motivation to participate in creating an equitable, humane, and sustainable world. Fundamental to any reform effort, we argue, is the need for preparing all students for a changing world and enabling them to address current societal challenges. Essentially, we claim that the driving question of any reform initiative must be framed around the challenge of: *How can science education provide all learners with equitable opportunities for participating in communities of learners in an increasingly globalized world?*

Emphasizing the Culture of Science. Another issue raised in this chapter is the de-professionalization of teaching, goals that do not fit local realities, and problems of curriculum adaptation. We argue that in order to address this challenge, science education reform initiatives need to consider the *culture* of science as well as the characteristics of an era of globalization. Current global social trends such as migration and hyperconnectivity call for a re-visioning of science education, which involves a rationale that goes beyond national politics, economic and technological advancements. Instead, a conceptualization of science that goes beyond binary oppositions, and which moves beyond borders and boundaries is needed, given that borders are limiting and considering that the boundaries of science are constantly shifting. A response to consider regarding this issue lies at the intersections of local, national, and international goals and an emphasis on the *culture* of science. Research has shown that science education practices and curricula emphasizing professional or western science goals alienate underrepresented groups (Calabrese-Barton & Osborne, 2001). The premises of these practices and curricula are based upon rigid teaching strategies and uncreative methods; a view of science as a very technical field that is practiced by intelligent individuals who manage to leave their subjectivities outside the field; a context of practicing science that is detached from cultural relevance; and, perspectives in science that lack representation from diverse groups (Zembylas & Avraamidou, 2008). As William Stanley and Nancy Brickhouse (2001) argued, Western science needs to be taught as a part of a culture rather than as transcultural. Such an approach helps students understand about other cultural views of science as well as some of the assumptions embedded in western science that otherwise might be invisible. In other words, school science, as something different from professional science, needs to be constructed around a culturally sensitive and context-specific science curriculum. The assumption of the above vision is that there needs to be a shift from the priority that emphasizes content mastery and standardized tests—that implies a narrow view of western science as an elitist discipline for only a few—to a commitment for a science education that nurtures and capitalizes on different orientations to learning and prepares students to function in an increasingly globalized world in which different knowledge systems can co-exist and be respected (Zembylas & Avraamidou, 2008). In order to do this, scholars such as Vongai Mpofu, Femi Otulaja, and Emmanuel Mushayikwa (2014) argue that contemporary theoretical frameworks are needed that can be used to guide and inform the integration of indigenous knowledge and Western science at classroom science level.

Another Chance with Standards? The most recent reform efforts in the U.S. were initiated in 2012 with the publication of *A Framework for K-12 Science Education*, which details a conceptual framework for the *Next Generation Science Standards* (NGSS Lead States, 2013). As of February 2016, 16 out of 50 U.S. states have adopted the NGSS. The overarching goal of this framework is to:

> Ensure that by the end of 12th grade, all students have some appreciation of the beauty and wonder of science; possess sufficient knowledge of science and engineering to engage in public discussions on related issues; are careful consumers of scientific and technological information related to their everyday lives; are able to continue to learn about science outside school; and have the skills to enter careers of their choice, including (but not limited to) careers in science, engineering, and technology. (p. 1)

A major difference in NGSS than earlier efforts is the interdisciplinary approach and the notion that crosscuttings concepts that unify the study of science and engineering influence modern life and are essential for every American citizen. This emphasis on how science relates to everyday life is also evident in the most recent reform efforts in the EU, as described earlier. We are hopeful that the content and structure of the NGSS will disrupt the tendency to focus on goals that do not reflect local realities and do not address societal challenges.

Still missing, however, is an emphasis on the role of science education in addressing crucial societal challenges by promoting social justice. These ideas are historically connected with the STS movement in the U.S. and connected to the construct of scientific literacy, even though "scientific literacy" has been given wide variety of meanings throughout the years. According to DeBoer (2000) "scientific literacy has usually implied a broad and functional understanding of science for general education purposes and not preparation for specific scientific and technical careers" (p. 594). In a review of the historical meanings of scientific literacy and its relationships with science education reform, DeBoer (2000) described how the construct has been used since the 1950s and how it featured centrally in Project 2061 and the *National Science Education Standards*. However, a consensus definition of scientific literacy has been missing, and a result, has made a holistic assessment of the goal of scientific literacy difficult to achieve. Specific elements of scientific literacy that have been included in the vision of scientific literacy are the following: applicability of science to the social environment, relationship between science and technology, ethics of science, interest in science, the methods, integrity, the spirit and the results of science, habits of mind that come from the nature of science. Clearly absent in these elements yet crucial in re-envisioning science education is a view of science for society and social justice in an increasingly globalized world.

Better Use of Testing. While countries like the U.Ss and most nations in central and south Europe find themselves at the bottom of international tests scores in mathematics and science, other countries (i.e., Finland, Canada, Singapore, Japan and Korea) provide powerful examples of successful transformations in their educational systems. One could argue that comparisons are not easy or even fair to make due to each country's unique demographics. Finland, for example, has a small size and a homogeneous population, even though that has been changing in the past 5 years, with no changes in its scores on international tests. Others would argue that comparisons with East Asian models that rely on long hours, exhaustive studying and memorization are not fruitful. However, there are some useful observations to be made and lessons to be learned from the best-performing educational systems. The most prevalent observation, perhaps, is how the educational system resembles each country's economic model. For example, the Finnish economic model is collective, much like students' scores in international tests: there are no gaps between high achievers and low achievers. In addition, the focus of Finland's educational system is on equity (not on competition), it has no standardized tests and no private schooling or universities. As a result, *all* students do well but only about 15% fall within a top-performing category, unlike the U.S and China, which have huge educational achievement gaps, much like wealth disparities in their economies.

Quite a few researchers such as Linda Darling-Hammond (2015) have examined the nature of whole-system reform in Finland. These scholars emphasize the importance of making the entire system work well, not just its "output part"; for example, early childhood development, well-being of children in school, and professionalism within education craft. A key to Finnish educational success is its well-trained teachers. Teachers have higher academic education in Finland than in many other countries, but they are also involved in systematic and continuous professional development. Analogous examples of teacher effectiveness policies can be found in Ontario and Singapore, and which are among the highest-performing jurisdictions in international tests of student achievement and their tests are ranked among the most equitable in the world.

Linda Darling-Hammond and Robert Rothman (2011) argued that Finland, Ontario, and Singapore provide important lessons for policies that aimed at teacher and school leader effectiveness. As they summarized in a report, the right efforts to improve teacher effectiveness can lead to higher and more equitable student achievement and they include: (a) a systemic approach; (b) strong recruitment and preparation; (c) attractive teaching conditions; (d) continuous support for learning; and, (e) proactive leadership development. Another lesson that Finland has to offer is that a *systemic* approach can enhance conditions for teaching and learning for all, and provide equal opportunities within an educational system. A historical overview of Finland's educational system shows that Finland has transformed a very poor and traditional educational system into an exemplary model of an educational system, as measured in achievement as well as a narrow gap between high and low achievement. This success story can be attributed to changes to the infrastructure of schools, teacher preparation, teacher professional development, curriculum, balancing decentralization and centralization, and providing supports for special needs students.

Singapore provides another example of successful reform with its students (all students, including many minority groups) scoring first in both mathematics and science on TIMSS assessment in 2003, and scoring near the top of the ranking in mathematics and science in PISA exams in 2009. As Tan Lay Choo and Darling-Hammond (2011) described, these successes were the product of a long-term commitment to developing a high-quality educational system through a system wide reform called "Thinking Schools, Learning Nation" launched in 1997. This reform has similarities to the ones carried out in Ontario and Finland: the reform had a coherent conceptual vision that focused on educational and societal goals (instead of economic goals); the reform was systemic and included changes in teacher recruitment, preparation, retention, school infrastructure, curriculum, and leadership; and at the heart of its account was teacher preparation, professional development, and teacher life-long learning.

We are not arguing that the U.S. or Europe or any other nation should adopt the Finnish or the Singaporean or any other country's educational system. As many researchers have argued and given that education is deeply culturally embedded and what works in one context may not work the same way in another context, so caution is to be exercised when considering what to take from comparative studies of educational systems. As Robin Alexander (2012) has argued, the problem is not with the desire to learn from others, but with the import from distant shores of miracle cures advocated by school improvement experts. However, we are saying that there are big lessons to be learned from these countries' examples regarding the ways in which reform was conceptualized and enacted through a more global lens.

WHAT REFORM? WHERE IS THE REFORM? WHOSE REFORM? HOW TO REFORM?

In reflecting on the nature and effects of science education reform over the last 20 years, we have to ask: *What reform? Where is the reform? Whose reform? How to reform?* In concluding this chapter, we contend that: there is a lot written about the *what*; the *where* remains very narrow, un-inclusive and un-equitable; the *whose* remains attached to national, economic, and technological ambitions, and the *how* remains un-thinkable. The attempts for reform in science education in the U.S., one after the other, have not fulfilled the dream of the right to learn. This is not news. What has changed over the decades are the lessons learned, with perhaps one of the most prevalent being that a new and contemporary rhetoric and vision for educational reform is needed, one that is focused on *science for society* instead of *science for economy*. To do so, we need to conceptualize change as a process that takes much time, instead of as a product that occurs overnight. It is time to start conceptualizing this change, re-theorizing the reform rhetoric, shifting paradigms by using global frameworks situated beyond the national landscape and adopting global, systemic and bottom-up approaches. Central to this is the realization that science education reform is political and requires a strong vision and a shared paradigm, as well as governance support, leadership, expertise, funding and resources.

REFERENCES

151 Cong. Rec. 22769 (2005) (statement of Sen. Lamar Alexander and Sen. Jeff Bingaman).

Alexander, R. J. (2012). Moral panic, miracle cures and educational policy: what can we really learn from international comparison? *Scottish Educational Review, 44*(1), 4–21.

American Association for the Advancement of Science, Project 206. (1995). *Science for all Americans: Summary*. Washington, DC: American Association for the Advancement of Science, 1995.

Australian Education Act. (2013). *Australian Education Act 2013: An act in relation to school education and reforms relating to school education, and for related purposes*. Retrieved May12, 2016 from http://www.comlaw.gov.au/Details/C2013A00067

Bybee, R. W., & Fuchs, B. (2006). Preparing the 21st century workforce: A new reform in science and technology education. *Journal of Research in Science Teaching, 43*, 349–352. doi:10.1002/tea.20147.

Calabrese-Barton, A., & Osborne, M. (2001). *Teaching science in diverse settings: Marginalized discourses and classroom practice*. New York, NY: Peter Lang.

Choo, T. L., & Darling-Hammond, L. (2011). Creating effective teachers and leaders in Singapore. In L. Darling-Hammond & R. Rothman (Eds.), *Teacher and leader effectiveness in high-performing education systems* (pp. 33–42). Washington, DC: Alliance for Excellent Education and Stanford, CA: Stanford Center for Opportunity Policy in Education.

Crawford, B. A. (2000). Embracing the essence of inquiry: New roles for science teachers. *Journal of Research in Science Teaching, 37*, 916–937. doi.org/10.1002/1098-2736(200011)37:9<916::aid-tea4>3.0.co;2-2.

Dabla-Norris, E., Kochhar, K., Suphaphiphat, N., Ricka, F., & Tsounta, E. (2015). *Causes and consequences of income inequality: A global perspective*. International Monetary Fund.

Darling-Hammond, L. (2015). *The flat world and education: How America's commitment to equity will determine our future*. New York, NY: Teachers College Press.

Darling-Hammond, L., & Rothman, R. (Eds.). (2011). *Teacher and leader effectiveness in high-performing education systems*. Washington, DC: Alliance for Excellent Education & Stanford, CA: Stanford Center for Opportunity Policy in Education.

DeBoer, G. E. (2000). Scientific literacy: Another look at its historical and contemporary meanings and its relationship to science education reform. *Journal of Research in Science Teaching, 37*, 582–601. doi:10.1002/1098-2736(200008)37:6<582::aid-tea5>3.0.co;2-l.

European Commission. (2004). *Europe needs more scientists*. Luxembourg: Office for Official Publications of the European Communities.

European Commission. (2007). *Science education NOW: A renewed pedagogy for the future of Europe*. Luxembourg: Office for Official Publications of the European Communities.

European Commission. (2015). *Science education for responsible citizenship*. Luxembourg: Office for Official Publications of the European Communities.

Figazzolo, L. (2009). *Impact of PISA 2006 on the education policy debate*. Retrieved from http://download.ei-ie.org/docs/IRISDocuments/Research%20Website%20Documents/2009-00036-01-E.pdf

Froese-Germain, B. (2010). *The OECD, PISA and the impacts of educational policy*. Canadian Teachers' Federation: Virtual Research Center.

Gorur, R., & Wu, M. (2015). Leaning too far? PISA, policy and Australia's "top five" ambitions. *Discourse: Studies in the Cultural Politics of Education, 36*, 647–664. doi:10.1080/01596306.2014.930020.

Hurd, P. D. (1986). Perspectives for the reform of science education. *Phi Delta Kappan, 67*, 353-358.

Improving America's Schools Act of 1994. Pub. L. No. 103–382.

Indiana Department of Education. (2016). *Indiana academic standards: Science and computer science*. Retrieved from http://www.doe.in.gov/standards/science-computer-science

Institute of Medicine, National Academy of Sciences, and National Academy of Engineering. (2007). *Rising above the gathering storm: Energizing and employing America for a brighter economic future.* Washington, DC: The National Academies Press.

Kahle, J. B. (2007). Systemic reform: Research, vision and politics. In S. K. Abell & N. G. Lederman (Eds.), *Handbook of research on science education* (pp. 911–941). Mahwah, NJ: Lawrence Erlbaum Associates.

Kincheloe J., & Steinberg, S. (Eds.). (1996). *Thirteen question: Reframing education's conversation* (2nd ed.). New York, NY: Peter Lang.

Kyle, W. C. (1996). Shifting ideologies and science education. *Journal of Research in Science Teaching, 33*, 1043–1044.

Mpofu, V., Otulaja, F. S., & Mushayikwa, E. (2014). Towards culturally relevant classroom science: A theoretical framework focusing on traditional plant healing. *Cultural Studies of Science Education, 9*, 221–242. doi:10.1007/s11422-013-9508-5.

National Academy of Sciences, National Academy of Engineering, and Institute of Medicine. (2010). *Rising above the gathering storm, revisited: Rapidly approaching category 5.* Washington, DC: The National Academies Press.

National Commission on Excellence in Education. (1983). *A nation at risk: The imperative for educational reform: A report to the Nation and the Secretary of Education, United States Department of Education.* Washington, DC: The Commission.

National Research Council. (1996). *National science education standards.* Washington, DC: The National Academies Press.

National Research Council. (2000). *Inquiry and the national science education standards: A guide for teaching and learning.* Washington, DC: The National Academies Press.

National Research Council. (2012). *A framework for K-12 science education: Practices, crosscutting concepts, and core ideas.* Washington, DC: The National Academies Press.

NGSS Lead States. (2013). *Next generation science standards: For states, by states.* Washington, DC: The National Academies Press.

No Child Left Behind Act of 2001, Pub. L. No. 107-110, § 115, Stat. 1425 (2002).

Organisation for Economic Co-operation and Development. (2008). *Encouraging student interest in science and technology studies.* Retrieved from http://www.oecd.org/sti/sci-tech/encouragingstudentinterestinscienceandtechnologystudies.htm

Osborne, J., & Dillon, J. (2008). *Science education in Europe: Critical reflections.* London: The Nuffield Foundation.

President's Council of Advisors of Science and Technology. (2010). *Prepare and inspire: K-12 education in science technology, engineering, and math (STEM) for America's future.* Retrieved from: http://www.whitehouse.gov/sites/default/files/microsites/ostp/pcast-stemed-report.pdf

Project 2061 (American Association for the Advancement of Science). (1993). *Benchmarks for science literacy.* New York, NY: Oxford University Press.

Rutherford, F. J., & Ahlgren, A. (1990). *Science for all Americans.* New York, NY: Oxford University Press.

Schwarz, C. V., & Gwekwerere, Y. N. (2007). Using a guided inquiry and modeling instructional framework (EIMA) to support pre-service K-8 science teaching. *Science Education, 91*, 158–186. doi.org/10.1002/sce.20177.

Settlage, J., & Meadows, L. (2002). Standards-based reform and its unintended consequences: Implication for science education within America's urban schools. *Journal of Research in Science Teaching, 39*, 114–127. doi:10.1002/tea.10012.

Stanley, W. B., & Brickhouse, N. W. (2001). Teaching sciences: The multicultural questions revisited. *Science Education, 85*, 35–49. doi:10.1002/1098-237x(200101)85:1<35::aid-sce4>3.0.co;2-6.

U.S. Department of Education. (2009). *Race to the top executive summary*. Retrieved from http://www2.ed.gov/programs/racetothetop/executive-summary.pdf

Vygotsky, L. S. (1986). *Thought and language* (Rev. ed.). Cambridge, MA: MIT Press.

Zembylas, M., & Avraamidou. L. (2008). Postcolonial foldings of space and identity in science education: Limits, transformations, prospects. *Cultural Studies in Science Education, 3*, 977–998. doi:10.1007/s11422-008-9121-1.

SECTION TWELVE

Science Education as a Political Issue

What's Missing in the Public Conversation About Science Education?

CHAPTER THIRTY

Rethinking Science Education in Light of Motivated Reasoning

DAVID E. LONG

We live in an extraordinary moment for science education to be used to affect social change. We also live in a time when some of the most critical issues affecting humankind such as anthropogenic climate change will come to affect us all. Epidemiological crises prompted by the development of antibiotic resistant strains of bacteria due to over-prescription of antibiotics at the hands of an evolution-ignorant, demanding public has also emerged as a new problem affecting all of us. Where once herd immunity was thought enough to guard against a small outlier of anti-vaccine attitudes, those attitudes, in some communities, are no longer so rare. Additionally, while genetic engineering promises to increase our ability to feed the world's population and vaccination has clearly improved public health over the past century, both have become value-laden tokens within an increasingly polarized culture. Rather than consider the outputs of scientific inquiry with a reasoned eye and mind, our engagement with evolution, climate change, vaccination, and genetic engineering has increasingly devolved into a politics where our identity trumps understanding. Biotechnologies are increasingly the terms by which ideological fights between often well-informed sectors of the public. Many people react negatively as an identity-protective reaction to information perceived challenging to values of their cultural affinity groups.

For each of the perennially challenging topics such as evolution, climate change, vaccination, and genetic engineering, science education can be a leader. Science education can and should serve as a mediating force in leveraging social change oriented with individual, community, and state interests in mind. Currently, our public imagination and rhetoric used to mobilize it for science education at its best most often invokes the limited metaphor of the pipeline. As we know, we entrust pipelines to carry pure water for our various needs, but know pipelines also carry effluent. Our goals for science education needs refreshed past these metaphorically limited conduits to solve these problems.

In this chapter, I discuss one example of a sorely needed, interdisciplinary approach to science education and science education research in service of finding pragmatic paths forward to ameliorating our most wicked socio-scientific issues. It is, a connector across the ragged edge of what we currently consider appropriate science education engagement, and the broader domain of science, public understanding, and political action. A burgeoning research base in motivated reasoning, cultural cognition, science communication, behavioral economics, and the learning sciences are all converging on a point long known by educational anthropologists. Our learning of socially charged science is first and foremost motivated by identity protective interpretations of science. Whereas learning certainly opens new horizons of understanding with which we imagine the world anew, we also learn, through culture and affinity with our cultural co-conspirators, to protect our ways of life. Alienation from the known and comfortable is often quite unpleasant.

By articulating this research, my chapter frames an ethically actionable, socially just position, balancing the reality of diverse imaginaries of the good vis-à-vis commonly shared socio-scientific issues also defending science education as a social good unto itself, as a pragmatically strong, problem solving toolbox worth cultivating for all.

EXTRAORDINARY TIMES

Although the phenomenon of climate change has been well recorded for some time, at least within the United States, we continue to dwell within a political climate where a large group of voters and their elected representatives deny the science of climate change and refuse to take action in response to it. Like both nationally representative studies of inconsistent or absent evolution education amongst U.S. high school biology teachers (Berkman & Plutzer, 2010) and close analyses of U.S. college students grappling with evolution (Long, 2011), climate change education is currently spotty or absent. Like in the case of evolution education, a recent nationally representative study of climate change education (Plutzer et al., 2016) shows American teachers underprepared and often unwilling to teach the science of climate change.

One of the conventional ways to deal with problems of evolution and climate change education has been to frame them as a matter of science denial. Denial has been seen conventionally as a matter of knowledge deficiency, a general problem of science literacy and public understanding. But as we also know, the issue of ideologically based reactions to science do not stop with evolution and climate change. Many pockets of the U.S. population now choose to opt out of childhood vaccinations once thought a stalwart defense against debilitating diseases that, through our best epidemiological efforts, were thought of until very recently as more or less eradicated. Mumps. Whooping cough. Measles. All are seeing unfortunate resurgences. Genetic modification of organisms within our food supply and the purposeful manipulation of human stem cells both offer tremendous promise for improving human life while another portion of the population knows, unequivocally, that such efforts should be banned. A quick perusal of both pro- and anti-GMO websites will simultaneously convince you of both the efficacy and social evil of GMOs.

For each of these aforementioned points of contention, why do they exist as contentious? As broader social science research shows, engagement with these topics by citizens is a more complicated relationship than what is known about science by the public. Current knowledge of each of these topics is not only variable across the population, but is mobilized in differing ways to do ideological work supportive of individual, community, or often corporate interests. For each, there is an intersection of human knowledge and practice drawing on knowledge from science, from technology, and informed by human values. Insomuch as evolution, climate change, vaccination and genetic engineering prompts a certain portion of the population to contest it, each is in its own way a political issue. How do we learn about each of these? And what is science education's role in crafting responsible understanding and action about each? Moreover, who chooses what is responsible action regarding each?

We need to unpack and clearly understand a twofold problem within science education that firmly roots the subject as the site of intersectional politics, where each topic—climate change, evolution, genetic engineering, and vaccination (amongst others)—is a political issue. This is not to say that every element of scientific knowledge carries a political valence. Far from it. We're all basically okay with gravitation, the germ theory of disease, etc., and most of the corpus of science knowledge. But there are clearly topics within the orthodoxy of science that, as we have seen in the reactions by scientists to those who question or reject particular scientific premises, carry with them an implied entailment of assumed politically correct action.

For these topics, so it goes, gaining an understanding of the topic is, for some scientists and science educators, enough of a self-evident warrant such that it carries with it an obligation to clear, socially responsible action. It is as though, for these scientists and science educators, the descriptive work of science then compels, clearly and without other consideration, a clear pathway for all concerned to take social action. The problem is plain. Some actions based upon a warrant drawn from science knowledge come to violate certain ways of life (Callison, 2014). And more foundationally, no action based upon science knowledge alone carries, by itself, a command to self-evident action. Action taken in the name of social improvement depends on values and ideology for a world we wish to see, not one that we have. Across the political landscape, we neither agree on visions of idealized future worlds, nor the best method of setting up and operating schools within which science education would affect such change. For example, seeing that the earth's climate is changing by using the tools of science does not necessarily imply that one must do anything about it. In the popular film series Batman, directed by Christopher Nolan (2008), Bruce Wayne (the wealthy alter ego of the superhero "Batman") is consoled with this sentiment as to explain the nihilism of the Joker character by his butler and confidant, Alfred. Some men, in the words of Alfred, "just want to see the world burn."

As the intended outcome of this chapter, I ask the reader to more broadly consider the current limits of status-quo, depoliticized science education practices and science education research and treat this analysis with the analytic eye of Hume's Guillotine. Scottish philosopher of science, David Hume, is most attributed for pointing out the "naturalistic fallacy"—where one cannot necessarily (or ever) derive an "ought" claim about the world based upon an "is" claim, given the problems of definitively identifying causation and causal relationships between seeming discrete phenomena. For us, this will be where the "is" project of science describing the natural and social worlds is divorced from the normative "ought" project of making changes—in our case through education—to the social world. Our oughtsmanship of educational improvement in this view is seeing and naming educational change as a matter of ongoing, intentional, power laden socialization. This view is at its root anthropological, as the ways that we marshal knowledge and technology for improvement depend not on brute facts produced by science alone, but by the multivalent usage of science knowledge and forms of technology that simply may not carry the same semiotic weight across lifestyles and cultures. The kind of project at hand, begun a few decades ago in science education by Lorie Hammond and Carol Brandt (2004), needs refreshing. We do this work in two steps.

Firstly, despite seminal work in science and technology studies and other social sciences that have shown the practice of professional science and science education as a normative project that corresponds with neither idealized notions of a scientific method nor actors following a coherent "nature of science," the field proceeds with its normative efforts without nearly enough self-evaluation as to its ability to command; (1) the kinds of levers required to affect the social change it pines for through national and international policy setting and, (2) implementation in schools through science "literacy" efforts.

As I have repeatedly heard the lament amongst colleagues at science education research professional meetings, there is a deeply felt sense that the findings of science educational research are not taken seriously by conventional scientists. For this, science education's normative project is a cart way out in front of its horse. The horse in question here—the descriptive project—is clear within the literature across the social sciences that examine the acquisition and manipulation of science knowledge. People are neither the kinds of rational actors assumed for decades within the field's research, nor the kind of formal thinkers who easily and willingly slip into the measured analytic frame of mind depended upon to grant reasoned epistemological judgments such as those of the "nature of science" genre of inquiry so dominant in our field in the past thirty years.

The issue is clear enough when pointed out. Across myriad cultures, human societies have produced multivalent epistemologies that engage quite well with technologies for the improvement and sustainment of their respective ways of life. Seminal works such as Glen Aikenhead's (1996) border crossing model of science engagement started this work, but as I point towards, the same border crossing is not unique, as in Aikenhead's work, to First Nations Canadians. People worldwide in cities, suburbs, exurbs, rural, and remote environs variably engage with science in ways that enhance the projects that matter to their lives. Science they do seem to enjoy. They just may not be interested in giving credit to, nor much attention toward formal epistemological and pedagogical efforts such as the nature of science, argumentation, learning progressions, or whatever next behaviorist, instrumentalist, socio-educational change gimmick comes into vogue. For pointed topics such as climate change and evolution, people live in environments which, via individual interest and through affection to local authority, form identities which are equally or more informed by those things learned from cultural affinity groups (houses or worship or social groups) more than science education policy or the teachers who teach science. For topics such as genetic engineering or vaccination, moral authority over whether to tamper with essentialized notions of a biological "kind," or class-based nations of corporate conspiracy have dominated the recent decade.

For this, the normative project of dominant science education research is hampered strongly by a lacking engagement, ironically enough, with the best social *science* of human learning that accounts for engagement with the field. The normative project of instilling certain quick-fix pedagogies such as the nature of science, argumentation, learning progressions, etc. do not account for the disruptive agency of human learning that comes from other domains of our learned experience. The supposed production of the scientifically literate citizen is folly unless it accounts for and thinks clearly about the limits placed upon it in the political economy of changing the identities and material conditions of lived lives. For our most perennially troublesome science education topics, social resistance to them becomes much clearer, and much more reasoned when we account for the holism of lived lives, and how science knowledge will work within it.

HOW SCIENCE EDUCATION HAS POSITIONED ITSELF: THE NORMATIVE PROJECT

When we think about what science education is, for those of us schooled in the highly industrialized part of the world, our imagination often most commonly conjures pictures of laboratories or classrooms. The specifics artifacts and acts—beakers, explosions, and measurements—are engrained deeply in the consciousness of most. Advocates of the famous "draw a scientist" test of student science identity and imagination get consistently stereotypical results for good reason—knowledge of a more diverse scientific field of practice does not exist in a child's field of phenomenal experience alone. Images from the community and various forms of media all of equal if not more input than school learning in populating the conceptual and symbolic repertoire of a child's imagination.

In the formal sense, schooling in science has produced and keeps reproducing people with a specific imagination for the field as existing in discrete siloed areas of inquiry and fact. Our systems of state or national level accountability and testing underscore this. Chemistry looks like this, biology is like this, physics is like this, and so forth. Geoscience or environmental science, more often than not, is not part of the minimum bar of school courses usually required University science coursework, for the much greater number of non-STEM degree seekers that matriculate on university campuses, acts to further emphasize this conceptual siloing.

We also know that this description illustrates a picture of scientific inquiry quite askew from the actual practice. Past the smoke created between social scientists questioning the degree of idealized objectivity practiced by scientists, and those same scientists pushing back in the 1980s and 1990s via the "science wars," the very real issue of scientists committing the act of culture in the perpetration of their science has been well-documented. Bruno Latour and Steve Woolgar's (1979) seminal study of laboratory practice prompted decades of analysis into the nature of science practice and the questionable correspondence with claims of scientific methods by scientists. Our idealized form does not match our practical form, as so often is the case in how we educationalists present school subjects. In my own recent experience as an evaluation researcher on a large federal STEM education research project, the principal investigator more or less saw science through the lens and materiality of the chemist. This was perfectly fine and had its merits, but was in no way inclusive of all the material and investigative apparatus we might, on the whole, engage with under the banner of scientific inquiry.

Rightly so, research within our field has, at least since the 1990s, begun to ask more complicated questions of the kind of knowledge production we're engaged in, and whether this production is created and used by all people or just a select few with particular social advantages. Indigenous knowledge formed by endemic epistemologies, feminist critiques of science, and now a much broader array of analytic lenses have begun to be applied to the field in service of a more democratic form of scientific participation. We take these ultimately liberatory ends as part of an ethos where all students can and should engage science for improvement. They are ends to be championed.

Science educators must be an active and responsible body politic. The findings of science must be communicated for science to work as a social practice, and that communication forms narrative. As we know from linguistic anthropologists such as Shirley Brice Heath (1983), the narratives we learn and carry with us animate us and direct the projects important to sustaining our identity. The narrative that science produces about the world equally does social work. While less well articulated within science education, the purpose of our engagement with forward looking, equipped narratives and their entailed technologies depends on the important projects in what matters in our lives. We all are members of communities of varying sorts—some small and local, some with national, international, or transnational constituents—but those communities all have interests. Those interests, projected out into hopeful futures, are served by foundational moralities whether they are individualistic, communitarian, neoliberal, and authoritarian, amongst many others. How we position ourselves within these sometimes economic, spiritual, or nationalistic identity formations may vary and may intersect with others, but what remains in common is that they are moral projects directed toward ends we hope to achieve. The catch is that they, across the culturally and politically diverse human terrain, are not all in agreement. Science education then, insomuch as education changes who we are as we proceed through it, is about changing the boundaries of our knowing and doing and this process, as a matter of political economy, is a political project. Science education is a political act.

Nevertheless, as a field of professional practice, we often do not act as though education or science education is a political act. In my experience, which I believe is very common, we are often armed with very different conceptual tools in how we frame and approach educational problems. For example, for those educationalists who pursue reasoned inquiry into the nature of and systematic reproduction of inequality in education and the social institutions within which those practices are nested, supposed deficiencies between groups is as much or more a product of asymmetric power relations between actors within institutions. Such investigations, which interrogate this power-laden asymmetry, are often described as a form of critical theory. Especially important for seeing science education as a political act, analyses of social power are derived from social theory and social science—not the siloed natural sciences described above. For this difference, science education researchers are often talking past each other in much of our most impassioned professional discourse. My point is well encapsulated in the interchange below between eminent science education scholar, Ken Tobin, and a science education faculty member during a scene at an invited talk he gave while visiting as outside reader for my dissertation defense.

> Ken: "… we all need to consider the critical approaches that will steer science education to address sustainability … you've all read critical theory … what's critical theory?"
> [… very long, awkward pause …]
> Faculty: (timidly) "… really, **really** important theory …?"

To be fair, I was trained in my doctoral work as much in social sciences that amplify critical theory and social/structural analysis as in the natural sciences. This faculty member, working within another department, was a traditionally trained scientist who completed his/her science education Ph.D. program with very likely little or no coursework that would have explored critical theory. To understand the nature of power and its effect on our reasoned inquiry, and what affect it has on our most commonly employed pedagogical strategies, we need better tools. Those tools are at least a better understanding of how we reason with science for the projects we care about in our lives.

HOW SCIENCE EDUCATION CAN IMPROVE ITS SCIENCE: THE DESCRIPTIVE PROJECT

There is a large and converging body of research in motivated reasoning and cultural cognition, with a great potential to impugn and hopefully reset many current pedagogical conventions in science education. Our current moment of science in service of neoliberal economic competitiveness—to "win the future" with "21st Century Skills"—is inadequate for the wicked issues, which require science's most apt toolkit. It's also undergirded by an ethic that has little to say for non-capitalist enterprises as the reason for science. The current dialogue is almost entirely framed by limited and limiting neoliberal ends. Science is more than a tool for leveraging individual or corporate success in globalized markets. It's a tool we need all people to have ready at hand to solve increasingly global problems. We should think of our work, using Hume's Guillotine, to both lop off science education's head, clean out the rot of a misguided normative project, and then reattach it towards an ought project both more ethically responsible for our future action, and scientifically accurate for our current work.

The overarching issue at hand is simple, but sticky when imported into science education. Our reasoning, quite unlike the hopes of nature of science researchers, is quite motivated and quite attached to existing structures, which frame our experiencing of the world. Much more often than not, we use science to protect our identity—not disrupt it. Most often this tendency is buried from immediate purview, as we do not usually have intra-communal quibbles over the status of knowledge as used in the maintenance of our identities or life projects. People generally are not looking to revolutionize the social games they are playing. Additionally, power laden status hierarchies tend to keep social power structures in place. But there are areas of science knowledge that, when considered against the narrative content of existing lifeways, come to be an intersectional point of contestation. Evolution is the most familiar long-term example from within the U.S., but climate change is increasingly salient under this same lens. This is not a rejection of pedagogies such as the nature of science, but a cautioned deflation of their utility in these cases. Such pedagogies assume a sort of social power through authority in education to shape learners in ways that the evidence of how we accommodate identity-threatening knowledge does not support. We simply do not have that much power.

Much research has been conducted to articulate the nature of this phenomenon. So much so that there is compelling reason to dramatically turn the orientation of our approach. Since George Lakoff and Mark Johnson's (1980) pioneering work on the structure of use and semantic work done by the human employment of metaphor, it has become unavoidably clear that there is no such thing as value neutral language in a strict sense. Lakoff and Johnson's work extends earlier anthropological work by Edward Sapir (1929) and Benjamin Lee Whorf (1956) where it was postulated and later empirically shown that a person's language shapes their ontology, and therefore their worldview, and subsequently, the basal categories of experience and linguistics that are deemed real, or any concern. Even when we suppose we are striving to employ some sense of detached, reasoned inquiry, the political conditions by which one engages in such a cognitive act is governed by background conditions which make such open inquiry possible. Of the work that metaphor does, for the mobilization of scientific theory, our communication by metaphor—no matter how benign the example—is always value laden. Much the same has been known from the slightly differing approach by Antonio Damasio (1994) where his work on embodiment and somatic markers pointed out the foundational error carried forward from Plato through Descartes in separating emotion from reason. Reason, as we are now seeing, is always imbued with emotion—to the radical notion that the supposed detached, objective scientific view from nowhere—especially the inclination to adopt such a view for personal use to accomplish social work—is itself a specific type of emotional state. Like the problem of emptying even interstellar space of its substance ontology, we cannot detach ourselves from an uninvested point of view. People are not ambivalent for their interest in science. They are motivated by the utility the view affords them to look upon the world. This point, well developed in Nietzschean perspectivism, has been with us for a while now.

Lakoff (2008) has continued this work in recent years, employing this framework of language and ideology to interpret the current political intractability within U.S. political discourse and voter affinities. Adding to this view, Jonathan Haidt's (2012) work on what he terms "the righteous mind" points to findings that underscore the aforementioned work, in that our reasoning is primarily intuitive, with the type of distanced, detached reason we project for our analytic selves to be more of a veneer than a substance quality of our reasoning faculty. As Haidt advances more so than some others, the moral framework, which drives our intuitive faculty, is shaped socially and handed down culturally. He does not ascribe this tendency as a wholly individualist nor cognitivist phenomenon. Much the same, Daniel Kahneman's (2011) Thinking Fast and Slow codified our reason into two types; 1) the quick, intuitive, and emotional thinking self, and 2) the slow, deliberative, and more logical self. This work, while well received, has been rightly criticized for little consideration of the social domain of life and learning.

A large and quickly growing body of work has emerged from within research on judicial proceedings and the reasoned weighing of evidence in jury trials. Dan Kahan's (2012) work on what he terms "cultural cognition" has shown clear evidence that many of our supposed science literacy measures, particularly some of the American Association for the Advancement of Science's work, are flawed for their tendency to provoke ideologically based reactions. In his work with measures of understanding of both evolution and climate science, he has shown both that evolution understanding measures have been worded such that they cover over more valid statements of understanding, by instead forcing respondents to affirm membership in cultural affinity groups. While levels of understanding in item tests he has conducted have shown overall poor understanding of both topics across the U.S. population, he has shown that groups often labeled as evolution or climate change "deniers" have little or no statistically significant difference in their level of understanding (Kahan, 2012). This is perhaps best articulated in a humorous blog post (Kahan, 2015) in which he compares case studies of a Pakistani doctor who denies evolution as a matter of his religious identity and practice within his conservative Christian faith community, and believes evolution as a matter of his practice of contemporary medicine. Paralleling this case, he also presents a case of a Kentucky farmer who both uses the best science known from climate change to logistically prepare for best farming practice while simultaneously denying climate change as a matter of ideologically charged identity politics.

What we can take away from this brief walk through research on culturally grounded reason is that our reason first and foremost is motivated. Echoing the seminal sociology of Erving Goffman (1959) from the mid 20th century, are multivalent performers of our social selves. We present different selves to others depending on audience and social position. Our reason is motivated to protect our identity and the affinity groups with which we call ourselves members, and aid us in furthering the political projects of our lives. This is both political in the conventional sense of governance and citizenry, but also in the existential, phenomenological sense put forth by Martin Heidegger (1962). As beings that take a stand on the status of our own being, we care about the proximal and intentional content of our lives, and the future projects we see ourselves projecting out into, along with the co-conspirators who make of the circle of care within our cultural affinity groups. When we look across the demography of any population, and its varied educational ends that that population seeks, the points of intersection are often points of conflict. Science as a source of epistemological authority only carries so much weight to any one person in the world. This realization, when applied to educational practice has special significance. It detaches us from our feigned ability to see ourselves as being in the business of knowledge transmission. We are also political actors, often shaping an ought-world of future possibility, which often creates conflict for learners, across the lifespan.

AFTER THE GUILLOTINE: SCIENCE EDUCATION AS A POLITICAL PROJECT

You cannot shape the motivation and interests of a vast, variable, culturally diverse population to adhere to a person to such a formalized regimen of thinking. As the social sciences that document what is and has gone before in human experience and specifically with teaching and learning have shown, there is no one prescription to follow which will produce an equality of outputs regarding a population's willingness and interest to adhere to the formal thinking demonstrated by the formalist pedagogies such as the nature of science.

Such pedagogies do not seem to be situated in a cultural ecology wherein the formal thinking being cultivated by the strategy accounts for all of the inputs people experience in the clock hours experience of their lives, or the values represented by them. All the while one sits in a school or university science classroom learning such formal demarcations, later, other home and community values and practices constantly underscore or reshape the style of thinking by which a person comes to acquire their general habitus in their social world. This is not to say that formal science education pedagogies are an unworthwhile endeavor—far from it. But it seems increasingly clear that it is less of an effective research paradigm to somehow produce legions of standardized formal nature of science thinkers, and better a tool for gaining epistemological skillfulness. Even then, for some topics, and many people, it is likely bound to fail.

Our best and most ethically sound strategy for working through contested science terrain first requires us to consider our emphasis on certain science education discourses which are impugned by the research reviewed above. Lastly, I will offer a framework for future professional commitment in science education based on a seminal piece in the foundations of U.S. educational history, sociology, and philosophy of education.

Abandoning Science as Truth as a Warrant for Care. Making the importance of this clear requires, due to the deep embeddedness of science as an assumed good, comparisons of human variable interest with other social practices. As this brief positioning and examination of the limits of science education's normative project, and further exploration of the motivated nature of our human reason has shown, many of our dearest research foci and pedagogical strategies are due for revision in light of the political and value-rational nature of human thinking disclosed in this work. A more robust descriptive project is called for. One that would investigate, in an anthropological style, the varied epistemologies, materialities, and semiotic frameworks that stand for—especially in developing and "undeveloped" nations—the reasoning style and tool-equipped engagement with respective worlds.

As seen above, much of science education's goals still are motivated by some conception of human deficiency, if the human lifeways in question are misaligned with science "standards." Scientific fact does not have a direct relationship to the content or interests of people's varied ways of life. That scientific information is assembled and presented as a truth does not correspond directly to peoples' interests. It is true that the American National Football League holds many games. It is also true that I have both little interest in it, and correspondingly little detailed knowledge of the style and moves of the game that matter. The factual status of the game, its history, nor its style of social practice or even factual status of existing has little to say about my interest.

Grounding STEM from Its Floating Signifier Status. As has become clear in recent years, the practice of science education, at least at the level of national education standards policy, has become subsumed within a framework of intended teaching, which has greatly changed the field's focus. The inclusion of engineering and technology within the *Next Generation Science Standards* (NGSS) (NGSS Lead States, 2013) has both deflated the importance of the descriptive sciences such as geology and paleontology, but also distanced itself from any semblance of science for an ethical purpose. Alberto Rodriguez's (2015) work on this point underscores the abdication of NGSS to any commitment to science for social justice.

The politics of invoking STEM is neoliberal at its core. As is bandied about, many practitioners are confused as to the boundaries and intent of the STEM movement, such that the arts are insisted for inclusion to make STEAM. With the inclusion of social sciences and philosophy, aren't we simply resurrecting the liberal arts? STEM is both code for science in many circles, but also a floating signifier—in that it is used by the constituent STEM field acronym disciplines to advance their individual field goals over the nominally friendlier integrated approach it seems to advertise. NGSS is sorely deficient in its appreciation of work on identity and how value-laden reason as detailed above may shape engagement and interest. This is an area we, as a research community, can clearly do more. If science education is not to be subsumed within the neoliberal ends of current STEM discourses, our politics must be critical and engaged with this movement.

Past the Pipeline Metaphor. Science education policy rhetoric has, for many years, insisted that there is a crisis of science education production. As the rhetoric asserts, not enough science graduates are coming through a pipeline ready to work in skilled, science and STEM fields. While there is a broad literature on the shaky validity of this pipeline claim, our point for a politics of science education focuses on a subtler point. We do not have a clearly articulated sense of what science education is for, past common notions of science arming a population to work towards the utopian project of "science literacy." Literacy calls exist most often outside the material and cultural conditions of human political economy. We should do better at situating our work within this domain. It may be that the evolution denying creationist or the climate change denying market fundamentalist businessperson may, in fact, be quite adept and useful in their application of science and formal reason for achieving greater good in other domains of existence, contributing work toward ameliorating other commonly shared problems. Those problems may have nothing to do with maximizing U.S. economic competitiveness.

The pipeline as metaphor does great work to lop off the consideration of science to serve broader human interests. It co-opts most of science's potential to activate the best inquiry-minded sense of investigative wonder by commodifying the practice into an ever narrowing set of tasks to be mastered in service of corporate and national competitiveness discourses. This type of speech alienates many from genuine interest in science, and also leaves whole areas of science less useful for commodification discourses more and more marginalized. My first professional fields of archeology and paleontology have little to gain from the STEM revolution as currently construed.

Past the Crisis Drumbeat—Toward a Multivalent Use-Value of Science Education. Science, once again, must be seen as a good unto its own. We have spent at least half a century with our imaginary for the use-value of science increasingly framed as a tool for economic or geopolitical competitiveness (Rudolph, 2002). Science has other uses. We must insist on them. Science is being quashed from many young people's interests within this constricted frame. From a telling study on the affective nature of student (dis)engagement with science, Peter Appelbaum and

Stella Clark (2001) report on the all-too-common nature of why science prompts people to disengage.

> Was science fun for them? Did they wish to share such fun with their own future students? No. They hated science in school. Science was scary. Science was a place of embarrassment. Science made them feel stupid, made them feel bad and angry. (p. 584)

As we think first about the success of our most common pedagogies in light of the research on motivated reasoning, especially considering pressing issues of importance to us all such as climate change, a strong rethinking of what we are doing, as a research community is warranted. Our work still echoes of a perfectionist's intent, where we hope that our science educational projects will cure social problems. This utopian hope must be tempered by realities of material conditions. Problems will always be with us. In addition, we can and must take clear ethically sound, professional action. The reworking of our dearest pedagogies will then be rearmed to serve different goods.

David Labaree (2006), while not a science educator, wrote a couple of decades ago of the multiple uses of education in the U.S. to serve individual goods, public goods, and citizenry goods. We have served the individual good well and now, as is increasingly clear in this era governed by neoliberal practices, to a much too great degree. Community and citizenry goods need reemphasizing. This framework can and should be a lens with which to present science for our future teachers in training if we do that work—or our students, if we, ourselves, are classroom teachers. Common pedagogies I have demanded as deficient might come back into new service, but with different political ends. Wendel Berry's (1990) oft-cited short piece called "What are people for" is instructive. I feel, and I sense that I am not alone, that we cannot, as a professional community clearly answer what science education is for outside of the narrow political project I have outlined. We can and should do better.

REFERENCES

Aikenhead, G. S. (1996). Science education: Border crossing into the subculture of science. *Studies in Science Education, 27*, 1–52. doi:10.1080/03057269608560077.

Appelbaum, P., & Clark, S. (2001). Science! Fun? A critical analysis of design/content/evaluation. *Journal of Curriculum Studies, 33*(5), 583–600. doi:10.1080/00220270010023812.

Berkman, M. B., & Plutzer, E. (2010). *Evolution, creationism, and the battle to control America's classrooms.* New York, NY: Cambridge University Press.

Berry, W. (1990). *What are people for?: Essays.* San Francisco, CA: North Point Press.

Callison, C. (2014). *How climate change comes to matter: The communal life of facts.* Durham: Duke University Press.

Damasio, A. R. (1994). *Descartes' error: Emotion, reason, and the human brain.* New York, NY: Putnam.

Goffman, E. (1959). *The presentation of self in everyday life*. Garden City, NY: Doubleday.

Haidt, J. (2012). *The righteous mind: Why good people are divided by politics and religion*. New York, NY: Pantheon Books.

Hammond, L., & Brandt, C. (2004). Science and cultural process: Defining an anthropological approach to science education. *Studies in Science Education, 40*(1), 1–47. doi:10.1080/03057260408560202.

Heath, S. B. (1983). *Ways with words: Language, life, and work in communities and classrooms*. Cambridge: Cambridge University Press.

Heidegger, M. (1962). *Being and time*. New York, NY: Harper.

Kahan, D. (2012). Why we are poles apart on climate change. *Nature, 488*, 7411. doi:10.1038/488255a.

Kahan, D. (2015). *So you want to meet the "Pakistani Dr."? Just pay a visit to the Kentucky Farmer.* Retrieved from http://www.culturalcognition.net/blog/2015/1/7/so-you-want-to-meet-the-pakistani-dr-just-pay-a-visit-to-the.html

Kahneman, D. (2011). *Thinking, fast and slow*. New York, NY: Farrar, Straus, and Giroux.

Labaree, D. F. (2006). *Education, markets and the public good: The selected works of David Labaree*. London: Routledge.

Lakoff, G. (2008). *The political mind: Why you can't understand 21st-century politics with an 18th-century brain*. New York, NY: Viking.

Lakoff, G., & Johnson, M. (1980). *Metaphors we live by*. Chicago, IL: University of Chicago Press.

Latour, B., & Woolgar, S. (1979). *Laboratory life: The social construction of scientific facts*. Beverly Hills, CA: Sage.

Long, D. E. (2011). *Evolution and religion in American education: An ethnography*. Dordrecht: Springer.

NGSS Lead States. (2013). *Next generation science standards*. Washington, DC: National Academies Press.

Nolan, C. (Director). (2008). *The Dark Knight* (DVD). United States: Warner Brothers Home Video.

Plutzer, E., McCaffrey, M., Hannah, A. L., Rosenau, J., Berbeco, M., & Reid, A. H. (2016). Climate confusion among U.S. teachers. *Science, 351*, 664–665. doi:10.1126/science.aab3907.

Rodriguez, A. J. (2015). What about a dimension of equity, engagement and diversity practices? A critique of the Next Generation Science Standards. *Journal of Research in Science Teaching, 52*, 1031–1051. doi:10.1002/tea.21232.

Rudolph, J. L. (2002). *Scientists in the classroom: The cold war reconstruction of American science education*. New York, NY: Palgrave.

Sapir, E. (1929). The status of linguistics as a science. *Language, 5*, 207–214. doi:10.2307/409588.

Whorf, B. L. (1956). *Language, thought, and reality: Selected writings of Benjamin Lee Whorf* (John B. Carroll, Ed.). Cambridge, MA: MIT Press.

CHAPTER THIRTY-ONE

Sociopolitical Activism and Transformative Learning

Expanding the Discourse About What Counts in Science Education

LYN CARTER, CAROLINA CASTANO RODRIGUEZ, AND MELLITA JONES

> A little rebellion now and then is a good thing.
>
> —THOMAS JEFFERSON, *LETTER TO JAMES MADISON, 1787*

There has been a small but critical conversation within science education for decades around the place of sociopolitical action, where the relationship between science education, social enquiry and political action has been recognized (e.g., Bencze & Alsop, 2014). In spite of this and claims from some authors that there has been a "broadening of the aims of science education to include social justice and sociopolitical action" (Reiss, 2007, p. 21), these concerns are somewhat marginalized in most discourses and practices in the field (see, for example, Castano Rodriguez, 2016). Interestingly, authors such as Roderick Watts, Matthew Diemer, and Adam Voight (2011) argue that sociopolitical development is aligned to the psychological trait of positive social involvement like social competence, community service, and citizenship. We believe this has implications for science education and science educators to re-position activism as a possible aim of science education.

Sociopolitical development also derives from liberation psychology, the branch of psychology that extends Paolo Freire's (1970) notion of critical consciousness amongst other anti-oppression constructs (Watts & Guessous, 2006). As such, like critical consciousness, the central aim of sociopolitical development is to articulate the relationship between creating just societies and the advancement of personal wellbeing. Hence, sociopolitical development embraces a higher purpose that brings with it action against social oppression (Watts, Griffith, & Abdul-Adil, 1999). Understood in this fashion, sociopolitical action and activism as part of psychological development should be relevant to science education, indeed education in general, as it speaks to the healthy and holistic social development of young people, in which we all have a stake.

Science is closely aligned with a range of social development issues, particularly in the current milieu where there are global concerns around climate change, genetic modification, mapping of the human genome, and other socio-scientific and techno-science issues that have seen an unprecedented level of human intervention with nature. Such developments increase the political and ethical dimensions of science teaching and learning to maintain its concern with the healthy and holistic social developmental of young people. In spite of this, only scant attention has been given to any kind of interrogation on the nature of activism or, more importantly, theorization of the sociopolitical activism construct itself within science education. A notable exception here is Wolff-Michael Roth (2014) who describes his theoretical concern and an evolving post-constructivist perspective on activism as event-in-the-making. Also, Steve Alsop and Lawrence Bencze (2012) in their debate with David Burns and Stephen Norris (2012), argue that more work in all areas is needed and highlight what they regard as the pertinent and persistent questions still requiring attention:

- How should SMTE take up activism?
- What could the work of activism be within SMTE?
- In what ways are existing SMTE practices activist?
- How might SMTE contribute more effectively to new social movements?
- How is activism pedagogical?
- What might pedagogies of activism offer and entail? (Alsop & Bencze, 2012, p. 395)

(The acronym SMTE stands for science, mathematics and technology education. Alsop and Bencze (2012) were writing in the *Canadian Journal of Science, Mathematics and Technology Education*, which is why they chose to use that acronym. For our purposes here, we can read it as science education.)

We consider these concerns in discussing what is missing in the public conversation about science education. For this, we describe part of our journey as academics committed to transforming the lives and educating passionate individuals who are committed to the transformation of society towards one that is more just and ethical.

TRANSFORMATIVE LEARNING

Transformative learning, a term first applied by Jack Mezirow in the late 1970s, refers to learning that merges "critical thought with critical action to effect change" (Kitchenham, 2008, p. 108). Mezirow (2000), also heavily influenced by Paulo Freire's liberation psychology and critical consciousness (Freire, 1970), defines transformative learning as a process whereby "we transform our taken-for-granted frames of reference to make them more inclusive, discriminating, open, and reflective so that they may generate beliefs and opinions that will prove more true or justified to guide action" (p. 214). Transformative learning theory is not well known within science education apart from some work by Kevin Pugh and his colleagues (Pugh, Linnenbrink-Garcia, Koskey, Stewart, & Manzey, 2010) who described transformative experiences adapted from John Dewey. With its focus on action, we believe that transformative learning theory may offer unique potential to further the sociopolitical agenda within science education.

In this chapter, we re-examine activism in science education as we take seriously the proposition that sociopolitical activism, as part of holistic psychological development, is relevant for the positive formation of young people in our care. It is as Michael Reiss (2007) notes, a legitimate and worthy, to which we add, courageous outcome of science education at the very time our world faces profound challenges, so many of which are anchored to scientific and technological knowledge and practice. We begin with our brief overview of activism leading to our typology that in no way wishes to be essentialist, but rather offers our reading of one amongst many. We then describe how we have adopted transformative learning as a framework to engender activism in science teacher education students at an Australian university.

ACTIVISM

Activism can be thought of as a profusion of actions, agendas, ideological positionings, theories and epistemologies animated by the need to make social, political, cultural, economic, and environmental change. It can range from large-scale collective public engagements to small-scale or individual acts that aim to make a difference. Consequently, one does not get far with activism before encountering the political.

While there is great similarity of processes and mechanisms of activism (Tilley, 2004), nonetheless modes and expressions of activism are always tempered by sociocultural, economic and political access, age and other social locations, as well as by the conjunction of the historical moment and geographical place (Hodgson & Brooks, 2007). It is useful then, to conceptualize sociopolitical activism as a "semantic field" that can describe an incredibly broad range of practices—from the very radical to the utterly mundane. There have been efforts to categorize these activist practices (e.g. Tilly, 2004), but for our purposes, only a basic delineation is needed to further our discussion. Hence, we present a brief typology ranging from everyday activism through to violent resistance and, whilst acknowledging a somewhat heuristic separation, we also recognize the fluid boundaries of these categories as one bleeds into the next.

Everyday Activism. Everyday activism involves acts by individuals, which, although performed privately or on a small scale, nevertheless constitute acts of support for, or resistance against, particular ideological standpoints. Everyday activism is conducted under the auspices of personal freedom or choice. Taking public transport, buying organic food, refusing plastic packaging, or supporting a local community group are common examples of ways in which a great many citizens *enact* political choices in their daily lives.

Conventional Political Activism. Common forms of sociopolitical action include volunteering, lobbying, advocacy, petitions, fundraising, speaking out, protests, media campaigns or legal cases. A science class choosing to write to their local government sharing their experimental investigations into the health of a nearby waterway would be an example, as would deciding to take part in a local climate change march. There are some authors who promote conventional political activism in science education including Patricia White (2012), and Sarah Stitzlein (2012). While by no means is there universal agreement about the appropriate level of activism (and certainly no agreement about the issues themselves), it is a central tenet of most Western political systems that the citizenry are free (at least in theory) to agitate for change, and in that way, influence the direction(s) of society.

Direct Action. Non-violent action and civil disobedience are more radical forms of activism, which often exist at the fringes of lawful behavior and can include a policy of deliberate law breaking as a means of political expression. Strikes, obstruction, sit-ins (including the recent widespread phenomena of OCCUPY), sabotage, hacking (or hacktivism), and the theft and release of confidential government information (à la WikiLeaks) are all examples of direct activist practices. For practitioners of direct action, the cause often exists in a highly privileged position vis-à-vis the rights of others. Direct action like this almost always articulates a clear challenge to the status quo. The consequences for those that participate openly in direct action can, on occasion, be severe—as Edward Snowden and Julian Assange are no doubt aware. Often, direct action is met with mainstream social censure or recrimination.

Violent Resistance. With violent resistance, the notion of activism begins a semantic bleed into a slippery slope of other labels—freedom fighter, agitator, guerrilla, insurgent, anarchist, traitor, partisan, seditionist, zealot and so on—where the distinction between "activist" and "terrorist" becomes largely a question of ideology, ethics, values, perspective and characterization. While an effort is often made to hold separate and condemn the most extreme forms of political action, they too have a place on the roadmap of activities aimed at bringing about social and political change. Violent resistance may begin as non-violent resistance then escalate (for example, the Arab Spring), or it may involve quite deliberately calculated violence for political ends (e.g., the Irish Republican Army's bombing campaign or the obdurately violent Israeli/Palestinian conflict). The lines between violent resistance, civil war, revolution and activism are in most cases entirely artificial, where the aim is to completely overthrow an existing ideological position. Where violent resistance is practiced, activists are invariably seeking a fundamental change to some aspect of the social system as a whole. It is practiced wholly outside the bounds of law and tends to regard existing power structures as illegitimate.

ACTIVISM AS CLASSROOM PRACTICE—WHO DECIDES?

That activism is political without any compass, or indeed from the typology charted above, without any limits, raises issues about the place of sociopolitical activism within the classroom. There is an unchallenged assumption that social change sought by science educators is of inherent value, within the acceptable category and ideologically appropriate. Assume, for example, activist science teachers presenting a class with the food cycle in a way that highlights the ethical dimensions of factory-farmed meat, and a portion of the student cohort decides to become involved with radical animal liberation movement—is that *desirable*? If that same movement destroys the property and livelihoods of others, is that *acceptable*? Are activist science educators complicit in activist practices they may not personally endorse? *Does it matter? Who decides?*

These are urgent questions for activist understandings of science education. With the theoretical vacuum in sociopolitical science education scholarship, it is probably not surprising that such questions are not well addressed. Derek Hodson (2010) sympathizes and "urge(s) teachers and students to be critical, reflective, robust in argument, and sensitive to diverse values and beliefs, but above all to have the courage and strength of will to do what they believe is right and good and just" (p. 205). Certainly, but in the slippery semantic and perspectival field of activism shown in the typology above, just who are the arbiters of what is "right, good and just"?

Burns and Norris (2012) acknowledge these dilemmas and propose an enquiry into the *moral* dimensions of an activist science education, with questions regarding the nature of values, the act of teaching values, and the proper limits of pedagogical authority being central. There have been efforts directed towards this end like Alsop and Bencze (2012), responding to Burns and Norris (2012), who reiterate the call for additional enquiry, arguing that "the STS movement has not paid sufficient attention to ethical issues and the moral development of students" (p. 402). At this stage, it is apparent that a significant discussion in activist science education literature of ethico-moral implications remains under-explored and under-critiqued. In this chapter we contribute to this discussion by describing our experience in incorporating activism as an outcome of the science education we offer to primary school teaching students. We start by exploring transformative learning theory as a framework that provided us with theoretical and empirical insight in empowering our students to consider taking responsible action.

TRANSFORMATIVE LEARNING: A METHODOLOGICAL FRAMEWORK FOR ACTIVISM

Mezirow's theory of transformative learning and its various iterations (for example, Mezirow, 1991, 2000, 2003), is based upon Paulo Freire and Jürgen Habermas' emancipatory view of knowledge. Mezirow's transformation describes the process of experiencing a deep shift in perspective that results in a radical alteration in outlook, and subsequently, actions. For Andrew Kitchenham (2008, p. 108), it couples "critical thought with critical action to effect change," making it an ideal approach to promote activism.

Mezirow (2003) sees transformative learning as primarily a rational and cognitive process involving:

- recognition that an alternative way of understanding may provide new insights into a problem;
- context awareness of the sources, nature, and consequences of an established belief;
- critical reflection of the established belief's supporting assumptions;
- validating a new belief by continuing assessment of its justification;
- coping with anxiety of the change and taking reflective action on the validated new belief. (p. 124, reformatted for ease of reference)

Other scholars have elaborated transformative learning beyond the cognitive. John Dirkx (2006) for example, integrates experiences of the *outer world* that includes cognitive, epistemic and sociocultural dimensions of the process of transformative learning with the *inner world* involving emotional, moral and spiritual dimensions. There is a clear link here with Burns and Norris' (2012) moral issues within activist science education noted above.

Furthermore, Patricia Cranton (2011) and Michael Kroth and Cranton (2014) argue that critical reflection and emancipatory education practices are necessary but not sufficient conditions for transformative learning to take place. Individuals must also experience their own *triggering events,* that is, encounter perspectives at odds with their own that cause them to revise their worldview and reauthorize personal biographies. Variously referred to then, Cranton's (2011) *triggering events,* or Ted Fleming's (2011) disorienting *dilemmas,* or Kaisu Malkki's (2010) potent *unpleasant emotions,* and ultimately, to Knud Illeris' (2009) much more strident *crisis-like situation,* confrontation of some type is deemed a necessary condition for learners' transformation towards action. It is this formalized confrontation that is missing from much of the sociopolitical science education, limiting, according to transformative learning theory, the effectiveness of much sociopolitical science education. As such, confrontation was among the first dilemmas faced as we explored the use of transformative learning to engender activism in our science teacher education units.

CREATING SPACES FOR TRANSFORMATION

Three years ago, we, the authors and science teacher education academics at a university in Australia, decided to discuss how we could further develop our teaching to empower students to take action on issues of contemporary, science-related, global and/or local relevance. Conscious of the diverse complexities associated with empowering adults to take action in a field, science education—which has been historically perceived as neutral, we decided to create a unique learning space devoted to this aim. We designed an elective unit for 20–30 students from an undergraduate primary teaching course. This unit has run for the last three years as a four-day intensive course (four Saturdays from 9 am to 4 pm) with additional research expected outside class time. The main goal of the unit was to engage primary teaching students in transformational processes that encourage action.

Of the three of us involved in the planning of this unit, one took responsibility for the teaching and assessment and the other two conducted the collection of research data. This separation of roles helped to ensure we addressed any conflict of interest between the teaching/assessment process and the research component. At the commencement of the first day, we explained our use of transformative learning theory, which Karen Swanson (2010) argues is necessary to foster connections between students and teachers and create an appropriate learning space. The students were thus aware of the four-step pedagogical process and the research being undertaken. Ethics approval was obtained from the university prior to our commencing implementation of the unit.

Following transformative learning theory, we presented students with diverse local and global issues linked to developments in science/technology that remained understudied in Australian society. These issues aimed to confront students' current beliefs and actions, and empower them to consider taking more informed decisions and actions in their lives. In the first year the unit ran, after introducing the process, we allowed students to identify issues they considered confronting for their own investigation. However, we soon realized that students had limited experiences/spaces for developing their own agency, and they tended to hold views of science and society as neutral entities that always worked for the benefit of the community. Thus, our students were limited in their capacity to critically review society and science, which was reflected in the themes of choice. As a consequence, from the outset, we felt we would not fully accomplish the goals we established in this first delivery of the unit. Moreover, by the end of the unit it was clear that students were confronted by the issues we introduced, but not so much by those they selected themselves. Consequently, in year two, we decided to present students with a list of possible assumptions held by our society for them to choose from and which they then deconstructed following transformative learning theory. Themes ranged from "In a democratic society, everyone has freedom of speech, including environmentalists" to "Zoos are important learning and ethical centers that aim to conserve species." Since year two, the unit has promoted a higher level of student engagement with transformation of thought and action, probably as an outcome of this shift in our pedagogy.

A range of forms of data were collected each year including: researcher field notes; students' responses to class tasks; comments students made in class; responses to emotional charts asking students how they felt at certain points of learning; student ratings on how confronted they were; what, if any, form(s) of action they felt compelled to take for issues/topics covered; and interviews conducted at the end of the unit. These multiple forms of data were used to assist in providing trustworthiness to the analysis and findings. The process of transformation that students experienced, and our own journey in continually designing and re-designing the unit, is a testament to both the complexities and the potential for meaningful teaching of science with a focus on activism. Based on evidence collected, we describe below our reading of students' experience of transformational learning in this unit and the challenges we faced in incorporating activism and transformation as genuine aims for science teacher education.

Confronting Students: Triggering Events. Deciding on the issues that students would confront and be of relevance to the students was a major challenge in the conceptualization of the unit. Aware of the limited opportunities students have to be confronted and engaged with controversial issues (Castano Rodriguez, 2016), we introduced the transformative learning process through the diverse issues surrounding the production of eggs in Australia. We arrived at the egg production topic after considerable contemplation and discussion. Initially we thought to use the highly provocative animal rights documentary *Earthlings* as our confrontation. Mindful of the concerns raised earlier about the emotional or crises potential triggering events can have, and given there is no roadmap for going "too far," the graphic nature of the film saw us step back from its use and settle instead, on the "safer" topic of eggs. We expected this topic would still be confronting, as we suspected that students would have limited knowledge of how agribusiness meets the egg demand, and we believed it was straightforward enough that they would probably see immediate and possible actions. We began by challenging the students with the statement: *If you eat eggs you are contributing to animal cruelty.* We provided examples from media, industry, animal protection organizations, and academic journals for students to critically analyze, ensuring they had access to diverse and contrasting positions regarding the issue. Students were encouraged to pose their own questions during their analysis of each source of information, and to carry out further research of their own. During this initial session, students commented on how confronted they felt by the realization that many of the views they held were influenced by mainstream media. They also recognized that without a concerted effort to look beyond mainstream media, it was difficult to access all relevant information surrounding an issue in order to make informed decisions about their actions.

On the second day, another confrontation was introduced in the form of a panel of activists. Across the four years the unit has run, diverse areas of activism have been represented in this panel session, including: activists from a local not-for-profit sanctuary for rescued farm animals; a Chief Executive Officer for a not-for-profit organization promoting more humane, non-animal methodologies in research; and climate change activists, including speakers from Greenpeace. Each speaker provided some background about their cause and their personal journey towards becoming activist. Articulate and passionate about their respective causes, the panel participants' descriptions of their activities could be categorized within the everyday and conventional political activism typology described above, as well as within the more radical nonviolent direct action, occasionally at the legal edge.

In an attempt to individualize some "triggering events," the third and fourth days of the unit, and the four-week period in between, sees students working in small groups on sociopolitical scientific issues they have identified as personally confronting and which they have chosen from the provided list. Students engage in in-depth research of those aspects of society that reinforce and limit our understanding of the social-ecological and ethical issues under study. They are encouraged to interview friends and relatives, and visit activist organizations to collect and contrast diverse views held by different members of our society. As an outcome, students create an activist campaign video to confront their peers on their chosen topic, and engage them in debate and discussion about the levels of action they would or would not be willing to undertake. Additionally, students have to select and commit to their own personal action associated with their topic for a minimum three-week period, and reflect on this and their new understanding of the issue under study.

To date, every semester this course has run, we have more than 20 students wanting to enroll. Each group has a unique dynamic and presents different challenges. Over the three years we have worked towards better empowerment of students to contemplate issues and actions they have never considered before. One of the biggest challenges for us has been how strongly we should confront students. As considered by transformative learning theories, there is a fine line between the impact of a triggering event that compels some form of action and instilling feelings of despair which are dis-empowering and often cause the learner to re-accommodate previous conceptions to justify not taking action.

Along our journey we have also been challenged to re-think and re-consider our own actions. For example, not only has this process strengthened Carolina's passion and commitment towards a vegan lifestyle, but she has also extended her ethical lifestyle with regard to environmental issues such as the use of plastics in everyday life products. Moreover, Mellita was confronted not only by the discussion and consideration of the issues presented in the unit, but also by one of the students who, during a final interview, assumed she was vegetarian. Mellita felt ashamed at the time to admit that she was not and as result, decided to become vegetarian as she drove home that very evening. More recently (another 12 months on) Mellita has also moved to a vegan lifestyle. This demonstrates the power of transformational learning, and of personal triggering events, not only on the students, but also on the teachers/researchers themselves.

Activists as "Triggers" of Confrontation Towards Sociopolitical Action. Activists who have participated in this unit have shared their passion and commitment for the cause they believe in. Only a very few students had had exposure to activists prior to taking this unit. For most, this space to share conversations and learn about the life and passion of activists has been an inspirational and empowering experience. For example, during the first year, 14 out of the 18 students who recorded their views on this experience, indicated they found the activists reasonably to highly confronting, with six reporting confrontation by new information: "There is so much I did not know." Two students felt confronted by their own inaction on issues about which they shared similar beliefs as the activists, while another felt completely overwhelmed and as a result, disempowered. Four students with low confrontation scores indicated prior knowledge about the issues.

In terms of inspiring action, 8 of 18 students felt that although they found the panel confronting and inspiring, they could not make a similar level of commitment. It seemed an "all or nothing" proposition and four students believed the panel's more radical direct actions were "too extreme." Several though, cited convenience over conscience as a reason not to act.

During the last two years, we have placed greater emphasis on the process of becoming an activist and the personal journey that activists have undertaken. This appears to be aligned with an increase in students reporting feelings of being inspired to change. Moreover, when guest activists have been closer in age to the students, an even greater increase in reported intention to take action has resulted. For example, last year we had two guest speakers who volunteer for Greenpeace: one a university student and the other a recent graduate; both extremely passionate and committed vegans and environmentalists. Although similar in many ways to other activist speakers, these particular speakers were able to relate their quite recent process of transformation. They each illustrated how their choices and lifestyles were still progressing and shared the challenges they still faced in considering their own actions. Students reported that they identified with these two activists and that the examples of how passion can drive actions were more accessible to them. Each year, several students indicate that they would take the issues identified by the activist panel into their own thinking when teaching, which itself could be seen as everyday activism.

These findings suggest that activist speakers can be a good way to confront students and raise their awareness about sociopolitical issues and actions, particularly if they have little prior knowledge on the topic. However, it is also apparent that presenting more radical action on its own does not necessarily engender action. Exposure to activists who are more recent to their committed actions, and presenting activism as a journey with multiple levels of "buy in" rather than an apogee of extremism, seem to be two ways that encourage and better empower students in thinking about their own activist stance.

From Promoting Critical Awareness to Taking Action. As mentioned earlier, in year one when students were left to propose their own confronting issue/topic for investigation, many of the issues nominated did not actually confront them enough to inspire action. Those that did were used along with other topics we selected, to form a list of issues for students to select from in year two and beyond. Some of the characteristics we have identified that most empower students to consider action is when their research increases their awareness of how limited mainstream information is about the issue; and how political and/or other vested interests are hidden and yet underpin many issues. For example, during year one the students researching genetically modified (GM) foods were unexpectedly surprised by their findings, with one student noting during the interviews:

> When you come across stuff like big companies like Monsanto when they develop a corn that is tolerant to everything except weeds and they also sell weed killer that sort of stuff just doesn't leave you. So you've got to tell people about that.

Another committed to take action on GM foods as a result of their research:

> All this information led me to take the action. When shopping, I took the list of GM and non-GM foods. This gave me the power to make what I believed was more ethical purchases.

These comments are characteristic in revealing the themes that most confront students. In these cases students question the societal institutions that they feel have misled their views and limited their opportunities to learn about associated controversies. They are particularly confronted by their lack of awareness that any controversy even exists prior to their research. While most of the actions reported fall within everyday activism through individual lifestyle choices, like the GM food example described above, some have resulted in conventional political activism. For example, the topics of GM foods, zoos and puppy mills have inspired students to participate in protests and petition the government for greater regulation. These results suggest that the higher the confrontation, the more robust and politically orientated the actions willing to be undertaken. At the end of the unit, most preservice teachers identified transformative learning as a powerful pedagogy of which the following comment is typical:

> This unit was so interesting and engaging! Each session I came away with new information and ideas. It challenged my learning and beliefs in a safe and open environment which was refreshing. … This was different to anything I had done before and I was grateful for the opportunity to challenge myself.

Despite the positive outcomes of promoting activism through science education we identify here, caution is necessary as confrontation generates anxiety and discomfort that might encourage students to accommodate the new information within their previous beliefs, thereby avoiding unpleasant feelings. Moreover, it is not clear if students who felt highly confronted preferred not to take action or the level of disorientation was disempowering them to take action. This needs further exploration.

TRANSFORMING SCIENCE EDUCATION FOR ACTION—A FIELD FOR FURTHER INQUIRY

Our purpose in writing this chapter was to provide insight into promoting transformational learning and activism as a legitimate, worthy and courageous outcome of science education in a time that our world faces profound ethical challenges related to scientific and technological development. Sociopolitical development is, we believe, an important part of the positive social formation of all young people. In addition to raising some of the ethico-moral difficulties around types of activism, we sought to explore what transformative learning theory with its focus on action could bring to the sociopolitical science education agenda. More specifically, we reported on our experience with primary school teaching students to investigate transformative learning to encourage activism. We hoped to show that pedagogies informed by transformative learning theory could contribute towards some of the questions raised by Alsop and Bencze (2012) and Hodson (2010) that include for example, "what might pedagogies of activism offer and entail?" After all, as White (2012) and Stitzlein (2012) so aptly remind us, student activism does not just happen—it needs to be actively planned.

Our journey suggests that transformative learning theory may be a viable approach for privileging the sociopolitical within the science classroom. The transformative learning pedagogical process of firstly, critical reflection and reflexivity, secondly, the mobilization of conflict (as triggering events), thirdly, the embeddedness of the individual as the social emotional learning self, and finally, the focus on action that we derived from transformative learning theory (Carter, Castano Rodriguez, & Jones, 2014), would appear to be readily implementable within the sociopolitical science classroom.

Confrontation seems to be an essential pedagogical tool that takes transformative learning beyond other awareness-generating sociopolitical approaches, and will more likely lead to action. Confrontation though, is not without its challenges, and presents possibly the greatest challenge for science educators wanting to implement transformative learning. Moreover, consistent with the queries raised above, there are ethical issues around the selection and structuring of the triggers, discomfort or crises, essential to transformation. How much of an emotional challenge can be viewed as appropriate before, as Burns and Norris (2012 p. 385) ask, "the proper limits of pedagogical authority" have been reached or exceeded? And what level of activism can be considered—everyday life style choices, acceptable conventional political dissent, more radical direct action or is it okay to go further? While well used to providing students with cognitive or intellectual challenges, science educators are less skilled at and secure about provoking emotional challenges. Certainly, attempting confrontation to transform learners' acquired frame of reference and action engenders an ethico-moral responsibility to ensure adequate support structures are available to assist learners in managing the potential distress and challenge they face in the transformative learning process (Burns & Norris, 2012). It is an area that clearly requires more research that could lead to deeper insights into Alsop and Bencze's (2012) queries noted above about how activism is pedagogical and what might pedagogies of activism offer and entail.

In considering Hodson's (2010) concern about who decides what is acceptable and responsible action, our findings show that the students themselves bounded the types of activism in which they were willing to engage. Predominantly from the everyday lifestyle category of individual action, most students had an emphatic sense of what was "too much of a commitment."

In raising some ethico-moral issues around types of activism that remain under- explored and under-critiqued, we also argue for a greater theorization of sociopolitical activist science education. Only then will we be really able to grapple with Alsop and Bencze's (2012), Hodson's (2010) and others' concerns and move the field forward. Much is to be gained from scholarship outside science education like transformative learning theory that is salient to student activism. Certainly, there is a lot of work to be done in expanding the conversation and thinking about what counts in 21st century science education.

ACKNOWLEDGEMENTS

With thanks to Nicolas Walker, Philip Clarkson and Jenny Martin for their comments on earlier drafts of this paper.

REFERENCES

Alsop, S., & Bencze, J. L. (2012). In search of activist pedagogies in SMTE. *Canadian Journal of Science, Mathematics & Technology Education, 12*, 394–408. doi:10.1080/14926156.2012.732256.

Bencze, J. L., & Alsop, S. (2014). *Socio-political activism and science & technology education.* Dordrecht: Springer. doi:10.1007/978-94-007-4360-1.

Burns, D., & Norris, S. (2012). Activist environmental education and moral philosophy. *Canadian Journal of Science, Mathematics and Technology Education, 12*, 380–393. doi:10.1080/14926156.2012.732190.

Carter, L., Castano Rodriguez, C., & Jones, M. (2014). Transformative learning in science education: Investigating pedagogy for action. In J. L. Bencze & S. Alsop (Eds.), *Socio-political activism and science & technology education* (pp. 531–545). Dordrecht: Springer. doi:10.1007/978-94-007-4360-1_30.

Castano Rodriguez, C. (2016). Which values regarding nature and other species are we promoting in the Australian Science Curriculum? *Cultural Studies of Science Education, 11*, 999-1021. doi:10.1007/s11422-015-9675-7.

Cranton, P. (2011). A transformative perspective on the scholarship of teaching and learning. *Higher Education Research and Development, 30*(1), 25–86. doi:10.1080/07294360.2011.536974.

Dirkx, J. M. (2006). Engaging emotions in adult learning: A Jungian perspective on emotion and transformative learning. *New Directions for Adult and Continuing Education, 109*, 15–26. doi:10.1002/ace.204.

Fleming, T. (2011, May). *Recognition in the work of Axel Honneth: Implications for transformative learning theory.* Paper presented at the International Transformative Learning Conference, Athens, Greece.

Freire, P. (1970). *Pedagogy of the oppressed.* New York, NY: Continuum.

Gallagher, J. (1971). A broader base for science teaching. *Science Education, 55*, 329–338. doi:10.1002/sce.3730550312.

Hodgson, D. L., & Brooks, E. (2007). Introduction: Activisms. *Women's Studies Quarterly, 35*(3/4), 14–25.

Hodson, D. (2010). Science education as a call to action. *Canadian Journal of Science, Mathematics and Technology Education, 10*, 197–206. doi:10.1080/14926156.2010.504478.

Illeris, K. (2009). A comprehensive understanding of human learning. In K. Illeris (Ed.), *Contemporary theories of learning: Learning theorists in their own words* (pp. 7–20). Milton Park: Routledge. doi:10.4324/9780203463321.

Kitchenham, A. (2008). The evolution of Jack Mezirow's transformative learning theory. *Journal of Transformative Education, 6*, 104–123. doi:10.1177/1541344608322678.

Kroth, M., & Cranton, P. (2014). *Stories of transformative learning*. Rotterdam: Sense Publishers. doi:10.1007/978-94-6209-791-9_6.

Malkki, K. (2010). Building on Mezirow's theory of transformative learning: Theorizing the challenges to reflection. *Journal of Transformative Education, 8*, 42–62. doi:10.1177/1541344611403315.

Mezirow, J. (1991). *Transformative dimensions of adult learning*. San Francisco, CA: Jossey-Bass.

Mezirow, J. (2000). *Learning as transformation: Critical perspectives on a theory in progress*. San Francisco, CA: Jossey-Bass.

Mezirow, J. (2003). Transformative learning as discourse. *Journal of Transformative Education, 1*, 58–63. doi:10.1177/1541344603252172.

Pugh, K., Linnenbrink-Garcia, L., Koskey, K., Stewart, V., & Manzey, C. (2010). Motivation, learning, and transformative experience: A study of deep engagement in science. *Science Education, 94*, 1–28. doi:10.1002/sce.20344.

Reiss, M. (2007). What should be the aim(s) of school science education? In D. Corrigan, J. Dillon, & R. Gunstone (Eds.), *The re-emergence of values in science education* (pp. 13–28). Rotterdam: Sense Publishers.

Roth, W.-M. (2014). From-within-the-event: A post-constructivist perspective on activism, ethics, and science education. In S. Alsop & L. Bencze (Eds.), *Activist science and technology education* (pp. 257–254). Dordrecht: Springer. doi:10.1007/978-94-007-4360-1_14.

Stitzlein, S. M. (2012). The right to dissent and its implications for schooling. *Educational Theory, 62*, 41–58. doi:10.1111/j.1741-5446.2011.00434.x.

Swanson, K. (2010). Constructing a learning partnership in transformative teacher development. *Reflective Practice: International and Multidisciplinary Perspectives, 11*, 259–269. doi:10.1080/14623941003672428.

Tilly, C. (2004). *Social movements, 1768–2004*. Boulder, CO: Paradigm.

Watts, R. J., Diemer, M. A., & Voight, A. M. (2011). Critical consciousness: Current status and future directions. *New Directions for Child and Adolescent Development, 2011*, 43–57. doi:10.1002/cd.310.

Watts, R. J., Griffith, D. M., & Abdul-Adil, J. (1999). Sociopolitical development as an antidote for oppression—theory and action. *American Journal of Community Psychology, 27*, 255–271. doi:10.1023/a:1022839818873.

Watts, R. J., & Guessous, O. (2006). Sociopolitical development: The missing link in research and policy on adolescents. In S. Ginwright, P. Noguera, & J. Cammarota (Eds.), *Beyond resistance! Youth activism and community change: New democratic possibilities for practice and policy for America's youth* (pp. 59–80). New York, NY: Routledge.

White, P. (2012). Making political anger possible: A task for civic education. *Journal of Philosophy of Education, 46*, 1–13. doi:10.1111/j.1467-9752.2012.00837.x.

SECTION THIRTEEN

Science Education Visions

What Is School Science for and What Should We Be Doing in the Name of Science Education?

CHAPTER THIRTY-TWO

Breaking Down Enlightenment Silos

From STEM to ST²EAM Education, and Beyond

ELISABETH (LILY) TAYLOR AND PETER CHARLES TAYLOR

DIFFERENCE AND DANGER

> *"So you have no idea what my aptitude is?" "Yes and no. My conclusion," she explains, "is that you display equal aptitude for Abnegation, Dauntless and Erudite. People who get this kind of result are ...," she looks over her shoulder like she expects someone to appear behind her, "... are called ... Divergent ... Divergence is extremely dangerous. You understand?"*
>
> —ROTH (2011, PP. 22–23)

This excerpt from the Divergent book series by author Veronica Roth (2011)—now made into movies—describes the struggle of young Tris who lives in a society that, as the story reveals, has been set up as a social experiment to assess whether a caste system that assigns people to factions according to their aptitudes in life, can create a peaceful and harmonious society. These so-called natural aptitudes are determined through simulations during which a young person is injected with a serum that induces an hallucination, conjuring up a challenging and sometimes life-threatening scenario in which the person has to find a solution to survive. The type of solution strategy chosen determines the faction to which the young person is assigned. On "choosing day" most young people are steered into making a life commitment to a particular faction.

In the story, people are separated into five factions: Abnegation (Selflessness), Candor (Honesty), Dauntless (Courage), Amity (Peaceloving) and Erudite

(Knowledge). These factions live under strictly separated conditions. However, it soon transpires that evolution has stepped in and created a mutant variety—Divergent—that combines several or all of the aptitudes. Divergents are feared by other factions and considered dangerous because they cannot be pigeonholed. As the story unfolds it becomes increasingly clear that Divergents hold the key to solving complex problems that other factions individually cannot solve because only they can access a variety of aptitudes. You might ask, "What does this Divergent book series, which is written for young adults, possibly have to do with overcoming science education Enlightenment silos?"

We see a strong parallel between the Divergent story and current developments in education systems worldwide that are exploring the benefits of integrating the factions of STEM (Science, Technology, Engineering, Mathematics) education with arts education to create exciting new STEAM curricular structures and pedagogies. Arts education is perceived (often uncharitably) by STEM educators to be a hothouse of divergent thinkers. In education systems worldwide, curriculum structures have been premised historically on the notion that science and the arts are akin to incommensurable paradigms (Kuhn, 1962) that have nothing to offer each other. However, in today's hi-tech digital world, this curriculum divide is closing rapidly with the uptake by both the sciences and the arts of exciting new technologies, such as augmented and virtual reality devices. Further breaking down of this divide is being advocated by business and industry leaders who argue that the IT workers of the future need to be creative and innovative thinkers. One of the Australian government's peak advisory bodies on higher education and training for the 21st century recently argued that arts education should be an integral component of the preparation of STEM graduates:

> IT leaders should add an "A" for fine arts to the science, technology, engineering, and math charter-STEAM, not STEM. Designing engaging solutions requires creative talent; creativity is also critical in ideation-helping to create a vision of reimagined work, or to develop disruptive technologies deployed via storyboards, user journeys, wire frames, or persona maps. Some organizations have gone so far as to hire science fiction writers to help imagine and explain moonshot thinking. (Deloitte, 2015, p. 126)

Perhaps surprising to many, the Australian Curriculum recognizes the importance of developing the higher-order abilities of young people for the 21st century, especially innovative and creative thinking. Our national curriculum requires teachers of all learning areas, including science education, to address a set of General Capabilities and Cross Curricular Priorities that serve to "contribute to the creation of a more productive, sustainable and just society" (Australian Curriculum Assessment and Reporting Authority [ACARA], 2016).

It may also be surprising to learn that the divide between the arts and sciences has not always existed. It comes as a direct result of the Enlightenment movement of the 18th Century that created a silo mentality amongst adherents of the separate factions. Boundaries between the factions of science, art and religion are defended "religiously" and divergents who choose to cross the boundaries between the silos often are regarded suspiciously. In the Divergent story, the faction system that created harmony is incapable of solving complex problems in isolation—this requires divergent thinking that crosses faction boundaries. We argue that whilst the Enlightenment silos have served us reasonably well for 300 years in creating harmony between the proponents of science, religion and the arts, the complex (or wicked) problems we are facing in the 21st Century require nothing less than a profound rethink of the silo boundaries and recognition that a full range of human abilities and skills is needed.

In this chapter we analyze reasons for having Enlightenment curriculum silos in science education, consider opportunities and provide suggestions for overcoming them in an age where out-of-the-box solutions are asked for and the standard "this is how we have always done it" response will no longer do.

WITHIN THE SAFETY OF THE ENLIGHTENMENT SILOS

We notice that when we visit schools in different countries and on different continents, 21st Century classrooms have not changed much from the previous century, when we were students. We still find students of equal age grouped together in most schools. Perhaps they are streamed according to similar ability or gender. Forms or year groups are amongst the most common physical representations of Enlightenment thinking.

But the most entrenched example of Enlightenment thinking is the way curriculum content or subject matter is divided into distinct disciplines, such as mathematics, humanities, art, geography, languages, and, in some schools, religious education. There is a hierarchy whereby science, mathematics and English are on top of the intellectual pecking order, and humanities, arts and religious education are viewed as less important by parents, students and the general public—sometimes this distinction exists within professional education circles. Commonly, students are not only organized into manageable units of size, age, gender and ability, they are, similar to the young people in the Divergent story, also required to choose a restrictive range of subjects to study during the course of their schooling—at least here in Australia. Their choices ultimately determine what they can study at university, which jobs they can aim for—in short, what they can subsequently do with the rest of their lives.

Every year, authorities warn that not enough graduates choose science, technology, engineering and mathematics (STEM) careers which, we are told, our nation desperately needs to boost the economy. Recent national reports on future-proofing Australia's high-tech, digital workforce call for STEM graduates with creative and innovative abilities (e.g., PricewaterhouseCoopers, 2015). Australia's Chief Scientist has called for education reforms to better engage students in STEM-related career pathways (Office of the Chief Scientist, 2013). There are national initiatives by business and industry to get more students into STEM subjects, such as the 21st Century Minds (21C) Accelerator Program which aims to prepare students to "think smart and creatively, solve problems, persist and take risks, have strong digital skills, and know how to collaborate effectively" (PricewaterhouseCoopers, 2016). Some are specifically targeting girls. Yet, whilst STEM has been around for a while internationally, the STEM movement in Australia has only recently attracted attention, mainly because it has just entered the awareness of (electorally competitive) politicians that STEM is necessary for Australia to remain competitive and economically viable.

We now live in an age of digital technologies, artificial intelligence, DNA mapping, robotics, nanotechnology, 3D printing, biotechnology and the "internet of things." Preparing students for jobs in these emerging hi-tech sectors, many of which currently do not exist, is essential to the nation's economic wellbeing and is one of the primary challenges for education systems worldwide. Graduates with *liquid skills* that enable them to adapt to a fluid working landscape throughout their lives are being called for by business and industry leaders (Infosys, 2016). Liquid skills include the ability to work with others, verbal communication, creative and critical thinking, active listening and active learning, and a disposition towards life-long learning. According to Infosys, these capabilities are deemed to be more important than high academic achievement for IT workers in the unfolding "fourth industrial revolution."

AUTOMATIZATION, RAPIDIFICATION AND THROWAWAY CULTURES—AN UNEXPECTED CALL FOR APPLIED LIQUID SKILLS

The problems facing our existence as human beings exceed by far anything that has confronted previous generations in terms of complexity and immediacy: climate change, repeated global and local financial crises, migrant and refugee crises, social disintegration of nation states, environmental destruction, coral bleaching, sea level rising, increasing automatization leading to job-loss, and many more. All of these problems have arisen despite, simultaneous with, or because of the stunning scientific, medical, and technological progress that has been occurring at lightning speed. Some argue that these problems are due to unbridled economic growth and blame free-market idolatry.

"There they go again!," we hear you say, "left-wing Marxist, anti-free market ideology!" The above alerts and analysis, however, are not ours to claim—far from it. The author who, for now, shall remain unnamed has called for the whole human family to seek sustainable and integral development in order to face the urgent challenge to protect our common home—the Earth. He has stressed that, "everyone's talents and involvement are needed to redress the damage caused by human abuse …" (p. 11). Moreover, he has blamed the "rapidification" of the pace of life and work for adding pressure to human lives—a development that has coincided with a period of irrational confidence in progress and human abilities, whilst "… the Earth, our home, is beginning to look more and more like an immense pile of filth" (p. 15). He has associated some of the current problems with our throwaway culture and has called for ecological approaches that should always include consideration of social perspectives, "… without excluding or favoring anyone" (p. 58), adding that, "there can be no ecology without an adequate anthropology" (p. 74). This author reminds us that our immense technological development has not been accompanied by a concomitant development in human responsibility, values and conscience. "Each age," he has stated, "tends to have only a meager awareness of its own limitations" (p. 65), causing humanity at present to use technology "according to an undifferentiated and one-dimensional paradigm" (p. 66). Analyzing the role that environmental education has been playing, the same author has concluded that:

> … in the beginning it was mainly centered on scientific information, consciousness-raising and the prevention of environmental risks, it tends now to include a critique of the "myths" of modernity grounded in a utilitarian mindset (individualism, unlimited progress, competition, consumerism, the unregulated market). It seeks to restore the various levels of ecological equilibrium, establishing harmony within ourselves, with others, with nature and other living creatures (p. 129).

By now, we hope, you might be curious as to who the author of this frank analysis of the status quo might be, and you might be surprised to find, as were we, that the above quotes originate from the head of a global think tank that is not usually known for calling into question so openly environmental, social, and economic politics in such an eloquent and direct way. It was in fact Pope Francis whose thoughts we have quoted above. His frank assessment, especially of the politics of climate change, was published in 2015 as "Encyclical on Climate Change and Inequality: On Care for our Common Home." The then Australian Prime Minister, Tony Abbott MP, who often described himself as an avid Catholic, was also a fierce critic of the science of climate change. When journalists confronted Abbott with the Pope's assessment of climate change he dismissed Pope Francis's ideas and advised the Pope to stick to what he is good at and to what he knows something about: religion. Never mind that the former prime minister trained as a journalist whilst Pope Francis studied chemistry and religion.

After this brief excursion into the borderlands between science and religion, we would like to ask you to take a step back and reflect on your first reaction when you learned that the author of the text above was not a philosopher, not a scientist, but the current Pope. He is, after all, the current leader of the organization—the Catholic Church—that in the past caused so much trouble for scientists many of whom, such as Galileo, were struggling to survive in the anti-science climate of the 16/17th Century. We imagine that some readers might have experienced a level of alarm, perhaps even strong resistance, on learning the author's identity. If that is what happened to you, do not be alarmed—this reaction is perfectly aligned with Enlightenment thinking that emerged over the past centuries as a consequence of the separation of science, religion, and art (Wilber, 1998). The creation of the Enlightenment silos has led to a sense of safety but also to resistance amongst adherents of the individual silos when there appears to be a breach of the boundaries. Boundaries create a safe space, but they also create limitations.

THE POWER OF CONVERGENCE: LIVING THE ENLIGHTENMENT SILOS

When I (Lily) was 17 years old, I had many discussions with my parents about my professional future. My father was a medical doctor and researcher, and he was adamant that, "art is something you can always do in your spare time! Do you want to end up as yet another unemployed, struggling artist?" What he said in the 1980s reflected a type of thinking that is still widespread today amongst the general public and in science-oriented academic circles—art serves pleasure, enjoyment, and entertainment. My father would have argued that the arts and religion are not to be taken too seriously whilst a serious career choice, for him, could only have been within the science domain. However, the distinction between the three domains has not always been in place: prior to the Enlightenment era, science, religion and the arts were parts of one unified domain (Wilber, 1998) with considerable overlaps. This meant that there was the potential for both mutually stifling and fruitful conditions. All this was to change through the Enlightenment movement. According to Oxman (2016, para 1), one of the hallmarks of the emerging Enlightenment era was:

> [...] cataloguing and concretizing the boundaries between the disciplines, which emerged from the "long eighteenth century" of the Enlightenment. For the next quarter of a millennium, we remained indoctrinated to the shibboleths of this relic, operating within discrete silos-of-thought.

Some people have since described the Enlightenment silos as driven by a culture of scientism (Smith, 1992), whereby science has become the surrogate religion that has taken on some of religion's former functions by creating and perpetuating belief and value systems—based on facts as they may be—often clashing with other fundamentalist belief systems. One example that comes to mind is the ongoing conflict, in the USA in particular, between defenders of evolution and believers of the Biblical creation story. It is equally concerning that some in the science community still argue that science is, and can be, value–free while many now recognize that … "values emerge from science, both as a product and process, and can be redistributed more broadly in the culture or society" (Allchin 1998, p. 1083). Values and morals, therefore, even in the absence of religion, continue to play a major role in human life—they are supposed to keep us safe. As a result of the Enlightenment era, in the public consciousness science now stands for progress, whereas religion and spirituality tend to represent a lack of realism, naïveté, and conservatism, and the arts have been largely relegated to their aesthetic functions and entertainment value.

In short, the Enlightenment period brought an end to the dominance of religion over the arts and science: finally, scientists, religion and the arts could go about their business unhampered by the others' ideologies. "Good riddance!" many might exclaim at this point. And that may be so were it not for one uncomfortable problem, as Ken Wilber in his book "The Marriage of Sense and Soul" (1998) points out: unfortunately, the baby was thrown out with the bathwater. In this book Wilber illustrates how when science and religion parted ways the arts, which for most of humanity's history had been intricately linked with religion and spirituality (Armstrong, 2010), found themselves equally sidelined from the Enlightenment endeavor. In short, the Enlightenment movement created freedom and isolation from each other paired with suspicion of, and thus a lack of collaboration amongst, the three domains. This divisiveness proves to be a constant obstacle for anybody envisaging cross-silo initiatives.

Some have lamented that the separation of science, religion, and the arts has led to a science and mathematics education that promotes the left-brain hemisphere over the right-brain. This bias has been criticized as narrow, reductionist, and too analytical. According to Sousa and Pilecki (2013), this situation presents a fundamental problem: "there is no neuroscientific evidence to support the notion that one hemisphere consistently dominates a person's thinking, learning, or behavior" (p. 39), and thus there is no good, scientific justification why this should be the case. Whilst our left hemisphere is good at understanding literal meanings of words, numbers and letters, details in visual processing, patterns in time and sequence, our right hemisphere is more adept at gathering information about form, images, and visual patterns, context, spatial perception, places, faces, relational operations in mathematics and geometry we are trying to overcome in this chapter.

The question arises, therefore, why does this artificial focus on the left brain and on critical thinking dominate education systems worldwide? The answer, we think, lies (again) in the deeply ingrained Enlightenment silos which have attached different "value" to the three domains and their associated disciplines: science and mathematics subjects are held in extremely high esteem; humanities and religion—not important; the arts—nice but not essential. This type of thinking is reflected also in how we design curricula, how we advise children regarding their future, and how governments distribute education funding. The Enlightenment silos have become entrenched in our values system, influence how we design education programs and envisage student learning. With regard to science education the silos have served their purpose by allowing us to teach science, religion and the arts separately—except in circumstances where there is a struggle between evolutionists and creationists. What has been lost in the case of science education is the explicit connection to values—formerly firmly represented by religion—and to creativity, which is still almost exclusively linked to the arts. "Why would science education need to focus on promoting creativity?" some might ask. Whilst creativity is grounded in the arts and covered in art classes, a) not all students take art classes and b) everybody including the most prominent scientific minds use creativity when arriving at innovative solutions (Root-Bernstein et al., 2008).

Eliot Eisner (2008) explained that the arts are concerned with expressiveness, evoking emotion, generating empathic understanding, stimulating imagination that disrupts habits of mind and creates open-mindedness, and eliciting emotional awareness. In sum, the arts enable us to discover our humanity. This is supported by what arts educators, Ramon Bucheli, Merryl Goldberg, and Ann Philips (1991, p. 250) say about what the arts offer:

> The arts can be, for both students and teachers, forms of expression, communication, creativity, imagination, observation, perception, and thought. They are integral to the development of cognitive skills such as listening, thinking, problem-solving, matching form to function, and decision-making. They inspire discipline and dedication. The arts can also open pathways toward understanding the richness of peoples and cultures that inhabit our world, particularly during this period of global change. The arts can nurture a sense of belonging, or community; they can foster a sense of being apart, or of being an individual. The arts give rise to many voices. By acknowledging the role of the arts in our lives and in education, we acknowledge what makes individuals whole.

In summary, both morals (linked to religion) and the arts have much to offer yet, as Wilber (1998) highlighted, science, religion and the arts now operate in almost separate universes—with few exceptions, we might add. The Enlightenment silos are the result of this split. Separation into silos has offered advantages over the past centuries allowing the individual domains to flourish independently, however, it has also led to losses particularly affecting how we relate to the world and to each other: scientific and economic progress unbridled by morals and values has led to many of the world's environmental, economic, and social problems we see today.

Currently we lack the elegance and aesthetics of diverse and creative approaches to complex problems, whose solutions need to draw on divergent and creative thinking, because we are so obsessed with the idea that progress and solutions can only come out of the scientific paradigm. Based on our analysis of the Enlightenment silos and what they mean for science education we argue that STEM, which represents only one of the three domains and which is currently hailed as the solution to all our problems, is unlikely to provide us with the necessary skills, knowledge and the essential shift in consciousness. Rather than staying with the limiting boundaries of only one domain, in order to overcome the global issues facing us today, bridges need to be built between the separate domains to enable collaboration, mutual enrichment, and exchange of ideas and problem-solving methods and visions.

THE POWER OF DIVERGENCE—RECONSIDERING THE SILO MENTALITY

Given the complex and unprecedented problems facing humanity today we require a complete rethink of the Enlightenment silos since, to paraphrase Pope Francis (2015), what we need now is to draw on all of humankind's strengths and skills to solve the problems so as to ensure our survival as a species. Science education undoubtedly has a major role to play in this struggle, however, we argue that it too needs to adapt and change in line with the environmental and social challenges ahead. Business as usual seems inappropriate at this time. To quote from Sir Ken Robinson's 2014 TED Talk, titled "How to Escape Education's Death Valley?":

> … one of the effects of "No Child Left Behind" has been to narrow the focus onto the so-called STEM disciplines. They're important. I'm not here to argue against science and math. On the contrary, they're necessary but they're not sufficient. A real education has to give equal weight to the arts, the humanities, to physical education.

Real education, to us, has to show connections to the everyday life of students which, except in schools and universities, is not separated into neat, distinct disciplines and domains. The introduction of STEM into the science education space is great and promising in terms of "better than nothing." However, to solve the world's problems it is unlikely to succeed because it fails to include the other two thirds of the human endeavor in times when we desperately need them: the humanities (including morals/spirituality/religion) and the arts. We concur with Zeidler (2016) that what is needed is an ability to contextualize scientific issues in responsible ways which requires prudence, morality, and character. By introducing STEAM (science, technology, engineering, arts, mathematics) as an approach that crosses disciplinary boundaries we can draw on more than "just" the resources of the left side of the brain. We can tap into everything that has helped humanity survive for millennia: experiential knowledge, values, and creative ingenuity.

Interdisciplinarity Is Not New. In education there have been attempts over the years to bring about border-crossings between individual disciplines, often expressed as "integrated curricula." In Australia, integrated curriculum approaches have been largely restricted to primary and lower secondary school, that is, middle year contexts which "lend themselves to a curriculum that takes account of the whole person rather than just the academic sphere or domain" (Groundwater-Smith, Mitchell, & Mockler, 2007, p. 46). Ken Robinson (2014) explained that, "education is not a mechanical system. It is a human system. It is about people, people who either want to learn or don't want to learn … they may find it boring … they may find it irrelevant. They may find that it's at odds with the life they are living outside of school." Robinson raises concerns that many "new" approaches in the USA stress the fact that they are "personalized." He says, "What is interesting to me is these are called 'alternative education'." He adds, "… and all the evidence from around the world is, if we all did that, there'd be no need for the alternative."

Antidisciplinarity—The Next Step. Could it be that even the term interdisciplinary is too narrow in focus given the enormity of what we are facing? Some have argued that what is needed is an approach that goes beyond an interdisciplinary approach like STEM—or even STEAM for that matter. Ito (2016, para 4) argues that what is called for is nothing less than an "antidisciplinary" approach. In his article "Can design advance science, and can science advance design?" Ito made a distinction between the two terms that are so relevant for our discussion on overcoming Enlightenment silos. He stated that, "interdisciplinary work is when people from different disciplines work together. But antidisciplinary is something very different; it's about working in spaces that simply do not fit into any existing academic discipline—a specific field of study with its own particular words, frameworks, and methods" (Ito, 2016). Ito reasons that hitherto unknown problems require new, creative solutions—not yet thought of due to a lack of cross fertilization across disciplines.

In his book, Artscience, David Edwards' (2010) makes a similar argument, namely, that creative breakthroughs in the arts and sciences occur in an "intermediate zone" of human creativity where neither art nor science is easily defined. For Neri Oxman (2016), the solution to the longstanding Enlightenment problem has arrived through the appearance of the antidisciplinary meme, "yanking us out of Aristotle's shadow and into a new 'Age of Entanglement'" (para 1). As part of her inaugural essay published in the online Journal "Design and Science", she proposes a map for four domains of creative exploration—Science, Engineering, Design and Art outlining her antidisciplinary hypothesis which reads, "knowledge can no longer be ascribed to, or produced within, disciplinary boundaries, but [it] is entirely entangled" (para1). Instead she proposes a "tentative, yet holistic, cartography of the interrelation between these domains, where one realm can incite (r)evolution inside another; and where a single individual or project can reside in multiple dominions" (Oxman, 2016, para 2).

She introduces us to the metaphor of the "Krebs Cycle of Creativity" (KCC) "recycling" the well-known biological metaphor of the Krebs Cycle which represents the process of cellular respiration during which energy is generated in a cell by giving it new meaning. The steps in the KCC are similar to those in the biological process. Oxman describes the KCC as a map outlining the perpetuation of creative energy (creative ATP or "CreATP"), analogous to the Krebs Cycle proper (Oxman, 2016, para 8). Using analogical thinking she compares four modalities of human creativity—science, engineering, design and art—to the Krebs Cycle's compounds whereby each modality produces "currency" by transforming into another. She explains that:

> The role of Science is to explain and predict the world around us; it "converts" information into knowledge. The role of Engineering is to apply scientific knowledge to the development of solutions for empirical problems; it "converts" knowledge into utility. The role of Design is to produce embodiments of solutions that maximize function and augment human experience; it "converts" utility into behavior. The role of Art is to question human behavior and create awareness of the world around us; it "converts" behavior into new perceptions of information, re-presenting the data that initiated the KCC in Science. (Oxman, 2016, para 9)

We would like to add the role of morals (drawn from the domain of religion), which is to safeguard against harm from excesses. The role of Humanities is to identify how the excesses caused by scientism, free market-idolatry, and rapidification of growth and progress affect human society and living conditions. Moreover, the role of the Humanities as the storehouse of thousands of years of human knowledge and wisdom is to provide modern humans with access to the insights and experiences gained by those who have gone before us. Our "need" for the inclusion of the Humanities is shared by other authors such as Kahn and Zeidler (2016) who suggested to use the term HARTSS*—standing for Humanities, ARTs, and Social Sciences. We believe that the term STEAM already includes the humanities and values. Researchers in the field are trying to—quite literally—come to terms with the complexity of the issue by inventing ever more inclusive acronyms. Whatever the acronym—it is impossible to consider any of these domains or modalities without including language—the role of language/s is to enable communication, cultural exchange, and the solving of complex problems, which in isolation seem impossible to overcome.

For the purpose of solving the world's problems we not only need communication between individual problem solvers in one domain but also across all domains and modalities, with each providing a variety of skills and ways of thinking. In other words, similar to overcoming factional loyalties in the Divergent book story, we need to abandon an exclusively convergent (one domain) outlook on life and the world's problems in favor of divergence. Divergent thinking draws on a multitude of aptitudes and skills to create a multitude of approaches to highly complex problems. STEAM education is a good start but it is essential to stress that when we say STEAM we are not just referring to science, technology, engineering, the arts and mathematics—we include values and morals, language and the humanities, all of which can offer to help with the solution of our human problems (Sousa & Pilecki, 2013).

ANTIDISCIPLINARITY IN SCIENCE EDUCATION—IS IT POSSIBLE?

In climate change science there is much talk about adaptation—exploring how systems and individuals learn to survive under changing conditions. For us, a similar question arises with regard to science education which is not exempt when it comes to living up to global challenges: how can an antidisciplinary approach help science education become better adapted to the rapidly changing conditions and be more adequate for the 21st Century context?

Science education has made progress in the area of socio-scientific approaches that have attempted to reintegrate values into mainstream science learning (e.g., Zeidler, 2016), yet more work needs to be done. These promising developments are also reflected in national curriculum documents for science in Australia that outline the aspect of "Science as a Human Endeavour" as one of the three dimensions of science education, together with "Science Understandings" and "Science as Inquiry" (ACARA, 2016). Whilst the latter two dimensions of this curriculum come as no surprise, the former has afforded science teachers opportunities to address values and ethics explicitly in their teaching. It has opened the door to let the Humanities in should a science teacher wish to do so. There is also a strong push to integrate literacy into all subject areas, including science. Again, this link to language is promising but requires further development.

At this point in time science teachers are still struggling with the interpretation and implementation of all dimensions of the Australian Curriculum, which has been introduced in stages since 2010. What is an exciting development is that the national science curriculum is directly linked to the "General Capabilities" and "Cross-Curriculum Priorities" dimensions which apply to all subject areas, to varying degrees. The General Capabilities focus on developing *critical and creative thinking, personal and social capabilities, ethical understanding and intercultural understanding*; and the Priorities provide compelling contexts for students to develop a moral conscience: *sustainability, Aboriginal and Torres Strait Islander histories and cultures, Asia and Australia's engagement in Asia.* These dimensions of the national curriculum provide science teachers with the justification they need to develop their students as global citizens and teach a form of science education that helps address pressing global needs.

However, since all of this is still new, the prospect of designing teaching and learning activities to develop students' higher-order abilities can be daunting for science teachers. Understandably, many are likely to focus primarily on teaching the "tried and true" dimensions of science understandings and inquiry, perhaps with a modicum of science as a human endeavor added to improve student engagement. This conservative (Enlightenment silo) standpoint is reinforced by assessment systems that privilege the science understandings and inquiry skills dimensions of the curriculum, especially for Years 11 and 12. Whereas a STEM curriculum focus allows teachers to bring in technology, design and engineering to enrich their science classes through robotics, coding, etc., that leaves the arts—how can they be helpful to science education and what would an integrated STEAM curriculum look like?

STEAM RATHER THAN STEM—A VIABLE ALTERNATIVE FOR SCIENCE EDUCATION

Combining science and the arts in the form of STEAM education has its origin in the USA where it was conceived for the goal of producing a citizenry and workforce for the 21st Century that is not only scientifically literate but also creative and innovative, curious and self-motivated, communicative and collaborative (Land, 2013).

Yet the idea is by no means new: back in the 1950s Snow (1998) had already argued for a rapprochement of the cultures of science and the arts. These days, the USA and Korea are at the forefront of producing STEAM curricula for their respective nations. Acknowledging that their existing curricula reinforced limitations in the development of students' higher-order abilities for the 21st Century, visionary science educators are now teaming up with their colleagues in the arts learning areas to design innovative interdisciplinary STEAM curricula and teaching approaches (e.g., Sousa & Pilecki, 2013).

Anna Feldman (2015) describes the uptake of STEAM programs and facilities in 27 school districts in Pennsylvania that is helping spark students' imagination, enabling them to apply creative thinking and design skills in their STEM projects. She also mentions the Boston Arts Academy where arts and STEM combine in "makerspaces" in which students engage in digital fabrications, electrical engineering, industrial design, and architecture using 3-D printing, laser cutters, circuit boards, etc., and where geometry students code to produce digital animations and create digital lighting plans for the college theatre. She adds, "STEAM uses project-based teaching to holistically foster students' skills in creativity, design thinking, tech literacy, collaboration, and problem-solving."

Guy Boy (2013), a space educator affiliated with NASA, argues that STEAM offers a human-centred education for preparing graduates for a new era in industrial design characterised by "holistic non-linear design of complex systems, such as the Internet, air traffic control management and nanotechnologies" (p. 1). He sees STEAM education as fostering essential creativity and collaboration skills for developing cross-disciplinary approaches required for designing complex and interconnected systems.

Other STEAM education programs and initiatives in the USA include a multidisciplinary program to foster creative thinking at the State University of New York that combines the arts, humanities and STEM fields. And the State University of New York, is exploring development of a new curriculum to enhance STEM education by integrating arts education. The driver is to produce graduates with the interdisciplinary critical and creative thinking abilities of innovation leaders (Madden et al., 2013).

In Australia, we have been involved in several STEAM-related projects—some completed, some ongoing.

- *A Big Picture school in metropolitan Western Australia*: a science/mathematics teacher integrated fictive stories about everyday ethical dilemmas into her Earth Science lessons, and demonstrated that her at-risk students engaged in ethical decision-making while developing scientific knowledge and inquiry skills (Taylor, Taylor, & Chow, 2013).
- *A Catholic secondary college in the north of Western Australia*: for the past 3 years, the Head of Science at the school has led an interdisciplinary team of teachers to develop a 7–10 STEAM curriculum. Learning areas represented in the project include science, drama, religious education, humanities and social sciences (HASS), and English. The Head of Science is now enrolled in a PhD program at Murdoch University and is investigating the impact of STEAM teaching on student engagement.
- *A government senior high school in metropolitan Western Australia*. The State Government of WA recently awarded Cecil Andrews Senior High School $4.8 million to build new STEM labs for the school. Under the visionary leadership of the principal, an arts educator, the school has embarked on a 7–10 STEAM curriculum development project. The Fogarty Foundation

has awarded Peter Taylor and Peter Wright (Murdoch University) a 3-year grant to support Cecil Andrew's STEAM curriculum project.

- *A Christian secondary college in Queensland.* The Head of Science is building on his previous doctoral research into socially responsible science, which investigated ethical dilemma pedagogy (Settelmaier, 2009) for engaging science students in higher-order thinking, and is working with colleagues in other learning areas to plan a STEAM curriculum for the school.

BREAKING DOWN THE SILOS—FROM STEM TO ST^{2}EAM, AND BEYOND

As educators promoting STEAM education we often encounter genuine interest but often also a sense of trepidation based on the way we have been taught to think within the limitations of our respective Enlightenment silos. Often the question is raised: does allowing the arts into the science education picture have the potential to dilute content knowledge and to distract from teaching science inquiry and skills (Feldman, 2015). The same concerns are often raised in relation to the inclusion of values education in science classrooms, although this has been remediated to some degree in recent years. These concerns are warranted when we consider how deeply ingrained Enlightenment (silo) thinking has been over the past three centuries. As a result, everything that seems to threaten the apparent safety zone of science-related domains sends shivers down the spines of practitioners. The silos took a long time to build—it will take some time before everybody feels comfortable with breaking down the silo walls and allowing for true collaboration and mutual inspiration.

At this moment in history, we suggest that STEAM offers a viable alternative to the Enlightenment curriculum thinking in the context of science education. In particular, we suggest not to stop at STEM but to go the whole way to STEAM education. However, a crucial word of warning! If STEAM education is to facilitate young people's development of higher-order abilities for the 21st Century, as discussed here, then it needs to be accompanied by a powerful theory of education. Otherwise it runs the risk of being just another fashionable but toothless acronym that attracts government funding while reinforcing business as usual Enlightenment silo thinking. So how do we inoculate STEAM education against possible recidivism?

Our work with science teachers in the schools outlined above has been underpinned by an established theory of transformative science education (Taylor, 2015) that fosters five interconnected ways of knowing, being and valuing: *cultural self-knowing, relational knowing, critical knowing, visionary and ethical knowing, knowing in action.* We believe that transformative learning can serve as a powerful theory of education for ensuring that STEAM education realizes its fullest potential. Perhaps a good way to ensure that this happens is to emphasize the central role of transformative education by modifying the STEAM acronym to ST^{2}EAM, where T^2 represents technology and transformative learning.

For innovative science educators who are being drawn to STEM education initiatives but are skeptical of embracing post-Enlightenment thinking, we wish to point out the following attributes of ST^{2}EAM education.

- ST^{2}EAM education does not oppose STEM education; rather it enriches and expands the scope of STEM.
- ST^{2}EAM education is a curriculum philosophy that affords teachers a creative design space across different learning areas allowing for the development of richly integrated school-based curricula.
- ST^{2}EAM education actively involves teachers in developing a humanistic vision of a 21st Century education and their roles as professionals.
- ST^{2}EAM education can be designed on a modest scale and implemented by an individual innovative teacher.
- ST^{2}EAM educators are being inspired by project-based learning programs (e.g., Holm, 2011).

To us, ST^{2}EAM has the potential to overcome the Enlightenment silos that have kept us limited to our own factions for so long, to use Divergent story language. We have kept to our own factions to allow for harmony and development within the individual domains. Yet some people have started to move beyond convergent silo thinking towards rich, varied, innovative and creative approaches built on divergent thinking. Some people fear divergent thinking since one can never be sure what comes of it since it requires divergent thinkers to come up with innovative solutions that nobody else can think of. Young Tris's comment in the Divergent book sums it up nicely, "I am Divergent. And I can't be controlled" (Roth, 2011, p. 442). Yet, rather than fear lack of control over divergent thinking we should wholeheartedly welcome its potential, especially as science educators. The human species needs nothing less to solve the problems of this world at this historic point in time.

REFERENCES

Allchin, D. (1998). Values on science and in science education. In B. J. Fraser & K. G. Tobin (Eds.), *International handbook of science education* (pp. 1083–1092). Dordrecht: Kluwer Academic Publishers.

Armstrong, K. (2010). *The case for god: What religion really means.* London: Vintage Books.

Australian Curriculum Assessment and Reporting Authority (ACARA). (2016). *Australian curriculum: General capabilities.* Retrieved March 10, 2016 from http://www.acara.edu.au/curriculum/general_capabilities.html

Boy, G. A. (2013, August). *From STEM to STEAM: Toward a human-centred education.* Paper presented at the European Conference on Cognitive Ergonomics, Toulouse, France. Retrieved May 10, 2016 from http://ntrs.nasa.gov/search.jsp?print=yes&R=20130011666

Bucheli, M. R. J., Goldberg, M. R., & Philips, A. (1991). Symposium: Arts as education. *Harvard Educational Review, 61*, 249–251. Retrieved April 10, 2016 from http://0-search.proquest.com.prospero.murdoch.edu.au/docview/212286241?accountid=12629

Deloitte. (2015). *Tech trends 2015: The fusion of business and IT.* Retrieved May 10, 2016 from http://www2.deloitte.com/au/en/pages/technology/articles/tech-trends-2015.html

Edwards, D. (2010). *Artscience: Creativity in the post-Google generation.* Cambridge, MA: Harvard University Press.

Eisner, E. (2008). Art and knowledge. In J. G. Knowles & A. L. Cole (Eds.), *Handbook of the arts in qualitative research.* Thousand Oaks, CA: Sage.

Feldman, A. (2015). *STEAM rising: Why we need to put the arts into STEM education.* Retrieved May 10, 2016 from http://www.slate.com/articles/technology/future_tense/2015/06/steam_vs_stem_why_we_need_to_put_the_arts_into_stem_education.html

Groundwater-Smith, S., Mitchell, J., & N. Mockler (2007). *Learning in the middle years: More than a transition.* South Melbourne, VIC: Thomson.

Holm, M. (2011). Project-based instruction: A review of the literature on effectiveness in prekindergarten through 12th grade classrooms. *InSight: Rivier Academic Journal, 7*(2), 1–13. Retrieved March 11, 2016 from http://bie.org/objects/cat/research

Infosys. (2016). *Amplifying human potential: Education and skills for the fourth industrial revolution.* Retrieved May 10, 2016 from http://www.experienceinfosys.com

Ito, J. (2016). Design and science: Can design advance science, and can science advance design? *Journal of Design and Science.* Retrieved May 14, 2016 from http://jods.mitpress.mit.edu/pub/designandscience

Kahn, S., & Zeidler, D. (2016). Using our heads and HARTSS*: Developing perspective-taking skills for socioscientific reasoning (*humanities, arts, and social sciences). *Journal of Science Teacher Education, 27*, 261–281. doi:10.1007/s10972-016-9458-3.

Kuhn, T. S. (1962). *The structure of scientific revolutions* (2nd ed.). Chicago, IL: The University of Chicago Press.

Land, M. H. (2013). Full STEAM ahead: The benefits of integrating the arts into STEM. *Procedia Computer Science, 20*, 547–552. doi:10.1016/j.procs.2013.09.317.

Madden, M. E., Baxter, M., Beauchamp, H., Bouchard, K., Habermas, D., Huff, M., … Plague, G. (2013). Rethinking STEM education: An interdisciplinary STEAM curriculum. *Procedia Computer Science, 20*, 541–546. doi:10.1016/j.procs.2013.09.316.

Office of the Chief Scientist. (2013). *Science, technology, engineering and mathematics in the national interest: A strategic approach.* Canberra: Australian Government.

Oxman, N. (2016). The age of entanglement: An inaugural essay for the Journal of Design and Science (JoDS). *Journal of Design and Science.* Retrieved May 14, 2016 from http://jods.mitpress.mit.edu/pub/AgeOfEntanglement

Pope Francis. (2015). *Encyclical on climate change and inequality: On care for our common home.* Brooklyn, NY: Melville House.

PricewaterhouseCoopers. (2015). *A smart move: Future-proofing Australia's workforce by growing skills in science, technology, engineering and maths (STEM)*. Retrieved March 11, 2016 from http://www.pwc.com.au/stem.html

PricewaterhouseCoopers. (2016). *21st century minds: Accelerator program*. Retrieved March 11, 2016 from http://www.pwc.com.au/stem.html

Robinson, K. (2014). *How to escape education's Death Valley?* Retrieved June 12, 2016 from https://www.ted.com/talks/ken_robinson_how_to_escape_education_s_death_valley?language=en

Root-Bernstein, R., Allen, L., Beach, L., Bhadula, R., Fast, J., Hosey, C., … Weinlander, S. (2008). Arts foster scientific success: Avocations of Nobel, National Academy, Royal Society, and Sigma XI members. *Journal of Psychology of Science and Technology, 1*(2), 51–63. doi:10.1891/1939-7054-1.2.51.

Roth, V. (2011). *Divergent*. London: HarperCollins.

Settelmaier, E. (2009). *Adding zest to science education: Transforming the culture of science education through ethical dilemma story pedagogy*. Saarbrücken: VDM.

Smith, H. (1992). *Forgotten truth: The common vision of the world's religions*. New York, NY: HarperCollins.

Snow, C. P. (1998). *The two cultures*. Cambridge: Cambridge University Press.

Sousa, D. A., & Pilecki, T. (2013). *From STEM to STEAM: Using brain-compatible strategies to integrate the arts*. Thousand Oaks, CA: Corwin/Sage.

Taylor, E., Taylor, P. C., & Chow, M. L. (2013). Diverse, disengaged and reactive: A teacher's adaptation of ethical dilemma story pedagogy as a strategy to re-engage learners in education for sustainability. In N. Mansour & R. Wegerif (Eds.), *Science education for diversity: Theory and practice* (pp. 97–117). Dordrecht: Springer.

Taylor, P. C. (2015). Transformative science education. In R. Gunstone (Ed.), *Encyclopedia of science education* (pp. 1079–1082). Dordrecht: Springer.

Wilber, K. (1998). *The marriage of sense and soul*. New York, NY: Broadway Books.

Zeidler, D. (2016). STEM education: A deficit framework for the twenty first century? A sociocultural socioscientific response. *Cultural Studies of Science Education, 11*, 11–26. doi:10.1007/s11422-014-9578-z.

CHAPTER THIRTY-THREE

Science Education in the Key of Gentle Empiricism

LEE BEAVINGTON AND HEESOON BAI

Let's begin this thinking about science education at the beginning: with the very meaning of *science education*. Science education may mean instructing in or learning about scientific information, knowledge, and methods in various disciplines. The goal here would be to equip our students with requisite knowledge and skills to become scientists, technicians, and science and technology educators. Our current system of science education is replete with this way of teaching science. And we need this form of education to cultivate future generations of scientists, technicians, and workers in science-related careers.

Science education may also mean educating people to have ways of understanding and approaching the world with worldviews based in or influenced by science. When our school children protest to us that they are not going to become scientists or technicians and so they don't want to and don't need to study science, we might appeal to this second meaning and tell them that for them to function in today's world, they have to have a certain level of scientific knowledge and understanding of the scientific method. If they press us further and ask us why we have to know science to function in today's world, we will point to the degree to which the construction of our world depends on science and technology. These responses on our part would constitute a justification for not only including science and technology education but also underlying its central importance.

Now comes the troubling part. What if our students, well trained to ask critical questions, ask us this question: "So, you say that in today's world everything is constructed by the modern scientific knowledge and technologies. But look how terribly troubled the world is today. You have been, in fact, telling us that according to many scientists, our world may not continue and cannot be sustained beyond the next few decades, at the rate we humans are consuming and destroying the world. If modern science and technology are what constructed such a world, then why should we continue to study them?" What would be our answer? "Oh, come on! Stop trying to avoid studying science. That won't get you far." Or, "Back to your work! Don't ask such silly questions!" Or, "Your mom and dad want you to be successful, and for that, you need to study science (and math, etc.)." Answers such as these do not address the serious point raised by our students. However, our chapter shall do just that.

The starting point of our discussion on science education is the acknowledgment that science is a human enterprise. What does that mean? It means that what we think science should be bears our metaphysical (literally, going beyond the physical) assumptions about the world: what the world looks like to us humans or how it might look differently when we adopt a different lens of understanding and investigation, and what we value or ought to value. This is apparent when we examine the history of science. Historians and philosophers of science would readily confirm this. Even just in the history of Western civilization, we can point to a diverse set of scientific worldviews, closely aligned with or based on ontological or metaphysical and epistemological views of the world (Pepper, 1970). Moreover, when we add such views from other civilizations, we have a cornucopia of scientific worldviews, notwithstanding the debate, sometimes violent, about which ones are true of reality. We shall also note that our general tendency is to discredit the scientific views of other cultures, especially from the distant times, as untrue. Our main point of making these observations here is to emphasize that science is not independent from human values and views, commitments, sensibilities, aspirations and tastes of a given culture and people.

Of particular interest to us (the authors of this chapter) are the exploitative and even violent views, attitudes, and resulting conduct that define humans' power relationship of domination over nature, integral to the West's Enlightenment and Renaissance preceding it and Modernity proceeding it (Merchant, 2008). A central figure in this way of thinking about the relationship between nature and humanity is Francis Bacon who *invented* the modern scientific method of experiment that is the hallmark of empirical research to this day. Merchant states: "Distributed among many of Bacon's works and often reappearing in slightly altered phraseology, these terms (such as, 'penetrate,' 'shake,' 'shape,' 'squeeze,' 'subdue,' 'wrest,' etc.) all connote some degree of violence toward nature" (p. 749). For emphasis, let us repeat: violence toward nature is built into the foundation of modern experimental science. So normalized is this understanding that we take it for granted and we think nothing of it. How else do we do science?

The question of "ought" is always a wake-up call for the mind that thinks descriptively as in "that's how it is," or "that's how it goes and how it's done." A master car thief may teach his apprentice, "Let me show you; this is how you steal a car." The apprentice, suddenly waking up from her trance, may ask herself, "Wait a minute. *Should* I be stealing cars? *Ought* I to continue to live a life of crime?" Similarly, many thoughtful people since Bacon's time woke up from the Baconian trance and have been asking: *Ought* we to think of nature as a female that needs to be "penetrated," "squeezed" and so on to have "her" yield, give up "her secrets"? Secrets that will enable us to manipulate, alter, further violate "her"?

Equally disturbing is the reduction of nature to the order of the inanimate. In recent decades, many thinkers (e.g., Erazim Kohak and R. D. Laing) from diverse but interconnected fields of philosophy, psychology, ecology, and health have all spoken about contemporary humanity living out the worldview of the mechanical universe that the 17th century Western philosophers like Thomas Hobbs and Rene Descartes championed. If the universe, including our planet, is basically made up of inanimate matter ("stuffs"), then there is nothing that would prevent us from hacking and destroying it to satisfy our insatiable desires for consumable goods and services.

The role of science, however, goes deeper than supporting the enactment of the mechanical universe. The process of rendering the universe to the order of the inanimate could not be accomplished without simultaneously numbing our psyche so that we don't feel the pain of snuffing life out of the animate universe. Neil Evernden (1993) speaks of the practice in which lab scientists would cut the vocal cord of animals that they were vivisecting, for the pain of hearing the animals howling in pain is extremely disturbing. By shutting down our senses and sensibility, we dehumanize our selves. It's only in hearing, seeing, and feeling the pain of others that our empathy and compassion are exercised. With all this in mind, we may continue to consider how contemporary science education in North America (and all over the world as the modernist worldview has spread globally) plays a vital two-fold role of continuing to render the world to be an order of the inanimate and simultaneously accomplishing psychic numbing and, hence, dehumanizing. And, against this backdrop of considerations, we may also re-envision a science that heals our psychic numbing and releases us from the practice of dehumanization.

SCIENCE THAT HEALS: THE GOETHEAN APPROACH

The gift of postmodernism—the era we have been in for the last few decades—is its massive questioning of the *grand narrative* and replacing it with many narratives. Monologue is being replaced with dialogue; fundamentalism, by pluralism;

hierarchical ordering by rhizomatic networking, and so on. As part of this invitation to seeing in multiplicity and ecological complexity, we can also entertain different paradigms of science. Since, as we saw earlier, our interest in doing science differently ultimately stems from ethical values, such as wanting to see less suffering and more flourishing, less pain and more joy, and in general more support for all beings and their wellbeing, we are searching for a science that heals and vivifies all beings, including, of course, human beings. It is in this context that we would like to introduce the Goethean "gentle empiricism," zarte Empirie.

Johann Wolfgang von Goethe was a celebrated writer, statesman, and scientist, and prospered during the Romantic era where holism, wonder, science and philosophy could belong to a single discipline (Richards, 2002). He wrote poetry and prose, inspired morphology as a new scientific discipline, and devised an alternative approach to scientific methodology. His "conscious-process-participation" (Wahl, 2005, p. 59) involved prolonged looking, direct attention, and deep empathy, with an "emphasis on the metamorphosis of the scientist" (Robbins, 2005, p. 115). This holistic and participatory approach, when adapted to the contemporary context of science education, offers an alternative way to understand and be immersed in the world's phenomena. Whether studying climate change, biodiversity, plant growth, or other biological or physical entities, zarte Empirie engages the scientist in a relational yet still precise scientific methodology.

Incorporating Goethean science in today's science classrooms requires us to step beyond the limits of the conventional reductionist science methodology that follows the predominant subject-object-separation epistemology and move toward a conscious-process-participation epistemology. In this Goethean approach, the learner provides space for the phenomenon to show up and for the learner to receive from the phenomenon under study, mobilizes his/her imagination as a tool to understand the varied qualities of a phenomenon, and employs intuition in order to more fully comprehend the object's outer and inner contents. By incorporating a science of qualities rather than *only* a science of quantities (Reason & Goodwin, 1999), we engage in a reciprocal relationship more inclined to reveal our mutual interdependence with other life phenomena. The alternate paradigm of Goethean science recognizes the importance of two-way perception where the scientist both perceives and receives and helps us to "recover the life that once pulsed through knowledge" (Reynolds, 2007, p. 160).

Today's scientific method that formulates questions, gathers data, interprets results, makes conclusions, and then formulates new hypotheses, is a precise methodology often documented in a lab report. The most coveted section of a lab report is the results section, "the centerpiece of your report" (Pechenik, 2016, p. 151), providing key data findings, including numbers, statistical parameters, and other relevant observations shown in figures, tables and written text. This is deemed the most important section of a lab report because it presents findings without "reflecting the author's biases, hopes and opinions" (p. 158), without interpretation or analysis, and certainly without subjectivity. In other words, it lacks the qualities that make us human. To a Newtonian scientist, this logical-rationalistic approach removes human folly from the equation and all the subjectivities that might infiltrate the conclusions made in the report. Their observations through one or more of their senses, as well as through their higher faculties—and all of the prejudices and partialities therein—are not to be trusted, because they are all human-generated. The data is seen as eternal, unchangeable, and what future researchers will reference. Ethically, scientists should report any bias involved in their experiments. In reality, the pressure to publish may increase scientific bias (Fanelli, 2010). To a Goethean scientist, however, prizing raw data so highly neglects the fact that all scientists are human and disregards the reciprocal relationship between the observer and the observed, whether the latter is electrons or electric eels.

In her paper on Goethean science education, Sherrie Reynolds (2007) asks: "Is it necessary to eliminate spirit in order that children may learn science?" (p. 160). The use of the word "spirit" will give many scientists consternation. For, how can you measure or analyze such an ethereal entity (if indeed it's an entity) objectively? But her question is pertinent, for her use of the word "spirit" is connoted with passion, ethics, curiosity and attentiveness in an educational context rather than a religious one. Spirit is a phenomenon, not a substance. When science is taught through a mathematical, positivist, formula-driven lens, this can alienate some learners whose primary inquiry in science is qualitative observation, curiosity with how things work, cognitive problem-solving on a meta level, or a fascination with the natural world.

Conventional K-12 science education occurs inside a classroom and focuses largely on the classification, measurement, and manipulation of phenomena. Kindergarten students are encouraged to explore using multiple senses, but by grade 7, in a slow progression that removes creative expression and any degree of open exploration, the dominant focus becomes variables, controls, and hypotheses. In each lesson a clear learning outcome is usually being sought, one that brings clarity for understanding a certain phenomenon, which can discount the pursuit of science as an ongoing and holistic journey where phenomena are examined over time and from multiple frameworks. In later grades, creativity and purposeful unstructured exploration is supplanted by a strict and rigorous scientific methodology that wants us to forget our humanness and the reality of all of the other beings with which we share our planet. Cartesian-Newtonian science does afford us an important way to gather unbiased data and understanding about the world, and all of the atomistic mechanisms that underlie phenomena, but there is an opportunity for its limitations to be complemented by Goethean science.

The four stages of the Goethean approach, as outlined by Isis Brook (1998) and developed by The Life Science Trust, can be listed as follows: (1) exact sense perception, (2) exact sensorial imagination, (3) seeing is beholding, and (4) being one with the object. These stages, detailed below (with examples of how to implement and facilitate each stage), cultivate a relational engagement with the phenomenon being observed over a prolonged spatial-temporal study. The distinct features of Goethean science include "rigorous attention to direct experience, empathy, intuition and imagination" (Wahl, 2005, p. 60), fostering a sense of wonder, incorporating philosophy and ethics in science, using emotional awareness and creativity to fully engage with phenomena, and contemplating the myriad interrelations among phenomena. The observer also becomes the observed, and is immersed, at different times, in direct sense engagement, creative expression, contemplative practice, and finding "unity in variety" (Bronowski, 1956/1972, p. 20) by tying together common themes that underlie seemingly divergent realms; in other words, the observer employs many of the most prized qualities of his/her humanness. "Goethe asserted that the human body is the greatest instrument to evaluate empirical experience because it permits an understanding of nature from the inside out" (Mason, 2014, p. 62), and this holistic understanding is what we are after.

Prior to the numbered steps below is a preparatory stage where the observer makes a note of first impressions and shares these preconceptions about the object of study. If "[w]e all have a history as observers and have formed ideas about the world, which influence what and how we perceive" (Wahl, 2005, p. 62), then acknowledging these formed ideas places us in an important context that emphasizes how our personal history and current state of being may influence our observations. This preparatory stage serves as an individual foundation often absent from current scientific study.

Exact Sense Perception. During this stage the observer steps back from the object of study and, by way of his/her ordinary senses, lists all of the facts and specific data perceivable from the object. Like Newtonian science, the observer is objective in his/her approach, doing his/her best to suspend judgement, evaluation, and personal interpretation. In encountering the phenomenon as it is, the Goethean scientist is present to its patterns, colors, shapes and other details. As Reynolds (2007) contends, "for many natural phenomena, the measuring, mathematical approach is blind to quality" (p. 163); during exact sense perception, qualities are recognized as pertinent observations.

We can apply this to elementary school students by having them go outside and use the five primary senses to perceive a single phenomenon, such as a fern. After the preparatory stage where they can write and share about who they are, how they are feeling right now, and their first impressions of the phenomenon, students will "remain within the realm of phenomena and concentrate on form, color, pattern, and behavior" (Reynolds, 2007, p. 168). The teacher will avoid classifying the phenomena (e.g., the fern) and let students use their own words to describe it. Drawing is an appropriate method here, both in the moment and later from memory (Brook, 1998), the latter as a way to emphasize the parts of the fern the student may have forgotten. Students can also make note of other phenomena interacting with the fern, such as a spider scurrying underneath the fern frond; such interrelations will be focused on in a later stage.

Another example, more suited to middle grade students, would be to have students observe an amoeba under the microscope. While the visual sense would obviously dominate here, this does not preclude learners from smelling the pond water from which the amoeba came, or using their imagination to make observations on how the amoeba touches with its pseudopodia, and what we might hear if the amoeba were large enough to make audible sounds. Students would record cellular patterns, shapes and structures in written form, and also make multiple drawings of their amoeba at different times or under different conditions (e.g., light intensity, slide mounting medium, time of day). They could also carefully observe, under the microscope, the way an amoeba ingests food, known as phagocytosis: how it surrounds and engulfs, with its pseudopodia, live prey like a protist or bacterium. It is a most fascinating process that students may enjoy drawing and even dance to enact its shape-altering ability.

Exact Sensorial Imagination. This stage involves shifting from a static view of phenomena toward an understanding of transition and metamorphosis. Now, the imagination can be employed to understand the temporal dimension of phenomena. We come to see that the object has a history. It is important here to move beyond the exact sense perception stage, which provided useful observations that are now frozen in time. Imagination is used to bring movement, development and transition to the phenomena, so that, as Daniel Wahl (2005) explains, "we can wilfully imagine a different sequence of transformation than the one that emerged based on our engagement in stage one" (p. 63).

To continue with our fern example, we can have students draw the fern frond and its leaflets as they might appear when the particular frond they examined was younger, and when this particular frond will be older. In the first stage the drawing provided a static image of the fern; now, students imaginatively draw a sequence of development. This alternate epistemology, rather than showing students a textbook figure or a time-lapse film, allows for students to imagine and reason for themselves.

Seeing is Beholding. In this stage the observer enters a state of "receptive attentiveness" (Brook, 1998, p. 56) and receives from the phenomenon with an open mind. The phenomenon is allowed to "articulate in its own way" (p. 56) while the observer closes down his/her own perception. As the recipient, the observer gives the phenomenon some of our humanness and conscious awareness, and therefore a way to express its gesture. Since "Goethe does not study nature from an observer's view seeking a reason for its existence outside itself, but as an entity that has its own reason to be" (Reynolds, 2007, p. 166) this stage is vital in revealing the phenomenon's purpose and intention in existence.

The observer expresses his/her observations in emotional language. This practice can lead to unexpected insights and the development of "new organs of perception" (Goethe as cited in Brook, 1998, p. 56). We would like to point out that this way of observing the world and coming to know it has a long, time-honored tradition, known as what we now call "wisdom traditions" of the world. These traditions bear ancient roots, reaching back thousands of years into antiquity and beyond, and still bearing abundant fruits on contemporary branches. One of the well-known wisdom traditions—there are many, throughout the world—is yoga in Hinduism, and here, the practice of devotion (bhakti) is central to yoga practice and philosophy (Scott & Bai, in press). It is through the practice of devotion that our organs of perception become attuned precisely and clearly to the exacting beauty and wonder of the world. Devotional practice enlarges and deepens our love for all, enabling us to participate in the phenomenology of life and cosmos. Through devotion, we become lovers, not destroyers, of life. Through devotion, we become part of, not apart from, the world. Our point here is that Goethean science education, too, engenders this profound love of and for the world.

To return to the seeing-is-beholding stage of science education: When students engage with the fern during this stage, they can do so in an extended period of silence. Questions such as "What does the fern possess that you haven't noticed before?" and "How would the thing studied describe itself if it had the ability to speak?" (Reynolds, 2007, p. 168) can help guide this practice for younger learners. Afterward, students can paint, write poetry, or partake in other arts-based expression that fosters an alternative understanding of the phenomena, all of which perhaps lead to a to a "morally responsive obligation to the observed" (Robbins, 2005, p. 113).

Being One with the Object. This last step of science education is really the fruit-bearing stage. Through this stage, we come to taste the fruit of our labor: full integration with and participation in the world. Thus, we come to realize the truth of ecology: that we humans are interconnected through and through, and that our presence interpenetrates all other phenomena. In knowing this, not just theoretically, but in our senses and heart, in our body, we truly become ethical beings. The manner of our being in the world and our actions we take would be in alignment with the welfare of the world. We would become healing, protecting, and nurturing agents: not the rampant consumers and destroyers that we witness in today's mainstream humanity.

Here, too, this notion of being one with the object has time-honored roots in the wisdom traditions of the world, and is well-known in contemplative arts and science, attesting to the integrative function of arts, science, and spirituality. The habit of separating science from arts and spirituality is typically modernist. Hence, Goethean science education offers an integrative function that can unite all disciplines of learning. Our science students, educated in a Goethean paradigm of science, would be polymaths, well versed in both arts and sciences. (Later, in the concluding section, we will also make the point that specialization in separate branches of sciences is not incompatible with our Goethean science education approach.)

Phenomenologically speaking, this idea of being one with the object is not some spiritualist talk. The phenomenon is eminently verifiable by undertaking the practice and seeing how it works in our consciousness. Anyone who has spent enough time with an object in concentrated attention can verify how a sense of intimacy grows with respect to the object. This phenomenon, known in some circles as engrossment or in others as participation or in Buddhism as non-duality, goes beyond the disciplinary boundaries and belongs squarely in the phenomenology of human consciousness. Human beings are so constituted that, given an opportunity of modeling and practice, they will experience being-one-with, or in contemporary neurobiology, attunement and resonance (Siegel, 2010). It is also known as intersubjectivity in philosophy and psychology (Gunnlaugson, Scott, Bai, & Sarath, in press).

When we spend attentive time with an object, we come to appreciate its intrinsic value, that is, we come to appreciate its being for its own sake, not as a mere means to meeting our desire and design: "When we spend time in deep contemplation of the structure of a plant, for instance, we come to appreciate the plant as an end in itself rather than a mere means" (Robbins, 2005, p. 123). Again, we can see right away how this kind of science learning is one-piece with ethics learning and practice. Ethics, unlike considerations of prudence, demands us to approach and treat the world intrinsically. The world is not there just to serve us and do our bidding. We, educated in Goethean science, would not exploit, violate, and damage the world. For, the world is our beloved, and we are the lovers, caring about and being devoted to its wellbeing and flourishing.

Goethean science education, being aligned with contemplative arts and sciences traditions, can be supported by learning and practices in arts just as equally as in science. Mindfulness practice, which is frequently being introduced now in schools in North America, can be harnessed to support Goethean science education. Mindfulness, as is practiced in schools, has become an anxiety relief and behavior management tool, which, while beneficial, is a limited practice, and not doing full justice to its rich and important educational potential. Since there is a great deal written about contemplative practices in support of "being one with object," we will not go into its scholarship here but simply refer the interested reader to such scholarship (e.g., in Zen tradition, Bai, 1997, 2002, 2003).

CONCLUDING THOUGHTS

Goethe's approach to science offers a foundation of ethical thinking, mutuality, and "a more appreciative, qualitative, meaningful and participatory engagement with nature" (Wahl, 2005, p. 60). As Robbins (2005) acknowledges, "the observation of nature is always also a process of self-discovery" (p. 125). In this way the scientist is also changed or transformed. Goethe (1790/2009) saw the need for imagination as complementary to empiricism, allowing the observed to see nature as "complete and unified as both creator and creation" (p. 106). This is also what we propose: that is, a pluralistic approach to K-12 (especially K-9) science education.

Goethean science, with its sensorial engagement, creative expression, and holistic exploration, fits well with elementary school science curricula. Instead of students dissecting frogs, the classroom can be moved outdoors, where frogs are part of the place-based learning environment. The students given a scalpel and dead frog learn to atomize and dissect, while those who share the environment with this living amphibian become part of the ecosystem, and gain intimate knowledge about this frog through delicate empiricism. Rather than focusing on anatomy and physiology, there is a shift in focus to ecology and environment, and a similar shift from anthropocentric to ecocentric beliefs. In later grades, dissection can be a topic of conversation rather than a compulsory exercise left unquestioned, and Cartesian-Newtonian science can supplement Goethean science with more technical training.

The pertinence of disciplines working together in science research (Van Noorden, 2015) is well suited to the holistic and unifying approach of delicate empiricism. "The problem," Reynolds (2007) contends, "is not that positivism has not led to useful science, but that it has substituted for all other forms of knowledge" (p. 164). The wide-ranging implications of climate change, for instance, necessitates an interdisciplinary study, while the Goethean approach ensures ethics and relationality are recognized as not only important but vital ingredients in this study. The largely abstract, and seemingly invisible, science fields such as molecular genetics or quantum physics may seem incongruent to Goethean methodology. How can you use the sense of smell to sequence genes or use intuition to understand atomic particles of matter? However, delicate empiricism is still essential and relevant here for instilling wonder, doing ethical science, and witnessing the interrelation and interdependence among varied phenomena. Goethe himself applied his approach to physics in his study of light and color.

There are core science disciplines best suited to Cartesian-Newtonian science, such as those involved with specialized medical practice. Cartesian-Newtonian science helps us gather data with limited bias, ensure that our findings are statistically significant, and has provided a prolific and profitable database of information on universal phenomena. Such empirical knowledge has produced innumerable rewards for human health and safety. However, this does not preclude Goethean science from supplementing the more technical scientific fields. If delicate empiricism can "help restore spirit to science education by reuniting a world that we have artificially fractured" (Reynolds, 2007, p. 166), then we need to seriously reconsider how we teach our children and young adolescents science. We can look to Goethean science, and other holistic science education models, such as Indigenous Science (Cajete, 1994), which are increasingly practiced in schools, as models that are compatible with most, if not all, branches of science.

REFERENCES

Bai, H. (1997). Ethics and aesthetics are one: The case of Zen aesthetics. *Canadian Review of Art Education, 24*(2), 37–52.

Bai, H. (2002). Zen and the art of intrinsic perception: A case of haiku. *Canadian Review of Art Education, 28*(1), 1–24.

Bai, H. (2003). Learning from Zen arts: A lesson in intrinsic valuation. *Journal of the Canadian Association for Curriculum Studies, 1*(2), 1–14.

Bronowski, J. (1956/1972). *Science and human values.* New York, NY: Harper and Row Publishers.

Brook, I. (1998). Goethean science as a way to read landscape. *Landscape Research, 23*(1), 51–69.

Cajete, G. (1994). *Look to the mountain: An ecology of indigenous education.* Durango, CO: Kivakí Press.

Evernden, N. (1993). *The natural alien* (2nd ed.). Toronto, ON: University of Toronto Press.

Fanelli, D. (2010). Do pressures to publish increase scientists' bias? An empirical support from U.S. States data. *PLoS ONE, 5*(4), e10271. doi:10.1371/journal.pone.0010271.

Goethe, J. W. (1790/2009). *The metamorphosis of plants.* Cambridge, MA: The MIT Press.

Gunnlaugson, O., Scott, C., Bai, H., & Sarath, E. (Eds.). (in press). *The intersubjective turn in contemplative education: Shared approaches for contemplative learning & inquiry across disciplines.* New York, NY: State University of New York Press.

Mason, D. M. (2014). Holism and embodiment in nursing: Using Goethean science to join 2 perspectives on patient care, *Holistic Nursing Practice, 28*(1), 55–64. doi:10.1097/HNP.0000000000000010.

Merchant, C. (2008). The violence of impediments: Francis Bacon and the origins of experimentation. *The History of Science Society, 99*, 731–760.

Pechenik, J. A. (2016). *A short guide to writing about biology* (9th ed.). Boston, MA: Pearson Education.

Pepper, S. C. (1970). *World hypotheses: A study in evidence.* Berkeley, CA: University of California Press.

Reason, P., & Goodwin, B. C. (1999). Toward a science of qualities in organizations: Lessons from complexity theory and postmodern biology. *Concepts and Transformations, 4*, 281–317.

Reynolds, S. (2007). A Goethean approach to science education. *Journal of Curriculum and Pedagogy, 4*, 160–171. doi:10.1080/15505170.2007.10411632.

Richards, R. J. (2002). *The romantic conception of life: Science and philosophy in the age of Goethe.* Chicago, IL & London: University of Chicago Press.

Robbins, B. D. (2005). New organs of perception: Goethean science as a cultural therapeutics. *Janus Head, 8*, 113–126.

Siegel, D. J. (2010). The mindful therapist: A clinician's guide to mindsight and neural integration. New York, NY: W. W. Norton & Company.

Scott, C., & Bai, H. (in press). Bhakti yoga as intersubjective contemplative practice. In O. Gunnlaugson, C. Scott, H. Bai, & E. Sarath (Eds.), *The intersubjective turn in contemplative education: Shared approaches for contemplative learning & inquiry across disciplines.* New York, NY: State University of New York Press.

Van Noorden, R. (2015). Interdisciplinary research by the numbers. *Nature, 525*, 306–307. doi:10.1038/525306a.

Wahl, D. C. (2005). "Zarte Empirie": Goethean science as a way of knowing. *Janus Head, 8*, 58–76.

CHAPTER THIRTY-FOUR

When Holism Meets Democratization

Re-centering Science Classrooms to Support Students' Feelings of Agency and Connectedness

ISABEL MARTÍNEZ-CUENCA, AUTUMN JOY FLORÊNCIO-WAIN, AND ALANDEOM W. OLIVEIRA

> [Rather than becoming] dependent on given knowledge of the outside world which we gladly take for granted ... [students should opt for] experiencing life and engaging it critically, of looking at life as an open-ended question, full of possibilities.
>
> —ILDEFONSO (2011, P. 55)

Captured in the above quotation is the gist of our present effort to answer the very complex question of what science education is for and what should be done in its name. As emphasized by the quote, we believe that science education should prepare students to live a wholesome life wherein tentativeness and openness are celebrated, and critical engagement and questioning are valued as tools for the pursuit of unity and meaning. Far from constituting a straightforward endeavor, such a pedagogical task, we argue, requires balanced hybridization of educational principles or tenets drawn from diverse educational paradigms, including holism and democratization. The text is organized as follows. First, attention is given to the ideological tenets and curricular features of holistic educational paradigms. The focus then shifts to instructional approaches emphasizing democratization (learner- and society-centered learning). We then conclude by arguing that an amalgamation of these progressive paradigmatic insights and ideological commitments can lead to a more coherent vision of what should be done in the name of science education.

HOLISTIC SCIENCE EDUCATION

From a holistic pedagogical standpoint, science education should be, from its introduction in the earliest grades to the most specialized level of professional study, a crucial bridge between individual learners' developing sense of self and their understandings of the world around them. Classroom science should serve as a conduit through which learners come to see themselves as both empowered stewards of the natural world and an integral part of it. For holistic educators, the goal is for students to see science as a vehicle for self-knowledge and empowered engagement with the environment. As such, science education needs to concern itself with cultivating four Cs, namely *caring*, *curiosity*, *collaboration*, and *connectedness*.

The Four Cs of Holism. Holistic educators consider the forces of physics, chemistry, biology and even earth sciences as being inseparable from our form and existence as human beings. The same forces that have given rise to the entirety of the known universe also shape our human selves. As such, science teachers can utilize knowledge of scientific facts and principles to forge a profound *connection* to the natural world a part of their identity and invite their students to do the same. In making our shared origins and interdependence foundational elements in science teaching across scientific disciplines, we set the stage for creating an intellectually stimulating, personally empowering and socio-affectively supportive classroom setting. As Givanni Ildefolnso (2011) writes, "the educated person needs to have the capacity to see the whole, not fragments of the world and of him/herself" (p. 53).

By adopting the philosophical stance that scientific understanding reveals us to be connected to all that is through common origin and essential similarity in composition and governing scientific principles, we set the stage for our students to care for and feel curious about all aspects of the natural world. Cultivating *connectedness*, from the holistic perspective, means locating the learner as an active agent within an interconnected and interdependent universe. This includes giving thoughtful consideration to the relationship of the self to local cultures, the community, nature, and the universe (Miller, 1997). John Miller (2011) suggests that students should be guided to acquire a sense of profound connection to knowledge, community, the environment, and to the cosmos from a non-dualistic understanding. From the holistic perspective, cultivating students' sense of connectedness is a means for achieving other ends of progressive education (such as developing empathetic relationships, socially responsible civic engagement, love of learning, strong knowledgebase, personal happiness, etc.) as well as being an end unto itself.

The type of connectedness aspired to in holistic education is reflected in the philosophy of Heesoon Bai (2015) and in the lived experience of Jill Bolt Taylor (2006). Bai describes the ontological convention of describing the universe as consisting of two categories—the animate and the inanimate—as a problematic and unnecessary bifurcation. Likewise, educators who adopt a holistic paradigm favor teaching methods and curricula that foster students' sense of profound connectedness to the whole of the universe, both animate and inanimate. In doing so, holistic educators seek to foster what Bai (2015) describes a felt and lived aliveness in which one can "experience all parts of reality to be interconnected and to interpenetrate each other, arising together, co-emergently, moment by moment" (p. 137). In a similar vein, Taylor (2006), a neuro-anatomist who studied her own experience of having a stroke in the left hemisphere of her brain, describes the experience of seeing and vividly experiencing her interpenetration with her surroundings. Taylor attributes the temporary loss of the logical reasoning abilities of her left-brain with the life transforming capacity to see no important boundary between the atoms in her arm and those of the air around. Her newfound capacity to conceive of herself as part of a continuity that includes the totality of the universe convinced her that our left-brain dominated culture had something to gain from more development of right-brain ways of perceiving and conceptualizing. In alignment with Bai and Taylor, the holistic paradigm can empower science educators to shine a light on how the current state of scientific understandings can be both utilized and problematized so as to underscore connections and inspire students to feelings of respect, wonder, and awe when thinking about the relationships between themselves and the universe that surrounds them. As Bai (2015) puts it, this way of presenting the natural world draws forth "an ethic that will enable us to live in communion, respect, peace, and harmony with the earth" (p. 138).

In her classic essay about teaching and learning, "The Having of Wonderful Ideas," Eleanor Duckworth (1973) notes that there is significant agreement about the incredible rate of intellectual development and uninhibited curiosity in early childhood and a notable decline in children's investigative impulses as they get older. She then goes on to ask, "What happens to children's curiosity and resourcefulness later in their childhood?" (p. 6). The answer, she suggests, is that whereas young children's "wonderful ideas" are celebrated, older children's ideas are dismissed as trivial or discouraged because adults find them silly, embarrassing, or destructive. As science teachers who embrace holism, we need to orient ourselves to celebrate and support students' *curiosity*. Of course, a significant part of science education involves providing information and introducing our students to concepts and practices, nonetheless, we must not lose track of that fundamental component of education, stimulating and supporting curiosity—that love of learning that goes beyond any single discipline and facilitates true engagement with all disciplines.

A particular type of curiosity that educators can find challenging to support is that of questioning the premises on which scientific claims are made and questioning the authority of the power-holders in schools and classrooms. As we discipline ourselves to honor rather than dismiss or trivialize student curiosity, we must also support them in engaging in epistemological questioning. As Joe Kincheloe (2008) stated, "Anytime teachers develop a pedagogy, they are concurrently constructing a political vision. The two acts are inseparable" (p. 9). It therefore stands to reason that questions of epistemology and critical pedagogy concern every aspect of knowledge building and can and should be incorporated, at appropriate levels of complexity, in all stages of science education. According to holistic paradigm, responsible science education must promote critical awareness of the contexts in which scientific understandings are constructed and the epistemological underpinnings of what is understood as true.

Most teachers have been faced with some version of the questions, "What is the point of learning this?" and "How do I know this is true?" from their students. While it is tempting to meet these questions with as succinct an answer as possible and return to our planned curricular objectives for the day, holistic educational practice requires a different approach. From the holistic paradigmatic perspective, the goals of education are better served by treating queries that challenge conventional epistemology and authority as having paramount importance to student learning. Educators are tasked with finding a way to honor such questions with time and attention in the spirit of genuine inquiry. Taking the time for such inquiry has several benefits. First, a teacher's decision to honor students' questions that challenge authority and epistemology conveys to our students the message that we hear them and care about what they think. Second, valuing such questions will encourage students to continue questioning and thinking. Third, validation of these questions can serve to demonstrate to students that our classroom is a safe environment for taking risks and asking difficult questions. Finally, when teachers demonstrate that they see these kinds of questions as valuable and worthy of class time, students are placed squarely in the role of active participants in inquiry, *collaborators* in their learning.

In allowing time and space in our classrooms for our students' difficult questions we give students the experience and support they need to see themselves as empowered meaning-makers. Our students are the present and future stewards of the planet and inhabitants of a world connected through ecology, economics, politics, social relationships and a shared humanity. Making time in curricular mapping to take difficult questions seriously is needed in order for students to come to know themselves as critical thinkers with the capacity and responsibility to raise their voices and engage on issues that matter to them. It is our responsibility to educate students to understand themselves to be active collaborators in the future of our planet and all of its inhabitants. To achieve this end, it is imperative that we are careful not to treat our students as passive recipients of knowledge doled out by a dominant authority.

Emphasizing our connectedness, honoring student curiosity, viewing students as collaborators in their learning and our collective treatment of one another and the planet are all demonstrations of and invitations to *caring*. Nel Noddings (2012) has written extensively on the ethics of caring and the role of caring in education. She describes the roles of the carer and the cared for as bidirectional. Not only must we hold an attitude of caring for our students, we must also attend to how they receive our caring intentions and direct our actions to their specific needs. In science classrooms, we are accustomed to concerning ourselves with our presentation and our students' reception of the subject area content. Noddings and other writers on the subject of caring point out that the best way to ensure that our students are receptive to the content of our lessons is to care; care about them personally, care for them and institute a culture of caring in our classrooms.

Through recognizing and valuing our interconnectedness, sustaining and supporting curiosity throughout life, caring about and for one another, and accepting the responsibility of collaborating with each other in our learning and engagement outside classrooms, we may even learn to peacefully share and care for our planet, while we honor, care for and support one another. This is what education is for and what science education should be about.

Other Ideological Commitments. Holistic educators are ideologically committed to a confluence of aspects of the instructional process: intrapersonal needs, interpersonal relations, extrapersonal setting, and the transpersonal (Miller, 1986). The *intrapersonal* deals with the cultivation of a kind of deep personal self-awareness in students. The *interpersonal* involves students' development of awareness of others, the cultivation of caring, and the development of good communication practices throughout the school setting. The *extrapersonal* refers to the social structures, physical school environment and connections to the larger community. The final aspect is the *transpersonal,* which relates to "profound truths" about the nature of existence, connections to something bigger (Baquet, 2011), and spirituality. While referring to this transpersonal aspect of education, Ron Miller states:

> The very act of perceiving and naming and knowing one's environment is a spiritual act. (p. 81)
>
> Bringing spirituality into education does not … mean injecting religious teachings into the curriculum; rather it means encouraging students to engage their world with a sense of wonder through exploration, dialog and creativity. (p. 80)

Holistic ideology fundamentally compels us to re-envision science education as a deeply transformative endeavor that attends to the inner lives of students. According to this paradigm, science education should invite and facilitate deep and meaningful connections on multiple levels. Science education should invite and allow for the discovery of the ineffable. Perhaps not proscribing that which is to be found can render students' quest for understanding yet more alive, more exciting, and more inviting. Perhaps choosing to believe that there is room for all sorts of understandings of life's "profound truths" is what we should be doing in the name of science education.

Holistic Curricular Features. The above ideological commitments of holistic ideology are exemplified particularly well in the book *Cosmic Education* (Duffy & Duffy, 2014), a K-6 curriculum based on the work of Maria Montessori. This curriculum integrates history, literature, and the sciences (geology, biology, astronomy, chemistry) into a series of origin stories within stories: Story of the Universe; Story of Stars/Solar System; Story of the Earth; Story of Life; Story of Humans; Story of Civilizations (ancient times); and, Story of One Nation (modern times). The units are organized in a chronological progression of detailed scientific and historical explorations of life and the universe; progressing both in time (from the Big Bang to the present) and in complexity (from primary to upper-elementary level). The guiding principle of the curriculum is that all of existence has a common origin and is profoundly interconnected and valuable, pointedly including the learner him or herself. As stated by the book authors:

> Cosmic education is intended to help each of us search for our cosmic task as a species and as individuals. To do this, we must understand ourselves in context. It is only against the background of our place in the Universe our relationship to other living organisms, and our understanding of human unity within cultural diversity, that we can attempt to answer the question: "Who am I?" (Duffy & Duffy, 2014, p. 6)

The question "who am I" and its correlates "where do I come from?" and "why am I here?" are meant to guide students through a journey of self-discovery and personal meaning-making and encourage learners to make personal connections to the content of the curriculum. Students are expected to learn predetermined bodies of knowledge (standard subject area content) in a specified order via a fixed methodology. At the same time, *Cosmic Education* seeks to cultivate independence and inquiry, peace, respect for the universe, reverence for the natural environment and unity of all things. The idea is to use the whole (i.e., the story of the universe) to better understand the parts (i.e., science, history, arts and other components of the universe) and initiate students into a deeply personal and meaningful voyage of self-discovery.

A second example of enactment of holistic ideals can be found in the Waldorf model (Ullrich, 2014). Waldorf education has at its base the ideology that students must be provided for in their development as whole beings. In Rudolf Steiner's (1985) terms, this means attending to students' "head, heart and hands"—that is, their thought, feeling and will. There are no specific methodologies for cultivating deep and caring relationships between the teacher and student and meaningful connections between students and the natural world. The goal is to provide students with experiences that feel fresh and alive, hence inspiring their curiosity.

Connections to the natural world are fostered in Waldorf education through curricular integration. Themes are investigated over the course of several weeks and then returned to in a cyclical fashion. Students record their observations and understandings of scientific concepts in notebooks using their own words and drawings. In addition to being expressed in writing, ideas are also enacted theatrically, rendered artistically, used in writing and so on, in what David Nicholson (2000) calls “layers of experience” (p. 587). Waldorf educators try to create ideal conditions for students to develop relationships with the subject matter by systematically evoking imagery and imagination (Uhrmacher, 1993), promoting aesthetically rich knowledge formation, and bridging epistemologically between artistic and scientific approaches to understanding nature. As Edvin Østergaard, Bo Dahlin, and Aksel Hugo (2008) point out, the intention of Waldorf educators is “to teach science in such a way that a sense of meaningful wholeness of nature grows in the students; a wholeness from which the human being is not separated or alienated” (p. 22). Careful modeling of respect for nature and thoughtful preparation of learning materials and environments so as to reflect and cultivate harmony with nature are foundational elements of the curriculum. The physical environment of the school grounds and classrooms are carefully attended to and natural materials are favored to the exclusion of plastic and machine made objects as thoroughly as possible (Woods, Ashley, & Woods, 2005).

Teachers' sense of caring for each of their students is achieved through specific practices that guide the teacher to direct his or her attention to each student throughout the school year. The teacher prepares a detailed written evaluation of each student's academic, social and personal development as evidenced in school, maintains regular contact with parents, and visits students' homes at least once a year.

Our intention in introducing Waldorf and Montessori methodologies for science education here is not to advocate the whole-cloth adoption of these particular approaches but simply to provide a window into the two real-world enactments of holistic ideals. The Montessori and Waldorf approaches to science education embody just two possible methodologies wherein qualities such as curiosity, caring, connectedness and collaboration are valued and cultivated, not subordinated to skill and fact mastery. This type of education involves acknowledging our shared humanity and coming together with the fullness of our whole selves to investigate and experience life as fully as possible. This, we believe, is an essential part of what we should be doing in the name of science education.

DEMOCRATIC SCIENCE EDUCATION

A second educational paradigm that we consider essential to answer the question of what science education should be about is democratization. Two ideological commitments are central to this educational paradigm, namely "learner centeredness" and "society centeredness." Both have historical roots that can be traced back to the progressive educational movement of the early 20th century (Ellis, 2004).

Learner-Centeredness. From this particular educational perspective, science teaching should become more learner-centered and support democratic pedagogical enactment by eliminating asymmetrical roles and countering control imbalances. In Freirean terms (1970), science educating must reach a dialectical equilibrium by which teacher and students engage in the subject matter horizontally, as catalysts and agents of learning. This democratization can increase opportunities for student interest, for meaningful self-realization, and for guided empowerment.

Traditionally, educational theorists and practitioners separate two fundamental factors in the educative process: the child and the subject matter. At the enacted curricular level, the traditional, knowledge-centered science lesson normally consists of a vertical relationship, a top-down, undemocratic imposition to recipient learners through the teacher. When visually depicted, based on David Hawkins' (1974) *I-Thou-It* theoretical framework, traditional science teaching consists of a linearly shaped pedagogical process, wherein unproblematic content (the *It*) is channeled by a teacher (the *I*) to a student (the *Thou*) (see Figure 34.1a about traditional teaching below). Fortunately, the spirit of transmission has been increasingly questioned due to prevalence of unidirectional relationships, students' infrequent direct contact with the subject matter, and their lack of agency and control in the learning process. Yet, even to this day, performance goals and the transmission of canons make the direct experience of the subject matter a rather sporadic affair.

Recent constructivist efforts to make science teaching more democratic have been shown to often engender pedagogy with a slight pyramidal shape, shifting the interplay of the elements of the practice from traditional top-down imposition to a pseudo-democratization (see Figure 34.1b about pseudo-democratic teaching below). In these pseudo-democratic lessons, science teachers attempt to create direct interactions between students and the subject matter, and foster learning through inquiry and group work activities. However, the science classroom still shows a predominance of teacher-centered implementations with asymmetrical participatory frames. A defining feature of this pedagogical mode is the occurrence of "pseudo-dialogism" in the form of IREs or *triadic dialogues* (Lemke, 1990)—teacher-led verbal exchanges wherein students' oral contributions are limited to isolated words aimed at guessing the answer being sought after by the teacher

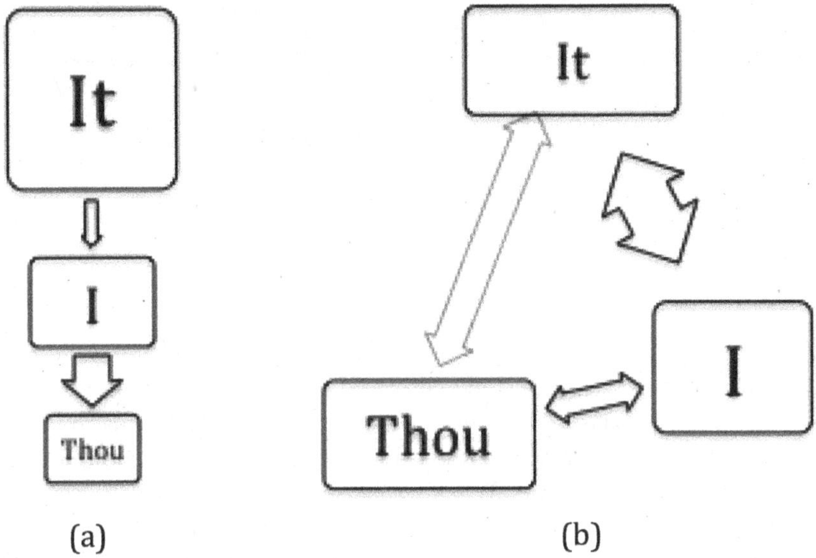

Figure 34.1: I-Thou-It Representations of (a) the Traditional Science Classroom, and (b) Pseudo-Democratic Science Teaching Approaches.
Source: Author

(far from a "true dialogue"). Despite the surface-level reorganization of the traditional imposition model, there are still some limitations in terms of the autonomy of the individual elements (particularly students) as well as the balance of their relationship.

Dialectical Classrooms and Horizontal Interactions. The future of progressive science education must be rooted in the past, in Deweyan tradition, finding ways to foster the participation of the learner in the learning process and the co-construction of the content she is studying. To this end, we believe science education needs a democratic enactment of the lesson: first, leveling the asymmetries of the elements of the practice to create a more dialectical lesson; and second, judiciously shifting the locus of control from the teacher to the learner.

To begin with, the current science practitioner must design dialectical lessons, promoting a horizontal interaction between child and subject matter. Paulo Freire (1970) defined dialogue as "the encounter between men, mediated by the world" (p. 88). He extended his metaphor to the educational plane, arguing that "authentic (science) education is not carried on by 'A' for 'B' or by 'A' about 'B', but rather by 'A' *with* 'B', mediated by the world" (p. 93). We visualize this mediation as a horizontal continuum between the *I* and the *Thou*, facilitated by the *It* (see Figure 34.2 about horizontal interactions below). The logical consequence of this leveling is an

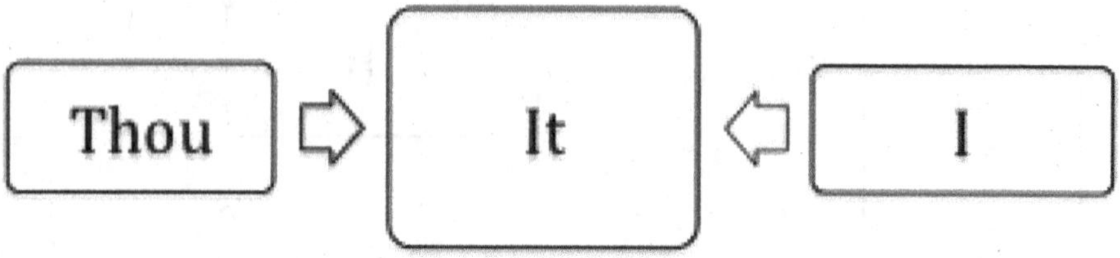

Figure 34.2: Horizontal Framework, Inspired by Freire (1970).
Source: Author

equilibrium between teacher and student, both engaging in the subject matter on the same plane, as catalyst and agent of learning, respectively. In this conception, the leveling is necessary because without it, there is not possible communication between "dialoguers" (i.e., students and teachers), which is the basis for true education (pp. 92–93). Particularly, because, as Freire suggested, the *I* cannot think for the *Thou* or without the *Thou*, nor can the *Thou* think without the *I*. Thus, practitioners should trust students' capacities to directly engage in the exploration of science topics to be able to actively connect incoming facts and experiences in the lesson with their own.

Furthermore, while symmetrical relationships are conducive to better learning environments, leveling is not sufficient in itself. Dialectical participation does not necessarily translate into learning opportunities unless we redress subject matter control imbalances. We are not suggesting that practitioners should completely relegate control of the subject matter to students (like proponents of extreme learner-centered traditions often do). However, it seems to us that science teaching can only become a true democratic enactment when the *initiative* to engage in the subject matter is at least partially delegated to the learner. The more latitude in negotiating and making executive decisions with respect to the enacted science curriculum, the more possibilities for students to pursue their interests (i.e., deep-seated inclinations). Allowing student initiative increases the opportunities not only for successfully and collaboratively deciding how to interact with the subject matter, but also for creating their own learning experience (see Figure 34.3 about a democratic practice below).

Addressing Dissenting Voices. The democratization of the science classroom does not come without its detractors. The symmetry of roles in the classroom may be criticized because the demotion of the teacher may result in epistemological relativism. Historically, the promotion of the teacher was mostly present in traditional knowledge-centered models, which prioritized regurgitation, and the exercise of idleness and silence in the classroom. As we know, this combination has proved to

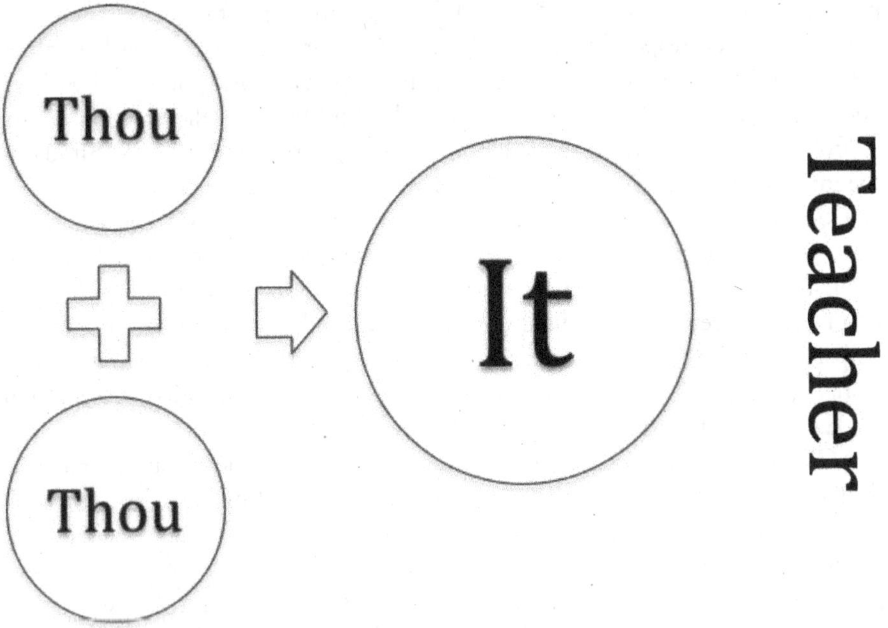

Figure 34.3: Democratic Science Classroom.
Source: Author

be a rather unsuccessful recipe for the self-realization of most students. The future of science education requires symmetries to foster both a passive and an agential dimension of learning (Roth, 2007). Passivity, in Wolff-Michael Roth's terms, refers to a state of non-intentionality, and yet of noticing, of being attuned, or of "open[ing] up … to being impressed by that which is other" (p. 6). We agree with these two dimensions and further argue that leveling hierarchies in the classroom creates a space for students to "actively orient and attend to" (p. 6) learning the content, while also passively noticing and being emotionally alert.

Secondly, skeptics of the democratization of teaching may also have reservations as to whether student self-direction and sharing subject matter control may translate into anarchy, abundance of mistakes, and curricular imbalance. Nevertheless, it is an oversimplification to assume that unstructured, unpredictable enactments result from student control. In practice, the implementation of lessons is a fluid, evolving, emerging product inevitably unpredictable, as it is affected by the dynamics between teacher, learners, subject matter, as well as other socio-contextual aspects. Furthermore, teachers should foster student self-direction and risk-taking as mistakes are a byproduct of learning and the exercise of responsibility.

Paradoxically, learning to be responsible and assuming the consequences of decisions are indispensable for social and moral maturation. Thus, we envision the process as a gradual exercise at the enacted level, scaffolding students' self-management, and the corresponding repercussions, as a function of age. In earlier grades, control could be narrowed to the options in a common task, such as choosing a book to read about living organisms. In following grades, self-direction can be scaffolded through learning centers from which students would choose. Similar to the interest center approach in Exploratory Experiences Curriculum for Elementary Schools, in a topic, such as rocks and minerals (Ellis, 2004), learners might select to read about rocks (reading center); measure them and observe their characteristics (science center); engage in artistic expression (art center); write about them (writing center), etc. Finally, in more advanced grades, learners might participate in science topic selection; engage in an open-ended manner with the topic; work on an individualized final product (e.g., mural, portfolio, video, presentation, etc.), and create their corresponding assessment criteria.

Given the benefits of role symmetry, and balancing subject matter control, as well as the fluidity of the practice, and the importance of mistakes and responsibility, we might as well explore more democratic realizations where students can create and manage their own guided learning opportunities in this empowering process we call learning. Ultimately, we argue that science education must be theorized and practiced *within* a more "cooperative" and "democratic" framework of engaged pedagogy, dialectally relating teaching to learning, favoring opportunities for meaningful self-realization and guided empowerment.

Society-Centeredness. Our pragmatist, knowledge-centered, performance-obsessed educational tradition is jeopardizing our democracies, Martha Nussbaum noted in 2012. Both Plato's idea of the democratization of education as well as Aristotle's character education have been poorly executed in pursuit of profit and the maintenance of the status quo. On the one hand, the expansion of education seems based on an efficient division of labor model common in factories (Tyack & Cuban, 1995). On the other hand, the obsession with prescriptive school conduct limits children to "first learn right conduct and later be allowed to question, analyze, and criticize" (Noddings, 2012, p. 13). Naturally, the product is the manufacture of lethargic citizenry with a particular understanding of the world. These epistemological misrepresentations of education may inadequately reproduce hierarchical stratifications, not only restricting access to epistemological and symbolic resources (cultural capital), but also cultivating conformity and limited democratic participation.

As we described in the previous section, democratization at the classroom level affords students not only to voice their ideas, but also to take the initiative to engage in the subject matter. Clearly, though, democratization should move beyond the classroom and the individuals, ultimately, positively transforming social realities and local communities. Here, we propose three notions and practices by which science educating can reclaim a true democratization of the purposes and the model of education: content relatability, reflective practices, and action or active engagement. Ultimately, we argue that these practices can create informed, well-educated citizenry; promote social mobility; and lay the foundation for the transformation of societies.

Content Relatability and Reflective Practice. First, following John Dewey (1902), the science teacher can help students to access symbolic practices and epistemological traditions by "psychologizing" materials. We argued that since students are creators of knowledge, direct interaction and exploration of the subject matter lay the foundation for learning. However, as Freire (1970) and many contemporary scholars noticed, students do not come to the science classroom with the same symbolic tools (language mastery and discourse practices) and cultural capital. Given such reality, directly engaging students with the subject matter might not be sufficient, especially if the content is inaccessible, whether due to language barriers (Bialystock in Wellington & Osborne, 2001), inflexible representations (Willingham, 2002), or the obscurity of academic and scientific language (Wellington & Osborne, 2001). In this line, science educators must design lessons where students can autonomously connect the new material in relation to their prior experiences; a process Dewey (1902) called "psychologizing." Similarly, Freire (2009) urged teachers to embrace and respect children's abilities and prior knowledge and employ "unconventional" teaching methods (e.g., calculating lengths with a kite instead of conventional trigonometry) to make the discourse practices and the content of science accessible. By eliciting connections between prior and incoming knowledge, not only can students relate to materials, but also gradually gain access to the tools and the cultural capital necessary to create an informed, well-educated citizenry.

Second, while granting access to the content of science should be the premise of good teaching practices, we propose that science education must additionally cultivate reflection. Dewey (1902) observed that students' experiences are operating forces in the incorporation of more knowledge. Yet, even if students' prior knowledge is brought to bear in the classroom, science might just become an inert mental collection of facts. In this line, science teaching should spur a holistic process of active student reflection, threading distinct parts/events together without fractures, purposefully selecting and rejecting what is integrated into their thought processes, blending prior to new ideas, and creating new cohesive conceptual units (Dewey, 1934/1980, in Jakobson & Wickman, 2007). By actively reflecting on the ideas introduced in class, students can create continuity across experiences, weaving meaning among initial and incoming threads (Rodgers, 2002) find relationships; transform those into more complex clusters of associations; and draw conclusions. As Dewey (1938) observed, reflective experiences are a "form of growth that creates conditions for further growth … attitudes and habits which alone open up avenues for development in other lines" (p. 13). In the short run, reflection not only supports genuine individual growth and self-realization but also epistemological empowerment. In the long run, when an educational institution provides students with opportunities for reflection and a continuum of growth-oriented learning experiences, it is playing its part in fostering social mobility.

Two Levels of Active Engagement: The Classroom and Beyond. Finally, together with access and reflection, the science practitioner must include two levels of "praxis" in the classroom. As Freire (1970) says, action, together with reflection, not only distinguishes humans from animals, but also has a potential to transform reality. On the one hand, a first level of action in the classroom might refer to engaging in reasoning activities, and hands-on experimentation to explore the conventional and canonical body of pre-existing scientific knowledge. As such, active participation in exercises, problems, disambiguating language, or making sense of scientific experiments, and theories allows students opportunities to take part in epistemological traditions. Most importantly in this level, students can uncover the scientific, political, economic, ecologic, social, and epistemological underpinnings of global issues, such as global warming, fracking, food sustainability, and the extinction of species, among others. While, initially, science teachers might need to model experiments and reasoning procedures; verbalize steps in problem understanding; evaluating competing theories/perspectives; and scaffold the learning process; ultimately, by "doing" in the classroom, students become independent problem solvers and "critical consumers" of knowledge.

On the other hand, on a second level, action extends beyond consumption to creation of knowledge and transformation of local and social communities by means of dialogue, discussion, and critical awareness. In fact, except for some sensory schemata, all of our knowledge is made of socio-empirical conceptual categories (Silverstein, 2007). As such, these categories are not pre-existing in a vacuum but created in communicative events. Indeed, following socio-constructivist as well as sociocultural premises, knowledge is socially constructed in each community and such epistemology is interactively re/negotiated by their members through discourse. In this line, science practitioners can help students become competent participants in science discourse. If so, when a conscientious teacher incorporates dialectical moments, these could support future participation in real life (scientific) discourse practices. Most crucially, in the classroom, knowledge is created by "talking (something) into being" and "talking (something) out of being as well" (Crawford, 2005, p. 141). Thus, habitually engaging in creating knowledge at the classroom level can, ultimately, be a precursor of empowered transformative action, participating in discourse that "talks changes into being" in our students' surrounding communities. Independently of the level, then, by fostering active engagement, critical awareness, and praxis, students can address obstacles, co-construct situations differently, and dialectically transform realities from the inside as a process of liberation rather than "salvation" (Freire, 1970, p. 94). Ultimately, praxis is essential in order to nurture responsible members of communities, drive social advancement, and contribute to the democratization beyond the classroom.

All in all, thoughtfully crafted experiences wherein 1) students' prior and incoming knowledge are related, 2) active student reflection is encouraged, and 3) engagement and critical awareness are cultivated are all essential to the task of stimulating interest and inextricable from the larger task of educating democratic citizens. On the one hand, since learning is mediated through complex interactive and interpretative processes, granting access to science materials, symbolic tools, and science practices (i.e., cultural capital) can help foster an informed, well-educated citizenry. On the other hand, students who are trusted to engage in reflection and agency as part of their own learning process will find themselves as empowered actors and critical thinkers inside and outside the classroom, not passive recipients of knowledge. Paraphrasing Thomas Sergiovanni, Ellis (2004) reminded us that shaping critical thinkers and informed citizens today, science educators can contribute to "tomorrow's caring and active citizens" (p. 69).

TOWARD A MORE COHERENT PARADIGM

Articulating the educative purpose of science is an extremely complex endeavor nowadays, given the legacy of conflicting and conflating orientations in the educational field. The polarization between content and pedagogy (i.e., its methodological implementation) reached a deadlock (Egan, 2003) with the emergence of diametrically opposed traditions (Page, 2010; Young, 2013). Educators with knowledge-centered views criticize holism and democratization as unproductively prioritizing students' subjectivities. According to Michael Young (2013), celebrating student subjectivities is becoming an increasing source of "national incoherence" (p. 113), trumping the diffusion of "powerful knowledge" and a nationally coherent curriculum (p. 107). On the other side of the battlefield, supporters of learner- and society-centered views of education counterargue that knowledge-centeredness is an infelicitous orientation and a destabilizing force that is culturally limiting and homogenizing (Ellis, 2004), and thus highly problematic in our pluralistic society. Further, rote learning and memorization can result in knowledge inflexibility, that is, reduced transferability to real-life problems (Sinatra & Pintrich, 2003; Willingham, 2002). Knowledge-centered education runs the risk of "producing generations of useful 'machines,' rather than complete citizens who can think for themselves, criticize tradition, and understand the significance of another person's sufferings and achievements" (Nussbaum, 2012, p. 1).

Rather than taking sides on this debate, a more productive approach to answering the question of school science's educative purpose is to argue for a paradigmatic composite that more coherently takes into account the three main components of philosophical underpinnings and visions of education (Ellis, 2004): knowledge, society, and the learner. We must go beyond static definitions, or unproductive dichotomies, and center on a compromise. Ideally, any curriculum should combine a sustainable (Van den Akker, 2003) balance of components as well as an integration of several paradigms. Yet, any practitioner is aware that balance is seldom achieved in any enacted curriculum (Ellis, 2004), even in the most structured exemplars. To limit the likelihood of curricular imbalances, we argue for more fluid, emergent, and dynamic conceptualizations of the purpose of education. We agree with Carl Leggo (1998) and Herbert Kliebard (2004) that this alternative conceptualization of education and schooling offer opportunities for dialogue across paradigms, and thus be more productive than polarized debates. Further, dynamic amalgamation of paradigms and ideologies, we believe, is likely to produce more moderate views of the purposes of science that are inclusive of all three key components of any science education effort, namely the learner, knowledge, and society.

Live in Fragments No Longer, Only Connect. We would like to conclude by quoting Edward Foster (1910) who, in the classic novel *Howards End*, makes the plea to "live in fragments no longer, only connect." A similar call for connectedness can be made to science educators. For too long, our science classrooms have been characterized predominantly by fragmentation (epistemic, social, and ontological). Coherent connections must be meaningfully forged among self, science, society, and democracy in order for school science to truly serve its educative purpose. As emphasized by the German philosopher Josef Pieper (1963), "Education concerns the whole man; an educated man is a man with a point of view from which he takes in the whole world. Education concerns the whole man, man *capax universi*, capable of grasping the totality of things" (p. 39). This is precisely what science education is for: helping students fully flourish as human beings capable of transcending situational fragments in the immediate context of the science classroom and seeing the whole.

REFERENCES

Bai, H. (2015). Peace with the earth: Animism and contemplative ways. *Cultural Studies of Science Education, 10*, 135–147. doi:10.1007/s11422-013-9501-z.

Baquet, J. (2011). *YOU ARE THAT foundational essays: The perennial philosophy and neo-perennialism: An introduction*. Retrieved October 20, 2014 from http://www.youarethat .org/foundations/neo-perennialism.htm

Crawford, T. (2005). What counts as knowing: Constructing a communicative repertoire for student demonstration of knowledge in science. *Journal of Research in Science Teaching, 42*, 139–165. doi:10.1002/tea.20047.

Dewey, J. (1902). The child and the curriculum. In S. M. Cahn (Ed.), *Classic and contemporary readings in the philosophy of education* (pp. 221–228). New York, NY: Oxford University Press.

Dewey, J (1934/1980). *The art of experience.* New York, NY: Perigree Books.

Dewey, J. (1938). *Experience and education.* New York, NY: Touchstone.

Duckworth, E. R. (1973). The having of wonderful ideas. In M. Schwebel & J. Raph (Eds.), *Piaget in the classroom* (pp. 258–277). New York, NY: Basic Books. doi:10.2307/3120071.

Duffy, M., & Duffy, D. (2014). *Children of the universe: Cosmic education in the Montessori classroom.* Hollidaysburg, PA: Parent Child Press.

Egan, K. (2003). What is curriculum? *Journal of the Canadian Association of Curriculum Studies, 1*, 9–16.

Ellis, A. K. (2004). *Exemplars of curriculum theory.* Larchmont, NY: Eye on Education. doi:10.4324/9781315855318.

Foster, E. M. (1910). *Howards end.* New York, NY: Vintage Books.

Freire, P. (1970). *Pedagogy of the oppressed.* New York, NY: Continuum.

Freire, P. (2009). *Teachers as cultural workers: Letters to those who dare teach.* Boulder, CO: Westview Press.

Hawkins, D. (1974). *The informed vision: Essays on learning and human nature.* New York, NY: Agathorn Press.

Ildefonso, G. M. (2011). Not a laughing matter: The value of leisure in education. *Curriculum Inquiry, 41*(1), 48–56. doi:10.1111/j.1467-873X.2010.00524.x.

Jakobson, B., & Wickman, P.-O. (2007). Transformation through language use: Children's spontaneous metaphors in elementary school science. *Science & Education, 16*, 267–289. doi:10.1007/s11191-006-9018-x.

Kincheloe, J.L. (2008). *Critical pedagogy* (2nd ed.). New York, NY: Peter Lang.

Kliebard, H. M. (2004). *The struggle for American curriculum, 1893-1958* (3rd ed.). New York, NY: Routledge.

Leggo, C. (1998). Living un/grammatically in a grammatical world: The pedagogic world of teachers and students. *Interchange, 29*, 169–184. doi:10.1023/A:1007566325713.

Lemke, J. L. (1990). *Talking science: Language, learning and values.* Norwood, NJ: Ablex.

Miller, J. P. (1986). Atomism, pragmatism, holism. *Journal of Curriculum & Supervision, 1*, 175–196.

Miller, J. P. (2011). Chapter 3 worldviews, educational orientations, and holistic education. *Encounter, 24*, 55–69.

Miller, R. (1997). *What are schools for? Holistic education in American culture.* Brandon, VT : Holistic Education Press.

Nicholson, D. W. (2000). Layers of experience: Forms of representation in a Waldorf school classroom. *Journal of Curriculum Studies, 32*, 575–587. doi:10.1080/00220270050033637.

Noddings, N. (2012). *Philosophy of education.* Boulder, CO: Westview Press.

Nussbaum, M. (2012). *Not for profit: Why democracy needs the humanities.* Princeton, NJ: Princeton University Press.

Østergaard, E., Dahlin, B., & Hugo, A. (2008). Doing phenomenology in science education: A research review. *Studies in Science Education, 44*, 93–121. doi:10.1080/03057260802264081.

Page, R. N. (2010). Struggle: A history of "mere" ideas. *Curriculum Inquiry, 40*(2), 205–220. doi:10.1111/j.1467-873X.2010.00478.x.

Pieper, J. (1963). *Leisure: The basis of culture*. San Francisco, CA: Ignatius Press.

Rodgers, C. (2002). Defining reflection: Another look at John Dewey and reflective thinking. *Teachers College Record, 104*, 842–866.

Roth, W.-M. (2007). Theorizing passivity. *Cultural Studies of Science Education, 2*, 1–8. doi:10.1007/s11422-006-9045-6.

Silverstein, M. (2007). How knowledge begets communication begets knowledge: Textuality and contextuality in knowing and learning. *Intercultural Communication Review, 5*, 31–60.

Sinatra, G. M., & Pintrich, P. R. (Eds.). (2003). *Intentional conceptual change*. Mahwah, NJ: Lawrence Erlbaum Associates.

Steiner, R. (1985). *An introduction to Waldorf education*. London: Rudolf Steiner Press.

Taylor, J. B. (2006). *My stroke of insight*. New York, NY: Penguin Group.

Tyack, D. & Cuban, L. (1995). *Tinkering toward utopia: A century of public school reform*. Cambridge, MA: Harvard University Press.

Uhrmacher, P.B. (1993). Coming to know the world through Waldorf education. *Journal of Curriculum and Supervision*, 9 (1), 87-104.

Ullrich, H. (2014). *Rudolf Steiner*. New York, NY: Bloomsbury Publishing.

Van den Akker, J. (2003). Curriculum perspectives: An introduction. In J. Van den Akker, W. Kuiper, & U. Hameyer (Eds.), *Curriculum landscapes and trends* (pp. 1–10). Dordrecht: Kluwer Academic Publishers. doi:10.1007/978-94-017-1205-7.

Wellington, J., & Osborne, J. (2001). *Language and literacy in science education*. Buckingham: Open University Press.

Willingham, D. T. (2002). Inflexible knowledge: The first step to expertise. *American Educator, 26*(2), 31–33.

Woods, P., Ashley, M., & Woods, G. (2005). *Steiner schools in England*. Bristol: University of the West of England.

Young, M. (2013). Overcoming the crisis in curriculum theory: A knowledge-based approach. *Journal of Curriculum Studies, 45*, 101–118. doi:10.1080/00220272.2013.764505.

Contributor Biographies

Martha Allexsaht-Snider completed her doctorate in Crosscultural Education at the University of California at Santa Barbara (1991) and is an Associate Professor in the Department of Educational Theory and Practice at the University of Georgia. Her research interests include: family-school-community interactions in diverse settings, including Latino/a communities and rural México, and professional development and equity in mathematics and science education. Currently, she is collaborating with colleagues in a four-year National Science Foundation grant titled Language-rich Inquiry Science with English Language Learners through Biotechnology (LISELL-B) that involves Latino/a middle and high school students, their families, and their science and ESOL teachers.

Leanne M. Avery is an Associate Professor in Science Education in the Department of Elementary Education and Reading at SUNY Oneonta. Her research focuses on STEM photodocumentation work with rural children, which provides key insights into how these children learn, where they learn, and from whom they learn. Via collaborative place-based teacher development with universities, schools and families, she also works to promote ways in which to value local/intergenerational knowledge in school science. By anchoring student learning to place and connecting this knowledge to global science, her work enables rural children to navigate and succeed in a rapidly changing world.

Lucy Avraamidou is Associate Professor of Science Education at the University of Groningen in the Netherlands. She received her Ph.D. from the Pennsylvania State University and worked as a Research Associate at the Center of Informal Learning and Schools (CILS) at King's College London, and as an Assistant Professor at the University of Nicosia in Cyprus. Her research explores how young people develop science identities through time and across various contexts. Her latest edited book volume is entitled: *Studying Science Teacher Identity: Theoretical, Methodological, and Empirical Explorations.*

Heesoon Bai is Professor of Philosophy of Education in the Faculty of Education at Simon Fraser University in Canada. She researches and writes in the intersections of ethics, ecological worldviews, contemplative ways, and Asian philosophies. She understands philosophy's task for today's troubled world to be, in the words of Raimon Panikkar, "to know, to love, and to heal." She brings this three-fold task of philosophy into her teaching and research. Her co-edited and co-authored books include: *Fields of Green: Restorying Culture, Environment, Education* (2009); *Speaking of Teaching: Inclinations, Inspirations, and Innerworkings* (2012); *Speaking of learning: Recollections, Revelations, and Realizations* (2014); *Contemplative Learning and Inquiry Across Disciplines* (2014); *The Intersubjective Turn in Contemplative Education: Shared Approaches for Contemplative Learning & Inquiry Across Disciplines* (forthcoming). You can find Professor Bai's published works here: http://summit.sfu.ca/collection/204.

Laura Barraza holds a Ph.D. on Education from the University of Cambridge, United Kingdom and currently working in Community Conservation Education programs in Africa and Mesoamerica. For the last 30 years she has dedicated her professional life to work on education in the fields of Indigenous knowledge, Environmental values, science education, socio-scientific issues, cultural heritage, and capacity building. The general goal of her research has focused on the analysis of environmental knowledge, attitudes and perceptions about the environment in different cultural settings; cognitive processes involved in the acquisition and incorporation of environmental knowledge in different regions of Mexico, the United Kingdom and Australia.

Angela Calabrese Barton is a professor in science education and teacher education at Michigan State University. Her research and teaching focuses in the areas of equity and social justice in science education, with a particular emphasis on the urban context. Drawing from critically-oriented research methods (multi-sited ethnography, collaborative design-based research, and case study), she investigates youth learning and identity work across settings and over time. She also works closely with teachers to design and adapt curriculum/pedagogy towards incorporating youths' cultural knowledge and experiences. She also engages in curriculum research and development that in two middle grades focal areas: a) engineering design and identity work and b) nutrition and science literacies. She has designed and taught after school and community-based science/engineering over two decades.

Lee Beavington is a doctoral candidate in Philosophy of Education at Simon Fraser University. He is also an author, photographer, and instructor for Kwantlen Polytechnic University's Amazon Field School, and teaches Ecology, Genetics, and Advanced Cell and Molecular Biology in the lab and field. His interdisciplinary research explores wonder in science education, poetic inquiry, and environmental ethics. Find Lee reflecting in the forest, mesmerized by ferns, and always following the river. More about Lee at www.leebeavington.com.

Ana M. Becerra has worked in the field of education for over 40 years, beginning as an instructional assistant in a reservation school in Montana, where became a classroom teacher. Through the years, she taught pre-K through middle school in mostly bilingual programs in Nevada and California. Ana directed equity projects at the University of California-Santa Barbara and was a director of Compensatory Education Programs and school administrator in Oregon. Ana has a Bachelor in Education (Montana State University), a Master in Science Education (University of Wyoming), and a Doctorate in Educational Leadership and Organizations (University of California-Santa Barbara).

Larry Bencze is an Associate Professor in Science Education at the Ontario Institute for Studies in Education, University of Toronto, where he teaches in the graduate studies programme. In addition to his Ph.D. and B.Ed. in education, he holds B.Sc. and M.Sc. degrees in biology. Prior to his work as a professor, he worked as a teacher of science in elementary and secondary schools and as a science education consultant in Ontario, Canada. His teaching and research emphasize history, philosophy and sociology of science and technology, along with student-led, research-informed, socio-political actions to address personal, social and environmental issues associated with fields of science and technology.

Purita P. Bilbao is a curriculum specialist for both basic and teacher education at West Visayas State University in Iloilo City, Philippines. She chairs the Technical Panel for Teacher Education in the Philippines. She is a co-author of the recently released book *Mother Tongue: For Teaching and Learning*. Over the last two decades, in her role as a faculty member of West Visayas State University, she has been a strong advocate for community-centered science education in the Philippines.

Lynn A. Bryan is a Professor at Purdue University where she holds a joint appointment in the Department of Curriculum and Instruction and the Department of Physics and Astronomy. She is also the Director for the Center for Advancing the Teaching and Learning of STEM (CATALYST). Lynn began her career in science education as a high school physics teacher, and since then has devoted her career to working with science teachers to bring innovative practices and cutting-edge science content into pre-college classrooms. Her current research interests focus on science teacher education, particularly teachers' development and enhancement of knowledge and skills for teaching through the integration of science, mathematics, and engineering; teaching science at the nanoscale; and teaching science through modeling-based inquiry approaches.

Cory Buxton is Athletic Association Professor of Education at the University of Georgia and a former high school science and ESOL teacher. His research aims to foster more equitable and engaging science learning opportunities for all students and especially for emergent bilingual learners. His most recent work is on creating spaces where students, parents, teachers, and researchers can engage together as co-learners while strengthening their academic relationships, their knowledge of science and engineering practices and careers, and their ownership of the language of science. His research has been funded by the National Science Foundation and the U.S. Department of Education.

Lourdes Cardozo-Gaibisso is currently completing her Ph.D. at the University of Georgia's College of Education in the Department of Language and Literacy Education. She is a research assistant for the National Science Foundation funded Language-Rich Inquiry Science with English Language Learners through Biotechnology project (LISELL-B), and President of the Georgia Association of Multilingual and Multicultural Education.

Lyn Carter is a very experience science educator in the tertiary sector. Her current research focuses on the broader influences of globalisation and policy development on science education, and the use of cultural theory and environmental sustainability as ways of expanding science education reform. She has published extensively in prominent international science education journals. Lyn is on the editorial board of *JRST* and *CSSE* and is an immediate past Assistant Editor of *Research in Science Education*, Australia's major science education journal.

S. Megan Che is an Associate Professor of Mathematics Education in the department of Teaching and Learning at Clemson University. Her research interests include social justice and mathematics education, with a particular focus on gender and race in mathematics teaching and learning. Dr. Che also conducts research on international mathematics education, specifically mathematics teaching and learning in postcolonial contexts.

Rowhea Elmesky is an associate professor in the Department of Education at Washington University in St. Louis. For over a decade, she has been involved in research activities that are focused upon understanding how education, and specifically, science education can be a transformational force in the lives of culturally marginalized and economically disadvantaged children. Rowhea works directly with school districts to engage in research that is responsive to stakeholder concerns and catalytic in affecting real change. Rowhea's award-winning work includes the co-edited book, *Improving Urban Science Education: New Roles for Teachers, Students and Researchers.* In addition, she is lead editor of the forthcoming volume, *The Power of Resistance: Culture, Ideology and Social Reproduction in Global Contexts.*

Abiola Farinde-Wu is an Assistant Professor of Urban Education in the Department of Leadership in Education at the University of Massachusetts Boston. In her previous position, she was a Visiting Assistant Professor in the Center for Urban Education at the University of Pittsburgh. Farinde-Wu's teaching and service focus on preparing urban pre-service and in-service teachers for diverse student populations. Her research interests are the educational experiences and outcomes of Black women and girls, teacher retention, and urban teacher education. She has co-authored numerous studies published in journals, including *Teachers College Record, Urban Education,* and *Teaching and Teacher Education.* In addition, she is co-editor of *Black Female Teachers: Diversifying the United States' Teacher Workforce.*

Roberto Gómez Fernández is a post-doctoral researcher affiliated with the Applied Educational Sciences Institute at the Université du Luxembourg. Roberto is currently collaborating in the project SciPol:Lux: Science Education, Innovation, and Policy in Modern Luxembourg which studies the teaching and learning of sciences in Luxembourg's primary schools combining a historical analysis with an ethnographic approach. He holds a Ph.D. in Education from the Université du Luxembourg and in Linguistics from the Universidad Autónoma de Madrid for his thesis on identity construction, learning and school newcomers. His research interests revolve around multilingualism, identity, and language use and science education in schools.

Autumn Joy Florêncio-Wain is a student, teacher and practitioner of holistic education, currently pursuing her PhD at the State University of New York at Albany. A passionate advocate for what she terms the four C's of holistic education: *curiosity, caring, collaboration and connectedness,* Autumn holds a BA in anthropology

(Bard College, 2000), an MS in TESOL (SUNY New Paltz, 2007), and is fluent in English, French, Spanish and Portuguese. Autumn has taught Pre-K through college in a variety of holistic and conventional settings around the world. Her work can be found in books and presentation archives, including that of the American Educational Research Association.

Anna Maria Gorges is a doctoral candidate at the University of Luxembourg. She holds a Master's degree in secondary school education from Trier University in Germany. In her research project, MultiSci, Anna investigates student interactions in a multilingual secondary school science classroom in Luxembourg. She is interested in how students use their full linguistic and semiotic repertoire as a resource for co-constructing knowledge in sustainability education. She explores the ways a multilingual and multimodal classroom approach can empower students for more agency in an alternative educational program.

Nidhal Guessoum received his M.Sc. and Ph.D. from the University of California at San Diego and is currently Professor of Physics and Astronomy at the American University of Sharjah, UAE. In addition to his technical papers in astrophysics, he has written extensively on issues of science, education, culture, Islam, and the Arab world. He has authored or co-authored several books (most notably *Islam's Quantum Question: Reconciling Muslim Tradition and Modern Science*—London: I.B. Tauris, 2011) and more than 150 articles (in Arabic, English, and French) in various magazines and newspapers, including *Nature*. He has been interviewed in numerous international media, including Al-Jazeera, BBC, NPR, Le Monde, and Der Spiegel.

Anita Hussénius is Associate Professor in Organic Chemistry at the University of Gävle and Director of the Centre for Gender Research at Uppsala University, Sweden. Her main research interest is about gender and feminist perspectives on science and science education. More specifically, her research is focusing on cultural aspects, i.e. discipline specific values and underlying assumptions of what is regarded as important knowledge and skills, and how such values and assumptions are mediated explicitly as well as implicitly. Another more action oriented interest concern science teachers' gender awareness and how an increased awareness of gender issues influence their teaching of science.

Sophia (Sun Kyung) Jeong is a doctoral student in the Department of Mathematics and Science Education at the University of Georgia. She is passionate about fostering creativity, encouraging inquisitive minds, and developing sociopolitical consciousness through science education.

Mellita Jones' research is concerned with effective teacher education where her focus has been on school-university partnerships, reflective practice and transformative learning in teacher education. Her recent work has involved school-based approaches to science teacher education and practicum partnerships for rural and regional teacher education courses. She also has an interest in working with teachers in the Pacific region and has had significant involvement in the Solomon Islands. Mellita has won a number of teaching excellence awards including an Australian Learning and Teaching Council Citation for Outstanding Contribution to Student Learning and has reviewed and written teacher education units for UNESCO's teacher education program for the Pacific region.

Shakhnoza Kayumova is an Assistant Professor in the Department of STEM Education and Teacher Development, and the Research Scientist at the Kaput Center for Research and Innovation in STEM Education, at the University of Massachusetts, Dartmouth. Her research focuses on conditions of possibility for enacting equitable science teaching and learning with culturally and linguistically diverse children.

Jiong Li has a master's degree in TESOL and World Language Education from Department of Language and Literacy Education at the University of Georgia. His research interest is in bilingual education and teacher-student interaction within bilingual classrooms. As a graduate assistant of the LISELL-B project, he has co-taught in middle and high school science classes, which enriched his experience in bilingual science education from a perspective of a language educator.

David E. Long is assistant professor of STEM education at Morehead State University. As an educational anthropologist, his primary research is ethnographic documentation and analysis of science learning at the social intersection of differing ideological, economic, and moral projects. He focuses on perennial problem of U.S. science education and communication, including the cognitive, social, and philosophical dimensions of student and teacher understanding of evolution, climate change science, and genetic engineering, with emphasis on how political and religious ideology mediates science education implementation in schools, universities, and in the civic discourse.

Sonya N. Martin is Associate Professor in Science Education at Seoul National University in Seoul, Republic of Korea. Sonya holds master's degrees in Elementary Education and in Chemistry Education from the University of Pennsylvania in the United States and a doctoral degree in Science Education from Curtin University in Australia. Prior to moving to Korea in 2011, Sonya was a tenured faculty member at Drexel University in Philadelphia, PA, USA. Her research focused on identifying science teacher practices that promoted language learning for culturally and linguistically diverse students. Currently she is examining the challenges facing teachers serving immigrant students in Korean science classrooms and she is examining the impact of Confucian ideology on interactions and discourse in science classrooms in Korea and Taiwan. Her research focuses on equity issues in science education.

Isabel Martínez-Cuenca is a Ph.D. student at the State University of New York at Albany. She earned a BA (1999) and ABD (2004) in TESOL and Applied Linguistics at Autonomous University of Barcelona (Spain). Weaving her passion for EFL/ESL teaching and TESOL teacher development, she studies the effects of an enrichment instruction program on immigrant L2 learners' longitudinal conceptual development. Under Dr. Istvan Kecskes' supervision, Isabel's research is designed to find ways to better support immigrants' transitioning into L2 science classrooms. Her research interests include learner-centered L2 pedagogy, bi- and multilingualism, L2 conceptual development, and L2 discourse and language use.

Maxine McKinney de Royston is an Assistant Professor in the department of Curriculum & Instruction at the University of Wisconsin-Madison. McKinney de Royston was recently a Ford Foundation Postdoctoral Fellow at the Center for Urban Education at the University of Pittsburgh. She is concerned with how discourses and practices related to race, identity, and equity influence Black students' experiences and opportunities to learn in mathematics and science learning environments, especially middle school classrooms. Her current work examines how math and science teachers understand the sociopolitical context of their work and how they position Black students' racial and domain-specific identities in ways that shape these students' classroom learning experiences and potentially their longer-term learning trajectories. Her work has been published in journals such as *Cognition and Instruction*, *Journal of the Learning Sciences*, *Teachers College Record*, and the *Journal for Research in Mathematics Education.*

Luis C. Moll is Professor Emeritus in the Language, Reading and Culture Program of the Department of Teaching, Learning and Sociocultural Studies, College of Education, University of Arizona. His main research interest is the connection among culture, psychology and education, especially as it relates to the education of Latino children in the US. His most recent book, *L. S. Vygotsky and Education*, was published by Routledge Press in 2014. Among his honors, he was elected to membership in the National Academy of Education (1998), and named Fellow (2009) of the American Educational Research Association.

Meshach B. Ogunniyi is a Professor Emeritus (Retired) of Science Education at the School of Science and Math Education (SSME) of the University of the Western Cape (UWC), Cape Town, South Africa. He has published extensively in the field within Africa and beyond. His focus during his tenure at University of Ibadan, University of Botswana and UWC has been on teaching, research and academic capacity building. He is an acknowledged member of the Southern Africa Association for Research in Mathematics, Science and Technology Education (SAARMSTE), the inter/National Association for Research in Science Teaching (NARST) and recently the Past president (2014-2015) of the African Association for the Study of Indigenous Knowledge Systems (AASIKS). His recent book (co-edited with Femi S. Otulaja), *The World of Science Education: Handbook of Research in Science Education in sub-Saharan Africa,* was published by Sense Publishers, in June 2017.

Alandeom W. Oliveira is an associate professor of science education at the State University of New York at Albany. He earned a Master's degree in science education at Southeast Missouri State University (2002) and a Ph.D. degree in science education at Indiana University Bloomington (2008). He has taught science education courses to teachers in Brazil and the US and has coordinated multiple professional development programs for school teachers, including Science Modeling for Inquiring Teachers Network, and Technology-Enhanced Multimodal Instruction in Science and Math for English Language Learners. His research interests include cooperative science learning, inquiry-based teaching, and classroom discourse and language use.

Fernanda Ostermann has a degree in Physics and a Ph.D. in Physics Education. She is a professor at the Physics Education Research Program of Universidade Federal do Rio Grande do Sul, Porto Alegre, Brazil and is a researcher of CNPq (National Council for Scientific and Technological Development). Her research activity covers sociocultural studies of physics teacher education and high school physics.

Femi S. Otulaja is an Academic Staff Development Advisor at the Science Teaching and Learning Centre (STLC) in the Faculty of Science at the University of the Witwatersrand, Johannesburg, South Africa. He holds a Ph.D. from the City University of New York (CUNY). He is a researcher with international scope in the field of Science Education, science teacher preparation and development. His research interests focus on sociocultural aspects of science education, integration of science and indigenous knowledge, and the scholarship of teaching and learning. He is the current President (2016-2017) of African Association for the Study of Indigenous Knowledge Systems (AASIKS) and an active member of SAARMSTE and NARST. His most recent book (co-edited with Meshach B. Ogunniyi), *The World of Science Education: Handbook of Research in Science Education in sub-Saharan Africa,* was published by Sense Publishers, in June 2017.

Maurício Pietrocola is a science educator and professor at The University of São Paulo in Brazil. He received his doctoral degree from The University of Paris 7 (Denis Diderot) in 1992 and has since been the author of numerous publications. His areas of work include curriculum development, pedagogical knowledge and innovative strategies of teaching and learning. His current focus is on connections between innovative education and risk taking, which contributes to an understanding of the failure of new educational innovations. In 2013–2014, Mauricio was a Fellow at the Advanced Research Collaborative at the Graduate Center of CUNY.

Cassie F. Quigley is an Associate Professor of Science Education in the Department of Teaching and Learning in the School of Education. She received her doctorate in Curriculum & Instruction at Indiana University in 2010. Her research focuses on broadening the conception of science and participation in science. In this way, my scholarship consists of two lines of inquiry, both focused on equitable practices in science education. The first line of research is exploring community-based methodological approaches to research that seeks to broaden the way science knowledge is documented and who participates in the construction of science knowledge. Her second line of research is examining equitable science teaching practices. In this line of research, she analyzes how teachers implement equitable practices, and ways to support teachers in making such pedagogical shifts.

Giuliano Reis is an Associate Professor of Science Education at the Faculty of Education at the University of Ottawa (Canada). He continues to see himself as a science teacher—and one who likes to believe that his work contributes to making the learning and teaching of science more captivating to all (unrealistically ambitious?). His main research interest is in the study of language in/for science and environmental education in various learning contexts. He is married to Juliana Reis and they have two daughters: Ana-Julia and Maria-Luiza.

Michael J. Reiss is Professor of Science Education at UCL Institute of Education, University College London, Visiting Professor at the Universities of Kiel, Leeds and York and the Royal Veterinary College, Honorary Fellow of the British Science Association, Docent at the University of Helsinki and a Fellow of the Academy of Social Sciences. The former Director of Education at the Royal Society, his academic interests are in science education, bioethics and sex education.

Flavia Rezende has a degree in Physics Education and a Ph.D. in Education. She is a professor at the Physics Education Research Program of Universidade Federal do Rio Grande do Sul, Porto Alegre, Brazil and researcher of CNPq (National Council for Scientific and Technological Development). Her research interests focus on high school and college science teaching and learning under sociocultural perspectives.

Carolina Castano Rodriguez is a Biologist and Educator from Colombia. She has lead several projects in Australia, Ecuador, Argentina and Colombia. Her main research interest is in the link between science education, social justice and ecojustice. Carolina has particular interest and expertise in working with marginalised and socio-economically disadvantaged communities in the field of critical pedagogies, transformative learning, peace education and pedagogies of care. She is a member of the Science Learning Transformative Group (SLTG). Currently she is leading two projects aiming at rethinking science education for the construction of peaceful and sustainable societies.

Kathryn Scantlebury is a Professor in the Department of Chemistry and Biochemistry, Director of Secondary Education in the College of Arts & Sciences at the University of Delaware and a visiting research professor at Uppsala University's Center for Gender Research in Uppsala, Sweden. Her research interests focus on feminist and gender issues in various aspects of science education, including urban education, preservice teacher education, teachers' professional development, and career paths in science. Scantlebury is a Fellow of the *American Association for the Advancement of Science* and co-editor of *Gender and Education.*

Ajay Sharma is an associate professor in the Department of Educational Theory and Practice, at the University of Georgia, Athens, United States of America. His current research centers on explorations of representations of nature in science education, and neoliberalism's impact on education. In his past research, Ajay Sharma has focused on understanding classroom discourse in K-12 science classrooms and science teacher education programs from the perspectives of individual agency and equity. Before becoming a university based academic, Ajay Sharma worked in the Hoshangabad Science Teaching Program in Madhya Pradesh, India, on middle school science curriculum and professional development.

Christina Siry is a Professor of learning and instruction at the University of Luxembourg, where she is also head of the Institute of Applied Educational Sciences. Her research is grounded in sociocultural theoretical foundations, and her work focuses on early elementary students' interactions as they engage in science investigations. She has published in a variety of international journals and edited book volumes, and she is currently a co-editor of the journal *Cultural Studies of Science Education.*

Anna Stetsenko is Professor in the Ph.D. Programs in Psychology and in Urban Education at The Graduate Center of the City University of New York. She came to CUNY with years of experience in a number of leading research centers and universities in Europe. She is currently working on topics at the intersection of human development, education and social theory focusing on activism, agency, and social change. With her interdisciplinary expertise in psychology, philosophy, and education combined with her international background, her works cut across many fields and connect cutting-edge developments and insights from a variety of frameworks.

Aik-Ling Tan is an associate professor at the Natural Sciences and Science Education Academic Group at the National Institute of Education (NIE), Nanyang Technological University. She is also the assistant dean for professional development in the Office of Graduate Studies and Professional Learning at NIE. Her current research interests lies in the area of student-teacher interaction in science classrooms and the professional development needs of science teachers. Aik Ling enjoys working with science teachers in schools to better understand the practical issues and tensions of teaching and learning science.

Tang Wee Teo is an Assistant Professor in the Natural Sciences and Science Education (academic group) at the National Institute of Education (NIE), Nanyang Technological University, Singapore. Her research work focuses on marginalised groups such as lower track students, who are less research about and understood locally. The book chapter draws on data collected from a large-scale two-year study (funded by Office of Education Research, NIE, under project no. OER51/12TWT) about lower track science students. Cultural sociology theories are applied to examine and unpack the complexities of the lower track students' science learning experience with the aim to inform policy changes and future research agenda. Currently, she is studying the subcultures of lower track science classrooms.

Elisabeth (Lily) Taylor is Adjunct Senior Lecturer at the School of Education at Murdoch University, Perth, Australia. Lily's research program has focused on ethical dilemma pedagogy in science education, education for sustainability, education against violent extremism and socio-cultural issues in education. She has been a co-investigator on several national competitive research and teaching grants. Lily has published a book, book-chapters and journal articles. She completed a counseling degree (MCouns) in 2013 and has been engaged in studies of Transpersonal Art Therapy. Her current research interests focus on the nexus between science education, art, transpersonal psychology and counseling.

Peter Charles Taylor is Professor of STEAM Education and Director of the Transformative Education Research Group at Murdoch University, Western Australia. Peter's humanistic vision for education involves integrating the Arts and Sciences to develop young people's higher-order abilities, such as creativity, critical thinking and compassion, which are essential for living and working in the complex and diverse world of the 21st Century. One of his main research and development projects involves assisting teachers to develop, implement and evaluate school-based, inter-disciplinary STEAM curricula that prepare students (i) for an emerging 21st Century *high-tech workforce* and (ii) as *ethically astute citizens* willing and able to engage with *sustainable development issues* such as climate change, loss of biodiversity and loss of indigenous knowledge systems.

Deborah J. Tippins is a Professor in the Department of Mathematics and Science Education at the University of Georgia, where she specializes in science for young children. Her scholarly work focuses on encouraging meaningful discourses around environmental justice and sociocultural issues in science education. Specifically, she draws on anthropological and sociocultural methods to study questions of relevance and justice pertaining to science teaching and learning.

Kenneth Tobin came to the Urban Education doctoral program at the Graduate Center of CUNY in the fall semester of 2003. Presently he is coordinator of the Learning Sciences strand. Prior to his position at the Graduate Center, Tobin had positions as a tenured full professor at Florida State University (1987–1997) and the University of Pennsylvania (1997–2003). Also, he held university appointments at the Western Australian Institute of Technology (now Curtin University), Mount Lawley College, and Graylands College (now Edith Cowan University). Before Tobin became a university science educator in Australia in 1974, he taught high school physics, chemistry, biology general science, and mathematics for 10 years. He began a program of research in 1973 that continues to the present day—teaching and learning of science and learning to teach science. Tobin's current research interests focus on mindfulness and wellness as components in a context of learning along the birth-death continuum.

Sara Tolbert is associate professor of science education and teacher education at The University of Arizona. Through critical qualitative, feminist, and participatory / interventionist research methods, she investigates how teachers can be prepared to engage youth with sociopolitical and community issues in science and create caring, contextually authentic science learning experiences that leverage minoritized students' lived experiences. Prior to becoming a professor of education, she taught science with/for teachers and youth in economically oppressed communities in the U.S. and abroad. She currently teaches elementary, secondary, and bilingual science methods courses as well as graduate courses on equity, social justice, and traditional indigenous knowledge in math and science education.

Maria Varelas is Professor of Science Education at the University of Illinois at Chicago's Curriculum and Instruction Department, and the Director of UIC's Center for the Advancement of Teaching-Learning Communities. Her research, supported by NSF funding totaling over 9 million dollars, focuses on science learning and teaching in urban schools, as she works with teachers to reform science education aiming at equity and social justice. Along with being recognized as a University of Illinois Scholar and Distinguished Researcher of the Year, she has earned numerous teaching awards from students and peers, and held leadership positions in the science education field.

Matthew Weinstein coordinates the Secondary Science Program at University of Washington-Tacoma. His scholarship focuses on issues of ethics, culture, science, and social justice. He is author of *Robot World: Education, Popular Culture, and Science as well as Bodies out of Control* (with Nidaa Makki), He is also the author of TAMS Analyzer (a qualitative coding and analysis package for Apple Computers).

Sara E. D. Wilmes received her doctorate from the University of Luxembourg. Her research explores science education in multilingual spaces and the use of postmodern research methodologies, such as collaborative autoethnography, to tease apart intersections between science learning and language learning. She holds a Master's degree in Primary Education and Science Education from the State University of New York at Buffalo.

Yu Xia is a master's degree student in World Language Education in the Language and Literacy Education Department at the University of Georgia. Her studies and research focus on bilingual education, ESL, and Chinese teaching. She is an experienced ESL teacher with interests in language, literacy, and linguistics. As a member of the LISELL-B project team, she works with science and ESOL teachers to explore new pedagogies, teaching strategies and methodologies for science investigation through language-immersed content-based bilingual science education.

Subject Index

Names Index

COUNTERPOINTS

Studies in Criticality

General Editor
Shirley R. Steinberg

Counterpoints publishes the most compelling and imaginative books being written in education today. Grounded on the theoretical advances in criticalism, feminism, and postmodernism in the last two decades of the twentieth century, Counterpoints engages the meaning of these innovations in various forms of educational expression. Committed to the proposition that theoretical literature should be accessible to a variety of audiences, the series insists that its authors avoid esoteric and jargonistic languages that transform educational scholarship into an elite discourse for the initiated. Scholarly work matters only to the degree it affects consciousness and practice at multiple sites. Counterpoints' editorial policy is based on these principles and the ability of scholars to break new ground, to open new conversations, to go where educators have never gone before.

For additional information about this series or for the submission of manuscripts, please contact:

Shirley R. Steinberg
c/o Peter Lang Publishing, Inc.
29 Broadway, 18th floor
New York, New York 10006

To order other books in this series, please contact our Customer Service Department:

Or browse online by series:

www.peterlang.com